CHASING *the* SUN

RANDOM HOUSE * NEW YORK

CHASING *the* SUN

The
EPIC STORY
of the STAR
THAT GiVES
US LIFE

...

Richard Cohen

Published in the United States by Random House,
an imprint of The Random House Publishing Group,
a division of Random House, Inc., New York.

RANDOM HOUSE and colophon are registered
trademarks of Random House, Inc.

Excerpt from "Notes Toward a Supreme Fiction" by Wallace Stevens,
from *The Collected Poems of Wallace Stevens*, copyright 1954 by Wallace Stevens,
and renewed 1982 by Holly Stevens. Used by permission of Alfred A. Knopf,
a division of Random House, Inc.

Excerpt from "Time," words and music by Roger Waters, Nicholas Mason,
David Gilmour, and Rick Wright, TRO-© copyright 1973 (renewed) Hampshire
House Publishing Corp., New York, NY. International Copyright Secured.
Made in USA. All rights reserved including public performance for profit.
Used by permission.

"In Praise of Limestone," copyright 1951 by W. H. Auden, from
Collected Poems of W. H. Auden by W. H. Auden. Used by permission of
Random House, Inc. Used in the e-book edition by permission of
The Wylie Agency (UK) Ltd.

Excerpt from "Telescope" by Ted Kooser, from *Delights & Shadows*.
Copyright © 2004 by Ted Kooser. Reprinted with permission of
Copper Canyon Press, www.coppercanyonpress.org.

"Mad Dogs and Englishmen" copyright © NC Aventales AG 1932 by
permission of Alan Brodie Representation Ltd. www.alanbrodie.com.

Excerpt from "The Poems and Plays of Oliver St. John Gogarty" by
Oliver St. John Gogarty, reprinted with the permission of Colin Smyth Ltd
on behalf of V. J. O'Mara.

LIBRARY OF CONGRESS CATALOGING-IN-PUBLICATION DATA

Cohen, Richard.
Chasing the sun : the epic story of the star that gives us life /
Richard Cohen.
p. cm.
Includes bibliographical references and index.
ISBN 978-1-4000-6875-3
eBook ISBN 978-1-58836-934-5
1. Sun—History. 2. Astronomy—History. 3. Science and civilization.
I. Title.
QB521.C625 2010
523.7—dc22   2010005885

Printed in the United States of America on acid-free paper

www.atrandom.com

9 8 7 6 5 4 3 2 1

FIRST EDITION

Book design by Barbara M. Bachman

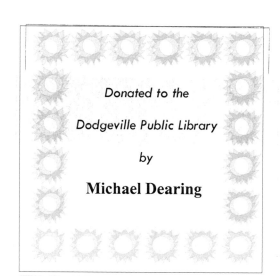

For Kathy

L'amor che muove il sole e l'altre stelle

THE LOVE THAT MOVES THE SUN AND THE OTHER STARS

CONTENTS

List of Illustrations   *xiii*

Preface   *xix*

SUNRISE: MOUNT FUJI   *xxv*

PART I · THE SUN BEFORE SCIENCE

CHAPTER 1 TELLING STORIES   3

Solar myths; Max Müller's discoveries;
Children of the Sun; an Inca festival

CHAPTER 2 CELEBRATING THE SEASONS   14

Solstices and equinoxes; worshipping through
the seasons; North American sun dances

CHAPTER 3 THE THREE THOUSAND WITNESSES   28

The Earth's standing stones; Stonehenge;
the Pyramids; Newgrange; and New York

CHAPTER 4 TERRORS OF THE SKY   42

Auroras and eclipses: their respective histories,
from 2000 B.C. to Shakespeare and Milton

PART 2 · DISCOVERING THE SUN

CHAPTER 5 THE FIRST ASTRONOMERS   59

The earliest stargazers; the Sumerians;
the Babylonians; the ancient Egyptians

CHAPTER 6 ENTER THE GREEKS   69

From Homer and Hesiod to Pythagoras,
Plato, and Ptolemy

CHAPTER 7 GIFTS OF THE YELLOW EMPEROR   *86*

*China's separateness; modes of thought:
yin and yang, the "Way" and the Sun*

CHAPTER 8 THE SULTAN'S TURRET   *100*

*Arabia's great astronomers; India and
Western Europe, A.D. 600–1543*

CHAPTER 9 THE EARTH MOVES   *116*

*The great four: Copernicus;
Tycho; Kepler; and Galileo*

CHAPTER 10 STRANGE SEAS OF THOUGHT   *136*

*Newton on gravity and color; the transit of Venus;
Newton's successors to 1850*

CHAPTER 11 ECLIPSES AND ENLIGHTENMENT   *153*

*From Captain Cook to Einstein's
great moment; Antarctica 2003*

CHAPTER 12 THE SUN DETHRONED   *168*

*Solar physics, 1800–1950; the Copenhagen gang;
making the atomic bomb*

PART 3 • THE SUN ON EARTH

CHAPTER 13 SUNSPOTS   *187*

*Their makeup explained; cycles, flares, and eruptions;
their influence on weather*

CHAPTER 14 THE QUALITIES OF LIGHT   *202*

*Light in war, in sport, in crime;
shadows, sunsets, and twilight*

CHAPTER 15 BENEATH THE BEATING SUN   *214*

*Heat and its effects; skin color fashions;
from sun-avoidance to sunbathing*

CHAPTER 16 SKIN DEEP   *231*

*Skin cancers; how melatonin works;
sungazing; seasonal affective disorder*

CHAPTER 17 THE BREATH OF LIFE   244
Photosynthesis and its discoverers; plants and the Sun; hibernation and migration

CHAPTER 18 THE DARK BIOSPHERE   259
Creatures of the deep; currents and tides

PART 4 • HARNESSING THE SUN

CHAPTER 19 THE HEAVENLY GUIDE   275
Navigation and cartography as aided by the Sun

CHAPTER 20 OF CALENDARS AND DIALS   288
Reckoning the years from before Julius Caesar to after Gregory VII

CHAPTER 21 HOW TIME GOES BY   307
Greenwich Mean Time, the creation of Standard Time, and Daylight Saving Time

CHAPTER 22 THE SUN IN OUR POCKET   322
Solar energy, from Archimedes' mirrors to solar-powered cars

PART 5 • INSPIRED BY A STAR

CHAPTER 23 THE VITAL SYMBOL   345
The Sun King; the metaphors of gold, blondes, and mirrors

CHAPTER 24 DRAWING ON THE SUN   358
The Sun in art, from the Renaissance through Turner, Van Gogh, Matisse, and Hockney

CHAPTER 25 NEGATIVE CAPABILITIES   376
The Sun in photography, film, and architecture

CHAPTER 26 TALK OF THE DAY   388
The Sun as reflected in classical and pop music from Mozart to the Beatles

CHAPTER 27   BUSIE OLD FOOLE   399
          The Sun in literature: Homer and Dante,
          Chaucer and Donne, Lawrence and Nabokov

CHAPTER 28   THE RISING STAR OF POLITICS   418
          Solar imagery and the Nazis; Japan's divine
          emperor; the Sun of Mao and Benedict XVI

          PART 6 · THE SUN AND THE FUTURE

CHAPTER 29   OVER THE HORIZON   439
          Solar astronomy, 1950 to the present;
          the coming "golden age" and what it may bring

CHAPTER 30   UNDER THE WEATHER   460
          The ozone layer; the Sun and global warming;
          will the Earth get too hot or freeze over?

CHAPTER 31   THE IMPOSSIBLE AND BEYOND   476
          Science fiction from ancient Rome to Swift, Verne,
          H. G. Wells, Isaac Asimov, and Arthur C. Clark

CHAPTER 32   THE DEATH OF THE SUN   490
          Solar catastrophes; asteroids and meteors;
          what will happen when the Sun burns out?

          SUNSET: THE GANGES   499

          Acknowledgments   513
          Notes   519
          Index   553

LIST OF ILLUSTRATIONS

p. xxix   An 1857 painting of the sun goddess Amaterasu, with Mount Fuji in the background. Minneapolis Institute of Arts, Gift of Louis W. Hill, Jr.

p. 1   The heroic archer Houyi, who shot down nine of the ten suns. © Mona Caron.

p. 5   The sun chariot of Surya. Michaud Roland et Sabrina/Rapho/Eyedea Illustrations.

p. 8   Max Müller. Walker & Cockerell, ph. Sc.

p. 9   A map of the distribution of solar cultures around the world. *Graphis* magazine, 1962.

p. 11   Audrey Hepburn and Gregory Peck in *Roman Holiday,* standing in front of the Bocca della Verità in the Church of Santa Maria in Cosmedin, Rome. Paramount Pictures.

p. 19   Celebrants jumping a bonfire during summer solstice festivities at St. Jean, in Alsace. Cabinet des Estampes et des Dessins de Strasbourg. Photo Musées de la Ville de Strasbourg. M. Bertola.

p. 21   A Mexican sun-ball game. Reuters/Henry Romero.

p. 25   The Okipa, the dance ritual of the Mandan Indians, as depicted by the artist George Catlin. Copyright © The Library, American Museum of Natural History.

p. 29   The massive pillars of Stonehenge. Photo courtesy of Madanjeet Singh.

p. 32   The passageway at Newgrange burial mound in Ireland. Courtesy of the Department of the Environment, Heritage and Local Government, Ireland.

p. 35   The pyramid at Chichén Itzá. Damian Davies/Getty Images.

p. 37   The Great Pyramid at Giza. Carolyn Brown/Photo Researchers, Inc.

p. 41   The sundial bridge at Turtle Bay in Redding, California, by Santiago Calatrava. Alan Karchmer/ESTO.

p. 43   Aurora borealis, March 1, 1872. Étienne Léopold Trouvelot. Science, Industry & Business Library, The New York Public Library, Astor, Lenox and Tilden Foundations.

p. 45   Eclipse in Antarctica, 2003. Photo by Woody Campbell.

p. 50   Eclipse at the Battle of the Medes and Lydians, May 28, 585 B.C. Camille Flammarion, *Astronomy for Amateurs* (New York and London: D. Appleton and Company, 1910). Painting by Rochegrosse.

p. 57   Petroglyph from the Camonica Valley in northern Italy. Sketch by Paul F. Jenkins.

p. 62   Babylonian astronomers observing the heavens. From John F. Blake, *Astronomical Myths* (London: Macmillan and Co., 1877), p. 187.

p. 67   The family of Akhenaten offers sacrifice to Atum the sun god. Erich Lessing/Art Resource, N.Y.

p. 72   Pythagoras in Alexandria. Sheila Terry/Science Photo Library.

p. 80   Hipparchus. SPL/Photo Researchers, Inc.

p. 89   Xi Chu with gnomon. HIP/Art Resource, N.Y.

p. 94   The water-driven astronomical clock tower of Su Song. Drawing by John Christiansen, from Joseph Needham, *Science and Civilisation in China* © Cambridge University Press. Reprinted by permission.

p. 96   Eclipse in Siam, 1688. Bibliothèque Nationale, Paris, France/Lauros/Giraudon/
         The Bridgeman Art Library.

p. 98   The Imperial Observatory in Beijing in the late seventeenth century. SPL/Photo
         Researchers, Inc.

p. 107   Indian astronomers, early seventeenth century. HIP/Art Resource, N.Y.; British
         Library, London, Great Britain.

p. 108   Hindu devotees pay homage to the Sun. Michaud Roland et Sabrina/Rapho/
         Eyedea Illustrations.

p. 110   Indian folk drawing depicting Surya Mandala with a cobra. From Madanjeet
         Singh, *The Sun in Myth and Art* (London: Thames and Hudson, 1993).

p. 113   The astrolabe of Lansberg. From Philips van Lansbergen, *Philippi Lansbergii Opera
         omnia* (Middleburgi Zelandi: Z. Roman, 1663).

p. 123   The castle and observatory of Uranienborg (constructed by Tycho Brahe). Biblio-
         thèque Mazarine, Paris, France/Archives Charmet/The Bridgeman Art Library.

p. 126   The planetary spheres as depicted in Kepler's *Cosmographic Mystery* (1596). Mary
         Evans/Photo Researchers, Inc.

p. 129   Nicolaus Copernicus. Science Source/Photo Researchers, Inc.

p. 129   Tycho Brahe. Hulton Archive/Getty Images.

p. 129   Johannes Kepler. Science Source/Photo Researchers, Inc.

p. 129   Galileo Galilei. SPL/Photo Researchers, Inc.

p. 131   Galileo's visit to the Vatican in 1611. Louvre, Paris, France/Peter Willi/The Bridge-
         man Art Library.

p. 138   Sir Isaac Newton. Science Source/Photo Researchers, Inc.

p. 140   Isaac Newton using a prism to break white light into a spectrum. Image Select/
         Art Resource, N.Y.

p. 145   *An Alchemist in His Workshop.* Courtesy of the Chemical Heritage Foundation Col-
         lections.

p. 154   Dancers during eclipse in Antarctica, 2003. Photo by Woody Campbell.

p. 159   Tintin uses an eclipse to outwit his Inca captors. Hergé/Moulinsart 2009.

p. 161   Parisian women view eclipse, 1892. From *La Vie Illustrée: L'Eclipse Solaire du 30
         Août, Rue de la Paix* (1905).

p. 162   A partial eclipse seen against a crescent on Islamabad's Faisal Mosque. AP
         Photo/B. K. Bangash.

p. 165   *Fifty Crows.* Photo by Antonio Turok.

p. 175   An advertisement for Tho-Radia cream. SPL/Photo Researchers, Inc.

p. 180   Albert Einstein and J. Robert Oppenheimer. USIA/AIP Photo Researchers, Inc.

p. 183   XX-28 George, a 225-kiloton atomic bomb, exploding over the Pacific Ocean on
         May 8, 1951. U.S. Department of Energy/Photo Researchers, Inc.

p. 185   Sunbathing. Illustration by Lt. Col. Frank Wilson from Stephen Potter, *Lifeman-
         ship* (New York: Henry Holt and Co., 1950).

p. 188   Galileo's sunspot drawings. Library of Congress.

p. 192   Christoph Scheiner mapping sunspots. SPL/Photo Researchers, Inc.

p. 195   Section of a Scotch pine showing the years of maximum sunspot activity. Cour-
         tesy of the Laboratory of Tree-Ring Research, University of Arizona.

p. 209   The great ball of Krishna, Mahabalipuram, on the southeast coast of India, 1971.
         Photo by Max Pam.

p. 210   A Cambodian farmer with a sunshade, 1952. Werner Bischof/Magnum Photos.

p. 212   Horizontal light can be used to discover the sites of lost villages. Ashmolean Mu-
         seum, University of Oxford.

p. 220   *Cleopatra Dissolving the Pearl*, by Joshua Reynolds, 1759. Kitty Fisher: Kenwood
         House: The Iveagh Bequest/English Heritage Photo Library.

p. 224   John F. Kennedy sunbathing, 1944. John F. Kennedy Library.

p. 226   English children undergoing sunlamp treatment. SVT-Bild/Das Fotoarchiv/Black Star.

p. 232   A mother and her albino child, Sudan, 1905. Michael Graham-Stewart/The Bridgeman Art Library.

p. 236   Protective clothing for the rare disease XP. Sarah Leen/National Geographic Image Collection.

p. 238   Max Schreck in *Nosferatu: A Symphony of Horror*, 1922. Photofest.

p. 251   Flower clock. Drawing by U. Schleicher-Benz.

p. 257   The Texas blind salamander. Dante Fenolio/Photo Researchers, Inc.

p. 260   A digital re-creation of a hydrothermal vent. David Batson/DeepSeaPhotography.com.

p. 264   The deep-sea angler fish. David Batson/DeepSeaPhotography.com.

p. 273   *Father Time Carrying the Pope Back to Rome*, 1641. © The Trustees of the British Museum/Art Resource, N.Y.

p. 285   Map of Commodore George Anson's attempt to circumnavigate the globe, 1740–44. Library of Congress, Rare Books Division.

p. 286   Jonas Moore's *A New System of Mathematicks*, 1681. Lawrence H. Slaughter Collection, The Lionel Pincus and Princess Firyal Map Division, The New York Public Library, Astor, Lenox and Tilden Foundations.

p. 289   Ethiopians celebrating the new millennium, September 12, 2007. Roberto Schmidt/AFP/Getty.

p. 297   William Hogarth, *The Election Dinner* (1755). Private Collection/Ken Welsh/The Bridgeman Art Library.

p. 302   Borneo tribesmen measure the length of the Sun's shadow at summer solstice with a gnomon. From Charles Hose and William McDougall, *The Pagan Tribes of Borneo* (London: Macmillan & Co., 1912).

p. 305   The Samrat Yantra, the giant sundial at the Jaipur Observatory. Science Museum/SSPL.

p. 309   An illuminated manuscript page from *Les Très Riches Heures du Duc de Berry* (1412–16). Réunion des Musées Nationaux/Art Resource, N.Y.

p. 312   The calendar used by an Inuk hunter in the 1920s. Revillon Frères Museum, Moosonee, Ontario.

p. 317   Atomic clock. National Physical Laboratory, Crown Copyright/SPL/Photo Researchers, Inc.

p. 323   Archimedes used mirrors to reflect the Sun's rays onto the Roman fleet. Archive Photos/Getty Images.

p. 326   Augustine Mouchot's solar-powered printing press at the Paris World Exhibition of 1878. The Granger Collection, New York.

p. 329   A revolving solarium at Aix-les-Bains, France, in September 1930. Fox Photos/Hulton/Getty Images.

p. 336   The Helios prototype flying wing, soaring on solar power, Hawaii, July 2001. Nick Galante/PMRF/NASA.

p. 343   Swastika sun symbols.

p. 346   Fireworks on the Grand Canal at Versailles, 1674. Réunion des Musées Nationaux/Art Resource, N.Y.; Chateaux de Versailles et de Trianon, Versailles, France.

p. 349   Indian woman stands naked in the water, facing the Sun. Michaud/Rapho/eyedea.

p. 350   Freewheelin' Frank Reynolds, secretary of the San Francisco branch of the Hells Angels. Photo by Larry Keenan.

p. 355   Five young Gypsies perform their salute to the Sun in the mid-1950s. *Tziganes,* Frans de Ville, 1956.

p. 360   Saint Benedict during a solar eclipse. Photo courtesy of the author.

p. 363   Turner, *The Fighting Téméraire.* National Gallery, London/Art Resource, N.Y.

p. 366   Monet, *Impression: Sunrise.* Erich Lessing/Art Resource, N.Y.

p. 373   Olafur Eliasson, *The Weather Project.* Tate, London, 2010.

p. 374   Pablo Picasso paints with light. Gjon Mili/Time Life Pictures/Getty Images.

p. 379   Brassaï, *Man with Umbrella* (1936). Brassaï. © Estate Brassaï—RMN: Agence photo RMN, 10 rue de l'Abbaye, 75006 Paris (France).

p. 381   Gustave Le Gray. *The Great Wave, Sète* (1857). The Metropolitan Museum of Art/Art Resource, N.Y.

p. 385   The Roman town of Bram, in France. Irène Alastruey/Agefotostock.

p. 386   The Walt Disney Concert Hall in downtown Los Angeles. Hufton+Crow/View/Esto.

p. 387   The Statue of Liberty. The Bridgeman Art Library

p. 389   The Sun and Moon dance in traditional Japan. Courtesy of the author.

p. 392   Judith Jamison as the Sun in the 1968 revival of Lucas Hoving's *Icarus.* Photo © Jack Mitchell.

p. 395   The set for the final scene of the second act of *The Magic Flute,* in an early-eighteenth-century production. Bildarchiv Preussischer Kulturbesitz/Art Resource, N.Y.

p. 400   Vladimir Nabokov. Philippe Halsman/Magnum Photos.

p. 411   D. H. Lawrence and Aldous Huxley. Topham/The Image Works.

p. 413   Norman Mailer with his sixth wife, Norris Church. Courtesy of Francis Delia.

p. 419   No. 10 Downing Street. ImageState/age fotostock.

p. 427   Charlie Chaplin in *The Great Dictator.* Photofest.

p. 431   Mao's Little Red Book. Courtesy of the author.

p. 434   Joseph Ratzinger, now Pope Benedict XVI. Courtesy of the Weltenberg Cathedral.

p. 437   Sketch from the seventeenth-century manuscript *De thermis.* NOAA Library Collection.

p. 443   Neutrino detector. Kamioka Observatory, ICRR (Institute for Cosmic Ray Research), The University of Tokyo.

p. 446   "The Suicide Squad" tries to send a rocket into space. Courtesy of NASA.

p. 455   NASA's Solar Probe Plus. JHU/APL.

p. 462   Eruption in Eyjafjallajökull. Photo by Sigurdur Stefnisson.

p. 467   Weather balloon. Novosti.

p. 480   Still from *Sunshine* (2007). Fox Searchlight/The Kobal Collection/The Picture Desk.

p. 482   Sunset on Mars. JPL/NASA.

p. 485   UFOs. From Harold T. Wilkins, *Flying Saucers Uncensored* (New York: Pyramid Books, 1955).

p. 494   The Sun heats a dying Earth, five billion years from now. Detlev van Ravenswaay/Photo Researchers, Inc.

p. 496   Artist's impression of a manned base on Mars. Julian Baum/Science Photo Library.

p. 503   Dasaswamedh Ghat, Varanasi, India. Courtesy of the Rare Book Division of the Library of Congress.

p. 506   Youths worshipping the Sun on the Ganges, India. Photo courtesy of Madanjeet Singh.

p. 511   Sunset on the Ganges. Art Wolfe/Getty Images.

p. 520   "They're keeping me from practicing my religion" cartoon. © Frank Cotham/The New Yorker Collection/www.cartoonbank.com.

p. 522   "Listen, pal, it's a free country" cartoon. Copyright © 2008 Leigh Rubin.
p. 524   From *Peanuts*. Peanuts © United Features Syndicate, Inc.
p. 528   "Narcissism: Thinks World Revolves Around It" cartoon. © Dolly Setton.
p. 533   "Astronomy Beyond the Visible-Light Spectrum" cartoon. © Dolly Setton.
p. 534   From *Peanuts*. Peanuts © United Features Syndicate, Inc.
p. 538   "Sunrise, Sunset, Noon" cartoon. Copyright © The New Yorker Collection.
         Jack Ziegler, 2007, from cartoonbank.com. All rights reserved.
p. 540   "Now that we can tell time" cartoon. Copyright © The New Yorker Collection.
         Tom Cheney, 2005, from cartoonbank.com. All rights reserved.
p. 543   "He's adjusting our solar panels" cartoon. © George Booth/The New Yorker
         Collection/www.cartoonbank.com.
p. 552   "That can't be good" cartoon. Copyright © The New Yorker Collection.
         Danny Shanahan, 2006, from cartoonbank.com. All rights reserved.

Insert
The Sun viewed in extreme ultraviolet light by NASA's SOHO satellite. SOHO
         (ESA & NASA).
Diagram of parts of the Sun. SOHO (ESA & NASA).
A diamond ring during a total eclipse. SOHO (ESA & NASA).
Coronal mass ejections. SOHO (ESA & NASA).
The dark eye of a sunspot. SOHO (ESA & NASA).
The Sun's magnetic loops. SOHO (ESA & NASA).
Active regions on the Sun. SOHO (ESA & NASA).
Coronal mass ejection. SOHO (ESA & NASA).
Progress of a coronal mass ejection (CME) observed over an eight-hour period on
         August 5–6, 1999, by LASCO C3. SOHO (ESA & NASA).
Auroras (top, center, and bottom right). Photos by Sigurdur Stefnisson.
Aurora (bottom left). Photo by Photo Researchers, Inc.
Sunrise seen from Mount Fuji. Courtesy of the author.
A time-lapse photograph of the Sun changing in color and brightness as it sets.
         Pekka Parviainen/Photo Researchers, Inc.
A pillar of light rises from the setting Sun. Courtesy of Michael Carlowicz.
The seashore at Futamigaura, near the holy site of Ise. Courtesy of Madanjeet Singh.
Louis XIV, aged fourteen, as Le Roi Soleil in *Ballet de la Nuit*, 1653. Photo: Bulloz Biblio-
         thèque de l'Institut de France, Paris, France. Réunion des Musées Nationaux/Art
         Resource, N.Y.
Three masks. Museum of the Sun, Riga, Latvia. Permission granted by Iveta Gražule.
Banco Central de Quito Ecuador/Gianni Dagli Orti/The Picture Desk.
Detail of ornamental Sun in Surya Chopar room in City Palace Museum, Udaipur, India.
         Richard l'Anson/Lonely Planet Images.
The Sun (2002), by Yue Minjun.
Girl with a Sunflower (1962), a painting from the Croatian naive school by Ivan Rabuzin.
Inti worshipping Sun, by Théodore de Bry. Bridgeman Art Library/Service Historique
         de la Marine, Vincennes, France/Lauros/Giraudon.
The legend of the nine suns. Bridgeman Giraudon. British Museum, London, U.K./The
         Bridgeman Art Library.
Sunset at the Crimean Shores, 1856, by Johannes Aivazovsky (1817–1900). © 2010 State
         Russian Museum, St. Petersburg.
New Year's celebrations at the Futamigaura seashore, Japan. From Madanjeet Singh,
         The Sun in Myth and Art (London: Thames and Hudson, 1993).

Heidelberg Castle by Joseph Mallord William Turner. (1840–45). Tate, London/
    Art Resource, N.Y.

Vincent van Gogh, *The Sower.* Erich Lessing/Art Resource, N.Y.

Roy Lichtenstein, *Sinking Sun.* © Estate of Roy Lichtenstein.

Edward Hopper, *Rooms by the Sea* (1951). Yale University Art Gallery/Art Resource, N.Y.

David Hockney, *A Bigger Splash* (1967). Acrylic on canvas, 96×96. © David Hockney/Art
    Resource, N.Y.

PREFACE

Once we begin to go back to a closer understanding of nature and man's relation-
ship to the Sun, we're going to start developing whole new concepts of who and
what we are, and why, and what our rightful place in the universe really is.

—HAROLD HAY, a pioneer of solar energy,
now in his 101st year[1]

The world is full of obvious things which nobody by any chance ever observes.

—SHERLOCK HOLMES in
The Hound of the Baskervilles

VOLUME 4 OF THE *ENCYCLOPEDIA OF ASTRONOMY AND ASTROPHYSICS*
defines the Sun as "a main-sequence, yellow dwarf of spectral type G_2V, mass
1.989×10^{30} kg, diameter $1,392,000$ km, luminosity 3.83×10^{26} W and absolute visual magnitude +4.82."[2] That is certainly one way of looking at the
star—although, as Bertie Wooster put it, not the sort of thing to spring on a lad
with a morning head. We describe the Sun in myriad ways and comprehend it
so variously, too. The Indian sages who wrote the Rig Veda between the eighth
and fourth centuries B.C. assumed a tone of awed reverence: "Only when illu-
mining Light shines, / Everything else shines; / The self-revealing Light illu-
mines / the entire universe"; while the philosopher-martyr Giordano Bruno
described the Sun with sixteenth-century extravagance: "Apollo, author of poetry, quiver-bearer, bowman of the powerful arrows, pythian, laurel-crowned,
prophetic, shepherd, seer, priest, and physician . . ."

Without the Sun, none of us would exist. Our star has been active for 4.6
billion years and has enough fuel for a further hundred billion at the present
rate, but continuing solar convulsions will shorten its life to a mere five billion
more. The Earth famously averages a distance of 93 million miles from it (more
accurately 92,955,887.6, but that rounding up by 44,113 miles seems an irrel-

evance). The Sun's core burns at a sustained 15,000,000°C (27,000,000°F). To give a sense of perspective: a single photon from the Sun's core takes 150,000 years just to reach the frontier of space. Every second, five million metric tons of mass are converted into nuclear energy, equivalent to the detonation of 90 billion one-megaton hydrogen bombs—as scientists express it, 3.8×10^{33} ergs (one erg being roughly the amount of energy a mosquito needs to take off). A constant blast of nuclear reactions pushes energy to the surface, releasing it as light and heat. Yet the Earth receives just one 2.2-billionth of that output, a figure so startlingly small that 120 years ago it was one of the talking points of science: Where was all that power going?

The Sun accounts for 99.8 percent of the mass of the entire solar system. As time goes by, it inexorably draws in upon itself and loses volume (although 1,300,000 Earths could still fit inside it), and some two billion years from now, a far cooler star, it will fade into a "red giant," then suddenly collapse into a "white dwarf" (alas, scientific vocabulary is rarely this poetic), all life on Earth having long since been extinguished. After about a further trillion years, it will have frozen to the ultimate cold.

IN THE EARLY 1990S, while running a publishing house in London, I decided I wanted to learn more about the Sun (no sudden epiphany, just the realization that I knew so little about what, above all other things, governs our lives), and went looking for an author who would write a wide-ranging account—but without success. Five years later, I left England for New York with the idea still in my head, and wondered whether in the intervening period someone had written a book that explained how that great ball of gases exercises its enormous and varied effects, and specifically how it creates and maintains life on Earth in all its diversity. A visit to the New York Public Library revealed 5,836 titles under "Sun," but none offered both the excitement of scientific discovery (Galileo first detecting blemishes on the Sun, or William Herschel identifying a new planet) and an exploration of the Sun's place in art, religion, literature, mythology, politics. I wanted to know about the great solar temple standing amid the ruins of Machu Picchu; Mozart celebrating the Sun in *The Magic Flute;* Charlie Chaplin mocking Hitler's love of sun imagery in *The Great Dictator;* and why for fifteen hundred years the civilized West stuck with the ancient notion of the Sun circling the Earth. But most volumes aimed at a specialized area, and the nonscientists who tackled a particular aspect either looked askance at the science or took minimal interest in it.

"There are two worlds," wrote the English essayist and poet Leigh Hunt (1784–1859), "the world that we can measure with line and rule, and the world that we feel with our hearts and imagination." Conan Doyle even has Sherlock Holmes assert that he doesn't want to know about the solar system, as that would take up brain space better devoted to more pertinent material.[3] Some people still find such a view sympathetic, maybe because the basic facts about the Sun are really beyond our understanding—they are too vast, too spectacular. As the science writer Ben Bova has observed, "Evolution hasn't adequately prepared us to easily understand things like quantum mechanics, curved space-time, or even the age of Earth, let alone that of the universe. . . . Metaphor helps, but it's really just a crutch, emphasizing more often than not the limits of our imagination when confronted with the overwhelming vastness of the cosmos."[4]

Back in 2003, I recall telling the biographer Richard Holmes, then just setting out on the group portrait of early-nineteenth-century British scientists that would become *The Age of Wonder,* that I intended to follow up my previous work, a three-thousand-year history of swordfighting, with one about the Sun. "Ah," he said, "first a study of honor, now one about belief." Maybe he is right, but I began with no particular agenda. The Sun is wonderful—literally "full of wonders"—and I wanted to know more about it.

This has meant that I have had to learn nearly all the science—for me a daunting task, as my high school was run by Benedictine monks who had little time for such disciplines, and the main memory I have of any scientific endeavor is of the black-robed, white-haired Father Brendan climbing through a window into our classroom carrying a large copper pan and a Bunsen burner. Proceeding to fry an egg, he told us that science should always have a practical end in mind. Many scientists and science-watchers would disagree—but I enjoyed my small portion of egg.

This book has taken eight years to complete, over which time I have visited eighteen countries on six of the seven continents. (During these years, my wife would frequently be asked about my whereabouts. "Oh," she'd reply, "he's out chasing the Sun.") Some readers may wish that I had ventured deeper into solar astronomy, but this book is not a rainbow—it has to end somewhere. What I came to realize was that, in the Western world especially, our relationship with the Sun has gone astray. Having cracked the code of our star's great power, the miracle of nuclear fusion, we have discovered new wonders in the Sun; but something of value has been lost in the process.

Once upon a time we thought that we were the center of the universe and that even the Sun (as well as the Moon and the planets) revolved around us. We were the center of everything. Yet not having mastered any of the primal forces that determined our lives, we were keenly aware of our own powerlessness. Hence our longing for mastery, as expressed so vividly in mythology, in the innumerable stories of men, or gods, or animals, trying to control the Sun: a longing for supremacy to be satisfied only through magical thinking, as if we had no other means at our command.

Thousands of years later, we know for a fact that our earliest, most basic idea about our place in the cosmos was false, and that that cosmos is vastly larger than we ever dreamed. We are mere specks. At the same time, we have created for ourselves powers far beyond any we could have imagined, and a decade into a new century, solar scientists believe we are on the edge of a golden age of discovery about our star, poised to answer such vexing questions as what causes sunspots and drives solar winds, how the Sun's rays can bring us usable energy, and how its magnetic particles and coronal ejections affect our climates. So, as our understanding of the Sun has increased along with our ability to harness it, we have stripped it of its primitive magical qualities.

The American astronomer John Eddy understood this:

> We have always wanted the Sun to be better than the other stars and better than it really is. We wanted it to be perfect, and when the telescope came along and showed us that it wasn't, we said, "At least it's constant." When we found it inconstant, we said, "At least it's regular." It now seems to be none of these; why we thought it should be when other stars are not says more about us than it does about the Sun.[5]

This book is therefore in part about the desacralization of the universe, using the Sun as the emblem of that ancient but ever stronger process, and about our sense of loss—sometimes conscious, sometimes not—as the wonder has been stripped away.

But it is more than that. Our star has also inspired artists across the centuries. Sometimes it is the direct subject of their creations, sometimes a symbol of what they have wanted to convey, infusing their work with an authority, even majesty, that no other force could match. Eight years into my subject, I have come to feel that the Sun has powers that are still well beyond our understanding—mythic indeed. We haven't at all managed to reduce it to just a "scientific little luminary, dwindled to a ball of blazing gas," as the fervent sun worshipper D. H. Lawrence put it. That ball of gas still has mysteries of a magnitude so cosmic that perhaps only the grandeur of myth can encompass what we now know about the Sun's world-destroying and world-birthing powers, and about its eventual death.

Consider global warming: we have been told that the science is settled, that greenhouse gases threaten the planet. I appreciate that to ask whether that is the prime cause precipitating the changes to our environment is to risk having this book branded a tract against common sense (as my three children repeatedly tell me). There is no doubt that we *are* witnessing extreme droughts, floods, epidemics, and food shortages, along with the accelerating disappearance of species and abrupt changes in animal behavior. Yet many of the world's leading solar physicists have told me that the coming "golden age" of discoveries about the Sun is likely to bear directly on how we understand what is happening to our planet's climate, and that in fact these revelations may dwarf what we currently know about man-made global warming. Will we ignore the research, as if even major cataclysms have to be our own doing and not that of our governing star?

To be sure, we are not blameless; but neither are we ruinously all-powerful. It is as if it offends us to think that we are still at the Sun's mercy. But we are. And of course the Sun has no mercy.

ONE LESSON I have learned in writing this book is that the Sun gets into everything. One evening in 2004, while sitting in the Carlyle hotel on Madison Avenue, I found myself in the midst of a half-dozen Masters of Foxhounds, men and women resplendent in hunting pink, there to attend a grand dinner. It transpired that true members of the blood dread days when the Sun is shining, as the heated atmosphere rises, lifting the fox's scent high beyond their pack's noses, and the quarry makes its escape. I cannot think how else I might have pulled such a sunny detail out of the air.

The U.S. Scouting Service Project's 2009 website promoted a game centered on the Sun (albeit one with scant regard for astronomical realities). "It is like musical chairs," its author writes.

There is one fewer chairs [*sic*] than boys. The extra boy is the Sun. The other boys are each assigned a planet (Mars, Jupiter, etc.). The Sun orbits around the chairs calling out the names of planets. The planet (boy) called gets up and walks (orbits) around the chairs with the Sun. When all the planets are orbiting, the den leader yells "Blast Off." All the boys scramble for a chair. The one left standing becomes the Sun.[6]

Well, please take your chairs. It is time for blastoff.

SUNRISE: MOUNT FUJI

You must become an ignorant man again
And see the sun again with an ignorant eye
And see it clearly in the idea of it.

—WALLACE STEVENS,
"Notes Toward a Supreme Fiction"[1]

You run and you run to catch up with the sun. . . .

—PINK FLOYD

My GOAL WAS TO GET TO THE TOP OF MOUNT FUJI IN TIME TO SEE the Sun rise on June 21, the summer solstice of 2005. However, I soon learned that it wasn't to be that simple: the five main trails used by the thousands of pilgrims and tourists who climb the mountain each year are off-limits until July 1, by which time most of the summit snow has melted, the strong winds have weakened, and the weather overall is kinder. The official season begins with a day of grand ceremonies, one at each of the several "entrances" to the mountain, ranging from a mass purification in one of the surrounding lakes to the *owaraji hono*, where a twelve-foot-high rope sandal made from rice stalks is offered to the gods in homage. The sandal, a symbol of safety, health, and strong legs, replicates the original peasant footwear that was worn (and worn out) on pilgrimages. I decided to stick with my walking boots.

From my apartment in New York City, I applied for a special police permit to make the climb off-season, but the official forms were daunting, demanding my blood type and contact details of my next of kin while warning that "climbing Mount Fuji is not easy, and in foul weather can be a nightmare." I opted to contact the Japanese Mountain Association instead, where a friendly Mr. Nakagawa said that he would raise no objection so long as I exercised extreme care. Temperature and oxygen levels could drop precipitously and suddenly, he cautioned, and there was real risk of high winds and rockfalls. Altitude sickness was another serious danger, and could bring on nausea and disorientation. "Only fools climb Fuji-san twice" runs the local saying, which wags in Tokyo have amended to "once." No matter: the chance to see the sunrise—*goraiko*—was incentive enough.

My actual journey started on June 20, from the small western town of Ise, seat of the two great shrines to Amaterasu, goddess of the Sun and mythical ancestress of the Imperial House. The day before, I had visited the shrines, the earliest of which dates back to the third century, although every twenty years each is torn down in an act of purification and renewal, then an identical successor raised. Now, at five o'clock in the morning, I boarded the train for Shin Fuji (literally "New Everlasting Life"), where I would pick up a bus for the three-hour ride to one of the midmountain stops before beginning the climb to the summit, 12,395 feet above sea level.

Since antiquity, Fuji has been revered as an embodiment of the nation. It rises from the Pacific coast of the great island of Honshu, about seventy miles west-southwest of Tokyo. Some twenty Buddhist shrines scatter across its slopes, to the very peak, Ken-ga-mine. Its majestic balance of beauty and power has for generations been invoked by nationalists as proof that the Japanese are the chosen people of the East—indeed, of the whole world. One sect, the Fujiko, believes the mountain itself to be divine.*

Fuji was formed by three volcanoes, the youngest of which became active some ten thousand years ago. Over millennia, lava and other outpourings from "New Fuji" boiled over the two older craters, creating a cone nearly two thousand feet across at their conjoint apex. To this day, geologists classify the mountain as potentially active—Krakatoa, after all, was dormant for two centuries before it blew in 1883—but Fuji last erupted on November 24, 1707. Eyjafjallajökull was five years away: I crossed this hazard off my list.

As the train pulled into Shin Fuji, the bus I needed was waiting about eighty yards from the station—and, but for the driver, empty. My stop was the fifth station of ten, just short of halfway to the summit and about 7,600 feet above sea level. I looked upward, out the window: the mountain was cultivated for the first fifteen hundred feet, then passed into grassy moorland, soon giving way to thick forest, in the midst of which we would find Station 5. After about twenty minutes, the bus took on a stockily built man in his mid-forties, with cropped black hair and a set, determined expression. My companion turned out to be a San Francisco accountant and computer analyst named Victor, who was also bound for the summit. I suggested that we might go up together. He took his time, scrutinizing me with some care, before inquiring what expe-

---

* Until the 1870s, no women were permitted to climb, and male pilgrims made the ascent in white robes. The first woman reached the peak in 1832, disguised as a man.

rience I had with this sort of climbing. I admitted I had none, and also acknowledged I was fifty-eight. He was clearly reluctant. "I guess we can try, but
if one of us is holding the other back, we split up. Okay?"

At Station 5, several tourists—there to take photos of the mountain and
visit the gift shop—were getting ready for home. Wandering off on my own, I
bought some clementines, a sturdy rosewood walking stick (complete with a
small Japanese flag and bear-discouraging bells—I promptly removed the bells,
not having heard of any bears on Fuji recently), and an inexpensive flashlight.
I soon found Victor, prodigiously well turned out in black climbing shorts,
black above-the-knee socks (not an inch of skin showing), black cotton gloves,
and a purpose-designed anorak with see-through canisters (each with its
different-colored liquid and plastic straw) strapped to either shoulder. In his left
hand he gripped two shiny steel pitons. "Use them for skiing?" I ventured. He
gave me a pained look before sitting down to put on black leather puttees "to
keep pebbles from getting into the boots." He checked his altimeter, temperature gauge, and compass, and made sure his three maps were ready for easy
consultation. In solidarity, I checked my jacket, hitched up my jeans, and wig-
gled my stick with its Japanese flag. And then we were off, quite on our own as
we strode up the rough dirt path toward the main route of ascent. Much of the
mountain was shrouded in heavy mist, but the outline of its still snowcapped
summit was clearly visible.

Victor and I exchanged life stories as we walked. After about forty minutes,
he pulled up abruptly. "Altitude sickness," he explained, taking a sip from the
bottle on his right shoulder. I was surprised, but slowed down. We continued
like this for a while, with Victor lagging several feet behind. What soon became
apparent was that he was more than just stocky—he was seriously out of
shape. Were I to make the summit in time to see daybreak, I would have to go
at my own speed, so with Victor's consent I set off by myself. It was just after
6 P.M., the Sun already quite low in the sky—for all that it was solstice eve. Perfectly safe, I told myself: thousands made this trip every summer. But the going
was getting harder, with the trickier sections lined with steel stakes linked by
chains. The signposts were in both Japanese and English: BEWARE FALLING
ROCKS AND DANGEROUS WINDS. Okay, I would look out for the rocks; but how did
one avoid winds? KEEP TO THE MAIN PATHS. That seemed simple enough—only it
wasn't, as I was about to find out.

When the Sun finally went down it did so abruptly, sending a chill through
me that was not to do with any loss of heat, but rather that rush of loneliness
one often feels at sunset. I turned on my flashlight, fixing it to my head with a
plastic band to leave my hands free. The terrain varied wildly: sometimes I
would find myself at the bottom of a great concrete stairway set into the

mountain, with maybe thirty high steps (steep for me, even steeper for the average Japanese). At other times, there were twenty feet of shale, which I would have to run up to gain any traction, or a series of gray-black boulders, of varying shapes and sizes, to clamber around or over. The angle of ascent changed too, from about twenty degrees to something far more acute, so I was ascending almost doubled over. The increasing unevenness of the terrain made it impossible to set up any rhythm, and keeping my balance was a constant battle—as was breathing. I paused to catch my breath, turned, and looked back down: a great vista of varied stone, then woodland, then moorland . . . but no Victor, not even the glow of his flashlight, nor anyone else. I felt an involuntary shiver.

As the ascent became more arduous, the signs became less frequent, so that more than once I wandered off the main path and had to retrace my steps. The shale and boulders were now broken by patches of rust-brown slag and knotted limestone crystals—beautiful, but treacherous to negotiate. Some of the pieces underfoot were full of airholes, like a sponge. I put a couple in my knapsack.

I began to feel light-headed, without the energy to take me farther. Altitude sickness in the form of heavy nausea was hitting me hard: even without Victor's altimeter, I reckoned I had hiked five or six miles—not always vertically. I counted the next fifty—seventy—seventy-five steps, then paused for thirty seconds before attempting the next seventy-five. I passed Station 6—an empty cabin surrounded by fresh scars of mineral diggings—and a half-hour later slowly climbed the steep concrete steps to Station 7, again empty. Station 8, I knew, just half an hour's further climb in summer conditions, could sleep more than a hundred people in season.

A few minutes after 8:15 P.M., I stumbled up to its bamboo porch. The door opened, and a dyspeptic Japanese man in his late sixties looked me over, hardly pleased to have a visitor out of season. "Take off shoes!" he barked. Behind him hovered a woman of similar age, his wife perhaps. She looked the more approachable of the two, and in pidgin English and sparrow Japanese she and I settled that I could stay for a few hours, so long as I carried my boots through to the dormitory and didn't make a noise when I rose to leave.

I followed her to a row of double-decker wooden bunks at the back, to find I wasn't the only visitor. Four reporters for the local television station, two of them women, were there to film the construction of a fast new downslope trail, close to the path I had used: the old route, the most dangerous of the three descents, had claimed too many lives. One of the men spoke English, and together we tiptoed outside, where he could smoke and I could share my clementines. Was I going on to the very top—on my own, in the dark? As it was still June, he reminded me, no part of the trail would be lit. Squinting at the thermometer that hung from the veranda roof, he worked out that it would be below freezing at the summit, and the chill factor from strong winds could bring it even lower. I thanked him for the warning, and the two of us made our way back to our bunks.

Sleep came quickly, all three hours of it—for my TV friend kept his promise and woke me at midnight so that I could stay on schedule. Sunrise was at 4:37 A.M., and I was determined to greet it at the summit. Altitude sickness had worn off, the path seemed clear, and I was making good headway when I looked up and almost lost my footing in alarm. No more than twenty feet ahead of me blazed a bright white light, with two serpentine strips of yellow right below it. "Hi there!" shouted a familiar voice. The glare, I realized, was no more than Victor's hand lamp, while the yellow zigzags were the phosphorescent strips sewn onto his lapels. He was sitting on a rock, his first stop since we had set out together over eight hours before. It was good to see him.

He took out his altimeter and estimated that we had more than two thousand feet to go, so we hurried on, talking only a little. I soon left him behind again, and glanced at my watch: nearly 2:30 A.M. The terrain was still erratic—pumiceous lava one moment, then shale, then great hunks of rock, then man-carved steps. On either side, snow-covered country glistened under my flashlight.

An 1857 painting of the sun goddess Amaterasu, with Mount Fuji in the background. Legend holds that she sent her grandson to pacify Japan, and his great-grandson Jimmu became the country's first emperor.

The thin air was draining me again, only this time more acutely, and now I had to stop every thirty steps and take a full minute to recover. I ate some chocolate and my last piece of fruit, and cursed my lack of preparation: Why hadn't I brought altitude pills? Well, the best course was to go ahead at a speed I could manage, but I was racing time—sunrise was in two hours, and it would take every minute to reach the summit. If I ever got there.

So much for the ambition to go "chasing the Sun," I found myself thinking. The phrase occurs in Samuel Johnson's preface to his famous dictionary. He is complaining about the endlessness of a lexicographer's task, seeing "that one enquiry only gave occasion to another, that book referred to book, that to search was not always to find, and to find was not always to be informed." He also speaks of the constricted vision of what could be done, and how pursuing perfection is like chasing the Sun, which, as soon as one reaches "the hill where he seemed to rest," seems just as far away as before. Not that I was after perfection; simply an end to the journey. Every step seemed to take an immense act of will, and I found the Pink Floyd song running through my mind: *You run and you run, to catch up with the sun.*[2]

Fifteen steps at a time now; then ten, taken slowly and at much shorter length than my earlier confident strides. I tried to beat out a tune with my walking stick, letting it bounce in pleasing rhythm off the stones. More than once I turned an ankle on the uneven surfaces. My flashlight dimmed before fading away completely, but luckily it was almost four o'clock—enough natural light to climb by. Then at last there rose above me a carved wooden arch, through which passed a stone stairway flanked by two immense statuary lions.[3] I had reached my goal, the summit of the land of sunrise. I was at the *torii*, the temple gate whose curved wooden beam is meant to invite the Sun, the King of Nature, to come birdlike and perch there.

I turned around slowly to take in all that lay below: a commanding view over the coastline, Fuji's southern face sweeping unbroken to the coast, while to the north and west stretched a line of lakes, where the rivers moving down Fuji's slopes are dammed by volcanic ash. It was just after 4:30 A.M. I sank to the ground and waited. The heavens were a mass of different-textured bruises; then all at once the beginning of the sunrise inched above the skyline, shedding its colors, each so vivid and changing so quickly as it moved upward that I found I was holding my breath, waiting for the next swath of indigo, or purple, or ochre. It was not yet so bright that I had to shield my eyes, but only seconds later the full glare struck home. Now I understood why we say something "dawns" on us, bringing a surge of recognition and inspiration, as if seeing something for the first time. The Sun crested the horizon like an infant newly born, and I savored the moment—atop Fuji on the longest day of the year, with the volcano, the whole panorama, seemingly mine. I couldn't feel possessive; just lucky.

For a while I wandered along the crater's edge, the whole mile and a half of it. A further half-hour went by until below me I spied a familiar figure. Victor had taken off his jacket and donned a kepilike construction. Sweating, but beaming widely from beneath its white flap, he, too, was about to gain the summit of Fuji-san, the sun goddess's own mountain.

PART ONE

The SUN BEFORE SCIENCE

The heroic archer Houyi
shot down nine of the ten
suns, so saving the world, and
is said thereafter to have
reigned over China from
2077 to 2019 B.C.

CHAPTER I

TELLING STORIES

I look upon the sunrise and sunset, on the daily return of day and night, on the battle between light and darkness, on the whole solar drama in all its details that is acted every day, every month, every year, in heaven and in earth, as the principal subject of early mythology.

—MAX MÜLLER,
the nineteenth-century
Oxford professor who
transformed the study
of solar mythology[1]

Man has weav'd out a net,
And this net throwne
Upon the heavens,
And now they are his owne.

—JOHN DONNE[2]

DONNE'S AWED YET MOCKING LINES WERE WRITTEN IN THE EARLY years of the Copernican revolution, but they could apply just as easily to man's attempt to make sense of the heavens—to make them "his owne"—by telling stories. Because all societies have myths about the Sun, their sheer variety is glorious—here it is a magician or trickster, there a ball of fire some figure must carry, another time a canoe, a mirror, or an amazing menagerie of beasts. In Peru and northern Chile, many tribes knew the Sun as the god Inti, who descended into the ocean every evening, swam back to the east, then reappeared, refreshed by his bath.[3] As soon as the horse became domesticated (early in the second millennium B.C.), the Sun was portrayed as guiding a chariot drawn by four flaming steeds. In ancient India, these were termed *arushá*, Sanskrit for "Sun-bright" (the Greek word "eros" shares that meaning, having evolved from the same root as "sun horse"). Birds are often invoked—a falcon, or an eagle, and of course the phoenix, which dies and is reborn from its own ashes. In Africa and India, the tiger and lion are solar animals, sunrise being represented by a young lion, noon by one in its prime, sunset by one in old age.

Where lions are absent, local communities adapt: in the pre-Conquest Americas, the eagle and jaguar are the chosen beasts.

Several cultures described the Sun in more than one way: to the Egyptians, the solar gods numbered not only Ra but Khepri, "the Self-Transforming One," and Harakhty, "the Far One." The Aztecs employed Huitzilopochtli (from *huitzilin,* a hummingbird) to mean both the rising Sun and the star at its zenith, and Tezcatlipoca, "smoking (or shining) mirror," for twilight or evening. The Sun is continually reborn; so that in all they had a jaguar sun, a wind sun, a rain sun, a rain-of-fire sun, and the god Nanahuatzin ("Full of Sores"), who became a fifth solar force, that of the earthquake. Yet whatever form the Sun takes—an eye, a wing, a boat, a dragon, a fish, a bird—there is a common core, a similarity to these tales that spring up in cultures often hemispheres, and millennia, apart.

Sometimes the Sun is seen as so overwhelming a threat that it must in some way be tamed. In ancient Chinese mythology, for instance, the goddess Xihi gives birth to ten suns, which rise simultaneously into the heavens, burning the harvests and all plant life—bar one huge mulberry bush, the *fusang,* on which the suns perch. Every morning the goddess bathes one of them, letting it fly up to her on the back of a crow. One day all the suns escape, and life on Earth becomes unbearable. A variety of monsters scour the land: the ogre Zuochi, with long teeth; Quiying, who kills with water and fire; a giant bird that unleashes the wind, Dafeng; the giant boar Fengxi; and the great serpent Xisushe. The wretched people below endlessly beg the suns to come down, but they refuse. Total destruction impends, until Houyi, a young archer, slays the ogre, the monster, and the giant bird, cuts the serpent in two, captures the boar and—his crowning act—shoots down nine of the suns. Ever since, the story concludes, there has been only the one last sun.

Aesop's fable "The Sun Gets Married" has a different plot but the same threat. One hot summer, word comes that the Sun is to marry. All the birds and beasts rejoice, especially the frogs, until a wise old toad calls for order. "My friends," he tells them, "you should temper your enthusiasm. For if the Sun alone dries up the marshes so that we can hardly bear it, what will become of us if he should have half a dozen little suns in addition?" Two stories, both teaching that one can have too much of a good thing.

Almost all ancient civilizations believed the universe to have existed for unknown ages without benefit of any human intervention. The same did not hold true for the Sun, which in a host of mythologies exists only by virtue of man's nurture. The Hopi of northeast Arizona, for instance, claimed they made the Sun by throwing up a buckskin shield along with a fox's coat and a parrot's tail (to make the colors of sunrise and sunset). But whatever form or character it took, the Sun was rarely cast as fully invulnerable (an old German custom forbade pointing at the star lest one do it harm), and it has been variously depicted as having been freed from a cave, or stolen, or having sprung into life through the self-sacrifice of a god or hero. Among the Inuit of the Bering Strait, all creation is attributed to a Raven Father, who is so annoyed at man's rapacity that he hides the Sun in a bag. The terrified people offer him gifts until he relents, but only to a degree, holding the Sun up in the sky for a time before removing it again.

Every early society personified the cycles of nature, but where the Sun is concerned, cultures have differed on its gender. In the Romance languages the star is male, but in the Germanic and Celtic it is feminine and the Moon masculine: in upper Bavaria the Sun is still spoken of as "Frau Sonne" and the Moon as "Herr Mond." For the Rwala Bedouin of Arabia, the Sun is a mean and destructive old hag who forces the handsome Moon to sleep with her once a month and so exhausts him that he needs another month to recover.[4] Other groups, such as the Eskimo, Cherokee, and Yuchi, also regard the Sun as female, while in Polish the Sun is neuter, the Moon male. These variations may have arisen from climatic differences: in some areas the day is mild and welcoming, hence the Sun tends to be termed feminine, whereas the Moon, ruling the chill, stern nighttime, is male. In equatorial regions, where daytime is searingly forbidding and the night mild and pleasant, the genders reverse. There

The sun chariot of Surya, driven by Arun, who is also personified as the redness that accompanies dawn and dusk

are exceptions: on the Malay Peninsula, Sun and Moon are both regarded as female and the stars as the Moon's children.[5]

Most creation accounts cast the Sun as paramount, both over the Moon and over the heavens. The Book of Genesis declares: "God made the two great lights, the greater light to rule the day, and the lesser light to rule the night."[6] The Egyptians referred to Sun and Moon as "the two lights," the right and left eye, respectively, of Ra—the left being described as weaker, because damaged. In Central and South America and among the Mundas of Bengal, Sun and Moon are man and wife. The Bengalis charmingly call the Sun "Sing-Bonga," believing him a gentle god who does not interfere in human affairs. Another myth of the same region fashions the star as a man with three eyes and four arms who is abandoned by his wife because his dazzle wearies her. She installs Chhaya (Darkness) in his place, but the Sun wins her back by reducing his effulgence to seven-eighths of his original brilliance (an interesting example of the spirit of compromise making companionship possible). Many stories are told about such marital troubles, it being a given that Sun and Moon can never live happily together.

It occurred to some of the more sophisticated ancient cultures to wonder why, if the Sun were indeed so powerful, he had to abide by strict laws rather than roam at will. Surely only a slave would perform so repetitively? Numerous legends were devised to explain this thralldom. The Sun was portrayed as erratic, sometimes hurrying too fast, at other times dawdling, coming too close to Earth one moment, the next moving too far away. The sixteenth-century poet Garcilaso de la Vega, one of the first bicultural Spanish Americans, tells the following story about Huayna Capac, greatest of Inca conquerors:

> One day this ruler stares directly into the rays of the Sun, and his high priest has to remind him that their religion forbids this. Huayna Capac replies that he is his king and pontiff. "Is there any amongst you who would dare command me to rise and undertake a long journey?"
>
> The high priest answers that this would be unthinkable.
>
> Huayna Capac continues: "And would any of my chieftains, no matter what his power or worldly estate, refuse to obey me if I should command him to travel to far-away Chile?"
>
> The high priest acknowledges that no chieftain would.
>
> "Then," says the Inca, "I tell you that this our Father the Sun must have a master greater than he, who thus commands him to journey across the sky day after day with never a respite, for if he were the Supreme Lord he would surely sometime cease traveling and rest."[7]

The Greeks, too, put the Sun in a somewhat less than exalted position; Homer does not even grant Helios a place among the Olympians. Nor is the Sun seen as always beneficent: in Mesopotamian myth, the solar god Nergal brings plague and war, his weapons being heat, parching winds, and lightning. Throughout history there remains a deep ambivalence: humanity cannot do without the Sun's power, but still wishes to tame or seduce it, to limit its hold over us.

WHAT IS THAT HOLD? In the latter half of the nineteenth century a remarkable scholar would make the Sun the focus of his research: Friedrich Max Müller. He would argue that the Sun lay at the root of language, and thus of all major myths, not just the obviously solar ones. Müller was born in 1823 in Dessau, then the capital of a small state within the German Confederation, the son of a poet. Initially he studied Sanskrit, which kindled an interest in philology and religion. He embarked on a translation of the Rig Veda, the sacred hymns of Hinduism, and in 1846 traveled to Britain to research the archives of its Indian empire, supporting himself by writing fiction—his first novel, *German Love*, becoming a bestseller. He stayed on, and in 1854 was appointed professor of modern languages at Oxford. Fourteen years later, he was made professor of comparative philology as well, and later yet the university's first professor of comparative theology. The breadth of his knowledge, along with the fact that he spent years preparing a massive fifty-volume English translation of *The Sacred Books of the East*, may well have made him the model for George Eliot's Dr. Casaubon—the pedant engrossed in his never-ending lifework *The Key to All Mythologies*—in her novel *Middlemarch*, published in 1871, when Müller's reputation was at its height.

In his time this German-born Oxford academic was a truly famous figure, his friends and acquaintances spanning two generations of the British intellectual elite: Macaulay, Tennyson, Thackeray, Ruskin, Browning, Matthew Arnold, Gladstone, and Curzon, among many others. Queen Victoria twice offered him a knighthood, which he declined as inappropriate. When he died, his widow received condolences from kings and emperors. In all, he wrote more than fifty books. His last words were, unsurprisingly, "I am tired."[8]

In his masterwork, *On the Philosophy of Mythology* (1871), he set about showing that the same kinds of stories, the same traditions and myths, could be found worldwide, and that the appearance and disappearance of the Sun and its worship as the source of life were the basis of most mythological systems. From the earliest times, man constructed his understanding of the world around the Sun.

Max Müller at age thirty, shortly after arriving in England

What we call the Morning, the ancient Aryans called the Sun or the Dawn. . . . What we call Noon, the Evening, and Night, what we call Spring and Winter, what we call Year, and Time, and Life, and Eternity—all this the ancient Aryans called *Sun.* And yet wise people wonder and say, how curious that the ancient Aryans should have had so many solar myths. Why, every time we say "Good morning" we commit a solar myth. . . . Every "Christmas number" of our newspapers— ringing out the old year and ringing in the new—is brimful of solar myths.[9]

More than a century on, we tend to take the validity of Müller's main arguments for granted. In his own time, however, he seriously overreached: his insistence that *every* myth derives from the Sun, as well as his emphasis on the primacy of Aryan mythology and his eagerness to converge *all* languages back toward a single common root, provoked a bitter battle between his camp and those who took different paths. The cause of solar panmythology lost its leading light with his death in 1900. But though Müller's work is now known to only a few, he remains a major figure in our understanding of solar myths.

THE SUN'S PLACE in the world's mythologies was taken up once again in 1923 by William James Perry (1887–1949), a cultural anthropologist at University College, London, and the anatomist Grafton Elliot Smith (1871–1937) when that year they coauthored *Children of the Sun,* which argued that during mankind's early history there were groups of people on most continents who believed themselves to be the progeny of a sun god. Unrepentant heliocentrists,

Perry and Smith contended that "the importance of this fact in the history of civilization, and especially in the study of mythology and tradition, cannot be exaggerated."[10]

They dated the first appearance of self-proclaimed descendants of the gods to around 2580 B.C. Claiming to be the actual progeny of Ra, the members of the pharaonic dynasties believed that at some point the Sun had come down to take the place of the king on Earth, thus making them his descendants. The subjects of the king were taught never to look directly at him; rain or sunshine were his to summon; he was master of magic and giver or withholder of the harvest.[11] The Egyptians took the divine nature of kingship further than any other society—although it was the Roman emperor Vespasian (A.D. 9–79) who joked on his deathbed, "Damn, I think I'm turning into a god."

Perry and Smith discerned similar belief systems among the Asuras of India, the Timurids of Indonesia, the Abarihu of San Cristobal in the Solomons, the inhabitants of many parts of Polynesia, New Zealand, and the Eastern Pacific, the Inca, the Mayans, and several North American tribes, and concluded, "Wherever it is possible to examine the ruling classes of the archaic civilization, it is found that they were what are termed gods, that they had the attributes of gods, and that they usually called themselves the Children of the Sun."[12] Like Müller, they finally pushed a valuable insight too far (the countries where such Children of the Sun held sway show up on a world atlas only in particular areas), yet they did identify a remarkable cultural pattern, which has enjoyed preeminence for thousands of years.

Perry and Smith then outlined the next cultural shift: as these societies

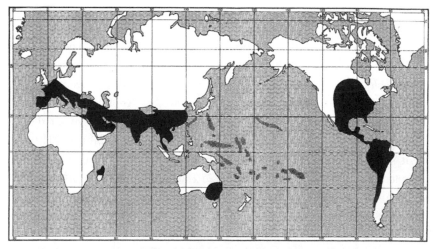

A map of the distribution of solar cultures around the world,
showing where Children of the Sun were to be found

sought to expand, they became more bellicose, and the Children of the Sun were transmogrified into war gods. For instance, as the Maya broke out of Guatemala into southern Mexico, their culture came to institutionalize cruelty and aggressiveness. The same applied to the Aztecs: "War, at first *defensive,* afterwards *offensive,* became the life of the tribe."[13] They found the same sequence in Melanesia, Polynesia, and North America. Like Müller, Perry and Smith asserted that their insights held true over enormous time spans. Just the major societies so far mentioned stretch over four thousand years: those in the Nile Valley run from 3200 B.C. until their conquest by Islam around A.D. 700. China's ancient culture can be argued to run from about 2800 B.C. until the end of the Tang Dynasty in A.D. 907—and so on.

In 1925, another man who would play a key role in reimagining solar mythology, the pioneering psychoanalyst Carl Jung, made two journeys that together shaped his philosophy. On his visit to the Pueblo Indians of Taos, New Mexico, the forty-nine-year-old traveler found himself deep in conversation with one of the tribe's elders. As they sat together on the roof of his house, the old man pointed at the Sun blazing overhead: "Is not he who moves there our father? How can there be another god? Nothing can be without the sun."

Jung in his turn inquired whether the Sun might not instead be a fiery ball shaped by an invisible god. "My question did not arouse astonishment, let alone anger," he wrote.[14] "His only reply was, 'The sun is God. Everyone can see that.'"

"We are a people who live on the roof of the world," continued the old man. "We are the sons of Father Sun, and with our religion we daily help our father to go across the sky. We do this not only for ourselves, but for the whole world. If we were to cease practicing our religion, in ten years the sun would no longer rise. Then it would be night forever."[15] Jung suddenly grasped the sun myth's central role in giving the tribe a sense of meaning and purpose.[16]

His own dreams had convinced him that human beings have an unquenchable longing for light;[17] and a trip he took to Africa several months later, to investigate the myths of the mountain peoples dwelling between Mombasa and Nairobi, bore out this theory. There he watched as they waited each day for "the birth of the Sun in the morning," for, as they explained, "The moment when the sun appears is God." For Jung, this daily celebration of the Sun in a place so far away from the Pueblo in Taos confirmed the universality of this primal longing (even if the Africans, unlike the Pueblo, saw the Sun as the *means* of creating light, not a supreme being).

For several decades after Jung's journeys, scant organized research was conducted on the religious practices of preliterate peoples, but in 1958, the great Romanian historian of religion, Mircea Eliade (1907–1986), took issue with all the heliocentric ideas that had been put forward, arguing that the Sun was "simply one of the commonplaces of what is in a vague sense religious experience,"[18] and that in the agrarian societies that followed the ancient hunter-gatherer cultures, in which farmers had already become dependent on the changes of the seasons for their crops and especially upon warmth and light, the solar deity turned from a god into a "fecundator" (his special word). For all the innumerable myths, Eliade wrote, only certain cultures (the Egyptian and many of the Indo-European and Meso-American civilizations) had developed true solar religions.

It seems to me, though, that Eliade was building on Jung's insights, not negating them. That so many cultures continued for hundreds, sometimes thousands, of years to hold on to myths of solar supremacy shows that such stories were central to their sense of self—and also tells us something about the Sun's enduring hold over mankind.

Audrey Hepburn and Gregory Peck in Roman Holiday, *standing in front of the Bocca della Verità ("Mouth of Truth"), the massive marble sun disk (second century B.C.) in the Church of Santa Maria in Cosmedin, Rome. Anyone who tells lies, Peck explains, and inserts their hand in the mouth of the god will suffer "the avenging justice of the Sun" and have it bitten off.*

IN JUST WHAT WAY, I decided to see for myself. In June 2004, I set off for Peru to witness Inti Raymi, the ancient festival that honors the Inca summer solstice. It takes place in Cuzco, over two miles up in the Andes, and opens in the northeastern quadrant of the town square. By the time I arrived, the skies had opened, a cloudburst was hammering down, and the crowd's red, blue, yellow, and green plastic rainwear rivaled the vividness of the actors—playing the pages of the long-slain Inca's court, the troops of orange-clad "court virgins," the Inca soldiery (Peruvian regulars, seconded for the occasion), and the royal entourage in all its splendor. After about an hour, we moved on some three hundred yards into Qorikancha, the Square of Gold, where messengers sounded conches to placate the Apu Inti (sun god), and the Willaq Uma (high priest) intoned exhortations in Quechua, the Indian language still spoken by some 20 percent of Peruvians. The woman next to me explained that local actors had been practicing for a month. Her husband, a jeweler, was playing one of the attendant priests; the high priest's acolyte was busily munching gum.

The third and final part of the day's events was held a short bus journey away, at Saqsaywaman—pronounced "sexy woman" but meaning "satiated eagle"—where a range of huge joined stones forms the most important Inca monument after Machu Picchu.* Their sheer size is remarkable—some weigh as much as ninety tons. Between 1533 and 1621 they were rolled on tree trunks greased with vicuña oil from a quarry more than half a mile away and arranged into the twisting form of a huge snake: a mighty presence of the old religion, yet somehow permitted by the Spanish. Every five yards, the blocks have been cut either "male" or "female," with prominences or sockets accordingly, and fit together so tightly that I couldn't get a sharp blade between them. This great wall formed the backdrop to the main celebration.

Our group's guide, Odilia, a small, well-organized woman in her late forties, mentioned that when she was a child her grandmother (now ninety-four) had told her that behind the stones lay a great earthen hill where once stood a palace and three astronomical observatories. Sure enough, in 1987 all four were discovered. Today excavations are going on all the time, and 328 stones used for solstice celebrations have already been found. Odilia next taught us how to pray to the Sun, leaning forward slightly, not so much out of reverence but to present the softest part of one's head to the great orb above. She removed her shoes and stretched out her arms, fingers splayed, to keep her body in touch with both Sun and Earth, the two great sources of life.

I wandered away from my group, to be swept up by hundreds of locals— women in ancient traditional costume, men decked out as conquistadors or matadors or draped in nothing but leopard skins, boys (no girls) sounding conches, horns, or cymbals, strumming guitars, or banging drums. Vendors

---

* In Machu Picchu still stands a stone column called an *intihuatana,* meaning "hitching post of the Sun" or, literally, "for tying the sun," the ceremony for which was to prevent it from escaping. The Spanish conquest never found Machu Picchu, but it destroyed all the other *intihuatana,* so ending the practice.

were offering everything from the ubiquitous tapestries of pumas and llamas to hats, coca tea, finger puppets, postcards, sodas, candies, city flags, ponchos, film equipment, chess sets. There was still a fine drizzle, so the gaudy plastic covers were being bought in numbers. Amid the clamor many were dancing, including an impossibly ancient crone who spun around as energetically as revelers a third her age.

In one corner, just off the main festival precinct, there bustled a traveling fair, complete with slides and merry-go-rounds. Jugglers and street artists performed their acts every few yards; sweaters, necklaces, and brooches in every imaginable style were being hawked; and chickens bubbled in huge black stewing pots tended by women in bright shawls, beads, and hats. The crowd was raucous, good-natured: this was their day.

The ceremony began at 2 P.M. On this great plateau a large rectangular stage had been constructed, with curtains painted to look like stones. Highland Indians from the four corners of the ancient empire trotted in to join their leader, who was poised to drink the sacred *chicha,* made from fermented corn. Some five hundred locals took part, roared on by an audience of forty thousand. Fires burst out at each corner of the plateau (the one nearest us sputtered fitfully and kept having to be relit); the conches were sounded yet again, yet more frenetically. The basic ceremony was much like a Christian Communion service, with girls decked out as Inca princesses offering up baskets of sacred bread (while trying to restrain their giggles). The blazing colors of their costumes took one's breath away. No Inca virgin had her heart ripped out to unveil the future, however, as was essential before the Conquest, and even the llama sacrifice was playacted—the animal was eventually set free to bleat its tale.

Something of genuine depth seemed to be asserting itself behind the commercial gloss: the pride of the long-conquered in their enduring ways, and even a sense that the star itself was, if no longer divine, still a very special power. To crown it all, just as the drizzle died away, the Sun came out and a rainbow magically appeared. The sun god, it seemed, had heard our prayers, and the solstice had been duly honored. As I walked away from the sacred field I wondered how anyone could doubt the Sun's control over us, in some essential sense still the prime organizing principle of our lives.

CHAPTER 2

CELEBRATING THE SEASONS

'Tis the year's midnight, and it is the day's,
Lucy's, who scarce seven hours herself unmasks;
The sun is spent, and now his flasks
Send forth light squibs, no constant rays;
The world's whole sap is sunk.

> —JOHN DONNE, "A Nocturnal Upon
> St. Lucie's Day, Being the Shortest Day,"
> c. 1611[1]

The summer solstice, the day the sun is said to pause, he thought. Pleasing,
that idea. . . . As though the universe stopped for a moment to reflect, took a day
off from work. One could sense it, time slowing down.

> —ALAN FURST, Dark Star[2]

AT ONE POINT IN HUCKLEBERRY FINN'S FLIGHT DOWNRIVER, HE COMES
across a copy of The Pilgrim's Progress (a book "about a man that left his family,
it didn't say why"), which he leafs through and decides, "The statements was
interesting, but tough."[3] Studying the Sun may provoke the same reaction, for
there is no easy way to explain how it works. We may think of it as rising daily
in the east and setting in the west, but it does so at different times, for different
durations, in different places. As Barry Lopez observes in his great book on the
Arctic, "In a far northern winter, the Sun surfaces slowly in the south and then
disappears at nearly the same spot, like a whale rolling over. . . . The thought
that a 'day' consists of a morning and a forenoon, an afternoon and an
evening, is a convention, one so embedded in us that we hardly think about
it."[4] The Sun is not a simple star, particularly in regard to the seasons.

During the course of a year the hours of sunlight vary from the maximum
(on the day known as the summer solstice, which augurs the beginning of
summer) to the minimum (the winter solstice, which signals winter). In summer, the Sun is brighter and reaches higher into the sky, shortening the shad-

ows that it casts; in winter it rises and sinks closer to the horizon, its light diffuses, and its shadows lengthen. As a hemisphere tilts steadily farther away from the Sun, daylight becomes shorter and the Sun makes ever lower arcs through the sky. On the first day of winter, the time of briefest day, the Sun rises at the point nearest the equator. On the year's longest day, it rises closest to that hemisphere's pole. On both occasions it seems to stop in its path before beginning to double back. ("Turnings of the Sun" is an ancient idiom, used by both Hesiod and Homer.) We see that effect at dawn, when, for two or three days, the Sun seems to linger for several minutes—hence the word "solstice," from *sol* (sun) and *stitium* (from *sistere*, to stand still: "armistice" translates as "weapon standstill").

In the Northern Hemisphere, the first days of spring and fall occur between March 19 and 21 and September 19 and 21. These are the equinoxes (from *aequa nox*, equal night), when the hours of dark and light are roughly even all over the planet and the Sun appears directly overhead at midday, after which it will keep advancing, northward from March 21 on and southward from September 21 on. Many astronomers, as well as mariners, believed that the equinoxes brought strong gales—a myth, or at least a misconception, possibly because of the sharp rise in the number of high winds in the second half of September, when the hurricane season does indeed reach its height in the Northern Hemisphere (but for reasons differently related to the Sun). They were not alone: Herodotus, recording the summer floodings of the Nile, speaks of the Egyptian Sun being "driven from his old course by storms" during winter, then "returning to the middle of the sky" in spring.

Societies that were organized around agriculture—such as those in early Egypt and the Americas—intently studied the heavens, carefully noting what happened each year and ensuring that the solstices and equinoxes were well charted. Despite their best efforts, however, their priests and stargazers came to realize that it was exceptionally hard to pinpoint the moment of solstice by observation alone—even though they could define the successive seasons by the advancing and withdrawal of daylight and darkness (albeit by another twitch of nature the earliest and latest sunrises and sunsets do not fall at the solstices).

The Earth further complicates matters. Our globe tilts upon its axis like a spinning top (orbiting the Sun at an angle of 23°40'), and this determines how much sunlight any part of the planet receives at any given time. Not only does it spin; its shape minutely alters and its axis wobbles, a process called nutation, "nodding," so that the Earth's orbit fluctuates between greater circularity and greater elongation. If the globe's axis didn't wobble and if its orbit were a true circle, then the equinoxes and solstices would quarter the year into equal parts; but our orbit being elliptical, the time between the spring and fall equinoxes in the Northern Hemisphere is slightly greater than that between fall and spring, the Earth moving about 6 percent faster in January than in July. The Sun takes (in round numbers) 94 days to go from the vernal equinox to the summer solstice, 92 days from there to the autumnal equinox, 89 days from there to the winter solstice, and 90 days back again to the vernal equinox. At high latitudes, the amount of solar energy received at midsummer can vary by 20 percent, depending on whether the Earth's various wobbles reinforce one another or cancel one another out. Tough, but interesting.

THE APPARENTLY SUPERNATURAL power manifest in solstices and equinoxes to govern the seasons has been felt as far back as we know, inducing different reactions in different cultures—fertility rites, fire festivals, offerings to the gods. Many of the wintertime customs in Western Europe descend from the ancient Romans, who believed that their harvest deity, Saturn, had ruled the land during an earlier age of rich crops, and so celebrated the winter solstice, with its promise of a return of summer, with a "Saturnalia," a high feast of gift giving, role reversals (slaves berating their masters), and general public holiday each year from December 17 to 24. They marked planting and harvesting with a ritual in which they sacrificed to the Sun a winning horse from one of the great chariot races. The historian Macrobius explained these celebrations:

> Certainly it is not empty superstition but divine reason that makes them relate almost all the gods—at any rate the celestial gods—to the Sun. For if the Sun, as men of old believed, "guides and directs the rest of the heavenly lights" and alone presides over the planets in their courses, and if the movements of the planets themselves have power, as some think . . . to foretell the sequence of human destinies, then we have to admit that the Sun, as directing the powers that direct our affairs, is the author of all that goes on around us.[5]

The transition from the Roman Empire and its pagan rituals to Christianity, with its similar rites, took place over several centuries. These were years of great upheaval—between A.D. 235 and 284, twenty-six different usurpers seized power in Rome—culminating in Constantine's great triumph at the Milvian Bridge in 312, which reunited the empire and ended half a century of civil war. Ascribing his victory to the Christian god, Constantine was later baptized in the new religion and enacted laws to advance Christianity. Many other customs were now appropriated and refashioned, as the Sun and God's Son became inextricably entwined in people's minds. Thus, although the New Testament gives no indication of Christ's actual birthday (early writers preferring a spring date), in 354 Liberius, Bishop of Rome, declared it to have occurred on December 25. The advantages of "Christmas Day" being celebrated then were obvious: as the Christian commentator Syrus wrote:

> It was a custom of the pagans to celebrate on the same December 25 the birthday of the Sun, at which they kindled lights in token of festivity. In these solemnities and revelries the Christians also took part. Accordingly, when the Church authorities perceived that the Christians had a leaning to this festival, they took counsel and resolved that the true Nativity should be solemnized on that day.[6]

In Christendom, the Nativity gradually absorbed all other winter solstice rites, and the co-opting of solar imagery was part of the same process.[7] Thus the solar disks that had once been depicted behind the heads of Asian rulers became the halos of Christian luminaries.* Then the question arose as to which day of the week should be given up to celebrating the Mass: the Church father Justin Martyr would shrewdly explain to the emperor Marcus Aurelius that Christians chose the day they did for celebrating the Eucharist because "it is on what is called the Sun's day that all who abide in the town or the country come together . . . and we meet on the Sun's day because it is the first day on which God formed darkness and mere matter into the world."

For a long time, the fixing of festivals remained something of a hit-or-miss affair. The Romans had been unsure when to celebrate the winter solstice. Julius Caesar officially marked the shortest day of the year as the twenty-fifth. Pliny, in the first Christian century, put it at December 26, while his contemporary, Columella, chose the twenty-third. By 567 the Council of Tours declared that the whole period from Christmas to the feast of the Epiphany on January 6 should form one cycle, and by the eighth century a system of twelve days of celebration was in place, with the Epiphany—day 13—starting a new se-

---

* Christian art at first avoided the halo, because of its pagan origins; but from about the mid-fourth century, Christ was portrayed wearing a crown of sun rays, as in effigies of the Roman emperor, then from about the sixth century on, the Virgin Mary and other saints were depicted with halos, and by the ninth the symbol had been completely assimilated. Again, some Christians were reluctant to use the pagan word "Sunday"; in Latin-speaking Christian countries, then in most Romance languages, the first day of the week became forever "the Lord's reign"—*dominica, dimanche, domingo, domenica*.[8] During Constantine's rule, the first day of the week was regarded by both pagans and Christians as its high point, and in 321, when he officially proclaimed Sun Day as the holy day of the week, some Christians believed it was for their sake, and began calling the day "Sunday" rather than the Lord's Day.

quence. Thus was created, in the words of Ronald Hutton, "a period between the winter solstice and the traditional (Roman) time of the New Year, in which politics, education, commerce were alike suspended, of peace, privacy, domesticity, revelry, and charity: a sacred period in a much broader sense than that associated with any one religion."[9]

Despite Christianity's apparent supremacy, many of the old customs survived. Not least, feasting during the festive cycle could become an end in itself—so much so that Church elders worried that the veneration of Christ was being lost. Saint Gregory Nazianzen (c. 276–374) urged upon his flock "the celebration of Christ's birth after a heavenly and not after an earthly manner," complaining they were "feasting to excess, dancing, and crowning the doors [i.e., decorating doorways]." More than a century later, both Augustine of Hippo and Pope Leo the Great again felt compelled to remind their people that Christ, not the Sun, was their proper object of worship; Saint Patrick, in his *Confessio,* declared outright that any sun worshipper would be damned eternally.

It should also be remembered that while Roman Christianity was the dominant culture in far the greater part of Western Europe, it was by no means the only one. In the eleventh century, the Danes conquered most of England, bringing with them Yule, their colloquial term for winter solstice celebrations, probably derived from an earlier term for "wheel." For centuries, the most sacred Norse symbol had been the "wheel of the year," represented by a six- or eight-spoked wheel or by a cross within a wheel, the arms of the cross signifying solar rays. The Norse peoples, many of whom settled in what is now Yorkshire, would construct massive solar wheels and place them on the tops of hills, while in medieval times processions bore wheels upon chariots or boats; and in some parts of Europe a taboo on using spinning-wheels during solstices lasted well into the twentieth century. The spinning-wheel on which Sleeping Beauty pricks her finger may exemplify this superstition, while the seven dwarves may represent the (then-reckoned) seven planets—as medieval lore had it, the seven greatest attendants of Earth.

An early-nineteenth-century miscellany elaborates on another connection between the wheel and the solstice. According to the authors, in much of Western Europe the wheel was an integral part of the summer solstice celebrations:

> In some places they roll a Wheel about, to signify that the Sun, then occupying the highest place in the Zodiac, is beginning to descend . . . we read that this Wheel was taken up to the top of a mountain and rolled down from thence; and that, as it had previously been covered by straw,

twisted about it and set on fire, it appeared at a distance as if the Sun had been falling [and] the people imagine that all their ill-luck rolls away from them.[10]

Dances celebrating midsummer might include spinning wheels around, and could go on for a whole month.*

A nineteenth-century engraving of celebrants jumping a bonfire during summer solstice festivities at St. Jean, in Alsace

Blazing wheels were often accompanied by festive fires, lit in sacred reenactment of the creation of the universe, a tradition at both summer and winter solstices. From at least the thirteenth century on, the lighting of fires on Midsummer Day became common across the whole of Europe, northwest Africa, Japan, and even Brazil. By the mid-nineteenth century, Cornish migrants in South Australia were celebrating the traditional European Midsummer with a bonfire on June 24—midwinter in the Southern Hemisphere. Up until about the same date, in villages of the north of England, whole communities would meet together at summer solstice "and be merry over a large fire,

---

* The feast day of Saint Vitus, the cult of whom took over from the sun god's in some countries, falls on June 15, close to Midsummer Day. In 1370, an epidemic of chorea broke out in Germany, especially along the Rhine Valley, its victims falling into involuntary convulsions. Their sad antics so resembled the annual celebrations that the disorder became known as "Saint Vitus' dance."

which was made for that purpose in the open street." This, of whatever materials it consisted, was called a Bonefire, because such fires were generally made of bones; there is a nice homily, of unknown date, *De festo sancti Johannis Baptistae:*

> In worship of Saint Johan the people waked at home, and made three maner of fyres: one was clene bones, and noo woode, and that is called a Bone Fyre; another is clene woode, and no bones, and that is called a Wode Fyre, for people to sit and wake therby; the thirde is made of wode and bones, and is called St Johannys fyre.[11]

Midsummer Day was a magical time and, far more than the winter solstice, an occasion for myriad superstitious doings. Shakespeare, inspired by the merrymaking and tomfoolery of the season, set *A Midsummer Night's Dream* at the solstice. Norwegian court records tell of how witches, fortified with home-brewed ale, would fly through the air on Midsummer Eve, often on the backs of cats,[12] while in Russia, Belarus, and the Ukraine it was believed that naked she-devils, climbing up through chimneys, would launch themselves upon the air. Francis Grose, a late-eighteenth-century original who began as a British officer of dragoons deployed to catch smugglers before switching to a career as an artist and antiquary, wrote: "Any unmarried woman fasting on Midsummer Eve, and at midnight laying a clean cloth, with bread, cheese, and ale, and sitting down as if going to eat, the street-door being left open, the person whom she is afterwards to marry will come into the room and drink to her by bowing; and after filling the glass will leave it on the table, and, making another bow, retire."[13] He adds that anyone who fasted on solstice eve and sat in a church porch at midnight would see the spirits of those parishioners fated to die that year knocking at the church door.

In some countries, the celebratory fires of Midsummer Eve were also meant to ward off evil spirits, and so were crowned with straw dolls, or witches' brooms and hats; while the encroaching darkness at the time of the winter solstice never failed to arouse fears that the Sun would burn out completely—hence the kindling of the Yule log and the Hanukkah lights (the latter tied to both lunar and solar calendars). The boar's head at Christmas feasts represents the dying Sun of the old year, a suckling pig, with the apple of immortality in its mouth, the new.

Outside Europe, customs varied widely. In Japan, young men known as "sun devils," their faces daubed to represent their imagined solar ancestry, would go from farm to farm to ensure the Earth's fertility. In China, the winter solstice honored the masculine (or *yang*) energies, the summer solstice the feminine (or *yin*); in early times, a sacrifice, often human, was offered to the Sun, the smoke joining Heaven and Earth. The Aztecs, who believed that the heart held fragments of the Sun's heat, ensured his continual well-being by tearing out this vital organ from hunchbacks, dwarves, or prisoners of war, so releasing the "divine sun fragments" entrapped by the body and its desires. Even sporting events were assigned a role: in 2005 I visited a sports field at Chichén Itzá in central Yucatán and was shown a fifteenth-century carving that depicted the bloody aftermath of one game the Mayans had played there. In a contest intended to guarantee the Sun's rebirth at the time of the vernal equinox, two teams, personifying light and darkness, played *ullamaliztli*, in which the ball represented the Sun's path on its nightly passage beneath the world and the players, using upper arms and hips, had to prevent it from hitting the ground or going out of bounds. Alas for the captain of the losing team, however, because he became the sacrifice whose blood revitalized Sun and Earth at the culminating ceremony.

In modern Mexico, this burning sun-ball contest reenacts games with ritual solar associations that were played for three thousand years in Mesoamerica up until Hispanic times.

Above all other rituals, reproducing the Sun by creating fires on Earth is the commonest solstice practice, at both midsummer and midwinter. (In Iran, families still observe the originally Zoroastrian winter solstice festival Zalda with nightlong vigils, fires blazing against the darkness; similar celebrations take place in Tibet and Muslim northwestern India.) Thomas Hardy, describing

Dorset villagers around a bonfire in *The Return of the Native*, offers an explanation for such a worldwide phenomenon:

> It seemed as if the bonfire-makers were standing in some radiant upper story of the world, as if these men and boys had suddenly dived into past ages and fetched therefrom an hour and deed which had before been familiar with this spot. . . . Indeed, it is pretty well known that such blazes as this the heathmen were now enjoying are rather the lineal descendants from jumbled Druidical rites and Saxon ceremonies than the invention of popular feeling about [the] Gunpowder Plot.*
>
> Moreover to light a fire is the instinctive and resistant act of men when, at the winter ingress, the curfew is sounded throughout Nature. It indicates a spontaneous, Promethean rebelliousness against the fiat that this recurrent season shall bring foul times, cold darkness, misery, and death. Black chaos comes, and the fettered gods of the Earth say, Let there be light.[15]

WHILE SOLSTICE CELEBRATIONS have always been the most common, other feasts to mark the changing of the seasons have taken place throughout the year. The French historian Emmanuel Le Roy Ladurie wrote an entire book about one carnival that took place in the town of Romans in 1580, and records the appearance of what has come to be known as the Candlemas bear. Every February 2 the beast would be prodded from its den to peer at the sky.[16] Were it cloudy, superstition held, the bear would be unable to see its shadow and so would behave as if winter were on the wane; but if it saw its shadow (because the weather was bright) it would be frightened back into its hole, and winter would endure for another six weeks. The bear even now is employed at such festivals in the Alps and Pyrenees, while elsewhere other hibernating animals perform a similar role—Ireland has a hedgehog for Saint Bridget's Day (Febru-

---

* On November 5, 1605, Guy Fawkes and his fellow conspirators were detected attempting to blow up king, Lords, and Commons, an event that became known as the Gunpowder Plot. The Parliament that so narrowly escaped destruction was postponed until the beginning of the following year, when an MP proposed a perpetual annual thanksgiving, whose service remained in the Anglican prayer book until 1859. By the 1620s it had become the most popular state commemoration in the calendar, surpassing by far even the king's or queen's birthday. In the last decades of the eighteenth century, by "ten o'clock, London was so lit up by bonfires and fireworks, that from the suburbs it looked in one red heat."[14] No longer witches' brooms but effigies of Fawkes, or sometimes of the Pope (often stuffed with live cats, who would shriek as the flames engulfed them) were paraded atop each fire.

ary 1), while Pennsylvania and other parts of North America celebrate Groundhog Day every February 2—a folk custom introduced by French and German settlers.

The Romans festival, along with so many others in France, came under attack during the Revolution of 1789. The leaders of that great convulsion were determined to abolish traditional Christian holidays. Sundays were to be workdays; a new calendar would be produced, a new order promulgated. How to begin the year was easy to settle, since the Republic had been declared on the autumn equinox of 1792. On that day (a revolutionary firebrand declared to the Convention in Paris), the flame of liberty illuminated the French nation. The new date—1 Vendémiaire, the first day of the month of gales—was given an explicitly "celestial" meaning, as is clear from the countrywide events that were planned to commemorate it. The municipal administration at Aurillac, in the Cantal (one of the eighty-three departments created by the Revolution) in south-central France, described how this glorious pageant should be celebrated:[17]

> Then will appear a superb, shining chariot drawn by twelve horses; this will be the chariot of the Sun; before it will be the signs of the zodiac and the hours according to their new division; beside it will walk young citizenesses, dressed in white, who will represent the hours of the day, while young citizenesses dressed in white and veiled in black will represent the hours of the night; on the chariot will be seen the still slumbering genius of France covered with a veil decorated with fleurs-de-lys and from time to time making simple movements.

For about a decade, such festivals were enthusiastically celebrated throughout France. But such zeal could not last, and soon the old festivals and calendars returned. The revolutionary system became part of history, with the life-bursting spring month of Germinal (March 21 to April 19) memorialized in the title of Zola's 1885 novel—and Thermidor (July 19 to August 17) in a lobster dish.

SUN CELEBRATIONS cut across time and culture, but possibly the most dramatic are the sun dances of the North American Indian, usually held in late June or early July before a buffalo hunt, or to recruit the power of the Sun cycle over crops. Most sun dances centered on a lodge, usually a circle of twenty-eight forked poles that represented the twenty-eight days of the lunar month. These would be joined together by more poles laid horizontally into the upright forks. A witness likened one such to "a circus tent, the top of which has been ruthlessly torn away by a cyclone."[18] Entrances to the lodges faced the sunrise. Sometimes a host of lodges—as many as seven hundred, enclosing more than ten thousand celebrants—would be arranged in one huge circle as much as six miles in circumference.[19]

For all these clans, the rite often took place under a full moon (whatever its other benefits, a sun dance was the occasion for energetic couplings). Each tribe had its own characteristics. The Shoshoni and Crow called their ceremony "the Thirst Lodge," the Cheyenne "the Medicine [in the sense of big magic] Lodge," and for the Lakota Sioux it was *wiwanyag wachipi*, "the Dance Staring at the Sun."[20] In Arizona the Papago hopped around a solar symbol in alternating directions, extending their hands toward the Sun, then stroking their chests to take in its power.

The spectacle reached its fullest development among the Teton Sioux. Participants congregated in a sweat lodge, a kind of sauna that served as a preliminary purification. The medicine men would pray for fair weather, and a dance leader would bring together an ornamented pipe, a perfectly formed buffalo skull, and some buffalo fat; then would follow the setting up of the "sun pole" (representing a path to the sky), a tobacco offering, and a dance of supplication. These preparations would take three to four days, during which the initiates would fast, their mounting hunger sometimes bringing spectacular visions. In more intricate versions, participants would raise tents in a circle representing the aurora borealis around a dance enclosure made of brush some fifteen feet west of a central pole symbolizing the Sun.

In *The Origin of Table Manners*, Claude Lévi-Strauss explores the ambiguous and equivocal nature of these dances:

> On the one hand, the Indians prayed to the Sun to be favorably disposed towards them. . . . On the other hand, they provoked and defied it. One of the final rites consisted in a frenzied dance which was prolonged until after dark, in spite of the exhausted state of the participants. The Arapaho called it "gambling against the Sun," and the Gros-Ventre "the dance against the Sun." The aim was to counteract the opposition of the intense heat of the Sun, who had tried to prevent the ceremony taking place by radiating his warm rays every day during the period preceding the dance. The Indians, then, looked upon the Sun as a dual being: indispensable for human life, yet at the same time representing a threat to mankind by its heat.[21]

The dance itself began at sunrise and continued for as long as five days. On the first morning, "pledgers," perhaps fifty strong, painted their bodies red for sunset, blue for sky, yellow for forked lightning, black for night, often including the form of a sunflower. Each wore an apron of deer or antelope skin, rabbit fur wristlets and anklets, and a downy feather in his hair, carrying in his mouth an eagle wing-bone whistle with porcupine quill or beadwork and eagle down decoration. The pledgers circled the phallic pole and saluted the Sun to drumbeats and intense singing. Faces fixed on the great solar orb, they danced on their toes or the balls of their feet. To "participate in the cosmic renewal of life by shedding their blood for the sake of all life," they would lash their arms and thighs as many as two hundred times and suspend themselves from eight-inch-long wooden skewers inserted by the *kuwa kiyapi* (holy man) into the small of their backs or into their shoulders. The skewers were attached to the central pole by tightly wound rawhide thongs at a height such that only the participants' toes touched the ground; alternately, some participants would suspend themselves upside down, their bodies hanging like corpses—the excruciating ordeal that Richard Harris undergoes in the 1970 Hollywood version of the ceremony in *A Man Called Horse*. The test of any dancer's honor was not to scream or cry out. Urged on by the rest of the tribe, each brave would pull hard on his wounds until he fell unconscious or tore himself loose, releasing him from his vow and signifying his willingness to suffer. Once freed, the

The Okipa, the dance ritual of the Mandan Indians, as depicted by the
artist George Catlin, c. 1835. The last full Mandan died in 1971.

liberated dancer would, after a prescribed interval, smoke pipes ("split the clouds"), enjoy a vapor bath, and at last drink and take food (customarily dog soup and buffalo meat) in a concluding feast.

At least ten plains tribes adopted the main elements of this ritual; others performed the ceremony without spilling blood. To an outside eye, such self-mutilation seems monstrous (Jesuits, keen to convert the heathen Indian, were particularly virulent in their condemnation); but the historian Joseph Epes Brown, who would collaborate with the Lakota chief Black Elk in his apologia *The Sacred Pipe* (1948), compares the sun dance with Hindu meditation techniques and the sun dance pole with the crucifix, equating the sufferings of the dancers with certain extreme forms of Christian penance. "When man in awful ceremony is actually tied to this Tree of the Center by the flesh of his body," Brown tells us,

> or when women make offerings of pieces cut from their arms, sacrifice through suffering is accomplished that the world and all beings may live, that life be renewed, that man may become who he is. The Sun Dance, thus, is not a celebration by man for man; it is an honoring of all life and the source of all life, that life may go on, and the circle be a cycle.[22]

In his moving account of the Kiowa, *One River,* the ethnologist Wade Davis writes that the sun dance was "by far the most significant religious event of their lives. It was a celebration of war, a time of spiritual renewal, a moment in which the entire tribe partook of the divinity of the Sun."[23] But its very existence was threatened by the white man. On the night of November 13, 1838, a great meteor shower blazed across the prairie sky, alarming the whole tribe and prompting the elders to predict the end of the world, a forecast that turned out to be accurate for the world of the Kiowa. Their first contact with white soldiers came the following summer. By 1890, East Coast administrators had banned sun dances altogether.*

At the end of his history of the Lakota Sioux, Clyde Holler asks simply, "Why on earth are they still dancing the Sun Dance?" Among the reasons, he offers this: "To attend the Sun Dance is to feel the power, to approach religious ecstasy, to experience for oneself the essence of a religion based on power. The

---

* Davis (who rejoices in the title of "explorer in residence" at the *National Geographic*) makes the point that modern celebrants, no longer allowed the excesses of old, chew on peyote as a substitute means of achieving enlightenment. A tribesman told Davis that peyote was like the buffalo of ancient times. "Both are children of the Sun. Peyote is the incarnation of the Sun. As are buffalo."

Sun Dance sweeps away any possible doubts about the existence of sacred power in a searing emotional catharsis."[24] He might have mentioned physical catharsis too. Dance promotes the uninhibited, the irrational—indeed, even sexual release, orgiastic excess. In *The Birth of Tragedy* Nietzsche set the Dionysian nature of dance against its Apollonian opposites, balance, reason, and discipline.[25] The bureaucrats in Washington were not alone: the Protestants of Tudor and Jacobean England, the inquisitors of the Midi, the conquistadors who savagely mastered South America, the gauleiters of National Socialism—all of them tried to regulate and contain solar-related dance rites, if not actually to outlaw them.

In those countries where ceremonial dances came under the sponsorship of the state, attempts were made to fix them to a particular time and place, rather than allowing them to arise spontaneously at some untried location. This made Sun celebrations easier to manage, and in truth worshippers yearned for central sites where they could congregate and pay homage. Material structures—Stonehenge (known throughout the Middle Ages as Chorea Giganteum, "The Giants' Dance"), the pyramids, those great erections of South America—all such sites came to embody belief, not least about the Sun.

CHAPTER 3

THE THREE THOUSAND WITNESSES

> *To lug those stones was quite an undertaking*
> *technology worth some remark*
>
> —PAT WOOD, "Dartmoor"

> *. . . The poet,*
> *Admired for his earnest habit of calling*
> *The sun the sun, his mind Puzzle, is made uneasy*
> *By these solid statues which so obviously doubt*
> *His antimythological myth*
>
> —W. H. AUDEN,
> "In Praise of Limestone"[1]

MORE THAN THREE THOUSAND STRUCTURES, NEARLY ALL OF STONE
and scattered worldwide, stand witness to man's obsession with the Sun. Most
are built in line with the direction of its rising or setting, while some align with
it just once a year, generally at a solstice—although the Great Sphinx at Giza
gazes out fixedly toward the vernal equinox. Whether one is looking at the first
pillars of Stonehenge* (dating from 2900 B.C.), the Mayan pyramids in the Yucatán (twelfth century A.D.), the megaliths of the Great Plains (down to the
nineteenth), or the Dancing Stones of Namoratunga, on the grassy plains of
Kenya (date unknown), here are the Sun's silent celebrants.

The majority of societies have erected monuments to the Sun with extraordinary care. A large percentage of mid-Neolithic tombs appear to have been
built to face the Sun at significant times, and in Britain alone more than nine

---

* A *henge*, while it means "hanging," is specifically a large Neolithic monument consisting of
circular banks of earth paralleled by an internal ditch, and frequently containing a circle of upright stones. Strictly, Stonehenge is not a henge, because the position of its bank and ditch are reversed.[2] A menhir is a solitary standing stone, a megalith a "great stone" (from *megas*, "great,"
and *lithos*, "stone").

hundred survive, above all Stonehenge, whose pillars, Samuel Pepys recorded after visiting them in 1668, were "as prodigious as any tales I ever heard tell of them, and worth going this journey to see." He then added, with typical honesty: "God knows what their use was."

I have visited Stonehenge a dozen times. On occasion, it can seem just a jumble of huge, squat blocks without any special allure. At other times, such

The massive pillars of Stonehenge, still the greatest of solar monuments

as Midsummer Eve, it is so thronged with worshippers that it could be a country fair. But when all but deserted, at either sunrise or sunset, it is magical, and one's imagination soars. Against the lateral sunlight, the stones gain in beauty and majesty. For men to have made such a place at such great effort shows how important the Sun was to them.

Present-day Stonehenge is the outcome of more than five millennia of building and accretion, from ditch banks of earth to chalk, to wood, and finally to stone. During its first six centuries, a round enclosure took shape, formed by two banks and a rough, "chalky, dazzling moon-milk white" ditch.[3] Within the circle fifty-six pits were dug, each roughly a yard wide and deep. There was a break in the circle at the northeast, with a pillar 256 feet inward from the entrance, the famous "heel stone" (a corruption of the Welsh *hayil* or Norse *hel,* both meaning "Sun"): 16 feet high (descending another 4 feet below ground), 8 feet in circumference, and some 35 tons in weight. On my last visit, one midsummer morning, I stood in the center of the circle and watched the Sun rise just to the left of the stone: easy to believe then that the site was as much observatory as place of worship.

Between 2300 and 2000 B.C. at least eighty-two upright slabs—known as bluestones from the tint they assume in wet weather—were floated, dragged, and hauled for 250 miles from the Preseli Mountains, in southwestern Wales, to form a crescent in the center of the circle. Over the next four hundred years, a further ring of uprights, consisting of thirty gigantic slabs of stone arranged in pairs that support lintels, was added, in a horseshoe-shaped curve, its opening in line with the Sun's midsummer rising and with the winter solstice sunset. Nearly seven yards high, these trilithons—"threestones" (two uprights supporting a lintel)—each weighing some forty-five to fifty tons, were cut from sarsen sandstone, a word suggesting pagan origin (from "Saracen").

Stonehenge first rates a mention in the *Historia Anglorum* (A.D. 1130). However, that is all there is—a mention; not until 1740 did an English antiquary, William Stukeley, ascribe an astronomical purpose to this mighty engineering feat, arguing that its principal axis pointed toward the dawn at summer solstice. After Stukeley (who also advanced the notion—not stupidly—that Stonehenge might have been a racetrack), theories came fast and curious, right down to the present day: it was a Bronze Age Lourdes; an ancient Druidic site; a prediction center for eclipses; a monument to the dead and part of a funerary and processional route (as many as 240 people are known to have been buried there); a tomb for Boadicea, the pagan queen of the Iceni, who led a revolt against the Romans; a mechanism for calculating the calendar; a solar and lunar observatory; a sophisticated forecaster of tides (curious, for a place so far inland); and a monastery or college for a superelite who were part of a worldwide prehistoric intelligentsia. All these centuries later, it is safe to say that scholars are firmly agreed on scarcely a single aspect of Stonehenge— although the weight of current research leans toward its having been created as a cluster of celestial observatories, some to track the movements of the Moon, others those of the Sun, and possibly to establish a calendar.* Maybe, as

* It is easy to take the accuracy of the alignments at Stonehenge for granted, perhaps because transporting and raising the stones was extraordinary achievement enough. But its builders

the archaeologist Jacquetta Hawkes once famously observed, "every generation gets the Stonehenge it deserves—and desires."[5]

WITH SO MANY different theories, how can we tell the purpose of such
buildings—not just Stonehenge, but other monuments, too? As Ronald Hutton wryly notes, "The people who built and used these great structures seem to
have taken a positive delight in making exceptions to every rule which we try to
discern in their behavior."[6] Point up in any direction, he suggests, and you will
most likely hit a star or phase of the Sun or Moon. Nor does the notion of monuments laid out according to a sophisticated geometry consort well with evidence on the ground: "A glance at the plans of Wessex superhenges will reveal
that the classic shape is that of a battered car tire."[7] Many prehistoric monuments clearly have no connection with the exact occasions of summer or winter.
For example, the stone figures, or *moai,* on Easter Island—more than nine hundred of them, averaging twenty feet in height—while they probably performed
some calendrical function, were based on lunar, not solar cycles.[8] One tomb on
the Isle of Arran faces the midsummer sunrise almost exactly, but nineteen similar tombs on the island point in quite other directions. As Bruce Chatwin reports
the conversation between Aborigine and rough Australian in *The Songlines,*

> The big man leered. "If all what them says was sacred sites, there'd be
> three hundred bloody billion sacred sites in Australia."
> "Not far wrong, mate!" called the thin aboriginal.[9]

However, overall the evidence is irresistible that by far the greater number
of structures are consciously connected to the Sun; and while Stonehenge may
be the jewel in the crown of such prehistoric monuments, many are the lesser
gems spread around the world. A particular favorite of mine is the burial
mound at Newgrange—Liamh Greine ("the Cave of the Sun") or "the Irish
Stonehenge"—which lies in the Boyne Valley about thirty miles north of
Dublin. A megalithic burial place dating back to 3200 B.C., it is roughly contemporary with the first Egyptian pyramid. From outside it is a mass of dull

---

used techniques that would keep predictions reasonably accurate. By contrast, modern mapmakers can make huge mistakes: in the early nineteenth century, with American–British Empire
relations under extreme stress, the United States erected two fortresses beyond Lake Erie, at the
huge cost (for the time) of $113,000. In October 1818, after three years' work, astronomers car-
rying out a boundary survey discovered that the border was three-quarters of a mile to the
south—both forts had been erected in Canada.[4]

stone atop a small hill; inside one sees a place of wonder. At precisely 9:02 A.M. at the winter solstice, sunlight shines through a small window at the end of a twenty-yard-long gallery, and for the next seventeen minutes creeps along the passage until its eerie glow falls upon a circular stone in the opposite wall of an otherwise dark little chamber.

Just north of the river Boyne in Ireland: the 62-foot passageway at Newgrange burial mound, down which the Sun flames each winter solstice

In his 2005 novel *Ireland,* Frank Delaney re-creates the lives of those who built the barrow. Inside the central chamber, directly in line with the passage-way, the tomb's creator ("the Architect") has placed a "round, smooth dish made of sand-colored stone," and now assembles the Elders within the tomb:

> At the farthest end of the passageway, red-gold colors tinged the edges of the rectangular aperture, making it look like a box full of light. Next, a shaft of sunshine slid through the box and lit the floor, just in-side the entrance.
>
> The Elders watched this glow and became transfixed. . . . Steadily, with no jumps, no lurches, the light filtered down the passageway. Its golden yellow beam, thick as a man's arm, flowed slowly, slowly—it was as though the sunlight was honey from a bowl that someone had tipped over and spilled into a long, pleasant stream. . . .

When the sunbeam reached the edge of the chamber, its yellow-orange light began to glow upward into the gathering, gilding the Elders' faces. The ray seemed to hesitate for a moment. And then it surged forward until it splashed into the great stone bowl—and it filled it exactly. Not a drop of sunshine slipped over an edge here, there, or anywhere. The final resting place of the sun for that brief moment was precisely within the full circumference of the bowl, leaving no area of the bowl's stone surface dark or cold. It lay there like a golden sphere.[10]

Stonehenge's greatest rival among the prehistoric monuments of Europe is not Newgrange, however, but the forest of alignments, stone circles, dolmens (burial places formed from great graven stones), and tumuli (earthen tombs) stretching beyond the remote seaside town of Carnac on the southern coast of Brittany. Some ten thousand megaliths—notably the Grand Menhir Brisé at Locmariaquer, a massive edifice that once rose sixty feet but now lies shattered into five parts, each of about 340 tons—huddle together like so many battle-weary giants. Local legend has it that the stones are Roman soldiers literally petrified by Pope Cornelius during the Roman schism of A.D. 251–53. The site boasts several burial chambers, deep caverns faced toward the rising Sun, a powerful index of the likely interest in things solar among those who dwelt here in the New Stone Age. It is among the great stones of Locmariaquer that Porthos meets his death (in *The Man in the Iron Mask*), covering Aramis's escape from a company of the king's guardsmen after being trapped inside one of the great caves.[11]

Most countries in Europe have at least one major site of ancient solar reverence. The most impressive constructions are hardly limited to the one continent, however. Halfway around the world, an extraordinary temple sits within the complex of buildings that makes up Angkor Wat in Cambodia. Constructed between 1113 and 1150, it forms part of the largest astronomically sited building in the world. The temple (Wat), almost a mile across, was designed both as a tomb for its creator, Suryavarman II, and as an observatory: nearly every measurement encodes calendrical information, and the bas-reliefs are all oriented west, to catch the setting Sun.

There are five particularly grand solar temples in India, all now in ruins: at Delhi; Konorak; Mudan; near Ranapur, in Rajasthan; and Modera, in Gujarat. In October 2006, I visited the one at Modera, built in A.D. 1026, about two hundred years before the next oldest, at Konorak. The shrine lies about six miles from the state's old capital, Patan, and its survival is all the more surprising given that it lies in an earthquake zone—badly hit in 1918, again in 1965, and once more on January 25, 2000, when the shock registered 7.9. My guide, a Brahmin who has converted to Buddhism, explained to me that unlike the peoples of Europe or the Americas, Hindus try never to face the Sun directly, which is why statues of Indian gods look eastward, a hand held up in blessing, so that anyone looking at a god catches the Sun's reflection. Temple domes are designed in the shape of massive lingams (a shape, I learned from several people, purposely imitated in the design of nuclear reactor cones, the latest tribute to the Sun's potency).

Solar-oriented monuments assume an amazing variety of forms. In North America, for instance, the Indians of the Great Plains position "medicine" (i.e., magical) wheels to register the Sun's course, each wheel consisting of a central circle or rock pile from which radiate spokes of stones. These creations can range up to a hundred yards in diameter, while the hub may be ten yards across and several yards high. They are found along the eastern boundary of the Rocky Mountains, from Colorado to Alberta and Saskatchewan. One, dubbed "the American Stonehenge," sits atop Medicine Mountain, in the Bighorn Mountains of north-central Wyoming. This sandstone edifice was held sacred by several plains tribes, and a line through two of its cairns (stone memorial mounds) lines up to within a third of a degree with the exact point of sunrise at summer solstice.[12] In Salem, New Hampshire, stands a rival "American Stonehenge"—a complex of cairns, chambers, walls, and huts between three thousand and four thousand years old encircled by notched "sighting stones," aligned on the sunrise and sunset solstices. In a series of markings, the site also celebrates "cross-quarter" holidays, which fall halfway between solstice and equinox.[13]

The Americas are particularly rich in such places. Northeast of St. Louis, Missouri, more than a hundred mounds dating from A.D 800 to 1550 and covering sixteen acres form the largest prehistoric earthwork in the world, many of its mounds arranged along the sunrise line of the winter solstice. As elsewhere, some are no more than elaborate burial tumuli, while others served as temples or great effigies of gods, or as fortifications.

The Mayas, Aztecs, and Incas all constructed such works, too. At Teuchitlán (literally, "Where Men Become Gods"), in the present Mexican state of Jalisco, arise some six hundred pyramids of various sizes, dominated by a great Pyramid of the Sun, one of three temples covering over an acre and aligned east-west, suggesting they too functioned as solar observatories. They were built between 300 B.C. and A.D. 200, one of the three being (initially) slightly misaligned, which fault so upset Montezuma that he had it razed and rebuilt—an act of intellectual integrity among a supposedly inferior people that so astonished Cortés he reported it to his king.

The pyramid at Chichén Itzá. On the spring and autumn equinox, at the rising and setting of the Sun, corners of the pyramid cast shadows in the shape of a plumed serpent along the side of the staircase, and as the shadow moves, the snake seems to slither down the stairs.

Further south, the entire Mayan city of Chichén Itzá in Yucatán was devoted to astronomical observation, the windows of its Spiral Tower being set toward solstices and equinoxes. While traveling in Peru, I saw temples of the Sun in Machu Picchu, Sacsayhuamán, and Cuzco all similarly disposed. Cuzco and Chichén Itzá both boast rows of pillars that divide the Sun's movements into periods roughly corresponding to months, helping the farmers to determine the time for sowing.[14] I have also seen solar markers on a hillside overlooking the Island of the Sun in Lake Titicaca. No wonder that in the last thirty years there has arisen the discipline of astroarchaeology, which combines archaeology with astronomy; most of the research so far published tells the same story, of sun worship united with practical scientific purpose and cultural celebration.

"To you, to me," wrote W. H. Auden, "Stonehenge and Chartres Cathedral . . . , are works by the same Old Man under different names: we know what He did, what, even, He thought He thought, but we don't see why."[15] The "why" is becoming clearer: the desire that places of worship should also serve a practical function. What is noteworthy is how societies living on the margin pour so much of their resources into such constructions, exacting the colossal common effort normally associated with major wars. (Around 1967, a NASA scientist declared that getting man to the Moon was the national integrative effort of modern times, parallel to Egypt's turning itself into one nation by building the pyramids.)

Worldwide, many ancient stone constructions have been so carefully crafted that they have lost nothing of their accuracy. But, because the Earth shifts its position over time (about one degree every seventy-two years), a temple or instrument originally aligned with the Sun will no longer face in precisely the right direction. Traveling in India in the fall of 2006, I visited the famous Jantar Mantar ("Instrumental Calculation") at Jaipur, one of five observatories erected by the Maharaja Sawai Jai Singh II (1686–1743) and the grandest of them all, consisting of sixteen huge instruments, in pink and yellow limestone and marble, set down in a vast park like a giant's playground. Throughout the observatory, the instruments still function effectively, although the one that measured the equinoxes was slightly off because of the Earth's precession:* the Sun's shadow should have crossed from one side of its wall to the other on September 21, but while I was there it did not do so for two more days.

SAMUEL JOHNSON'S HERO Rasselas describes the pyramids of Egypt as "the greatest work of man, except the wall of China."[16] To another visitor, they appeared "the Sun's rays made stone"; but however described, they are the supreme manifestation of a hierarchy whose kings were inextricably linked with the Sun, and who commissioned these astonishing structures as tombs for themselves and their families. Each tomb symbolized the primeval mound that grew up from the waters during the world's creation: it was believed to form a gigantic stairway to the heavens, where a pharaoh's soul would become one of the "immortal" stars. The two great moments of solstice, toward which the earliest pyramids were aligned, were seen as gateways for the royal souls on their journey into and out of life.

Just the empiric facts take one's breath away. The Great Pyramid of the

---

* The Earth's axis is engaged in a dance, and as it makes its longer loop, taking about 25,800 years to complete each circuit against the sky, this wobble created by the poles shifts the Earth's incline—only slightly over any human lifetime, but enough over the course of the millennia to change the apparent positions of the stars at any given season of the year. There are also seasonal shifts within the year. If a particular star or group of stars is directly overhead at, say, a given midnight, one finds that they gradually change position as the days, weeks, and months go by, until after a year has passed, they are directly overhead once again: a process known as "precession." In the same way, the Sun's position changes relative to the stars, until after a year it too returns to the same point. Its annual movement makes it appear to be slipping back from west to east compared with the stars, which move in the same direction—which is why the sidereal day, i.e., the time required for a complete rotation of the Earth in reference to any star, is about four minutes shorter than the solar.

Pharaoh Khufu at Giza, on the outskirts of where the ancient metropolis of Memphis once stood (Cairo now spreads over the site), is the oldest and sole survivor of the Seven Wonders of the Ancient World, and ranked for millennia as the tallest structure on Earth. Napoleon, fascinated by all things Egyptian, calculated that there was enough stone in the three great pyramids of Giza to build a wall three yards high (he neglected to say how thick) around the whole of France. Although there were imitations at Thebes and elsewhere, the first ones were built in Middle Egypt. The Egyptians knew next to nothing of engineering—they simply multiplied wedge, lever, and inclined plane by enormous manpower—yet the maximum inconsistency between side lengths of the Great Pyramid is, astonishingly, less than 0.1 percent. Herodotus asserts that the Great Pyramid was built by 100,000 slaves (a more credible figure is 25,000) fed on onions and garlic, who laid an average of 340 blocks a day over two cruel decades.

The Great Pyramid (the tallest) at Giza,
built c. 2800 B.C. out of 2.3 million stone blocks

The most intense period of pyramid building stretched from 2686 to 2345 B.C. Beyond the Great Pyramids, there were about eighteen primary sites and some twenty-nine others, each part of what is a single architectural complex. About a hundred of these monuments were built, put together from polished, pearly white limestone, red granite, quartzite, alabaster, mud brick, and clay-lime mortar. The largest rests on a base 756 feet (249 meters) square, and the tallest reaches 481 feet (158 meters), at an inclination of just over 50 degrees.*

The very first pyramid—which is also the world's oldest existing construction in stone—was built at Saqqara, sixteen miles south of Cairo, at the beginning of the Third Dynasty (2690–2610 B.C.). The cult of Ra reached its zenith under the Fifth Dynasty (2494–2345 B.C.), by which time the pharaoh (which means "great house," a reference to his final residence) had become identified as the son of the sun god. Neuserre, sixth king of the Fifth Dynasty, even had his burial pyramid at Abu Jirab, in Lower Egypt, made smaller than his sun temple, as a mark of respect—and possibly fear.

During the reign of Akhenaten and his wife Nefertiti, between 1379 and 1362 B.C., the Sun was the all-embracing deity of the Three Kingdoms. After Akhenaten's time, it was demoted from supreme godhood but was still an object of worship: Ramses II (1279–1213 B.C.) had two temples cut out of the living rock on the west bank of the Nile, south of Aswan, in what today is northern Sudan. The taller edifice is 119 feet from base to summit, with four colossal stone gods 67 feet high, atop which stands a row of carved sacred baboons, the Watchers of the Dawn, who as allies of Ra in casting back the forces of night are depicted with arms raised in adoration of the rising Sun. The temple is oriented so that on each February 22 and October 22 (interestingly, eight months apart, not six), the morning Sun fills the entire length of the elaborate passage chamber—just as at Newgrange—and illuminates four of the five gods against the back wall of the innermost shrine; but not that of Ptah, the primal creator, who dwells in the dark.†

The pyramids have been subject to almost as many theses as Stonehenge.

---

* The true derivation of the word "pyramid" has various contenders. The ancient Egyptian term *pir e mit* has been variously translated as "division of ten," "division of number," and "division of perfection"; also as "fire in the middle." The Greeks, possibly jealous, are said to have taken up a derisive term for the structures, *pyramides,* meaning either the little wheaten cakes sold on street stalls or storehouses for grain. There would have been some justification to the latter barb. Christian Europe during the Middle Ages based its idea of Egypt mainly on the Bible, assuming that the pyramids were Joseph's grain storehouses as described in Genesis 41–42, and calling them "the Barns of Abraham."

† Because they lay so far upriver, the temples were unknown until their rediscovery in 1813. With the enlargement of the Aswan Dam in the 1960s, they seemed doomed to being flooded under Lake Nassar. Between 1964 and 1966 UNESCO and the Egyptian government took both temples apart block by block and reassembled them on top of a cliff two hundred feet above their original site.

Like that great monument, they clearly had a powerful connection to the solstices. Their basic orientation, established by wooden stakes and ropes, was at first north to south (the solstice axis), then, with those built from the Fourth Dynasty on, east to west (the axis of the equinoxes), even though the entrances remained on the north face. The Egyptologist Martin Isler explains what had to be accomplished:

> The linkage between the rising Sun and a pyramid oriented by its sides to the cardinal points is straightforward. As the Sun travels from winter to summer solstice, it moves along the horizon in an arc of about 50 degrees from north to south at the latitude of Giza. To equate the king's pyramid with the Sun . . . required that the sides of the structure face due east and west. Due east is the midway point for the Sun as it moves between the summer and winter solstices. It is also the only position in which a structure square in plan can have one side face the birth of the Sun in the east, another face its death in the west, and still have its north side directed at the circumpolar stars.[17]

Most pyramids oriented to the equinox were aligned so that on that date, they seemed to swallow the setting Sun, which must only have increased the witnesses' awe. Their orientation also allowed them to be used to mark the Sun's course (an ancient text refers to "the shadow of Ra" and "the stride of Ra")—so that they served as both solar temples and seasonal clocks. Along with their adjoining temples, the pyramids are the surest and grandest evidence of how worship and astronomy have gone hand in hand.

Whole civilizations have married their concepts of time and space with their relationships to their gods, creating a bridge between Heaven and Earth. However, the religiously oriented commemorations, solstice points, and other solar phenomena are not the work only of ancient or precivilized societies. The great cathedrals of Europe face east to declare their sacredness. Although Pope Leo the Great (440–61) forbade Christians to worship the rising Sun, its veneration was deemed acceptable so long as the Sun was seen only as a symbol of Christ entering his church—a doctrine more easily decreed than enforced. To prevent sunlight (which was both distraction and rival) from streaming through open church doorways, Pope Vigilius (537–55) commanded that all places of worship face their apses, rather than their main entrances, east. Only occasionally was dispensation granted for a cathedral to have its doorway aligned with the sunrise of the winter solstice and the setting Sun of the summer solstice, or for it to lie with its main axis facing the rising Sun on a day of special significance, such as its patron saint's festival.*

No matter that the Church goes through such periodic renunciations of the Sun. The fact remains that its holy days are rooted in pagan solar festivities, and the astronomers who study the Sun have for centuries been in the employ of the Church. I have walked across a 1798 gnomon—a marked surface that shows the time by registering the shadow cast by the Sun—outside the main church of Bergamo, in northern Italy. Another gnomon, one that establishes the spring equinox, stands at the end of the brass strip, or "rose line," that runs across the nave of St.-Sulpice, in Paris, made newly famous in *The Da Vinci Code*. There is a long history of telescopes being placed in Italian cathedrals, and during the seventeenth and eighteenth centuries the cathedrals of Bologna, Rome, Florence, and Paris all served as solar observatories.[19]

Across the Atlantic, St. Mark's Church-in-the-Bowery was built and consecrated in 1799 at what is now Stuyvesant Street and lower Second Avenue in Manhattan. After a fire in 1978, restorers uncovered a rose window in its south wall aligned with the altar. In 1983 a stained-glass rose window succeeded it, so that at noon on the spring equinox the Sun strews the nave with streaks of colored light.

As with the sacred, so with the secular. Some two miles north, in the plaza of the McGraw-Hill building on Sixth Avenue at Forty-eighth Street, stands the fifty-foot steel "Sun Triangle," installed in 1973. The steepest side points to the Sun's zenith at highest summer, the lowest arm to its place at noon on the first day of winter, about December 21. Another recent example is the huge sundial bridge built by the Spanish architect, sculptor, and structural engineer Santiago Calatrava between 1996 and 2004: "Tilted like a catapult ready to be sprung,"[20] it spans the Sacramento River in Redding, California, its north-south alignment making the pylon's shadow indicate the time of day.

Twice a year, in the borough of Manhattan, the Sun is seen to perform a neat trick of its own. Because most of the island is laid out on a grid, and the grid is canted 28.9° east of true north, the rays of the rising and setting Sun are essentially at right angles to its great avenues. This meant that on May 30, 2009, the setting Sun could be seen shining straight down all the borough's east-west streets at once, from Fourteenth Street (where the city's grid proper

---

* Medieval man would talk of "walking widdershins," meaning that it would bring harm to walk around a church in opposite movement to the Sun; walking sunwise around free-standing megalithic monuments survived from the remotest past, and even in the seventeenth century, in parts of Scotland, local people walked sunwise around a stone called "Martin Dessil"—the Gaelic word for "clockwise" being *deasil*.[18]

The bridge by Santiago Calatrava at Turtle Bay, in Redding, California. It opened on July 4, 2004, and is the world's tallest sundial.

begins) on up. In the days immediately after, the point of sunset continued to shift to the north, right up until the summer solstice three and a half weeks later. Thereafter, that point migrated southward again, until sunset was once more lined up with Manhattan's cross streets on July 11 at 8:27 P.M. "Perhaps," Neil deGrasse Tyson, director of the Hayden Planetarium, suggests, "anthropologists in the distant future will presume the Manhattan grid to have some astronomical significance, just as we have found for Stonehenge."[21]

CHAPTER 4

TERRORS OF THE SKY

*On the twentieth day, an eclipse befalls. The King is slain, and a nobody seizes
the throne. On the twenty-first day, another eclipse. Devastation. Corpses are
found throughout the country.*

—Babylonian Table of Portents, 1600 B.C.

*Nothing can be sworn impossible
. . . Since Zeus father of Olympians
Made night from midday
Hiding the light of the shining sun
And sore fear upon man.*

—ARCHILOCHUS OF PAROS,
(714–676 B.C.), describing a solar eclipse

AURORAS AND ECLIPSES, THOSE GREAT CELESTIAL DRAMAS, ARE TWO
of the most awe-inspiring occurrences in nature, and for thousands of years
our fear of them has triumphed over whatever the scholars of the ancient
world and onward have told us about their causes.

Auroras are the flares that appear far over our heads, the most spectacular
and best known being the northern lights (aurora borealis) in the Northern
Hemisphere and southern lights (aurora australis) in the Southern. They often
start off as a mild glow, hugging the horizon; bright patches (called "surfaces"
by scientists) may follow. Next an arc appears, stretching "like a fluorescent
basket handle across the sky."[1] Finally, the entire heavens will be "wreathed in
celestial fire." Auroras typically swirl fifty to two hundred miles above the
Earth's surface, although some have come as close as thirty-five miles. The
flood of particles shot out by the Sun reacts endlessly with atoms and ions in
the atmosphere to generate photons—the basic particles of light—and when
enough of these collisions occur with sufficient intensity, they set off blazing
displays. Given the Earth's rotation (and because the arriving particles are
constantly changing direction as they strike the thin upper air), the auroras

An aurora borealis on March 1, 1872, at about half past nine in the evening

seem to move across the sky. They are visible mainly at latitudes within the Arctic Circle (most frequently during the equinoxes), and on the shores of the Great Lakes will appear a few dozen times each year; but on at least two occasions, auroras have manifested simultaneously in the Northern and Southern Hemispheres—first on September 16, 1770, when one was observed to the north while Captain Cook recorded another over the South Pacific; then again in October 2001.

Mercury and Jupiter, like Earth, are magnetized, and any planet with a magnetic field and an atmosphere will probably display auroras. But with the sole exception of Jupiter, which displays even more intense and powerful auroras than ours, none can compare with those found on Earth in amplitude, duration, luminosity, or brilliance. One observer has described them as "gargantuan, ghostlike arms, chasing and darting, appearing and disappearing spontaneously,"[2] while another recorded "different forms, wavering, many colours diffusing and changing, sometimes far away, sometimes filling the heavens around and above, plunging great dropping spears and sheets of colour earthward towards your very head as though a great hand were dropping colour like burning oil."[3] Robert F. Scott, who lost to Roald Amundsen in the race for the South Pole, wrote how overwhelmed he felt before an aurora: "By its delicacy in light and colour, its transparency, and above all by its tremulous evanescence of form . . . the appeal is to the imagination by the suggestion of something wholly spiritual."[4] Auroras even make themselves heard, giving off a "whistling, crackling noise,"[5] as in "the waving of a large flag in a fresh gale of wind."[6] The late-eighteenth-century American historian Jeremy Belknap compared their sound to a muted rustling or swishing, "like running one's thumb and forefinger down a silk scarf."

The giant plumes of gas that burst on the edge of our atmosphere were first given poetic life in the apocryphal Book of Maccabees ("horsemen charging in midair, clad in garments interwoven with gold").[7] In Norway, they were sometimes seen as Valkyries ("choosers of the slain"). The Inuit believe they are the dead reaching toward those left behind, or spirits playing football with a walrus skull. In the Middle Ages the lights were taken as the cause of plague or disaster, while as recently as 1802 Walter Scott wrote, "He knew, by streamers that shot so bright, / That spirits were riding the northern light."[8]

After the aurora borealis of November 14, 1837, one scientist in New Haven reported that at about six in the evening the sky was thick with falling snow when everything "suddenly appeared as if dyed in blood."[9] Trees, housetops, the entire atmosphere, were tinged with scarlet, so much so that people thought the town on fire.* The aurora lasted for half an hour before diffusing "like the light of an astral lamp." Elsewhere in the United States, great streamers appeared—bright red vertical columns alternating with white clouds—before fusing at their apex. Then at about 9 P.M. innumerable arches of brilliant scarlet shot up from the northern arc of the horizon, like the striped inside of a gorgeous circus tent. Similar phenomena were reported that night in Pennsylvania, Maryland, Georgia, and Ohio—sometimes accompanied by what observers called "merry dancers," great plumes of color liberated into movement. Simultaneously, auroras appeared over many parts of Europe, "a patch of the most intense blood-red colors ever seen." Although auroral glows have been known to reach as far equatorward as Italy and France, they are seen that close only about once every ten years. The vivid coloring has a simple explanation: While atmospheric nitrogen will make auroras purple and violet, oxygen makes them green and red (from interactions that take place at higher altitudes). They can also appear in reddish pink and greenish yellow, or with a violet sheen intermingled with pastel pinks and greens.

---

* In A.D. 37 the emperor Tiberius saw a vivid reddish light in the sky and, as Seneca reports, "battalions were hurriedly dispatched to Ostia [the vital seaport at the Tiber's mouth], in the belief that it was burning." Then, in London in January 1938, fire brigades were called out in the belief that fires were sweeping Windsor Castle. It could also work the other way around. In Daphne du Maurier's *Rebecca*, Maxim de Winter, driving frantically back to his Cornish mansion, sees the evening sky emblazoned in garish orange. "That's not the northern lights," he cries, "that's Manderley!"

We still do not know exactly what sets off these spectral light displays, although in 2007 NASA launched five spacecraft, each about the size of a washing machine, to try to solve the mystery. Initial findings show that they follow the snapping of magnetic fields—similar to the blowing of an electric fuse. The scientists involved hope to have more information before the next peak of solar storms arrives, between 2010 and 2012.[10] Meanwhile, auroras continue to fill us with wonder—which may be why so much superstition still surrounds them.

THE MOST ECONOMICAL definition of an eclipse comes from the Italian book *Eclipses: An Astronomical Introduction for Humanists,* and runs: "An eclipse occurs when three celestial bodies are in the same direction, so that one of them obstructs the reciprocal visibility of the other two."[11] What we now know to be the plane of the Earth's orbit around the Sun is called the "ecliptic," because eclipses can take place only when the Sun and Moon are both situated on a plane with the Earth's path. In short, a lunar eclipse is caused by the Earth coming between the Sun and the Moon, blocking the sunlight that usually reflects off the Moon and makes it visible. A solar eclipse occurs when the Moon passes between Earth and Sun, obscuring the Sun. Our Moon is unusually large—relative both to other moons in the solar system and to the planet it orbits—and for a total solar eclipse to take place it has to be in exactly the right place, and the Earth has to be exactly the right distance from the Sun. Sometimes, when the Earth and Sun are at their closest, making the Sun seem

Sun and Moon dance together on the horizon, Antarctica, 2003.

slightly larger, the Moon will mask only about 98 percent of it, leaving a circle of light visible and causing what is known as an annular eclipse (from the Latin *annulus,* "ring")—possibly an even more dramatic sight than a total darkening. Usually the Moon cuts across only a section of the Sun, causing a partial eclipse; on such occasions only about one-sixth of the Earth's surface encounters some part of the Moon's shadow.

One day long in the future there will be no total eclipses. Observed from Earth, the Sun and the Moon (in this epoch, at least) have almost the same apparent diameters, as their sizes are in the same ratio as their distances: the Sun is just over four hundred times broader than the Moon (866,000 miles in diameter as against 2,160), but just under four hundred times farther away—an extraordinary coincidence, which over hundreds of millions of years will not endure, because the distances between the Sun and the Earth, and the Earth and the Moon, are very slowly increasing. The Moon initially orbited close to the Earth, perhaps just 16,000 miles distant; today it averages 238,000 miles. The slowing effects of tidal friction on our planet will continue to press the Moon into an inexorable compensatory outward spiral, at the rate of 1.5 inches a year. Eventually, some hundreds of millions of years hence, it will be too far off to cover the Sun from our view.

For as long as we have records, eclipses have awakened fear and end-of-the-world imaginings, independent of mythology or culture. Eclipses, as well as other heavenly occurrences that were believed to be almost impossible to predict, such as auroras, comets, and meteor showers, are described in a Greek word meaning that the stars are aligned against you: *dis-astra*—"the stars are against." "Eclipse" itself comes from the Greek and means abandonment or forsaking. The Chinese word translates as "being eaten away" (*rìshí,* literally "Sun-eat"); in Spanish, the word means "to make dim or indistinct," and in Russian, "to black out." For the Cakchiquel tribe of North America, it glosses as "the Sun carries sickness"—a belief shared by many cultures. In Asian folklore, children born at the moment of deepest darkness will emerge mute or deaf. Even today, many Thais hoard black joss sticks and jelly and sacrifice black chickens to Rahu, the Sun-devouring demon, much as their ancestors did, and give eclipse-tainted goods as alms to beggars in the belief that their sins will be redeemed by their donations.

But it is in India that the most enduring superstitions are found. Only a little over a decade ago, in Jajai, a village in the Punjab, pregnant women were warned not to venture outside during an eclipse lest their babies be born blind or with a cleft palate. Food cooked before an eclipse must be thrown out as impure; anyone wielding a knife or ax during an eclipse is believed to be risking an accident. Cows are kept indoors—this being Hindu India—while children must either go to temple or remain at home. It has long been thus.[12]

On July 22, 2009, the longest solar eclipse of the century—six minutes and thirty-nine seconds—was best seen from Varanasi, the holy city in north-central India. Three years before I had traveled there and met a Sanskrit scholar from one of the city's five universities. We sat in his courtyard with cows and rabbits wandering about, and his children, all nine of them, peeping around corners to take in the strange visitor. "I was silly, and was away from home during the last eclipse," reflected Dr. Sudharka Mishra, pulling at his worn white dhoti. "Nobody took the cows inside, with the result that that one"—he pointed to a light brown three-year-old—"was born blind." Sure enough, great mounds of excess skin rose up from its lower eyelids, making it almost impossible for the poor animal to see.

Dr. Mishra had gone on to explain that for Hindus solar eclipses were *kalki*, "polluting." The obliteration of the Sun, even for a moment, was a terrible omen. When such an event was due, a special ritual known as a *homa* was held, which involved an offering to a consecrated fire, followed by the *suryagrahana* (solar eclipse) *puja* rite and the chanting of hymns; pregnant women had their bellies smeared with a mixture of ghee (clarified butter) and cow dung, as stipulated in the Vedic scriptures.* In Varanasi, after an eclipse, Hindus make their way to one particular pond in the center of the city and immerse themselves. I had seen the pond, about fifty feet square: it must have been packed. Ritual cleansings like this are still carried out all over India.

Because early societies experienced eclipses as a crucial disorder, they tried to incorporate them into their worldview as something that could be readily understood. In the Qur'an, they are punishments from God, and one must abase oneself to avoid further divine displeasure. North American Indians believed eclipses were the fault of dogs and coyotes; South American tribesmen blamed the jaguar; the Bakairi of Brazil believed an enormous blackbird spread its wings over the Sun. The list goes on.[14]

Following an eclipse, any setback—the death of a monarch, a defeat in battle—would be carefully noted, in the belief that any future darkening at the

---

* Ghee is known to have detoxifying properties, while cow dung contains a substance similar to penicillin. In ancient cultures, dung (the fresher the better) was used to combat a range of illnesses, and its particular power to absorb radiation is described in the Vedas. These beliefs persist. In 2005, after Pakistan tested a nuclear weapon, one Indian company advertised its brand of house paint by announcing that it had been mixed with dung: "Coat your walls with our paint and be free from atomic fallout!" Following the Chernobyl catastrophe of 1986, the Ukrainian peasantry sealed their huts with dung; and even NASA's manned space capsules were reputed to be coated with a treated layer of dung as a form of protection against the Sun.[13]

same point in the solar year would inflict a similar misfortune. Thus it became important to be able to predict when eclipses might occur, and the early glimmerings of science were put to work to shed light on what the future might bring.

It was the Babylonians who discovered that solar eclipses form a pattern. In the course of a year, the Earth's passage makes it appear as if the Sun is moving through a circle. When the Earth has gone 19 times around the Sun and the Moon 223 times around the Earth, a particular eclipse of exactly the same type will repeat itself; so had there been a lunar eclipse at dawn on (our) May 18, 603 B.C., say, one could have forecast that an almost identical eclipse would occur at sunset on May 28, 585 B.C. To recognize such a pattern over so extended a period of time was an extraordinary accomplishment for so early a civilization.[15]

By the end of the third millennium B.C., the Babylonians were maintaining a chart of sun and weather omens (such as lightning, thunder, and earthquakes). With this in hand, they tried to outwit the heavens: when an eclipse was forecast, some lowly individual would be selected to stand in for the king until the event had passed. To make it clear that it was he who was to suffer the impending evils, this "substitute king" would be forced to listen to a reading of the portents.[16] The true king remained out of sight, undergoing intensive rites of purification. Letters written to the real ruler would address him as "Farmer," but once the eclipse had passed, the ministers would inquire of the "Farmer" on which day the hapless decoy should meet his fate, to be mourned as if he were indeed a fallen sovereign.

The Babylonians would measure the eclipse's degree of totality by stretching out a hand at arm's length in front of the eye; this method was later adopted by the Greeks, whose word for "finger" or "finger's breadth" also meant a unit of measurement, twelve fingers (oddly) defining a total eclipse.[17] The first total eclipse for which we have records is set down in the Chinese *Ch'un-ch'iu* chronicles as having occurred on July 17, 709 B.C., a date agreeing exactly with modern computations.[18] Ancient China left behind a huge number of records; up until the Qing Dynasty took over in A.D 1644, there are about a thousand references to these heavenly darkenings. Exact forecasting was desperately important—hence those who managed it won considerable prestige, and people were not slow to hazard guesses, however wild.

But forecasting was no easy matter. The Babylonians may have achieved it to a certain extent, but few other cultures did. To predict a solar eclipse in Herodotus' day, for example, one would have had to know certain astronomical facts that came to light only in the third century B.C. In the second century B.C., Hipparchus—according to Pliny—in what one might call a solar legend, forecast the motions of Sun and Moon for six hundred years; three hundred years later, in about A.D. 150, Claudius Ptolemy's thirteen-book description of the universe, the *Almagest*, laid out a purportedly scientific system for judging eclipses. Such was his authority that his views, for all their inaccuracy, enjoyed wide and prolonged currency, and some fifteen hundred years later, in 1652, the volumes dealing with such matters were republished as "an easie and familiar method whereby to judge the effects depending on eclipses."

For hundreds of years, these dread occasions were predicted and recorded as warning, criticism, or confirmation, as appropriate. During times of civil war in medieval Japan, the court astrologers tended to overpredict; if nothing happened, they were rewarded for having prevented any occlusion. Chinese astronomers, on the other hand, would use predictions in highly political ways. If they disapproved of their ruler, they would meticulously record an eclipse that had occurred, which could be taken as a divine rebuke to the government because eclipses were considered warnings from heaven. They might even announce a totally fictitious one, as happened in 186 B.C., during the reign of the deeply unpopular empress Kao Tsu. However, if they approved of their rulers, astronomers might just as readily omit all record of an eclipse that had actually taken place. The belief that eclipses presaged misfortune kept Chinese astrologers close to the seats of power. One might imagine that a shortage of ensuing disasters would have devalued their status, but the fates threw up just enough occasions to support their authority. For instance, almost immediately after the eclipses of both March 4, 181 B.C., and January 18, A.D. 121, a dowager empress of China met her death; and less than a year after the eclipse of August 28, A.D. 360, the emperor Mu died, aged only nineteen.

Humankind seems to have an unquenchable thirst for superstition, and most societies have assumed that great events in the universe at large are conjoined with those in the human world; in the case of eclipses, these were often believed to be markers of the death of a famous person or of some other momentous occasion. "Every time after such an accident," Machiavelli says scornfully, "you get some such extraordinary event turning up."[19] Thus random acts of history seemed to confirm such superstitions. Pragmatism could sometimes triumph over folklore, however. In his *Life of Pericles*, Plutarch recounts that an annular eclipse (now identified as that of August 3, 431 B.C.) so terrified the Athenian sailors that they threatened to desert an expedition against the Lacedaemonians. Seeing that the pilot of his galley was filled with alarm, Pericles doffed his cloak and cast it over the man's head. "Is there anything so terrible in that?" When the man gave a muffled no, Pericles asked what the difference was between cloak and eclipse, except size. The fleet set sail.[20]

The emperor Claudius also deployed common sense before the eclipse of August 1, A.D. 45, which astrologers had told him would fall on his birthday. When they went on to say that their omens predicted rioting against him as a consequence, Claudius put out a public announcement of the imminent event, describing what kind the eclipse would be (annular), and explaining to the citizenry how such celestial happenings come about.[21] His birthday passed uneventfully.

Foreknowledge of eclipses could be used to exploit superstition as well as to defuse it. More than once, a ruler whose priests had predicted a solar eclipse would rely on the panic he knew would ensue to send troops to storm an enemy city. In July 1917, after a long desert trek, T. E. Lawrence and his Arab irregulars closed in on Aqaba, now Jordan's only port, which they planned to take from the Turks. Lawrence knew that a lunar eclipse was due on the sixth, and held back his troops until the garrison, alarmed by the loss of the Moon, set about frightening off the devouring monster with the usual assortment of drum hammerings, klaxons, and shouting. That gave Lawrence his moment to attack—and Aqaba fell.[22]

Sometimes, though, there really was a consequence on Earth for what happened in the heavens. In 1183, a brutal three-year civil war between the Minamoto and Taira clans was ravaging Japan, but just as the Minamoto were preparing for battle the Sun was put out. Having no rational explanation for

The armies of the Medes and Lydians were locked in battle on May 28, 585 B.C., part of a grueling five-year war, when "the day was turned into night" and the fighting stopped at once.

such a calamity, they took to their heels.[23] At other times, the connection between eclipse and event appears to have been fabricated. There is a famous case, datable to May 28, 585 B.C., when the Medes and Lydians, more than five years at war, made peace, supposedly because an eclipse put a sudden end to a major battle. Herodotus and Pliny both recorded such an event as taking place, the former writing that the great astronomer Thales (c. 624–c. 546 B.C.) had predicted a solar eclipse for that exact year (though he was no more precise than that). Even so, the conjunction is most likely mythical, with both historians guilty of political spin, combining an eclipse and a battle from two different dates. Herodotus also seems to have described an eclipse that never took place at all, describing how one shadowed Xerxes of Persia (519–465 B.C.) as he set out to conquer Greece; modern computations can find no eclipse at the time.*

Eclipses seem to have unhinged even normally authoritative sources. The Anglo-Saxon Chronicle, that register from the reign of Alfred the Great to the mid-twelfth century, records:

In this year [A.D. 1135] King Henry went across the seas at Lammas [August 1] and the next day, when he was lying asleep on board ship, it grew dark over all lands, and the Sun became as if it were a three-nights-old Moon. . . . People were much astonished and terrified and said that something important would be bound to come after this—and so it did, for during that same year the King died.

Henry I, whose old age the Chronicle paints as racked with misery, did indeed die exactly two months later; but the eclipse described actually took place two years before.

Eclipses have been used to date historic events—thus one is recorded eleven days before Alexander overthrew King Darius at Arbela, and by working out events retrospectively one can mark the battle as fought on October 21, 331 B.C. But such claims are made with varying degrees of accuracy. The great Johannes Kepler precisely calculated the world's creation day as July 23, 3999 B.C., and coupled it with an eclipse—indeed, these feature in similar ways in the creation stories of several cultures, and in attempts to date important events in religious history. "The birth of our Lord and Savior is itself quite positively fixed," contended one nineteenth-century divine. "It is said in the Bible

---

* An annular eclipse is said to have taken place on March 7, 51 B.C., the day Caesar crossed the Rubicon. Homer makes two possible allusions, once each in the *Iliad* and the *Odyssey*, but neither reference can be made historical;[24] he creates a third that is surely fictional when he has Odysseus kill Penelope's suitors during an eclipse. (Similarly, Aristophanes in *The Clouds* and Borodin in *Prince Igor* both use eclipses for dramatic effect.)

that Herod . . . died when Jesus was still a child, and probably a few months after his birth. Now, the Jewish historian Josephus tells us . . . there was an eclipse during Herod's last illness; and as this eclipse was known to have occurred in April, in a certain year of what is called the Julian period, it is quite clearly shown that our Savior must have been born just about the time the Bible states he was."[25]

An even more ominous darkening, the only one to shadow Jerusalem during Christ's public ministry, is said by Church historians to have occurred on the day of his crucifixion. They describe it as being total, beginning at noon, and passing across the Black Sea from near Odessa on to the Persian Gulf. Matthew, Mark, and Luke (probably using a common source) all invoke it in their narratives. According to all three, "From the sixth hour there was darkness all over the land until the ninth hour," with Luke adding, "while the Sun's light failed."[26] But Jesus was crucified at Passover, a spring full-Moon festival, when the Moon had to be on the opposite side of the Earth. And no eclipse lasts anywhere as long as the three hours that the Gospel narratives proclaim.

Everywhere in the world people have reacted to eclipses with profound fear, yet a simple explanation of what they were was put forward as far back as the sixth century B.C., when Thales proposed an answer. A century later, his fellow countryman and fellow astronomer Anaxagoras (500–428 B.C.) would write that "the Moon has not any light of its own but derives it from the Sun. . . . Eclipses of the Moon are due to its being screened by the Earth, or sometimes by the bodies beneath the Moon; those of the Sun to screening by the Moon when it is new."[27] For this, however, Anaxagoras was indicted for atheism and condemned to death (he chose exile instead), and the link between eclipses and misfortune continued unbroken for another two thousand years.

Certainly it was still going strong in the early ninth century, when the archbishop of Lyons appointed by Charlemagne witnessed three lunar eclipses in a single year and recorded that each time his terrified flock took to blowing horns, hammering on metal instruments, and screaming, "*Vince, luna,*" urging the Moon to fight back.[28] The Church officially forbade such behavior (as it would repeatedly do in the centuries to come), but it was no match for the strength of superstition—after all, the monster *did* always retreat. Even the great emperor himself was not immune: in 810, Charlemagne was just over seventy when solar and lunar eclipses fell within two weeks of each other (November 30 and December 14) and alarmed him into immediately preparing his will; by the end of January he was dead. Almost thirty years later, on May 5, 840, a solar eclipse lasting nearly six minutes so frightened Charlemagne's son Louis the Pious that he too fell ill and died before June was out.

(The bitter struggle for his throne led to the historic Treaty of Verdun, dividing Western Europe into the three major areas we now know as France, Germany, and Italy.)

Popular attitudes were little changed even into the seventeenth century, and the ominous conjunction of full solar and lunar eclipses within weeks of each other, albeit rare, was particularly alarming. When this happened in October 1605, Shakespeare used it as a powerful metaphor for a sense of impending doom, incorporating it into the first act of *King Lear*, in which Gloucester invokes the recent horrors:

> These late eclipses in the sun and moon portend no good to us: though the wisdom of nature can reason it thus and thus, yet nature finds itself scourged by the sequent effects: love cools, friendship falls off, brothers divide: in cities, mutinies; in countries, discord; in palaces, treason; and the bond crack'd 'twixt son and father. This villain of mine comes under the prediction; there's son against father: the king falls from bias of nature; there's father against child.

He wanders off, leaving his bastard son, Edmund, alone on stage. Gloucester is a good man yet a gullible one (as we are soon to see again with his other son, Edgar), and a creature of his time in his acceptance of superstition. Edmund, on the other hand, may be the play's villain, but he is also the supreme rationalist, and here has Shakespeare's tongue:

> This is the excellent foppery of the world, that when we are sick in fortune—often the surfeit of our own behavior—we make guilty of our disasters the sun, the moon and the stars: as if we were villains by necessity, fools by heavenly compulsion; knaves, thieves and treachers, by spherical predominance; drunkards, liars and adulterers, by an enforced obedience of planetary influence; and all that we are evil in, by a divine thrusting on: an admirable evasion of whoremaster man, to lay his goatish disposition to the charge of a star!

Shakespeare's skepticism was shared by his king. That same year, after a 90 percent eclipse, James I wrote a satirical letter to Lord Salisbury enumerating all the evils that his subjects might ascribe to the Sun's temporary occlusion. For James, the superstitions attached to forecasting were the devil's works, and he wanted them outlawed. Monarch and playwright alike proved singularly unsuccessful in changing opinion.

John Milton was born three years after the first production of *King Lear*. In 1638, he published *Lycidas*, which laments the death of his Cambridge contemporary Edward King, cast away when his ship foundered off the English coast:

> *It was that fatal and perfidious Bark*
> *Built in th'eclipse, and rigg'd with curses dark.*
> *That sunk so low that sacred head of thine.*

The same year that he published *Lycidas*, Milton took a continental tour during which he visited Galileo, under house arrest in Florence. The great astronomer was by then frail and blind, yet the two men spoke for some time, and one would like to think that the causes of eclipses would have figured in their conversation; but superstition held sway even over a mind as sophisticated as Milton's. In *Paradise Lost*, his description of the faded splendor of the fallen archangel captures the state of unease that most people still felt in the face of an eclipse:

> *As when the sun, new ris'n,*
> *Looks through the horizontal misty air,*
> *Shorn of his beams, or from behind the moon,*
> *In dim eclipse, disastrous twilight sheds*
> *On half the nations, and with fear of change*
> *Perplexes monarchs.**

By the time Milton finished his great work, mentioning eclipses in Book II, lines 663–66, and Book X, lines 410–14, he, too, was completely blind. He had struggled with poor sight since childhood, and as early as 1647 had already lost most of the vision in his left eye. For the next five years he tried various remedies to stave off his fate, but by March 1652, at the age of forty-three, he had to accept that both eyes had failed him, "without any hope of day."[29] In *Samson Agonistes* (1671), he portrays Samson railing in words that might have been his own:

---

* Book I, lines 594–99. *Paradise Lost* was not completed until 1665, when censorship was severe. One of the chaplains through whom the Archbishop of Canterbury gave or refused publishing licenses found this passage objectionable; following the death of Cromwell and in the turbulent days after Charles II returned to power, the lines were seen as dangerous, imputing to monarchs "perplexity" and "fear of change." The poem was allowed to see the light of day only through the intercession of an influential friend of the poet's.

O dark dark dark, amid the blaze of noon,
Irrevocably dark, total eclipse
Without all hope of day!
O first created Beam, and thou great Word,
Let there be light, and light was over all;
Why am I thus bereav'd thy prime decree?
The sun to me is dark
And silent as the moon.

Coincidentally, the twenty-ninth of March 1652 (by the old calendar)—the very same month that Milton finally lost all sight—saw "Black Monday," a solar eclipse that panicked all Britain. The rich loaded up their coaches and fled from London, and there was a thriving trade in a variety of cordials supposed to allay ill effects. At Dalkeith, in Scotland, the poor threw away their possessions, "casting themselves on their backs, and their eyes towards heaven and praying most passionately that Christ would let them see the Sun again, and save them."[30]

Not everyone reacted with such alarm. Samuel Pepys mentions that during a later (spring) occlusion, his doctor's daughter, having risen at dawn, was concentrating so hard on composing a letter that not until about nine o'clock—although she was in the midst of an almost total eclipse—did she even notice that "the light of the Sun look[ed] somewhat dim."[31] But she was the exception. In Paris, people flocked to confession in such numbers over the days leading up to an eclipse forecast for August 21, 1664, that one priest felt obliged to declare that the event had been put off for two weeks in order to manage his crowds of penitents.

In the cities of Western Europe, however, the days of superstitious ignorance were numbered, for the march of science was accelerating. Thus the great astronomer Edmond Halley (1656–1742) was able to predict the total solar eclipse of May 3, 1715 (which covered almost the entire south of England for more than three minutes), and so carefully observed it that he could estimate the width of its track. He had readings taken from fifteen different points along its path, which enabled him to calculate the Sun's diameter with greater accuracy than ever before. Thomas Crump, in his wide-ranging survey of the history of eclipses, declares that Halley's observations "can be taken to be the beginning of modern eclipse astronomy."[32] Fear and awe remained; but they were now augmented by knowledge.

PART TWO

DISCOVERING the SUN

Bronze Age petroglyphs from all over the world connect the Sun with fecundity, as in this drawing from the Camonica Valley, in the central Alps of northern Italy, of a male figure with the solar disk attached to his phallus.

CHAPTER 5

THE FIRST ASTRONOMERS

Of all tools, an observatory is the most sublime. . . . What is so good in a
college as an observatory? The sublime attaches to the door and to the first
stair you ascend;—that this is the road to the stars.

—RALPH WALDO EMERSON, 1865[1]

Astronomy: a fine science. Useful only to sailors. While on the subject,
laugh at astrology.

—GUSTAVE FLAUBERT, c. 1870[2]

IN 1894, SIR NORMAN LOCKYER (1836–1920), ONE OF BRITAIN'S FORE-
most living scientists—the man who had discovered helium in the Sun before
it was known on Earth—published *The Dawn of Astronomy*, in which he broke
down the work of his ancient predecessors into three distinct phases.[3] Civilizations first went through a period of worship, where celestial phenomena were
construed only as the actions, moods, and warnings of the gods; next, people
began putting astronomy to practical uses, such as agriculture or navigation;
and finally, the heavens were studied solely for the pleasure and purpose of
pure knowledge. Though Lockyer's divisions are seen now as too schematic—
for instance, we may question whether the reverential stage is ever completely
outgrown, and his definition of the third stage ignores the sense of wonder
that often drives astronomers—his categories still offer a useful yardstick.

Long before mankind gathered into cities, there was astronomy—or at
least there were people studying the Sun. We have evidence of observatories in
several parts of the ancient world; however, one city above all others deserves
to be called the birthplace of an organized and recorded science of astronomy:
Babylon.[4]

The country was settled by a people of unknown origin between the Tigris
and Euphrates rivers, in what is now the heart of modern Iraq. The surrounding land was called Sumer, although another name for this fertile flat area was
Eden. The Sumerians began farming about 9000 B.C.; and roughly 2,500 years

later they invented the wheel, cart, and plow, transforming every aspect of life—from travel and transport to warfare and industry—and reducing the number of workers required to raise a given amount of food. These were crucial steps in freeing some people to become full-time priests, scholars, and merchants. Around 3100 B.C. they invented a system of writing that incorporated astronomical concepts into two of its characters, one representing a star, the other "the Sun over the horizon."[5] By about 2350 B.C. the Sumerians had been subdued by their neighbors to the north, but it was the conquerors who were transformed: Sumerian remained the language of religion and scholarship, and divining the future by observing the stars became a preeminent intellectual discipline.

At first, the Sun was cast as subordinate to the Moon, and worshipped in only two minor cities, Larsa and Sippar. By 1800 B.C., the growing metropolis of Babylon (about seventy miles south of present-day Baghdad, taking its modern name from the Greek rendering of the Akkadian *bab ilani,* or "gate of the gods")[6] had become preeminent in the region. Ruled by the Amorite king Hammurabi, who encouraged the careful recording of the movements of the stars, the Babylonians believed the Sun to be a baleful divinity that spied for the high gods and was an unforgiving judge of mankind. A precursor of future images of the Jewish and Christian God, the Sun is depicted as an old man with a long beard, beams radiating from between his shoulders. This figure overtook the Moon in importance and metamorphosed into the Sun Royal, exercising the attributes of earthly divinity and kingship.

The next to subjugate the land, around 750 B.C., were the Assyrians, whose kings called themselves "Suns of the World." During their campaigns of conquest—in which they would line the walls of each fallen city with the skins of its inhabitants—Babylon sank into political and economic decay. Eventually, a series of revolts culminated in the Assyrians' overthrow in 606 B.C., and although the new masters, the Kaldi, or Chaldeans (a loose federation of five tribes of West Semitic origin, from among the southern swamps and coastal plains), held sway for one of the briefest ascendancies in Mesopotamian history, their great king, Nebuchadnezzar (who ruled from 605 to 562 B.C.), oversaw a renaissance that turned Babylon into the cultural center of a massive empire. The position of astronomy was assured; special schools flourished not only in Babylon but also in other large towns. Exchanges of ideas among Persians, Jews, and Greeks living in the same cities were eased by a lingua franca written in simple characters: Aramaic.

Initially, the Chaldeans—their name now all but synonymous with "Babylonians"—were content simply to watch the skies, but by around 550 B.C. they felt curious and confident enough to apply mathematical techniques to their records, dividing the heavens into zones, the most important being what we now know to be the celestial equator (in later years this would be identified as the path of the Earth's ecliptic journey). Along this they constructed the zodiac (from the Greek *zo idion,* "living creature"), a diagrammatic representation of the great figures that they imagined to be marking the Sun's movements. The following mnemonic lists them in order, starting with the vernal equinox in March:

> *The Ram, the Bull, the Heavenly Twins*
> *And next the Crab the Lion shines,*
> *The Virgin and the Scales;*
> *The Scorpion, Archer, and He-Goat,*
> *The Man who bears the Watering-Pot*
> *And the Fish with the glittering tails.*

The Chaldeans' advances were made amid great upheavals. Around 689 B.C. Babylon had been destroyed by the Assyrian king Sennacherib. The city was then rebuilt in even greater splendor (according to Herodotus, its walls were fifty-six miles in circumference and 335 feet high), but collapsed once more under the onslaught of the Medes and Persians, to be absorbed into the empire of Cyrus the Great. By the time the Chaldeans were overthrown in 539 B.C., they had become famous for their expertise in studying the stars. According to the Greek academic Strabo of Amasia (64 B.C.–c. A.D. 23), a special district of the kingdom was "set apart for the local philosophers, the 'Chaldeans,' as they were called, who were concerned mainly with astronomy; but some of these, who were not approved of by the others, professed to be writers of horoscopes." While most of what they recorded was purely descriptive, astrology became the business of interpreting the heavens—not for ordinary individuals, but for king and country. Several branches of this pseudoscience grew up, the first and most important being the art of foretelling a subject's life from the position of the stars, as well as the Sun, at his birth or conception. The Chaldeans would draw upon astronomical information to predict human destinies, work out where certain planets were in the sky at a child's birth, and from these circumstances draw conclusions.

Some aspects of the stars' influence were obvious: for example, the Sun's position on the ecliptic was tied to the seasons. If the heavens could tell you the most propitious times for planting or harvesting, why shouldn't they also indicate the best time to marry or to undertake an unavoidable but dangerous journey? Understandably, astrologers took to drawing such conclusions about the influence of the celestial bodies, even if they were without scientific basis.

Yet without the promptings of astrology, naked-eye astronomy would never have reached so deeply. At first there was no division between the disciplines: one practiced astronomy for the purpose of astrology. It was a fruitful partnership. By the fifth century B.C., astronomers had established a year ahead the ex-

A nineteenth-century depiction of Babylonian astronomers observing a meteor trail

pected intervals between sunrise and sunset for various days, the monthly apparent movements of the Sun, and much else. They developed the arithmetic required to predict the motions of the Moon and planets and from there to construct a calendar. Even more important was their discovery of a system to predict the passages of the Moon and the Sun, enabling them to forecast eclipses.

The Chaldeans maintained highly accurate and detailed records, but frustratingly we lack any explanation as to how they constructed their tables. They made their observations from step pyramids called ziggurats, which were attached to temples often the size of fortresses and painted the seven colors of the rainbow. The tower of Babel—portrayed in the Bible as a vainglorious attempt to build a stairway to the stars—is the best known, but the most commanding in all Babylon was another ziggurat that rose over four hundred feet in seven gleaming enameled levels, its zenith crowned by a shrine that held a table of solid gold and an ornate bed on which each night some young woman awaited the pleasure of the god, or his more assiduous earthly representative.

While much has been made of a climate allegedly perfect for observation—Plato lauded Mesopotamia's cloudless atmosphere, and Herodotus wrote that "the air in these regions is constantly clear, and the country warm through the absence of cold winds"—this was likely based on thirdhand accounts. Sandstorms and mirages would have been serious obstructions to following phenomena close to the horizon. Even in the best conditions, it was not easy to distinguish between a planet and a fixed star when gazing on the myriad points of heavenly light.

Observation by itself, then, was not enough. The Chaldeans began to analyze and interpret what they had seen, writing their commentaries in journals, almanacs, and star catalogues. By a conservative estimate, they recorded 373 solar and 832 lunar eclipses during their brief dominance. A wide range of daily happenings—meteorological events, river levels, fires ravaging the city, the deaths of kings, epidemics, thefts, conquests, plagues of locusts, famines, and—of particular importance—fluctuations in the prices of barley, dates, pepper, cress, sesame, and wool (always in that order) were scrupulously transcribed onto clay tablets.* The Chaldeans also inscribed their calculations of planetary positions on tablets that in A.D. 509 the Greek writer Heliodorus would name *"ephemerides"* ("matters specific to the day")—in effect, working journals, showing the consecutive appearances of the same objects in the sky. About seventy tablets from the main series survive; and of seven thousand notations, fully a quarter involve the Sun.

Divination, observation, calculation: in Babylon, anyone who could read studied mathematics. The average schoolchild knew far more about the heavens than do most of us today, who rarely examine the skies. In an age far re-

---

* See *The Cambridge History of Islam*, vol. 3, p. 534, and N. M. Swerdlow, *The Babylonian Theory of the Planets* (Princeton, N.J.: Princeton University Press, 1998), p. 26. The Assyrian capital of Nineveh stood only a few hundred miles from Babylon. At the great library established by Ashurbanipal (668–627 B.C.), at least twenty-two thousand astronomical tablets were collected. In 612 B.C., when the Assyrian Empire finally came crashing down, Nineveh was sacked and the entire library destroyed, except for the tablets, which were buried—and thus preserved—under the collapsed walls. Without such a disaster these records would not have survived to resurface during the 1870s, when a French official in Baghdad, his curiosity piqued by some large fragments of sculpture, began to dig there. It seems extraordinary that such vital sources of information were overlooked for so long. Muslims may have kept their distance because the Qur'an states that the tablets were bricks inscribed by demons and baked in hell, while Christians possibly thought it foolhardy to dig for the buried knowledge of a people so excoriated in the Bible. Another fifty thousand tablets, written between 3000 and 450 B.C., were excavated from the temple library of Nippur. There may be as many as half a million scattered around the museums of the world, only a fraction of which have ever been deciphered, while countless more still lie under the Iraqi sand. See Norriss S. Hetherington, *Ancient Astronomy and Civilization* (Tucson: Pachart, 1987), pp. 25–26.

moved from our modern benefits of gas or electricity, the lights of heaven had no competition, so that a horseman did not carry brands or tapers at night, but rode on under the light of the stars. Yet for all its accomplishments, organized astronomy remained an elitist activity. A college might contain just two or three people genuinely taken with the heavens—and these were hampered, as astronomers are to this day, by the limits of their technology.

One powerful mathematical tool gave Babylon much of its mystique in the sciences: the place/value numerical system. There were two versions: one used a decimal base and was the method for day-to-day business and for calculating lunar cycles; the other used a base of sixty and was for measuring solar cycles. The latter, although more cumbersome, would be employed by various societies for almost two millennia—until the time of Copernicus—and from it sprang the division of degrees into sixty minutes and minutes into sixty seconds.[7] Such a method was far superior to any other in the ancient world, and is curiously close to that of the modern computer.* Wonder was acquiring number as its servant, a crucial development for any science.

Babylonian mathematics—sexagesimal arithmetic, algebraic operations, geometric rules—spread over a far greater area than their astronomy ever reached, and indeed preceded systematic astronomy by more than a millennium. But major shortcomings remained. The early mathematical observers treated time not as a measure, but almost as a quality; nor were they interested in quantifying space—there was no concept of absolute distance. No trace appears in Babylonian astronomy of any desire to construct a comprehensive scheme to account for all heavenly phenomena, to investigate their nature and causes, or to take a synoptic view of the universe as a whole. The Babylonians had their lunar calendar and were the first to divide the day into twelve-part halves, but their divisions of daylight had to vary with the changing seasons, and they were unable to determine the Sun's exact apparent orbit around the Earth. This shortcoming in turn hampered attempts to work out the shape of the Earth; they never learned that it was roughly spherical. The gnomon (at its simplest, a vertical stick whose shadow is used to tell the time) was beyond them, not least because they had no concept of north, south, east, and west as cardinal points: anything approaching compasslike clarity had to await the third quarter of the fourth century B.C. They had not even established the se-

---

* In the 1790s, during the French Revolution, radical French mathematicians attempted to grade the circle by dividing each quadrant into a hundred parts, each then decimally subdivided; but the system never caught on. Less scientifically, but more enduringly, the French word for eighty—*quatre-vingt*, "four-twenty"—survives as a reminder that our ancestors counted on their toes as well as their fingers.

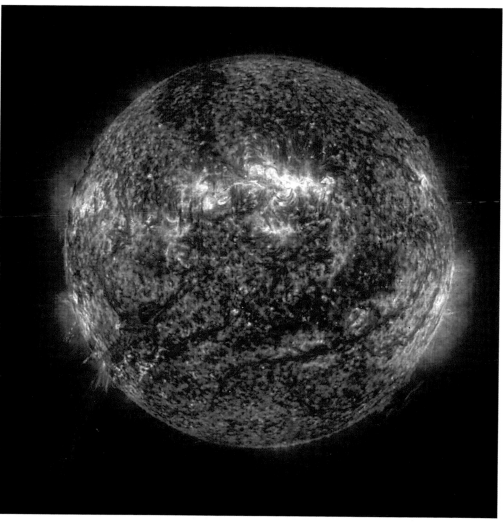

The Sun viewed in extreme ultraviolet light by NASA's SOHO satellite. Bright spots mark the most active regions, where sunspots, flares, and coronal mass ejections form.

The spectacularly varying colors in images of the Sun are explained by the fact that the EIT (Extreme ultraviolet Imaging Telescope) records images of the solar atmosphere at several wavelengths, and so shows solar material at different temperatures. In images taken at 304 angstrom, which show as orange, the bright material is at 60,000 to 80,000 kelvin; in those taken at 171 angstrom (blue), at 1 million K. Images at 195 angstrom (green) correspond to about 1.5 million K, those at 284 angstrom (yellow) to 2 million K.

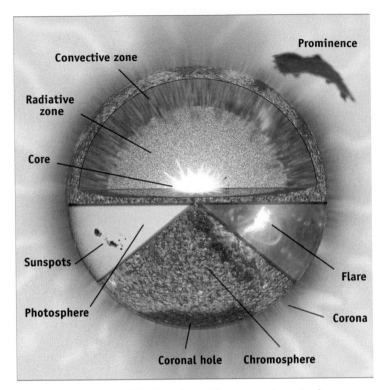

In this diagram of parts of the Sun, the artist's cutaway shows the three principal zones: the inner core, the transition zone (where particles begin their ascent toward the surface), and the convective zone, where energy circulates like porridge bubbling in a pot.

A diamond ring during a total eclipse, photographed in 1999; such rings occur as the Sun rises over the jagged limb of the Moon.

These coronal mass ejections (CMEs) are like rubber bands that can be stretched and twisted until they break, releasing massive doses of hot air and energy.

The coolest part on the Sun is the dark eye of a sunspot— 3,800° F, compared to 6,000° F for the rest of the surface. A sunspot this size could swallow the Earth with ease.

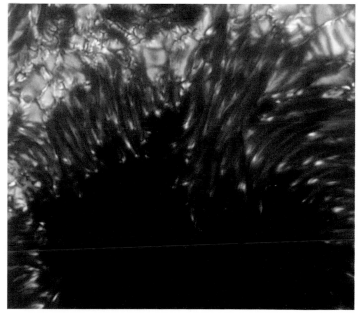

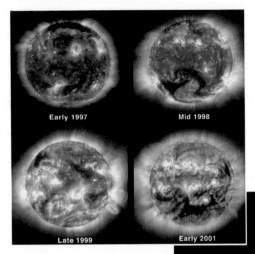

Early 1997   Mid 1998

Late 1999   Early 2001

The Sun changes from relative calm to a bright, tangled mass of magnetic loops in the years leading up to a solar maximum— the peak of the Sun's eleven-year cycle.

This LASCO C2 image, taken January 8, 2002, shows a widely spreading CME as it blasts more than a billion tons of matter out into space at millions of kilometers per hour.

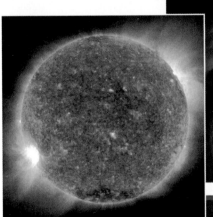

With two active regions (the brighter areas above) in profile almost diametrically apart, SOHO got a good view of the extensive areas above the Sun influenced by the powerful magnetic fields associated with the two regions.

A CME observed over an eight-hour period, as photographed by LASCO C3 on August 5–6, 1999. A dark disk has been employed to mask the Sun, so that the structures of the corona will be easier to see. The white circles represent the size and position of the Sun.

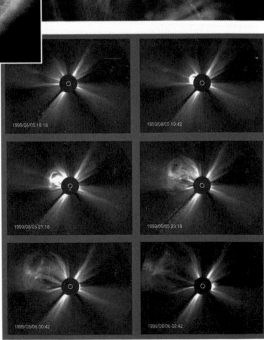

Like snowflakes, no two auroras are the same. In the fourth century B.C., Aristotle described "glowing clouds" and a light like flames of burning gas—one of the first scientific accounts of an aurora.

Dawn on Midsummer Day 2005, photographed from the top of Mount Fuji at 4:37 A.M.

*The seashore at Futamigaura, near the holy site of Ise,
where on New Year's Day homage is paid to the sun goddess,
Amaterasu. The "Wedded Rocks" in the foreground are believed
to shelter Izanagi and Izanami, the male and female couple
from whom all the islands of Japan arise.*

A pillar of light rises from the setting Sun in the North Atlantic, off the coast of Maine. Sun pillars will sprout from the top of the Sun when light reflects off the undersides of platelike crystals in the air.

quence of the planets outward from the Sun; but then, no one knew any better. That Babylonian astronomy has its current high reputation may be due to the fact that so many Greek records have perished: clay lasts better than papyrus. However, overall I share Noel Swerdlow's view: "That the complete development of this science took hundreds of years shows how difficult it was, far more difficult than any other science in the ancient world, both because of the magnitude and complexity of its subject . . . and because it was the first empirical science."[8]

Around 330 B.C. the region fell to Alexander, upon whose early death his marshal Seleucus took over the greater part of the conquered lands. Sixty years later, the mighty dynast Antiochus I forcibly moved most of the city's inhabitants to the newly founded metropolis of Seleucia, sixty miles to the north. For the astronomer-astrologers of the old city, this was an indignity too far. They kept up their practices in the temple of Bel almost to the end of the first Christian century, but it was their dying gasp. "Thus ended," writes the historian James McEvoy, "the two-thousand-year Babylonian cultural development of watching the heavens."[9]

EVEN BEFORE BABYLON was relocated, its learning had spread to Egypt, Greece, and Rome to the west, to India and possibly as far as China to the east. The Egyptians adopted the Babylonians' zodiac while rejecting their worldview, with its emphasis on portents and fateful predictions and the universe as bad news. Herodotus refers to Egypt as "a gift of the river": possibly the Nile's annual flood in a cycle ideal for growing crops encouraged a sense of well-being. Their priest-astrologers knew that the Nile would overflow its banks around the summer solstice, when bright Sirius rises, and would make their calendrical predictions accordingly.[10]

From the beginning, the Nile's yearly flooding yielded an unexpected benefit: as the government's main source of income was a land tax, there was a constant need to reestablish boundaries and settle disputes caused by the floods, giving rise early on in Egypt's history to a cadre of skilled surveyors.*

---

* See Martin Isler, *Sticks, Stones, and Shadows* (Norman: University of Oklahoma Press, 2001), pp. 55 and 135. Geometry, the pure mathematics of points and lines, curves and surfaces, takes its name from the Greek words for surveying: *ge*, "earth," and *metrein*, "to measure." "Algebra," on the other hand, derives from an Arab word meaning "compulsion," as in compelling the unknown *x* to assume a numerical value. A more colorful notion advanced by some historians is that "algebra" means "putting back into its proper place" or "bonesetting." The bonesetter in *Don Quixote* is known as *algebrista*, but whether because he resets or because he compels, Cervantes never makes clear.

The Egyptians, an exceptionally practical people (even though they remained 95 percent illiterate, they had invented writing slightly ahead of the Babylonians, around 3200 B.C.), devised sophisticated ways of measuring that would eventually be joined to their building skills to create the pyramids. But this was only one of many advances. They tried various ways, for instance, to gauge the Sun's diameter, using gnomons, water clocks, and even horses. In this last instance, the horse was galloped across level ground from the moment the Sun's upper edge appeared on the horizon until the entire disk had risen clear above it—a period of approximately two minutes, in which time the animal was reckoned to cover ten *stadia* (the term is a corruption of the Egyptian word for "racecourse"), and it was inferred that as the motion of the Sun must be equal to that of the horse, its diameter must be ten *stadia* too. This was hardly good geometry.*

The Egyptians had their own mathematical system at least as early as 2800 B.C., but it was crude, and unlike the Babylonians, they never progressed beyond a basic arithmetic—although they did manipulate this to produce both multiplication and division. With the exception of 2/3, the only fractions they used were in the form of 1/*n*. Although the mathematical papyri of around 1800–1600 B.C. bring off some fairly sophisticated calculations, such as the volume of a truncated pyramid, their geometry could not transact complex computations, nor could they manage the type of calculation necessary to progress from observation to prediction.[11] There is no evidence that they understood the concept of latitude or ever noted down eclipses with any regularity. What did emerge, in the intricate development of their religion and the pronouncements of their priest-astrologers, was a canon of times to be especially careful. And while they derived few predictions from what they saw, they did map stars, planets, and constellations and record their movements.

---

* A modern stade is a tenth of a nautical mile, about 600 feet; but lengths in the ancient world— and indeed on into the seventeenth century—were rarely uniform. A *stadion* could vary considerably, even within a single society: some say it represented the distance a man in ancient times could cover running flat out without taking a breath (reckoned at around 190 yards), and that this governed the length (and naming) of sports stadiums. Others say it originated as a standard length of a plowed furrow, 600 *pedes* (from *pes*, "foot"); the length of a *pes* varied, and in consequence so did that of a stade. But then, the measurements for pounds and pints were equally unreliable. Henry II of England was said to have established the length of a yard by measuring a line from his nose to his thumb. The French had hundreds of measures for nautical miles, the Germans a "mile" all their own. It was only between 1700 and 1900 that much energy was spent on ensuring equalities throughout Europe. Even now, we have yet to harmonize mile-kilometer differences, while goldsmiths remain a profession unto themselves, with an entirely different system of weights, using Troy ounces, which dates back to William the Conqueror and the trading city of Troyes.

Their understanding of what astronomy could accomplish might be summed up by an inscription on the pedestal of a statue of the astronomer-astrologer Harkhebi that dates to the early third century B.C.: while it tells us little about the man, it does describe his duties. He was to determine the risings and settings of the constellations of the zodiac; to establish the high points of each planet; and from this predict the risings of other bodies. He was also expected to foretell the solstices—"Knowing the northing and southing of the Sun, announcing all its wonders and appointing for them a time, he declares when they have occurred"[12]—the dates for celebrating the gods' feast days, and the exact relation of the spirits of the gods to the pharaoh.

Like most preindustrial peoples, the Egyptians continuously struggled to impose order on their universe—viewing the Sun itself as needing protection against those forces always lurking to disrupt the continuity of things. In a world characterized by disorder and the frequent calamities of nature, religion joined forces with science, its priests toiling to protect the people through ritual and cult and to create a body of anthropomorphic explanations for the

The family of Akhenaten (Amenophis IV) offers sacrifice to Atum the sun god. A relief from Amama, Egypt, 1350 B.C.

makeup and workings of the cosmos.[13] The sky was not an empty vault, but a goddess who each night reconceived the Sun—an early example of an immaculate conception—and each morning gave him birth. Even the dead space that separates the heavens from the Earth was a god. In this universe, creation and existence were the product not of impersonal forces but of individual wills and actions; and the various divinities the Egyptians created embodied a network of relationships, at the center of which basked the Sun.

Over time, their philosophy of creation became yet more complex. Atum, the Creator, the "pre-existing being in whom all existence was inherent and through whose self-realization all creation evolved," brought all matter out of himself—creation was the result of nothing other than his masturbation, a notion to us extraordinary, even ridiculous, but for the Egyptians the self-realization of Atum's own substance.[14] Often Atum would change places with the Sun: the star was his ultimate incarnation. At other times he was the conscious void within which the Sun first arose; the moment of orgasm was as useful a way as any to describe the way the universe came into being—a cosmology that suggests that the ancient Egyptians were, by several millennia, the first proponents of a Big Bang theory.

Such beliefs, and the Egyptians' first struggles toward scientific knowledge, developed in parallel. In these early centuries the Egyptians, like the Babylonians, never attained Lockyer's third stage: it would take millennia before the Sun was studied solely to gain knowledge. But close by was a society that valued knowledge perhaps more than any other in history.

CHAPTER 6

ENTER THE GREEKS

Astronomy? Impossible to understand and madness to investigate.
—SOPHOCLES, 496–406 B.C.

Don't confuse progress with perfectibility. A great poet is always timely; a great philosopher is an urgent need. There's no rush for Isaac Newton. We were quite happy with Aristotle's cosmos. Personally I preferred it. Fifty-five crystal spheres geared to God's crankshaft is my idea of a satisfying universe.
—BERNARD NIGHTINGALE,
the literary Byron-lover in
Tom Stoppard's *Arcadia*

THE EARLIEST REFERENCES TO ASTRONOMY AMONG THE GREEKS AP-
pear around 800 B.C. in the poems of Homer and Hesiod. By that time their countrymen, like the Babylonians and the Egyptians, had put names to various stars and quantified the risings and settings of the Sun.[1] However, they were not content with merely recording the movements of the heavens: once they took to astronomy in earnest, they inquired into what the skies were made of; the shapes and sizes of the Sun, the stars, the planets, and the Earth itself; how far these were from one another, and what revolved around what, in orbits of what form; and the number of stars, and whether they could be located with systematic accuracy. To this they added a raft of further questions about their immediate world: Exactly how long was the year, or the month? When were the equinoxes, and could one pinpoint the exact moment of the solstices? The one matter on which almost everyone agreed was that the Earth itself was stationary and central—a belief that would bedevil science for millennia.

Despite that crucial misapprehension, a formidable line of Greek students of the stars extends for over a thousand years, from the time of Hesiod until at least the death of Ptolemy. The muster of ancient Chinese philosophers may have stretched twice as far, but the Greek roll call is the more impressive. The great German scholar Otto Neugebauer lists 121 astronomers of note—and even that tally excludes figures such as Pherekydes of Syros (the teacher of Pythagoras), Zeno of Elea, Plato, and Epicurus, who, while not primarily astronomers, made significant contributions. Many familiar names appear—Aristotle, Euclid, Archimedes, Ptolemy—and others who were giants in their time but less well known now, such as Parmenides, Anaxagoras, Eudoxus, Heraclitus, and Aristarchus.

Of course, they built in part on the foundations laid by the Babylonians. Herodotus (c. 484–c. 425 B.C.), in his only book, *The Enquiries* (more often transliterated as *The Histories*), describes how his countrymen not only copied the Babylonian records and calculations but also borrowed several of their measuring devices, such as one for reckoning the motion of the Sun along the ecliptic. And the first glimpses of an astronomy that went beyond pure observation came from the Egyptians. The early Greek practitioners did not even seem to consider the prediction of heavenly phenomena particularly valuable: Plato—in the *Dialogue of Phaedrus*—blames his countrymen for not being interested enough in the planets. This mind-set did not change until the Greeks came into contact with the observation tablets of the Babylonians during the creation of the Macedonian/Greek Empire under Alexander, who dispatched literally camel-loads of Babylonian astronomical tablets to the Greek cities along the Adriatic coast.[2]

By around 650 B.C., the old aristocratic order had been overthrown, supplanted by a succession of tyrants (then meaning rulers who had seized absolute authority, not necessarily cruel despots) who encouraged trade and economic reform, so that for the first time there was an opportunity throughout the Greek states for a different style of life—enjoying the means to reflect, rather than having to be solely concerned with survival. It was from this point on that, to quote the great scholar of the period D. R. Dicks, "the Greek passion for rational investigation along mathematical lines . . . converted a mass of crude, observational material into an exact science."[3]

The first great practitioner was Thales (c. 625–c. 547 B.C.) of Miletus, a prosperous port serving the whole Aegean. A nobleman by birth, he was also a statesman, engineer, and merchant, canny enough, the story goes, quietly to corner the olive oil presses of Miletus and neighboring Chios just ahead of what he knew would be a bumper harvest, thereby amassing a considerable fortune: proof, he said, that philosophers could make money if they chose.

Thales had traveled to Egypt to learn the elementary principles of practical geometry, a lesson that would inspire him to make astronomy a deductive science. He was the first to posit that the Moon eclipses the Sun when it moves into a direct line between it and Earth, and the first Greek to maintain that it shone by reflected sunlight. He also is said to have established the times and sequence of the equinoxes. He is amply quoted in Plutarch, Pliny, Cicero, and Diogenes Laertius—which is fortunate, as none of his writings survive, and indeed both ancient and modern scholars have built him up into a scientific hero on very thin evidence.

According to Diogenes (born c. 412 B.C.), Thales was the first to gauge the Sun's diameter in relation to its apparent orbit around the Earth with any accuracy, estimating it as 1/720 thereof, a ratio that is an exceptionally close estimate of Earth's actual orbit around the Sun. Less impressively, he argued that water was the prime element of the universe, that the Earth floated like a cork on a great sheet of this water, and that the Sun was made up of a burning earthly substance. He was on firmer ground when postulating that the duration of the Sun's circuit between solstices was never even and that the solar year lasted 365 days; but again we know too little of his writings to be certain what he actually said. Ironically, he is best remembered for something he never did: predict the solar eclipse of May 28, 585 B.C.*

ONE GIANT OF THIS time was Heraclitus (535–475 B.C.), a mystic poet who rejoiced in the nickname of "He who rails at the people" for his arrogant misanthropy. The 130 fragments of his writings that survive emphasize two aspects of the cosmic order: its continuity and its periodicity, especially that of the Sun. Aspects of his thinking can still be seen in the writings of Darwin, Spencer, and other masters of the nineteenth century. He could be absurdly wrong: for instance, he suggests that the Sun is just a foot across, about the size of a shield, and that there is a new Sun every day. But his belief that the universe is composed of worlds that are being created and destroyed in a continuous cycle would have modern physicists nodding in agreement. He understood that nothing is constant—hence his dictum "One cannot step twice into the same river."

Another in the line of great Greek mathematician-astronomers was Pythagoras of Samos, who flourished during the second half of the sixth century B.C. He is said to have spent twenty-two years traveling in Arabia, Syria, Chaldea, Phoenicia, Greek Gaul (better known today as the French Riviera), and possibly India before settling in southern Italy when over fifty years old. (He also spent some time in Egypt studying astronomy, geometry, "and," as the

---

* Plato adds a less dignified anecdote of the astronomer in his youth being so caught up in his stargazing that on one occasion he tripped and fell down a well, to be promptly teased by "a clever and pretty maidservant from Thrace," who asked, "How do you expect to understand what is going on up in the sky when you don't even see what is at your feet?"[4] The first absent-minded-professor joke.

popular historian Will Durant puts it, "perhaps a little nonsense.")[5] An astronomer of exceptional originality, once in Italy he invested his energy in founding an influential religious brotherhood and studying mathematics and theoretical geometry, where he discerned mystical properties in specific numbers. He created a new word to cover such inquiries, which translates as "love of wisdom"—in Greek, "philosophy." In the sixth century, "philosopher" and "Pythagorean" were synonyms.

Pythagoras saw number as the essence of all things, harmony as the most beautiful, and the universe as fundamentally well organized. He pioneered the science of acoustics, which he extended to a hypothetical "harmony of the spheres," believing that, as with the strings on an instrument, each orbiting planet sounded a characteristic note, its pitch determined by its size and speed. The five planets then known, together with the Sun, the Moon, and the Earth, formed an octave. (No one seemed worried by the fact that the Earth, believed stationary, would not therefore be vibrating.)

He also believed that those bodies in the sky moved in two separate circular

A French artist's depiction of Pythagoras (c. 580–c. 490 B.C.) discussing the heavens with Egyptian priests during his prolonged studies in Alexandria

motions, so that the Sun circled the Earth every twenty-four hours in one revolution, but also made a second journey around the Earth along a different circle, at an angle to the first, once every year. He was attempting to rationalize the apparent motions of the stars in a universe deemed geocentric, and his theory broke down when applied to the planets; but this was the first serious attempt to explain their motions. All of modern science stems from Pythagoras' core insight: that there are patterns in nature and that those patterns can be described mathematically.

Pythagoras quickly passed into myth, and legends about him were already circulating by Aristotle's time. He had many followers, several of whom advanced our knowledge of the Sun. We know that one of them, Parmenides, worked out that the bright side of the Moon shone because it was turned toward the Sun and was also the first to pronounce that the Earth was spherical. Once this idea had established itself, Pythagoreans went on to reason that the heavens, too, must be spherical, and so spherical coordinates made their appearance. Philolaus (c. 470–c. 385 B.C., a contemporary of Socrates) even preempted Copernicus by moving the Earth from the center of the cosmos, making it a planet. However, in his system, it still did not orbit the Sun but rather circled an unseen "central fire." All else was spread out across space—including the Sun, which was not a body with its own light but a reflector of this central fire. The universe centered on fire because this was the noblest of the elements, and the center the most honorable place—a theory adopted in most of the leading schools in Athens.

Some of the finest solar astronomers, however, came from outside the main city. Among them were the three "Anaxes"—Anaximander, Anaximenes, and Anaxagoras—who, like Thales, all hailed from Miletus. As indicated by the prefix "Anax," which means "lord," each was wellborn. Anaximander's pupil was Anaximenes, who taught some thirty years before Anaxagoras, the greatest of the three.

Anaximander was already sixty-four when, around 560 B.C., he wrote one of the first works of natural philosophy. He may also, following the Egyptians, have introduced the gnomon, and appears to have marked down in some form the motions of the planets, the angle of the ecliptic, and the dates of solstices, equinoxes, and seasons. He engraved the first "world" map into a tablet of brass, and with his geometric model of Earth, Sun, Moon, and stars sought to depict the heavens in nonmythological terms.

Anaximenes (d. 528 B.C.) was less original, and is best known for his doctrine that air is the source of all things, in contrast to Thales' view that everything comes from water. Although eminent in his time, he had little to say about the Sun—only that it didn't travel under the Earth but circled around it, hidden by mountains at night—in sharp contrast to the third member of the group, Anaxagoras (494–428 B.C.). Although his name translates as "lord of the marketplace"—his parents evidently hoped he would turn out a merchant prince—when Anaxagoras moved to Athens from Miletus, at the age of twenty, he gave away his inheritance in order to devote himself to studying the skies.

One of his early pupils—and later close friend—was Pericles (c. 495–429 B.C.), who grew up to become the most influential person in Athens at the height of Greece's golden age. This period began after the Greeks had repulsed the Persians in 479 B.C. and endured until around 399 B.C. During this time, Athens became famous, evolving under the guidance of Pericles into the great intellectual center of the Mediterranean world, a place where the work of Anaxagoras and his fellow astronomers could flourish. Skywatchers of that era still had only the most primitive technology to help them, and even their observatories were probably no more than flat-roofed buildings. What they did have was mathematical geometry, even then closely allied with astronomy, and the same men were usually active in both fields. Yet solar astronomy continued to be a mix of the utterly wrong (as would come to be seen) and the formidably acute. For an instance of the latter, Oenopides of Chios (active around 460 B.C.) not only came up with the idea of the ecliptic, but went on to measure the angle of the Earth's axis against the ecliptic plane and figured it to be 23°45′—less than three-tenths of a degree off from the presently accepted value of 23°27′.

Among his many accomplishments, Anaxagoras was even able to explain lunar eclipses, but this insight was not taken up; he was too far ahead of his time. Around 440 B.C., a meteorite over a foot in diameter fell on Thrace in broad daylight. Anaxagoras, visiting and seeing it on display, concluded that it had fallen from the Sun, which must, he reckoned, be a mass of red-hot iron; thus heavenly bodies were not divine beings but material objects. The authorities charged him with atheism, and he retired into exile at Lampsacus, where he died, much honored by his new neighbors, in 428 B.C.[6]

About 432 B.C., Meton of Athens, a leading geometer and engineer, made a detailed study of the summer solstice, probably employing a gnomonlike device known as a *heliotropion* ("Sun turner"), a pillar fixed on a level platform, which enabled him to measure the ratio of the noon shadow. He also constructed a *parapegma,* or simple engraved list of astronomical events, and merged the traditional agricultural calendar into one for city use. He became famous—indeed, a celebrity of sorts, his enthusiasm for geometric astronomy such that he found himself publicly made fun of in Aristophanes' comedy *The Birds:*

PITHETAERUS: In the name of the gods, who are you?
METON: Who am I? Meton, known throughout Greece and at Colonus.

PITHETAERUS: What are these things?

METON: Tools for measuring the air. In truth, the spaces in the air have precisely the form of a furnace. With this bent ruler I draw a line from top to bottom; from one of its points I describe a circle with the compass. Do you understand?

PITHETAERUS: Not in the least.

METON: With the straight ruler I set to work to inscribe a square within this circle; in its center will be the marketplace, into which all the straight streets will lead, converging to this center like a star, which, although only orbicular, sends forth its rays in a straight line from all sides.*

The same summer that Meton was making his solstice calculations, Athens became embroiled in a struggle against a coalition led by its archrival, Sparta, a conflict that endured, with one brief interlude, for twenty-seven years. When it finally wore itself out, the political structure of the "tyrant-city" was in ruins. A plague that lasted for three years had killed over one-third of the population. Into this turmoil, around 428 B.C., a great figure was born. He was named Aristocles—"best and renowned"—but his prowess as a wrestler at the Isthmian Games led to the nickname Platon, or "broad." Unlike the fathers of Pythagoras and Socrates, the one a merchant seaman and the other a stonemason, both Plato's parents were wealthy, each coming from one of the most ancient and aristocratic families in Greece.

Athens was by now a state of about a quarter of a million people, half citizens and their families, the rest slaves and resident foreigners, its capital numbering about seventy-five thousand. The city was never more than a narrow margin away from serious hunger, and devoid of almost any medical services—a place where even the people who owned other human beings had a standard of living greatly below the working class of modern industrial democracies. (The term for "burglar" was "wall-piercer," because houses were so fragile.) Having survived his home city's crushing defeat and a brutal oligarchic revolution in 404 B.C., a confused restoration of democracy, then the trial and death of Socrates in 399 B.C., Plato fled to Megara, a trading port some twenty-six miles to the northwest, thence to Egypt. After various adventures—

* Meton got off lightly. Socrates, whose teaching contained a strong holier-than-thou element, so angered Aristophanes that he ridiculed the former's methods in *The Clouds*. In one scene an old gentleman finds his way to the "School of Very Hard Thinkers," where he finds Socrates suspended from the ceiling in a basket, engrossed in thought, while his students are bent down with their noses to the ground. He asks one: "Excuse me, but—their hind quarters—why are they stuck up so strangely in the air?" The student replies: "Their other ends are studying astronomy."

at one point he was sold into slavery, although quickly ransomed—by 395 B.C. he had returned to his home city. Nine years later, following further travels in Italy and Sicily, he borrowed money from friends to purchase a patch of land on the outskirts of Athens named after the local hero, Academos. This he would make his Academy, where he was to teach for forty years: little could he have imagined that it would remain the intellectual center of Greece for the next five hundred.

Plato made astronomy a branch of applied mathematics at his Academy, the inscription over its doorway famously reading, "Let no one enter who does not know geometry." No record exists of the Greeks' using algebra before the time of Christ.[7] Astronomy, which Plato sometimes called Spheric, occupies a comparatively small place in his writings, except in the dialogue *Epinomis*, which describes the work of an astronomer. Having embraced the concept of a spherical Earth, stationary at the center of a spherical universe, he encouraged his students to investigate this system more carefully, but this soon raised some potentially disturbing questions. Plato believed the stars and planets to be visible images of immortal deities whose movements were part of a transcendent order. However, some celestial bodies did not move with the same regularity as the rest, but "wandered" (the Greek word for planet, *planetos*, originally meant "wanderer"). How to explain these motions within the context of his belief in that transcendent order?

For Plato, the task of philosophers was to comprehend the reality underlying creation. If direct observation conflicted with understanding of that reality, he cautioned that things were not necessarily what they seemed—a valuable enough axiom in scientific study generally, but a cold shower for would-be astronomers in the Academy. But Plato's pupils were not put off, because even though he never understood the discipline, his Academy made their deliberations possible. When he died in 347 B.C., one of his students erected an altar to him and gave him a funeral almost befitting a god. The student's name was Aristotle. He didn't even like Plato very much.

ARISTOTLE (WHOSE NAME translates as "best objective") was born in Stagira, to the north of Greece, in 384 B.C. By the time he was ten both his parents were dead, and at seventeen he was sent to the Academy, staying for almost twenty years and becoming so much its star that Plato called him "the mind of the school." Possibly inspired by his father, who had been court physician to the king of Macedon, he first concentrated on medicine, biology, and zoology, soon advancing from pupil to teacher and lecturing on almost every subject. When Plato died, he was succeeded by his nephew Speusippus, whose intellect Aristotle disdained. Possibly piqued, maybe keen to widen his horizons, and certainly aware of the mounting animosity toward the Macedonians, who had so recently conquered Athens, he chose exile at a school on the Aegean island of Assos. When his patron, the ruler of Assos, was betrayed and crucified by the Persians, he fled to nearby Lesbos, until he was invited to tutor the unruly son of Philip II of Macedon, the future Alexander the Great. About 335 B.C. he returned to Athens and, possibly with money from Alexander, founded his own school, the Lyceum, near the temple of Apollo Lyceius, god of shepherds.

His followers were known as the "Peripatetics," taken from their master's habit of strolling up and down the covered walkway, the *peripatos*, while he lectured. The students came from prosperous families—merchant and land-owning families—and keen rivalries developed between the Lyceum, the Academy (with its largely aristocratic student body), and the school of the political theorist Isocrates, which catered to Greeks new to the city. Isocrates' strength was rhetoric, the Academy's mathematics, politics, and metaphysics, the Lyceum's natural science.

By this time, the Pythagorean notion of two sources of light, a central fire with the Sun its mere reflector, had been discredited, but its sister idea of several interconnecting crystalline spheres revolving around the Earth had been seized on and developed. Eudoxus of Cnidus (in what is now southwest Turkey), who was teaching in Athens when Aristotle arrived, argued that the universe comprised not two spheres but twenty-seven. This elaboration enabled him to explain why four of the planets sometimes came to a stop, then reversed course and moved westward (the phenomenon of "retrograde motion"). Each planet's path, he argued, was that of a hippopede—a name taken from the figure-eight-shaped rope used for hobbling horses' hooves. This theory also made sense of the Moon and the other planets' traveling along approximately the same path as the Sun but then deviating north and south.

Invited to lecture at the Lyceum, Eudoxus postulated that the stars and the planets were fixed to the shells of these twenty-seven spheres, which, transparent and thus invisible, encircled the Earth. One carried the Sun, which completed its transit every twenty-four hours, to bring about the day. Another slowly rotated the Earth about its axis, by which it was attached to a larger sphere, this rotation establishing the year. The axis of the newly posited sphere, tilted relative to its outer neighbor, was what moved the Sun down the sky in winter and up it in summer.

Aristotle went even further, proposing that the heavens were made up of *fifty-five* such bodies, all rotating around the Earth at different but constant speeds—which makes one think of some circus high-wire acrobat spinning plates. His idea was that since there could never be an empty space in the heavens ("Nature abhors a vacuum"), and the spaces between the spheres in this universe had to be filled, the greater number, each sphere nestling with those immediately above and below it, would do the job nicely. The Earth was completely surrounded by this rotating nest of spheres, like the core of an onion. Beyond all these spheres Aristotle envisioned yet one sphere more, the Prime Mover—not to be confused with the creator of the universe, but still the dynamic force that drove all things. This tortured web of elaboration satisfied the world's astronomers for centuries—until the invention of the telescope.

The whole structure was based more on a priori philosophic speculation than empirical observation, so Aristotle underpinned it with logic as best he could. The universe is spherical, he reasoned, because, of all shapes, a sphere, which looks the same no matter from where it is viewed, is the most perfect. Earth, too, has to be a sphere, as proved by the Moon's face being overrun by a curve of darkness during an eclipse. Furthermore, travelers north or south do not see the same stars at night as those who stay at home, nor do the stars appear to occupy the same positions in the sky: so terrestrial journeys must be over a curved surface.

Why the Earth should be at the center of the universe was explained by its uniqueness and immobility. While an understanding of gravity was millennia away, Aristotle argued that the Earth's heavier elements, rock and water, were of their nature drawn toward the center of the universe, where they came together in the shape of a sphere. The lighter elements—air and fire—intrinsically moved up and away. Echoing Plato, he added that heavenly bodies were made up not of these four elements but of ether, which he named (because it was the fifth) "quintessence . . . more divine, and prior to . . . the four in our uniquely changeable world that lies under the wandering Moon."[8] This ether was neither heavy nor light but "ageless, unalterable and impassive" (i.e., free from suffering), and—key to the construct of Greek astronomy—passed in an unceasing circular motion, so bringing each heavenly body back to its starting point, enforcing an eternal recurrence.

There were obvious flaws in this explanation. For example, each planet—and the Sun—attached to its sphere, must always be the same distance from the Earth; yet the Moon appeared to change in size by up to almost one-sixth. Did not this indicate that its distance from the Earth was changing? Another indication that the planets' distance from the Earth was not constant was that their brightness—especially that of Mars—was variable. Transient mutable phenomena such as comets and meteors were assumed to be atmospheric occurrences, taking place below the orbit of the Moon. Eudoxus and Aristotle for some reason ignored all this, as well as the fluctuation in the apparent velocity of the Sun. But what Eudoxus did discover, and Aristotle appreciated, was that the motion of the planets could be explained by combining the uniform rotations of concentric spheres on inclined axes. At least in respect of geometrical model-building, this marks the beginning of scientific astronomy.

Following Alexander's death in 323 B.C., anti-Macedonian agitation drove Aristotle from the city a second time, and he died the following year in Chalcis, on Euboea, leaving behind the prototype of a modern university library, a zoological garden, a museum of natural history, and an unparalleled body of work (he is said to have spent his honeymoon collecting specimens of marine life). His strengths were not in mathematics or physics, and he undertook no astronomic observations, yet his influence on this as on the other sciences was profound. He systematized the whole approach to learning and settled the principles governing research, collecting a mass of data from which deductions could be drawn.

Even as he advanced his beliefs, however, his almost exact contemporary Heracleides of Pontus (390–322 B.C.—they died within days of each other), who had studied under Plato but lived on the southern coast of the Black Sea, is said to have suggested that Mercury and Venus might revolve around the Sun, not the Earth, and also that the Earth might rotate daily on its axis, despite appearances to the contrary. He went on to assert that the Sun caused winds, which in turn brought about high and low tides. A younger astronomer, Theophrastus ("speaking like a god," a nickname given him by Aristotle) of Eressos, on Lesbos, discovered sunspots, though we have no idea how. Aristotle's will designated him his successor.

Early in the third century, one of the most remarkable of all Greece's astronomers, Aristarchus of Samos (310–230 B.C.), was the first to discover precession. He also calculated that the ratio of the Sun's diameter to the Earth's was between 19:3 and 43:6. How could so large a body revolve around one so much smaller? A few years later he made the world-shattering—or at least Earth-demoting—assertion that our globe must go around the Sun, not the other way around, and that, apart from the Moon, the other planets did too. The Earth, not the heavens, turned daily, and made a complete circle of the Sun in the course of a year. The Sun, along with the fixed stars, did not move.

A Stoic astronomer in Athens promptly branded him an atheist and circulated a tract charging him with having set "the hearth [sic] of the cosmos in motion." Otherwise, this new theory, so momentous in its implications, was buried. This may seem odd, given that Aristarchus' description not only built on discoveries and investigations that had been in train since the sixth century B.C. but also resolved a whole array of problems, such as retrograde movement. However, if the Earth truly moved, then Aristotle's theory of falling bodies would be undermined, without any compensating theory to take its place; and, if we were living on just another planet, what would become of the terrestrial celestial dichotomy implicit in the majesty of the heavens?

As Aristarchus himself had to admit, there was nothing to be seen in the world around us to suggest that the Earth moved: surely if it did it would hurl people against one another, clouds and birds would be left behind, things would just be pitched off. Common sense dictated that he was wrong, reinforced by the prejudice that man must surely live at the center of creation. Then there were the doctrines of astrology—at that time still respected as a science—which also required a fixed, central Earth. All in all, there seemed many good reasons for keeping the Earth at the center.

Whether or not the hypotheses of this solitary astronomer, working on his own, ever reached them, the so-called Hellenistic philosophers (i.e., those who practiced in the centuries after Aristotle's and Alexander's deaths) decided to leave matters (and the Earth) where they were. The Sun was important, but the Earth much more so. If difficulties emerged to which science had no answer, they should not threaten the status quo. One wonders whether some harbored doubts. For most astronomers of that time, science existed to support the conviction that humanity inhabited a universe of manifest order and design. But, as Durant writes,

Hipparchus (c. 190–c. 120 B.C.) at the Alexandria Observatory.
At his left is the armillary sphere he invented, while he is looking
not through a telescope but through a sector-defining tube.

Since the organization of a religious group presumes a common and stable creed, every religion sooner or later comes into opposition with that fluent and changeful current of secular thought that we confidently call the progress of knowledge. In Athens the conflict was not always visible on the surface, and did not directly affect the masses of the people; the scientists and the philosophers carried on their work without explicitly attacking the popular faith, and often mitigated the strife by using the old religious terms as symbols or allegories for their new beliefs; only now and then, as in the indictment of Anaxagoras . . . did the struggle come out into the open, and become a matter of life and death.[9]

The next figure of substance was Hipparchus of Nicaea, who has been called the greatest astronomer of antiquity (even though he stayed firmly within geocentric orthodoxy). Operating around 130 B.C. from his observatory on the island of Rhodes,* he determined the length of the solar year to within six minutes of accuracy and made numerous calculations of the Sun's diameter. He mapped some 850 stars (crucial, as without defining positions in the sky, mathematical astronomy is impossible), and devised a scale, divided into six ranges of intensity, that measured their brightness: even today, no one has been able to improve on it.

So accurate was his mapping that he had to accept Aristarchus' discovery of precession when his calculations indicated that the stars were not quite fixed in relation to the Sun. Ever the committed Aristotelian, he set out to explain the apparent circular motions of the main celestial bodies in a way that would make sense without upsetting the geocentric theory. Aware that the seasons were of unequal length, he had the Sun revolve at a uniform speed, but moved the Earth from the center of its path, then suggested that it was how the plane of the Sun's motion lined up with the Earth's axis that determined the timing of both the solstices and equinoxes. He remains the best early example of how astronomers tortured their reasoning to keep the Earth where the ancient world wanted it.

Hipparchus' long and productive life ended around 120 B.C., and with him the long Greek tradition of astronomical observation and speculation. Rome

---

* The center of a flourishing sun cult, Rhodes devoted to the Sun a quadrennial festival with athletic games, the most important being a chariot race, and erected the famous Colossus, one of the Seven Wonders of the World, in 284 B.C. This figure of the sun god towered over 105 feet (32 meters), only to collapse in an earthquake in 218 B.C. It took nine hundred camels to haul away the wreckage.

was in the ascendant, and uninterested in the heavens. It would take over two hundred years for another outstanding astronomer to appear.

LONG BEFORE HIPPARCHUS, the great successor states of the Macedonian Empire, from Greece to Iraq, had been conquered by the Roman Empire.

Most members of the Roman elite were suspicious of Greek scientific learning (its medicine aside). By the last years of the Roman Empire, a measure of astronomical knowledge was at least considered a basic aristocratic educational requirement, but its study was accepted only insofar as it could be applied to literature: Did it help toward a better understanding of a poem? In an empire of some fifty million people, the number of natural philosophers in Rome itself became too small for any fruitful collaboration,[10] leading to a decline in scientific knowledge generally.[11] There was virtually no solar astronomy of great value by an Italian-born scientist for over nine hundred years. The historian of science Timothy Ferris says of the Romans,

> Theirs was a nonscientific culture. Rome revered authority; science heeds no authority but that of nature. Rome excelled in the practice of law; science values novelty over precedent. Rome was practical, and respected technology, but science at the cutting edge is as impractical as painting and poetry. . . . Roman surveyors did not need to know the size of the sun in order to tell time by consulting a sundial; nor did the pilots of Roman galleys concern themselves overmuch with the distance of the moon, so long as it lit their way.[12]

IT IS AN IRONY of history that while Rome was turning its back on the heavens, one part of its empire was celebrating the last major figure of this period. Ptolemy—Claudius Ptolemaeus (c. A.D. 90–168)—was an Egyptian geographer and astronomer who flourished for some forty years at Canopus, some fifteen miles east of Alexandria, nicely rendered by that iconoclastic historian Dennis Rawlins as "an infamously licentious town which was an ancient combination in one of Hollywood, Lourdes and Las Vegas."[13] He left four works, any one of which would place him among the most important authors of the ancient world: *Syntaxis mathematica*, better known by its Arabic title, the *Almagest*, a thirteen-volume "book" of data about the stars; *Tetrabiblos* (the "Bible of the astrologer"), which he begins by distinguishing between the two modes of studying the heavens, mathematical astronomy and horoscopic astrology; *Harmonics*, which relates musical harmonies to the properties of mathematical proportions derived from the harmonies he considered to exist throughout the universe; and *Geographia*, a compilation of what was known about the world at that time.*

In the *Almagest*, Ptolemy suggested that the planets moved in circles within other circles, with the Earth at their very center, all the while assuming that their actual movements were unknowable. He concluded that the Sun was about 1,200 Earth radii from us (one-nineteenth of the actual figure), essentially the same estimate accepted during the whole Middle Ages. At one point he does consider a Sun-centered universe, but discards it: After all, where was the evidence? But he does hold that one of the two main composite motions of each planet is governed by its relation to the Sun, which made the conversion to heliocentricity much easier when it came.

When the *Almagest* appeared, critics charged that he had lifted large sections from Hipparchus and that several observations were fabricated—at one point he does unwittingly assign two quite different dates thirty-seven days apart to the same celestial event, and there are many other instances of sloppy research and borrowed ideas.[14] But if he was a fraud, he did a hopeless job of covering his tracks, and his achievement is better seen from a different perspective.

Colin Ronan, in his history of astronomy, points out that during these years, scholarship was "primarily in a reminiscent mood, collating and assessing the achievements of previous generations."[15] Ptolemy was the supreme collator; although the *Almagest* contains errors not corrected until the seventeenth century, the huge number of tables he created was sufficiently accurate to be taken up by Copernicus (never a sophisticated observer of the skies). The *Almagest* shares with Euclid's *Elements*—which laid down the basis for geometry—the distinction of being one of the two mathematical texts longest in use. It was Ptolemy's thinking that set the course of astronomy for the next fifteen hundred years.

Ptolemy regarded his work as part of an ongoing inquiry, but it was treated by his successors as definitive. If, as seems most likely, he came up with his theory of orbits within orbits in order to maintain the Earth at the center of the universe, then in turn had to shape his "observations" to fit his theory, he might almost not have bothered. Rulers and ever more powerful Church

---

* "Books" then were more like scrolls, of necessity brief affairs, corresponding to a modern-day substantial chapter. *Liber* meant a unit of writing, not a unit of creation (a book in our present-day sense), so it is not impossible for someone to have written many books, totaling several dozen in one lifetime. Livy authored one book of 127 chapters, Caesar's opponent Varro 490—yet in modern terms just 200 to 250 pages. By contrast, Henry James authored some forty books, few of them slim. But then every generation throws up at least one person who cannot stop writing.

dignitaries—their views reinforced by the rising importance of the Christian idea that doctrine counted for more than knowledge—were becoming skeptical of the value of looking into the heavens. With Ptolemy's death, as Ronan has it, "the light of astronomical research went out [throughout Western Europe] for a thousand years";[16] no astronomers from that civilization made any significant advance on what he had done. Conditions for scientists began to worsen, and not until far on into the Middle Ages were Ptolemy's ideas revived, as the philosophy of Aristotle was wedded to medieval theology in the great synthesis of Christian faith and ancient reason undertaken by such philosopher-theologians as Thomas Aquinas. The Prime Mover of Aristotle's universe (while never conceived by him as the cause of all creation) fed into the God of Christian theology; the outermost sphere of the Prime Mover became the cosmological embodiment of the Christian vision of heaven; and the central position of the Earth was interpreted as a sign of the Christian God's concern for mankind. Thereafter it served the Church not to study the heavens too carefully.[17]

Astrology was the leading discipline of the time; fortune-telling became an obsession that lasted through the centuries, and its connections to alchemy and number symbolism were to become important elements in both Christian and Muslim Arab thinking. As Nietzsche was to ask, "Do you believe that the sciences would ever have arisen and become great had there not been beforehand magicians, alchemists, astrologers and wizards, who thirsted and hungered after obscure and forbidden powers?"[18]

Rome's leaders were no exception. In the days before he accepted the sobriquets of Augustus ("the Increaser") and Emperor ("Victorious General"), Caesar's great-nephew and successor, Octavian, was converted to astrology when a leading practitioner examined his horoscope and immediately knelt in worship before the man he saw as his future sovereign. Although the adopted son of Augustus, Tiberius (42 B.C.–A.D. 37), banished all astrologers from the capital, he privately continued to depend on them. Nero (A.D. 37–68), too, was officially skeptical, but had one Barbillus read the heavens to discover his enemies, whom he promptly executed. These years and the centuries that followed were a period of regression, as astronomical observation and investigation became subservient to sun worship and false prophets abounded. Tertullian the Convert (c. A.D. 155–245), a lawyer from Carthage, wrote: "As for us, curiosity is no longer necessary."

Here astronomy came to theology's aid. Ptolemy's theories survived the collapse of the Roman West for their technical quality, their coherent picture of the cosmos, their usefulness to astrology, and their harmony with Christian teaching. The system outlined by Pythagoras, refined by Plato and Aristotle, given final form by Hipparchus, and recorded by Ptolemy promoted, as Franz Cumont says, "a divinity unique, almighty, eternal, universal and ineffable, that revealed itself throughout nature, but whose most splendid and energetic manifestation was the sun."[19] The way these astronomers described the Sun chimed in almost perfectly with how the Church, appropriating solar imagery, wanted its believers to see the Christian God.

This cumulative reading of the universe afforded far more to Christianity than merely a legacy of pagan festivals and symbols. "To arrive at the Christian monotheism," continues Cumont, "only one final tie had to be broken, that is to say, this supreme being residing in a distant heaven had to be removed beyond the world." The various cults devoted to the Sun not only made straight the roads for Christianity; they heralded its triumph. Little wonder that for the next fourteen hundred years Church and state should combine so effectively to keep man's understanding of the universe exactly as the Greeks had left it.

CHAPTER 7

GIFTS OF THE YELLOW EMPEROR

Among the Greeks, the astronomer was a private person, a philosopher,
a lover of truth . . . in China, on the contrary, he was intimately connected
with the sovereign pontificate of the Son of Heaven, part of the official
government service, and ritually accommodated within the very walls of the
imperial palace.

—ANDRÉ DE SAUSSURE, c. 1910[1]

The people of our humble land have always understood heaven.

—ZOU YUANBIAO,
writing to Matteo Ricci, c. 1597[2]

So MUCH OF OUR UNDERSTANDING OF CHINA IN GENERAL AND ITS
solar science in particular depends on the work of one Western scholar: the remarkable British polymath Joseph Needham, whose life spanned the twentieth
century (1900–1995). He has been compared to both Darwin and Gibbon in
bestriding his area of study. A contemporary went so far as to claim that one
would have to reach back to Leonardo to find so much learning in a single
individual—Needham's knowledge embraced mathematics, physics, history,
philosophy, religion, astronomy, geography, geology, seismology, mechanical
and civil engineering, chemistry and chemical technology, biology, medicine,
sociology, and economics.

He spoke eight languages with comfort, wrote poetry throughout his life,
and was a profoundly committed Christian and Communist, while other interests ranged from morris dancing to steam trains, accordion playing, and
nudism.[3] He spent most of his life in Cambridge, initially specializing in em-
bryology and morphogenesis, and in 1924 marrying a fellow don, the biochemist Dorothy Mary Moyle. Then in 1937 a Chinese postdoctoral student, Lu
Gwei-djen, arrived at the Department of Biochemistry: Needham fell in love,
and totally redirected his intellectual pursuits. He taught himself enough
Mandarin to get by, and became obsessed with two questions: Why were the

Chinese so unaware of their past scientific achievements? And why had not the scientific revolution that swept through seventeenth-century Europe taken place in China first?

Just before the outbreak of the Second World War, Needham signed a contract with Cambridge University Press to write a one-volume history "taking up the fundamental question of why modern science originated in Europe and not in China."[4] Going out in 1943, he spent two years with the British Scientific Mission in Chungking (Chongqing), the Nationalist Party capital of China, and for the next six years traveled the country by whatever conveyance was available—jeeps and junks, camels, wheelbarrows, litters and goatskin rafts—collecting rare works on traditional learning as he went. He returned home, his "short study" having grown into three volumes. The first appeared in 1954 and included a table of contents for the entire enterprise, which at that point he was projecting to seven volumes. Eventually those seven would become twenty-seven, mostly written by Needham, who worked at them almost until his death at ninety-five. The chapters on astronomy alone fill three hundred pages.[5] While his remarkable work has been augmented over the years, and sometimes corrected in its detail, it has never been replaced.

IN ONE CENTRAL myth of China's beginnings, the early king Fuh-hi and his four successors, together known as the "Five Sovereigns," are said to have founded the Chinese Empire, ruling from about 3000 to 1600 B.C. and expanding the kingdom from what is now north-central China to the eastern sea—a huge territory, with an 8,700-mile coastline. A competing myth presents a certain Xuan Yuan, or Huangdi, the Yellow Emperor, who became the leader of all the northern tribes after decisively defeating his enemies in 2698 B.C. He could move through the air at enormous speed, riding a dragon as high as the Sun— this beast having come from "the land where suns are born." Legend holds that he believed that numbers had philosophical and metaphysical properties and helped to "achieve spiritual harmony with the cosmos." Both Xuan Yuan and Fuh-hi, like the Babylonians, are said to have devised a numerical system based on sixty as part of a method of timekeeping known as the Heavenly Stems and the Earthly Branches.

Mythical though these figures were, there is some truth in what the myths describe. The first dynasty to appear in Chinese records is the Xia, running from 2033 to 1562 B.C. During the subsequent Shang or Yin Dynasty (1556–1045 B.C.) the Chinese developed a calendar, learned how to write, and became highly proficient in bronze metallurgy. This era was a critical time in Chinese intellectual history. We know that they were investigating the heavens by the fifteenth century B.C., and inscriptions suggest that by the fourteenth—well before the earliest known Sumerian calendars—they had established a solar year of 365 ¼ days, formulated systems for measuring daily time, incorporated the cardinal points of direction into their cartography, and mapped the paths of the Sun and the Moon accurately enough to predict eclipses.

Astronomy was of vital interest to the Chinese, and considerable resources were devoted to it. Early on in his account, Needham makes it clear that until at least the sixteenth century A.D. they connected celestial events with the performance of their rulers, whom they saw as divinely appointed. The Earth, its emperors, and the entire cosmos were one, and so long as a sovereign governed well, the celestial bodies would follow their appointed courses without deviation or surprise. Were his reign unjust or otherwise defective, comets or new stars would blaze across the sky.[6] Since unusual heavenly events either presaged disaster or pointed to inadequacies of administration, a ruler had good reason to keep such portents secret, and records were closely guarded. While major happenings were visible to all, many lesser ones were not noticed and could be passed over or publicized at will; thus many astronomers gained considerable power, even influencing policy, and were central figures of court life. They developed so-called techniques of destiny, which included inscribing predictions on the shoulder blades of ox or deer or on tortoise shells. These "oracle bones" are our main source of information about the period from the fourteenth to the eleventh century B.C.

Given the discipline's potential as an instrument of intrigue (independent astronomers might be tempted to calculate horoscopes to the advantage of a rival would-be dynast), the emperors made it a crime to leak astronomical information—just as Tiberius and Nero would later do in Rome. A typical edict declares:

> If we hear of any intercourse between the astronomical officials or their subordinates and officials of other government departments or miscellaneous common people, it will be regarded as a violation of security regulations. . . . Astronomical officials are on no account to mix with civil servants and common people in general. Let the Censorate look to it.[7]

Even to study astronomy in private was punishable by a two-year prison sentence.[8]

The most startling difference between Chinese and Western astronomy was the method the Chinese used to track the planets through the night sky. They named the ecliptic the "Yellow Path" (*huang-tao*), with a set number of

夏至致日圖

表竿

土圭

The astronomer Xi Chu (A.D. 1130–1200) using a gnomon to determine the summer solstice. The animals may represent the Chinese zodiac.

"houses"—the Ten Heavenly Stems—positioned along it. These combined with both the Twelve Earthly Branches and another sequence, the Five Elements, to create cycles of sixty hours, days, and years. To make the construct easy to remember, each branch was assigned an animal symbol.* "So far as the historian of science is concerned," Needham comments tersely, "whoever invented the animal cycle is welcome to it."[9] As he notes, a Western astronomer would hardly recognize a map of the constellations produced by a Chinese astrologer—a fair point, as this famous menagerie does tend to obscure the real advances that the Chinese made.

The recording of astronomical observations and cultural practices tended to go together. The Li Ki, or Book of Rites, for instance, a compilation of religious practices from the eighth to the fifth century B.C., describes the Son of

---

* The myth developed that the Buddha invited all the animals to celebrate the New Year, but only twelve turned up: mouse, ox, tiger, rabbit, dragon, snake, horse, sheep, monkey, rooster, dog, and pig, in the order they now take in the zodiac. Out of gratitude, he decided to name the years after them in the order of their arrival, people born in a given year inheriting the personality traits of that particular animal.

Heaven (the king/emperor) as a stargazer whose function was to predict the moment when his people were to participate in the rites of heaven and Earth by sowing the new season's grain. Stargazing (as with the ancient Greeks) was seen as part of everyday life—as the late-Ming-era scholar Gu Yanwu wrote, "In the Xia, Shang, and Zhoui dynasties, everyone was an astronomer." "Astronomy" was *tian wen*, literally "the pattern of the skies," and encompassed the systematic study of all heavenly events—the waxing and waning of the Moon, the movements of the planets, and the type and color of comets. The project of mapping the heavens was not unique to the Chinese, but, along with the Babylonians, they were the most persevering and accurate observers until the Arabs, and from the middle of the fourteenth century B.C. on they recorded nine hundred solar eclipses over the next 2,600 years. The Chinese of this era also noted the appearance of sunspots—they are mentioned at least 120 times in the official histories. Although observations of the shape and splitting of such spots did not come until much later—these were first noted in 28 B.C.— this was long before they were written up in the West. One wonders how the Chinese saw spots at all, when their skies were not especially clear, and even given good conditions a spot has to be more than fifty thousand miles in diameter to be visible to the naked eye.

The sixth century B.C. saw the standardization of weights, measures, and other practicalities, such as the width of roads. Even gnomons were officially set at eight *chi* (just under eight feet). By the fifth century the Chinese understood what caused eclipses, and were the first to record the appearance of what we call Halley's comet, in 467 B.C. Between 370 and 270 B.C., two of the greatest Chinese astronomers, Shih Shen and Kan Te, along with a colleague, Wu Hsien, drew up the first star catalogues; to give some idea of how advanced this was, Hipparchus would not produce comparable work for another two centuries.

In all this, the Sun operated as one force among many (Chinese painted screens depicting the universe often omit the Sun altogether): to give it primacy would have been to disturb the balance of nature. The Chinese saw the heavens as a bowl turned upside down, resting on a square Earth, with four cardinal points of terrestrial space, north, south, east, and west, and a fifth, the center, heartpoint of the Celestial Realm. The Sun was essentially a circumpolar star, illuminating first one zone of the Earth, then another. Sun and Moon were attached to heaven, which moved at great speed, carrying them along with it.

From around 200 B.C. on, the main observatory under the Ch'in Dynasty employed a corps of more than three hundred astronomical experts. They divided the sky into twenty-eight unequal sectors, radiating from the north celestial pole in the same way that lines of longitude radiate from the poles of a terrestrial globe, derived from the Moon's apparent motion set against the stars—which allowed them to calculate a "sidereal month" of 27.32 days. These astronomers employed a cosmograph, or *shipan*, with a rotating disk (heaven) above a square fixed plate (Earth). Around the disk's edges were marked the names of the twenty-eight constellations, while in the disk's center was a representation of the Big Dipper. The instrument enabled the user to locate a star at any time of the year, and also functioned as a clock. Once the astronomer had calculated what star was passing the meridian at sunset, he could tell which constellation had reached its zenith at noon or midnight—and, knowing that a particular constellation had risen at the vernal equinox, which stars were reaching their high point, or alternatively were on their way down. Some time before the second century B.C., this cosmograph would evolve into the compass, its function shifting from the celestial to the terrestrial.

Chinese astronomy was thus predicated on a very different system from that of the Greeks or later Europeans. The Egyptians and Greeks charted the Sun's risings and settings and those of the other stars near the ecliptic, so, for instance, the risings of Sirius, measured against successive sunrises, would announce when the Nile was about to flood. "Such observations," wrote Needham, "required no knowledge of pole, meridian or equator, nor any system of horary measurement. . . . Attention was concentrated on the horizon and the ecliptic."[10] The Chinese focused instead on the pole star and its circumpolar brethren, basing their system on the meridian—the great circle of the celestial sphere passing through the pole star at the observer's zenith—and calculating the culminations and lower transits of the pole star's immediate neighbors.* (Not that the two systems were entirely separate. Homer, for one, was aware of the significance of the pole star as a mark of the north; and the sentinels at the siege of Troy changed their guard according to the vertical or horizontal positions of the tail of the Great Bear.)

Such an approach had a clear scientific basis, but for most Chinese their understanding of the heavens—and thus of the Sun—was bound up in their overall way of looking at the world. Chinese cosmology affirms a life of balance, of virtue, of performing one's duty; and their philosophy of what occurs in the heavens reflects whether or not that balance is achieved, particularly as understood through their three key terms, *chi, yin,* and *yang.* Before about 300 B.C., chi in its narrowest sense means "life energy," but thereafter it was used to cover a multitude of phenomena: air, breath, smoke, mist, fog, the shades of

---

* A circumpolar star is one that, as viewed from any given latitude on Earth, never sets (that is, never disappears below the horizon), owing to its proximity to either of the celestial poles. Such stars are therefore visible from that location for the entire night on every night of the year—and would be continuously visible throughout the day, too, were they not overwhelmed by the Sun's glare.

the dead, and also cloud forms—more or less anything perceptible but intangible; the physical vitalities; cosmic forces and climatic influences that affect health and the seasons (many of the Sun's effects were manifest in chi); flavors, colors, and musical modes. Chi could be benign and protective or pathological, an agent of disease, and destructive.

One of the primary schools of Chinese thought, Taoism, is a philosophy of duality and balance—thus the doctrine of yin and yang, which embrace almost all opposable forces, yin being cool, wet, and female, yang hot, dry, and male. Other antitheses include day and night, light and dark, honest and deceitful.[11] Taoists propound the idea that, as yin and yang surge through the Earth, they cause the periodic changes that bring about tides and seasonal heats and colds. They believed that numbers were key to maintaining balance, five being particularly auspicious—each of the five prime locations, north, south, east, west, and center, having a corresponding color, animal, element, and flavor. Nine was also important, not so much auspicious as a number to be heeded, so that on the ninth day of the ninth month—the double yang—the people of Beijing would assiduously set out to climb two of the city's main hills (the Wall of Pure Metamorphosis and the Destiny of Trees Surrounding the Gate of Chi) in order to avoid disaster.[12]

Yin and yang modified chi: they were the complementary aspects of any configuration of space or time, what Needham calls the "two fundamental forces or operations in the universe." The Sun was of a fiery yang nature, the Moon and the Earth were yin, air simply a vacuum.* The distinction was between bodies that shone by their own light and those that shone by reflected light. Furthermore, there were five phases (wu hsing, the five powers, or the five travelers, because in continual movement) of five life processes: wood, fire, earth, metal, and water. By the second century A.D. no one was sure what the unifying principles of these materials were, but despite such uncertainty, a common language emerged to describe the cosmos and its analogues on Earth for both state and individual: a dynamic harmony existed, leading to one true Way. This classifying system originated in humanistic thought, then passed into state cosmology, and from there was taken up by scientists.

IN 221 B.C., the eastern kingdoms were forged together to become China, the great central state of Ch'in. "China" means "land of the house of Ch'in" but

---

* As in most cultures, the Sun had several synonyms, including Radiant Number (yao ling), Vermilion Luminosity (chu ming), Lord of the East (tung chun), Great Luminosity (ta ming), Crow of Yang, and Fluid Pearl. Needham comments that the Chinese elite used many synonyms "precisely in order to bewilder the uninitiated."[13]

also translates as "middle": Ch'in was the Middle Kingdom. The long distances and physical barriers that had isolated it from other civilizations gave its people the impression that their country was the center of the Earth, the sole source of civilization—an idea that would persist for millennia.

However, China's expansion into central Asia in the first century A.D. would expose its astronomers to Hindu and Persian ideas, provoking intense speculation in cosmology and astronomy. Some of these doctrines were almost too far ahead of their time: in A.D. 5, Wang Mang (45 B.C.–A.D. 23), regent and later emperor, argued from what he had learned from these foreign influences that the workings of the heavens happened "of themselves," and natural disasters were not sent to warn or punish, a thesis that went largely unheard. Another theory that won wide acceptance, associated with the Later Han Dynasty (A.D. 25–220), pictured the universe as unending empty space in which floated the Sun, Moon, planets, and stars—notable as the first time anyone from any country had suggested an infinite uncentered universe.

Much of Chinese mathematics (from, say, around 100 B.C. on) was devoted to calculating the calendar and predicting the positions of heavenly bodies. (As it was, Chinese calendric science produced a strong tradition of algebraic-arithmetical astronomy, in contrast to the Western emphasis on geometry.) In 132, China's greatest "Grand Astrologer," Zhang Heng (A.D. 78–139), basing his insights on such mathematics, invented the first seismoscope for registering earthquakes: a cylindrical device with eight dragon heads around the top, each with a ball in its mouth, and eight frogs, one directly beneath each dragon, on the bottom. When a shock wave came, a ball would clatter out of a dragon's mouth into a frog's. He also wrote that moonlight is only reflected sunlight, and that lunar phases are the result of the Moon's changing reflections of sunlight from different points on its course. He invented an astrolabe with graduated rings and a sighting tube.* While his fellow Chinese were reacting to lunar eclipses by pounding gongs, Zhang was explaining their real cause. Needham believes he set "the standard for excellence in astronomy," although even Zhang was partly conditioned by the thinking of his time: while he was the first in China to construct a celestial globe that actually turned, it bore little or no relation to reality, but rather reflected his vision of the universe as expressed in *Hun-i Chu:* "The heavens are like a hen's egg and as round as a crossbow bullet; the Earth is like the yolk of an egg, and lies along the center. Heaven is large and Earth is small." And, in a passage not translated by Need-

---

* Astrolabes show how the sky ideally looks at a specific place and time. This is done by drawing the constellations on their faces and marking them so positions in the sky are easy to find. Once set, the instrument's face has a complete representation of the sky, allowing many astronomical problems to be solved visually.

ham, Zhang adds: "Heaven takes its body from the *Yang*, so it is round and in motion. Earth takes its body from the *Yin*, so it is flat and quiescent."[14]

Between the second and eleventh centuries A.D., little changed in Chinese understanding of the Sun—although their astronomical literature was equal in extent to those of botany, zoology, pharmacy, and medicine taken together. By the twelfth century, the imperial library contained 369 books on astronomy and allied subjects, and by the thirteenth, the empire's astronomical instruments were far superior to anything in Europe; however, the nongeometrical nature of its mathematics left it still unable to chart the skies accurately and hampered further progress. This was one consequence of its isolation. Until about A.D. 400 the country was almost completely shut off from all but its immediate neighbors. Christian missionaries made the arduous journey overland as early as the following century, but there was little if any cross-fertilization of ideas, and the country did not become an object of general curiosity in the West until 1250, when Pope Innocent IV sent Franciscan monks into Asia—to discover a civilization that was in many ways remarkably advanced.

The Franciscan reports were augmented several decades later by Marco Polo (1254–1324), who, writing about a period toward the end of the thirteenth century, described a land "of immense extent, with the largest cities, widest rivers and greatest plains ever seen, where gunpowder, coal, and printing were in general use."[15] One might add to that an economy based on paper money, cities with more than a million people, a genuine bureaucracy, and

A reconstruction of the water-driven astronomical clock tower of Su Song (1020–1101) at Kaifeng, in northeast China. The clock tower employed 133 different clock jacks to indicate and sound the hours.

such items as silk, tea, porcelain, herbal medicines, lacquer, playing cards, rockets, astronomical clocks, dominoes, wallpaper, the kite, even the folding umbrella. Marco Polo was describing China as it was under Kublai Khan, during whose reign the Chinese Empire stretched from the Pacific to much of Eastern Europe, and China was open to foreign influence as never before.*

Under the Ming Dynasty (1368–1644), as the country again sealed itself off for more than a hundred years, its astronomical undertakings sank into the general deep sleep of science—a decline so precipitous that it is hard to believe it happened; but it may have been that the Chinese "believed that because the good life had been discovered their education need never change."[19] The situation would alter only with the coming of the Jesuits. They found a culture profoundly cut off from much of the knowledge of the West, including its

---

* How far, if at all, was ancient Chinese astronomy influenced by the West? In 1986 the Palace Museum in Beijing sent two horse bones to the British Museum asking for help. Both bore fragments of cuneiform script, and one was identified as a poor copy of a cylinder (of which the BM has an example) that describes the conquest of Babylon by Cyrus II of Persia in 539 B.C. At first it was presumed that the horse-bone script, with its square structure and pictographic aspects, was a forgery, but after prolonged study, experts decided it was genuine. How had it suddenly turned up in Beijing? In 1987 an article appeared in a Chinese historical journal by one Xue Shenwei, a doctor of traditional medicine who had died well into his eighties the year before. Traveling in eastern China in 1928, he had learned of the bones from a local scholar. In the years that followed he monitored their progress, and over a decade later bought them from an antiques dealer, who told him that his purchases came from the northwest of Xinjiang, in the desert around Gu'erban: the villagers there called such pieces dragon bones and used them as spindles. During the destructive excesses of the Cultural Revolution, Xue Shenwei buried the two bones; when he judged it safe he dug them up, and just before his death gave them to the palace museum.[16] Xinjiang probably lay on the line of an old trade route, so fourth-century-B.C. visitors from the West could indeed have carried the bones with them. I asked a senior member of the BM whether this discovery was not rather important. "I suppose," he mused, "we do tend to keep some things to ourselves."[17]

Six years after Beijing received its bones, about 113 mummified corpses dating back at least 3,800 years were excavated at Qizilchqa, known as "the Land of Irrevocable Death"—through which runs a key stretch of what would become the Silk Road, the four-thousand-mile network of interconnecting beaten paths and caravan tracks that served from around 500 B.C. to A.D. 1500 as the main highway between Europe, the Middle East, India, and China. The bodies, although their skin is parched and blackened, have long noses, long skulls, and deep-set eyes, typical of Caucasians—even blond hair. Qizilchqa, in the Taklamakan Desert of the Tarim Basin (modern-day Xinjiang), lying between the Gobi Desert to the northeast and the Himalayas to the south, is one of four sites where such corpses have been found, not in ones and twos but groups of thirty or more, suggesting permanent settlements and not just a few stray travelers. Victor H. Mair, professor of Chinese studies at the University of Pennsylvania, believes that there was a substantial two-way traffic between East and West beginning at least 3,980 years ago. Would such caravans have brought scientific knowledge with them? We do not know, and the present Chinese government has impeded further research, but at least two of the "Tarim mummies" were still wearing their original silk clothes—decorated with a star map and an ancient text, possibly in Phoenician, on astronomy.[18]

geography as much as its astronomy. When one of the missionaries hung up a map of the world in his hut and invited his Chinese guests to examine it, he reported:

> To them the heavens are round but the Earth is flat and square, and they firmly believe that their empire is right in the middle of it. They do not like the idea of our geographies pushing their China into one corner of the Orient. They could not comprehend the demonstrations proving that the Earth is a globe, made up of land and water, and that a globe of its nature has neither beginning nor end. The geographer was therefore obliged to change his design and . . . left a margin on either side of the map, making the Kingdom of China appear right in the center. This was more in keeping with their ideas and it gave them a great deal of pleasure and satisfaction.[20]

Chinese geography may have been backward, yet it was three Chinese inventions—the magnetic south–pointing compass, the true rudder, and ships that could sail to windward, with a compartmentalized hull to make them less likely to sink when broached—that made it more feasible for Europeans to sail eastward by sea. Even then, when in 1551 the first of the great missionaries, Francis Xavier, set out from neighboring Japan, it took him nine months of negotiations to establish himself on an island seven miles offshore, and a further nine weeks to set foot on the mainland. Some thirty years later, an Italian member of the Society of Jesus, the mathematician and astronomer Matteo

An eclipse of the Sun in 1688, viewed by Jesuit missionaries in the presence of the king of Siam

Ricci, took four months to reach the Portuguese outpost of Macao, enduring illness, shipwreck, and imprisonment before finally entering Beijing.

Ricci was the most notable of all the Jesuit missionaries to China, the first Westerner to be admitted to Beijing and then the Forbidden City. While baptizing many high officials, he acquainted his hosts with the advances of the Renaissance—although he and his fellow Jesuits also imported many of the current scientific errors of Europe, and indeed would ridicule the notion of "the emptiness of the heavens." Yet, as Needham observes, "The vision of infinite space, with celestial bodies at rare intervals floating in it, is far more advanced . . . than the rigid Aristotelian-Ptolemaic conception of concentric crystalline spheres."[21]

Soon after his arrival in 1584, Ricci set about expounding to his hosts what he had learned in his European studies—even providing them with a world map that proved, among other things, that the Sun is larger than the Moon.[22] The Chinese resisted, as to their minds they were the intellectual masters of the world, certainly in the sciences—until Ricci produced a chiming clock that he said replicated the movements of the stars. To test it, his hosts set about predicting a solar eclipse by their methods, while Ricci and his European assistants used their device. The Chinese-predicted hour came and went in the full light of the Sun; the Western-predicted eclipse fell precisely on time.

As Ricci's renown grew, he was invited to the country's most important observatory, in Nanking. Instead of the elaborate semimagical charts he had expected, however, he found four magnificent instruments of cast bronze, embellished with imperial dragons, more sophisticated than any he had seen in Europe: two spheres for determining eclipses marked with degrees according to the European system, but in Chinese characters; a gnomon; and, largest of all, a nest of four astrolabes. The eunuch in charge* explained that the instruments had been cast by a Muslim astronomer some two hundred fifty years before.

Ricci examined one of the spheres and pointed to the graduated arms. "Is that where you read the shadow?" The eunuch looked puzzled, then nodded, before trying to pull the visitor away, but Ricci continued his inspection, notic-

---

* "In the later periods of the Byzantine Empire," writes the historian Beverly Graves Myers, "the heavenly court found an earthbound reflection in the imperial court. The eunuchs who surrounded the emperor—who was considered God's representative—with their brilliant white clothing and glowing beardless faces served as the equivalent of angels."[23] Sun Yaoting, the last eunuch of the imperial court, died as recently as 1996, aged ninety-three.

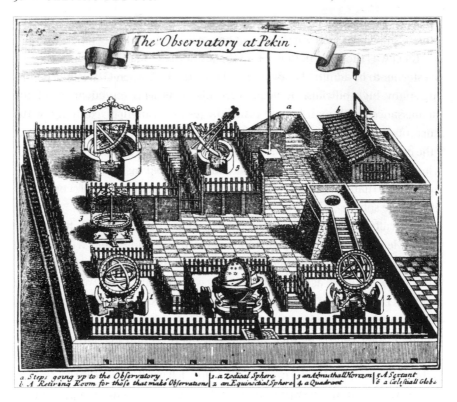

The Imperial Observatory in Beijing, as it appeared in the late seventeenth century.
Here sunspots were discovered and records of comets kept. Large bronze spheres (lower frame),
a quadrant (upper left), and a sextant (upper center) are all identifiable.

ing that degrees were indicated by a system of knobs that could be read at night by touch. Then he made a startling discovery.

"Surely Nanking stands at thirty-two degrees?"

The eunuch agreed.

"But these instruments are set for thirty-six degrees." His guide was silent, and tried again to divert Ricci's attention. Intrigued, his guest posed two or three simple questions about the gnomon—all of which received evasive answers, until at last the chief eunuch confessed.

"They're beautiful instruments, but we don't know how to use them. Two sets were made, one for Beijing, the other for Pingyang. These are from Pingyang."

Now Ricci understood. The town of Pingyang stood at 36 degrees, and the instruments had never been adjusted. They were no more than objects of magic. He gradually learned that some astronomical knowledge had come from the Arabs—so China had acquired its preliminary understanding of the

Ptolemaic system from the same source and at about the same date as Europe had. In the reviving West, such information had been enthusiastically received and elaborated; but in China, where science was not judged conducive to the good life or to good government (as opposed to astrology's value as the source of predictions), it was not included in the approved list of studies.[24]

Ricci did his best to dispel such ignorance, but considerable opposition to his teaching remained. Not until three years after his death, when the Chinese Board of Astronomers seriously mistimed an eclipse, did the throne finally decree that the calendar be brought up to date and European works on astronomy translated. However, China was almost immediately invaded by the Manchu, order broke down nationwide, and the country was once more closed off. Isolated it may have been, yet well into the eighteenth century it was held in awe by many Europeans as the most prosperous and sophisticated of any civilization on Earth. Samuel Johnson's first prose work (1733) refers to the Chinese as "perfectly, completely skilled in all sciences."[25] Not until 1850 was their astronomy merged with that of the rest of the world.

Today the Imperial Observatory in Beijing has its own subway station.

CHAPTER 8

THE SULTAN'S TURRET

Awake! For Morning in the Bowl of Night
Has flung the Stone that puts the Stars to Flight;
And Lo! The Hunter of the East has caught
The Sultan's Turret in a Noose of Light.

—Opening verse of the
Rubáiyát of Omar Khayyám

And when the night overshadowed him he saw a star and said, "This is my
Lord!" And when he saw the Moon beginning to rise he said, "This is my
Lord!" And when he saw the Sun beginning to rise, he said, "This is my Lord,
greatest of all."

—QUR'AN[1]

BACK IN 1959, THE NOVELIST AND CULTURAL HISTORIAN ARTHUR
Koestler published *The Sleepwalkers*, a remarkable account of scientific discovery through history. His summary of what Islamic civilization added to astronomy is harsh. The Arabs, he concludes, were

> the go-betweens, preservers and transmitters of the [Greco-Indian]
> heritage. They had little scientific originality and creativeness of their
> own. During the centuries when they were the sole keepers of the treasure, they did little to put it to use. They improved on calendrical astronomy and made excellent planetary tables; they elaborated both the
> Aristotelian and the Ptolemaic models of the universe; they imported
> into Europe the Indian system of numerals based on the symbol zero,
> the sine function, and the use of algebraic methods; but they did not
> advance theoretical science. The majority of the scholars who wrote in
> Arabic were not Arabs but Persians, Jews and Nestorians; and by the fifteenth century, the scientific heritage of Islam had largely been taken
> over by the Portuguese Jews.[2]

While his comments are correct in one sense, they do not take into account that during Europe's Dark Ages, any holding operation—if that adequately describes the Islamic contribution—had great value in itself. Just to retain knowledge and incrementally advance it was enormously important, not least in solar astronomy. Tellingly, the great Belgian-American scholar George Sarton (1884–1956) divides his magisterial five-volume *History of Science* into half-centuries, each of which he associates with one central figure. He begins in 450 B.C. with Plato, followed by Aristotle, Euclid, and Archimedes; but from A.D. 750 traces an unbroken line of Arab and Persian scientists—Jabir, al-Khwarizmi (who worked on calendars, the true positions of the Sun, spherical astronomy, and eclipse calculations), al-Razi, al-Masudi, Abul-Wafa, al-Biruni, and Avicenna—ending with Omar Khayyám (A.D. 1050–1100).

Ghiyath al-Din Abu'l-Fath Umar ibn Ibrahim Al-Nisaburi al-Khayyámi—the man the West calls Omar Khayyám—was largely unknown to it until 1859, when an effete Suffolk gentleman scholar, Edward Fitzgerald, published an anonymous translation of the four-thousand-line *Rubáiyát* (literally "Quatrains"—stanzas of four lines). By that time Fitzgerald was well into middle age, and his previous work had aroused little interest. He was probably unsurprised when this, too, seemed to meet the same fate. Two years after publication, however, a London bookseller sold off his remaining stock from a penny bargain box outside his shop near Leicester Square, and another poet, Dante Gabriel Rossetti, picked up a copy and was soon declaring the work a masterpiece; by the end of the century, aided by the long-running enthusiasm for Persian culture in nineteenth-century Britain, it would become the most popular poem of its time. Yet the *Rubáiyát* turns out to be only one of Khayyám's many accomplishments.

Khayyám lived in Naishapur (now in northeastern Iran), close to the Afghan and Turkestan borders. His last name means "tent maker," and in one of his verses he writes punningly of "Khayyám, who stitched the tents of knowledge"—which is precisely what he did. He was the author of the well-received *Treatise on the Demonstration of Problems of Algebra* and was the first to come up with a general theory of cubic equations; he also helped reform the Persian calendar, basing it on the solar year, not the lunar—the result, according to Gibbon, being "a computation of time which surpasses the Julian, and approaches the accuracy of the Gregorian style."[3] He measured the solar year at 365.24219858156 days, an extraordinarily accurate computation, given that at the end of the nineteenth century it was measured at 365.242196, now refined to 365.242190. He constructed a star map and helped build a major observatory.

Khayyám and his seven fellow scientists are Sarton's group of choice, but they were abetted by many others. Faced with Koestler's dismissal, one might ask what made Arab science, in particular its mathematicians and astronomers, so outstanding for more than 350 years, and what is a fair estimate of Islam's contribution to solar astronomy? These two questions take up the first half of this chapter; and it turns out that the answer to one is tightly bound up with that to the other, because Islam valued knowledge not only for its own sake, but also, in the case of astronomy—and specifically the study of the Sun—for its contributions to religious observance. Thereafter we will look at how, from around A.D. 750 through to the coming of Copernicus, studying the Sun was evolving in other cultures—most notably in India—as well as back in Europe.

It all started in A.D. 630, when the prophet Muhammad marched upon his birthplace, Mecca. The city surrendered without a fight and became the center of pilgrimage for Muhammad's new religion of Islam. In the moment of his next victory, over the Jewish settlement of Khaibar, he and some of his men were given dinner by a local woman who spiked their meat with poison. He took more than two years to die, finally succumbing eight days short of his sixty-third birthday.

By then, he had already established Islam as a social and political force on a par with Judaism and Christianity. Within a few decades, his successors would unite Iran, Iraq, Egypt (taking Alexandria and burning what remained of its great library), Syria, Palestine, Armenia, and much of North Africa. By A.D. 750, the rest of North Africa had followed, along with the entire Iberian Peninsula and much of central and large stretches of southern Asia. In the tenth century, Islam moved even deeper into Africa as well as spreading across what had been the mainly Hindu principalities east of the Indus in what are now Pakistan, India, and Bangladesh, then onward as far as the frontier provinces of China.

The Muslim theocrats could not have been more unlike their Christian or Chinese counterparts in that they tended toward tolerance—and so, during its golden age, Islam was eager to embrace learning of all kinds, its scholars regularly exchanging ideas with astronomers and natural historians from other countries. During the years A.D. 661–751 the caliphs,* ruling from Damascus, strove systematically[4] to increase the store of knowledge.

The reason for such passion for learning lay in the Muslim holy book. More than one-fifth of the Qur'an—more than eleven hundred verses—is con-

---

* "Caliph" comes from the Arabic khalīfa, meaning "vice-regent" or "successor," specifically the successor to Muhammad. It also defines what such a person's role should be: to govern according to the Qur'an and the practice of the Prophet. It is a word seldom used for anything other than the leader of the entire Sunni community.

cerned with natural phenomena.* Along with the teachings of Muhammad, the Qur'an forms the great motivating force for both individuals and the state to concentrate their energies on such subjects as mathematics and astronomy. As the Islamic scholar M.A.R. Khan has noted,

> Some of the most eloquent passages in the Qur'an refer to the grandeur of the stellar world, the regularity of solar and lunar movements among the constellations, the phases of the Moon and the dazzling brilliance of the restless planets. No wonder that the Arabs and the later converts to Islam from other nationalities and civilizations took to astronomy so enthusiastically and left their permanent mark on it.[5]

The successors to the caliphs of Damascus, known as the Abbasid Dynasty, came to power in A.D. 751 and ruled for more than five centuries, choosing Baghdad over Damascus for their capital and dedicating themselves to building on the scientific achievements of the first caliphs. Some twelve years after their move to Baghdad, they began to reconstruct that city, working from a design created by two men, one a philosopher and the other an astronomer. Within fifty years of its planning, this new metropolis became the preeminent cultural and scientific city of the world. At a time when the Church forbade the practice of intrusive surgery, Arabs were giving anesthetics and performing complex operations. Mathematics exploded: they introduced the tangent in trigonometry, posed and solved cubic equations, conducted extensive studies of cones, replaced chords with sines, and laid down the basic theories for the solution of trilineal figures (a rarefied area of higher mathematics). They brought the world its first international banking system, and also the pendulum and the windmill. Astronomy flourished, along with the rest of the sciences.

While the center of observational astronomy was Baghdad, the renaissance went far beyond one city, thriving in Muslim capitals from Spain to central Asia. The caliphs, particularly al-Mansur (who ruled from A.D. 754 to 775) were great patrons. In a typical example, a scientist from India arrived at court bearing a copy of the *Surya Siddhanta* (*Solar Principles*), written by the Hindu astronomer Aryabhatta four hundred years before. The caliph had it rendered into Arabic—Muslims were familiar with Indian and Persian writings some time before they read the great Greek texts. The caliphate of al-Ma'mun (A.D. 813–833) was even more progressive. Himself an astronomer, he founded a cel-

---

* A nice echo of the Qur'an's place in things appears in Zadie Smith's first novel, *White Teeth*, when the twins Millat and Magid Jones meet in a lecture room in a "red-brick university, South-West by the Thames": "Millat arranges the chairs to demonstrate the vision of the solar system which is so clearly and remarkably described in the Qur'an, centuries before Western science" (London: Penguin, 2001), p. 397.

ebrated academy, the House of Wisdom, and paid his savants famously well: one senior librarian, upon his retirement, was given gold equal to the weight of all the books he had translated. One of Sarton's eight, al-Khwarizmi, belongs to this period, as does al-Kindi, whose massive work *Optics* would become the basis for many of Newton's theories.

It was al-Ma'mun who introduced the best Greek works on astronomy to his people—Ptolemy's *Almagest* was translated at least five times in the late eighth and early ninth centuries, once into Syriac and four times into Arabic. He built two observatories, one in the desert outside Riyadh and the other at Qasiyaun, west of Damascus: over the next seven centuries a dozen observatories would be set up at royal expense, and many more were financed by the astronomers themselves. Much of this research was Sun-related: calculating rates of precession and the length of the tropical year, and advancing reasons for why the Sun appears larger near the horizon than at the zenith.

Their work was also specifically bound up with Islam. Muslims seek to pray facing Mecca (a mountain-set city about fifty miles inland from the Red Sea). It is not only Muhammad's birthplace: it is also revered as the first site created on Earth, where Ibrahim (Abraham) and his son Isma'il built the Kaaba ("cube") that Muhammad was to adopt as the focus of all Islamic worship, a physical revelation of the presence of God, which is tended by Muhammad's tribe, the Quraysh. Such ritual acts as the call to prayer or the halal method of slaughtering animals for food must also be performed facing the Kaaba; and Muslim graves and tombs were laid out so that the body lies on its side facing the holy wall. (Modern burial practice differs slightly, but the body remains oriented toward the Kaaba.)[6] Since Muslims direct their prayers and other acts of devotion toward Mecca, they need to know exactly where the Kaaba lies.

How do worshippers spread across such a vast part of the globe determine the *qibla*, the direction that they must adopt to be facing toward the holy city?* In the beginning, the faithful used the risings and settings of the Sun and the positions of the fixed stars—the earliest mosques faced Mecca only roughly. Out of a desire for a more faithful precision, the caliph al-Ma'mun commissioned a team of scholars to establish the coordinates of Mecca and Baghdad. Similarly inspired, the mathematician al-Biruni wrote a treatise on mathematical geography, *Tahdid*, expressly for the purpose of determining the *qibla* from Ghazna, the walled city along the route between Kabul and Kandahar. During the ninth century, according to the distinguished Arabist David King, "Differ-

---

* The Kaaba is actually a pre-Islamic pagan shrine of uncertain origin and date, standing about fifty feet high by thirty-five feet wide in the center court of the great mosque of al-Haram. Embedded in its walls is a large black stone, held by scientists to be a meteorite, that all pilgrims are expected to kiss.

ent schemes of sacred geography were developed . . . in which the world was divided into sectors about the Ka'ba[,] and the *qibla* in each sector was defined in terms of the rising or setting of the sun or a certain fixed star."[7] Muslim mathematicians compiled tables to express the *qibla* as an angle to the meridian for each degree of difference in terrestrial latitude and longitude, using geometric and trigonometric solutions. The extremely complex calculations involved led to the production of astronomical tables that made it possible for believers, wherever they might be, to know precisely which way to face. By the beginning of the tenth Christian century, mosques were being aligned toward Mecca with much greater accuracy: those in Egypt and Andalusia, west of the City of God, face the Sun as it sets at the midwinter equinox, while mosques to the east, in Iran, Iraq, and central Asia, face the rising Sun.

It was not just a question of the direction to pray, but when. Muslims required accurate knowledge of the Sun and Moon so that they could determine the exact hours for prayer. According to the twentieth-century historian of science Mohammad Ilyas, "so important was this role that only the major astronomers were assigned to the principal mosques in the different parts of the Muslim world during the medieval period."[8] The Qur'an does not specify particular times for prayer; rather, the faithful are encouraged to use the passage of the Sun for guidance. In the early years of Islam, the times were set by simply observing shadow lengths by day and twilight phenomena in the evening and early morning. Muslims begin their day at sunset with *maghrib*, moving on to their second prayer, the *isha*, as night falls; then come the *fajr*, or dawn prayer, and the *zuhr* at noon, to commence shortly after the Sun actually crosses the meridian. Last comes the *asr*, or afternoon prayer, which begins when the shadow of any object has gone beyond its midday minimum by an amount equal to its own length. In some Muslim communities, a sixth prayer is added, the *duha*, begun at the same time before midday as the *asr* begins after it.

While some early mosques were built without consulting an astronomer, young scholars were encouraged to study astronomy and mathematics, not just to prepare a calendar (the Islamic version is strictly lunar) but also as a crucial aid to religious practice. That the long line of great medieval scientists from the Muslim world coincided with the development of a dynamic and coherent Islamic liturgy is no accident.

THE INTRODUCTION OF the plane astrolabe—Greek for "star taker"—possibly as early as the eighth century by the Islamic mathematician and astronomer Muhammad al-Fazari, who died sometime between 796 and 806, only increased Arab interest in studying the heavens, particularly such spectacular manifestations as comets and eclipses. A highly versatile instrument, the astrolabe could measure altitude and answer so many of the key questions about spherical astronomy that European scholars dubbed it "the king of instruments," writing dialogues in its honor. The best craftsmen were specifically named *alasterlabi*—makers of astrolabes—such was the demand for the product. Most such devices were fashioned in the royal workshops (*buyutat*); to make and decorate one's own instrument was a significant achievement.[9]

When compared to the Greeks' speculative approach, the use of these great aids—astrolabe, meridian quadrant, armillary sphere, and celestial globe—transformed the practice of astronomy, providing the foundation on which scientists build the equipment and laboratories of today. In 992, the obliquity of the ecliptic was measured with a quadrant whose radius was 40 cubits—about 58 feet; at the great observatory at Samarkand, the gigantic meridian arc had a radius of over 120 feet. These magnitudes alone gave Arabian astronomers considerable advantages in accuracy and would be their greatest contribution to our understanding of the Sun.

By the thirteenth century the Abbasid Caliphate, harried by Mongol invaders, was in disarray, and east and west broke into semi-independent states. In 1220 Genghis Khan, at the head of some two hundred thousand horsemen, swept across central Asia and Afghanistan, overwhelming Samarkand, Bokhara, and Balkh, and pressing on into eastern Turkey. Genghis died at the age of seventy-two, by then having converted to Islam and changed his passion from destruction to building, but not before he had managed to erase, in the space of twenty years, one of the world's most advanced civilizations. Meanwhile, the Crusaders were again wreaking havoc on the western frontiers of the caliphate. The deathblow came in 1258 when the Mongol prince Hülagu Khan, Genghis's grandson, sacked Baghdad, destroying its palaces and public buildings. The libraries and collections of fine art that generations of caliphs had so assiduously assembled were pillaged and burned. Some believe that only one book in every thousand survived.

Arab astronomy, however, would live on. Three years before he sacked Baghdad, Hülagu had succeeded in taking the vital mountain stronghold of the Assassins.* Among the prisoners was the astronomer Nasir al-Din al-Tusi,

---

* This was a particularly terrifying and bloodthirsty religious sect founded in 1090 and violently opposed to the Abbasid Caliphate. Amin Maalouf, in his history of the Assassins, writes: "The serenity with which the members of the sect accepted their own death led their contemporaries to believe that they were drugged with hashish, which is why they were called hashashûn, or hashîshîn, a word that was distorted into 'Assassin' and soon incorporated into many languages as a common noun." *The Crusades Through Arab Eyes* (New York: Schocken, 1989), pp. 98–105.

who had been held as an honored astrological adviser by the Assassins' grand master. Al-Tusi now transferred his allegiance to Hülagu, and in 1256 a huge observatory was begun under his direction at Maragha, south of Tabriz, and continued to function for more than eighty years, becoming one of the most important of its time. Al-Tusi's criticisms of Ptolemy's theory of planetary motion would anticipate the Copernican revolution.

The Mamluk sultans ruled Egypt and Syria from the mid-1200s to the early 1500s. Moving northward to Turkey, westward to northwest Africa and southern Spain, and eastward to India, they effected a renaissance of sorts—new

Two Indian astronomers stargaze through a seventeenth-century telescope

centers of learning grew up around the great observatories at Maragha and Samarkand, the latter built under the patronage of Ulugh Beg (1393–1440), a grandson of King Tamur the Lame (Marlowe's Tamburlaine). Ulugh Beg ("great prince") devoted himself to astronomy and compiled star tables so precise that for generations of Arabic astronomers they superseded even

Ptolemy's. But the resurgence was transient—the only part of the Samarkand observatory that still survives is a giant underground marble sextant.

Various revivals were attempted in the years that followed, but none lasted long; after Ulugh Beg was assassinated in 1440, science in the then Muslim world withered away. But for 350 years, Muslims had put astronomy to work in the service of Islam, and by so doing kept it alive.

IN 1068, A scholar of the time picked out a group of eight nations according to their perceived contribution to science: "the Hindus [Indians], the Persians, the Chaldeans, the Hebrews, the Greeks, the Romans, the Egyptians [the scientists at Alexandria], and the Arabs."[10] The list is a simplification, ignoring the many interactions among them; it gives no credit to China, which, although it may have been a world unto itself, should not have been discounted; and imperial Rome is

Hindu devotees pay homage to the Sun, in Sanskrit called "Mitra" or "Friend" because of its warmth, light, and other life-giving qualities.

an unexpected and perhaps unwarranted inclusion. But that list will shape the rest of this chapter, which looks at what Hindu scientists gave to solar astronomy and assesses how Western Europe fared in the years before Copernicus.

The Vedic religion, the precursor to modern Hinduism, was one of the earliest to have its beliefs and rituals set down in written form (Sanskrit); the Vedas as a whole (the earliest of which date back to 1500 B.C.) are filled with hymns praising the Sun as the source and sustainer of all earthly life, and one of its associated texts, the Jyotisa Vedanga, refers to the Sun (Surya or Aditya) as "a storehouse of inexhaustible power and radiance."* The first Indian study devoted to astronomy, it was already an influential work by the fourth century B.C.[11]

Perhaps as a result of that early interest, the influence of Indian science was pervasive—and not only in China, where Indian astronomy made inroads within a century or so of the Vedas being written. That great authority on early science Otto Neugebauer writes that Indian astronomy constitutes "one of the most important missing links between late Babylonian astronomy and the fully developed stage of Greek astronomy represented by the *Almagest*."[12] It is possible that Indian astronomers exchanged ideas with their Greek and Babylonian counterparts at least from the time of Ptolemy, if not well before: key concepts already present in the Rig Vedic texts turn up in Babylonian astronomy around 700 B.C. We do know that Indian astronomy was a felt presence in China, along with Buddhism, during the Tang Dynasty (A.D. 618–907), when many Indian skywatchers took up residence in the Chinese capital. By A.D. 718 Indian astrology had also taken root, while the Chinese adopted the Indian calendar, organized around the twenty-eight lunar mansions. Next to be influenced was Persia, whose Abbasid caliphs were so enrapt by Indian science that they worked to spread its methods into central Asia.

From the fifth century A.D., Jainist doctrine (with its emphasis on rebirth and reincarnation, but also on supernatural forces issuing as karma)† began to assert itself through most areas of India. It decreed Mount Sumeru—supposedly situated in the Himalayas—to be the Earth's center, around which revolved the Sun, Moon, and stars. By day, the Sun was thought to travel south from the mountain, while at night the Himalayas hid it from view. Jains had no notion of daylight in one hemisphere and night in the other, but imagined each to have its own Sun.

The Indians' picture of the cosmos derived primarily from religious doctrines, and none of their ancient records of the Sun or of sunspots or eclipses has been found. Instead, Indian astronomers took the statistics compiled in Babylon and Egypt and adapted them to their own needs—calendar computation, timekeeping, the casting of horoscopes, and the prediction of eclipses. Yet

---

* One word for "bright" in the Jyotisa Vedanga is "citrus." It seems reasonable to suppose that the fruits we know as "citrus," such as oranges and lemons, gained a generic name from their sunlike color. When the strife goddess Eris set out to precipitate the Trojan War, she cast down not an apple but an orange, for which the Greek is *chrisomilia*—"golden apple."

† Jainism, or Jain Dharma, began in the sixth century B.C. as a revolt against Hinduism. It regards every living soul—even those of animals and insects—as potentially divine, and advocates nonviolence. Unusual in being an atheistic religion, it teaches, among other things, that one should not eat, drink, or travel after sunset. Currently there are 5.2 million Jains in India, found in all but one of the country's thirty-five states.

Indian folk drawing
depicting the Surya
Mandala with a cobra,
symbolizing the Sun,
water, and vegetation

it was the Indians who reckoned the age of the Earth as 4.3 billion years, when even in the nineteenth century many scientists were convinced it was at most a hundred thousand years old (the current estimate being 4.6 billion). Indian civilization most influenced Western astronomy not through its theories but through its development of mathematical tools: brilliantly innovative spherical trigonometry, for instance, and the invention of the decimal base, with the concept of zero functioning as a number.[13]

THE FALL OF ROME extinguished the essentially bilingual civilization of the West, resulting in the almost total disappearance of the Greek language outside the Eastern Empire, with most major works of Greek science meeting the same fate. Astronomy moved philosophically from a pagan to a Christian milieu, and geographically from one intellectual center in Alexandria to another in Constantinople; but for between three hundred and seven hundred years after Ptolemy, the discipline, both observational and theoretical, was almost entirely neglected. The vicissitudes of wars and faiths subjected learning to long hiatuses—disastrous for a highly technical science dependent on competent and consistent instruction handed down from one generation to the next. In 398, a Christian mob wrecked the library of the Alexandrian Museum. In 415, another mob lynched the leading astronomer, the Lady Hypatia. Any writings from 650 to 800 were negligible, partly the effect of Iconoclastic persecutions in the eighth century.

Throughout much of the old world, most people were living on the absolute margin, their lives threatened by barbarian invasions, calls to war, and ever more frequent migrations, famines, plagues, floods, and other disasters. In 526, Antioch was wrecked by an earthquake; in 542, Constantinople was scourged by bubonic plague; nine years later an earthquake and tsunami killed tens of thousands in Lebanon. The year 856 saw a major earthquake in Greece, 1069 a famine in Egypt. The tenth century alone brought twenty severe famines, some lasting two to three years. The eleventh century was no better: France endured at least twenty-six famines, England a famine every fourteen years or so. Earthquakes in 1138, 1268, and 1290 killed upward of half a million. In 1228, a great flood swept away a hundred thousand people and a substantial portion of the land in the Low Countries. Then, in the middle of the fourteenth century, the bubonic plague returned in the form of the Black Death, raging for six years and carrying off well over a third of the people of Europe as well as multitudes in Asia and North Africa, probably 25 million in all. It was achievement enough to survive, let alone map the heavens.

At the end of the ninth century, the West was just beginning to emerge from its Dark Ages. Arab astronomers had streamed into Spain, and by the tenth century, Córdoba, which boasted half a million inhabitants, several libraries, a medical school, and a large paper trade, was a substantial center of learning. While trade routes between the West and the Arab world were by this time long established, the exchange of scientific knowledge began only when Christian scholars visited Spain in the eleventh century, not only imbibing Arab ideas but also reclaiming their own heritage as a result. By the twelfth century, Toledo was the hub of translating from Arabic into Latin, Hebrew, and Castilian as well as many other languages. The *Almagest* was rendered by an anonymous Sicilian into Latin in 1160, although it remained a difficult work: some words were not understood but simply transliterated—such as "nadir" and "zenith"—with the result that a number of simplified and often inaccurate versions appeared. Like a game of Chinese Whispers, each translation became progressively more removed from the original.*

Even so, by 1200, thanks to the Arab translations in Córdoba and Toledo, reasonably accurate Latin versions of Ptolemy and the main works of Aristotle, Plato, Euclid, Archimedes, and Galen were circulating. Ptolemy's writings became almost synonymous with astronomy and were rarely questioned— although the leading Arab astronomers certainly had their doubts, discerning yet more errors in his system, particularly in his theories of planetary motion.

---

* The great scientist's influence endures. In the early months of 2008, a New York couple sent their young son to a small nursery school in Brooklyn. *Seven* of his fellow pupils were named Ptolemy.

How to explain, for example, the great variation in the brightness of Mercury and Venus by a Ptolemaic reading? The greatest of a new generation of Arab scientists, Ibn Rushd (1126–1198), known as Averroës, whose massive commentaries on Aristotle earned him the sobriquet "*the* commentator," wrote: "The astronomical sciences of our days offer nothing from which one can derive an existing reality. The model that has been developed in our times accords with the computations—not with the way things actually are."[14] King Alfonso X of Castile (1221–1284) is said to have remarked, on being introduced to the Ptolemaic model, that had he, Alfonso, been present at the Creation, he might have given the Almighty some better advice.

From the dissolution of the last vestiges of imperial authority at Rome in the late fifth century until the final years of the Renaissance, the Roman Catholic Church dominated the West. During the first half of this period, virtually the only schools were run by monks: pupils had to make do with limited and secondhand sources, their teachers offering little encouragement to original inquiry or even observations. Scientific knowledge was for the most part regarded as little more than an aid to understanding the Bible. Saint Augustine of Hippo, one of the largest-minded of Christian thinkers, observed:

> It is not necessary to probe into the nature of things, as was done by those whom the Greeks called *physici;* nor need we be in alarm lest the Christian should be ignorant of the force and number of the elements,—the motion, and order, and eclipses of the heavenly bodies; the form of the heavens. . . . It is enough for the Christian to believe that the only cause of all created things, whether heavenly or earthly, whether visible or invisible, is the goodness of the Creator, the one true God.[15]

Professor John North of Trinity College, Cambridge, a punctilious custodian of the period, adds: "Generally speaking, Christian man saw himself as a degraded and miserable creature whose only hope was in prayer and penitence, and as one to whom a rational understanding of the motions of the planets was supremely irrelevant."[16] It is hardly surprising, then, that astrology continued to be an integral part of everyday life, those few astronomers who remained being relegated to accumulating data and improving the precision of formulas and parameters. Astrology became the patron of astronomy, providing a market for treatises and tables and so contributing to the survival of works that would otherwise have perished.*[17] Gradually, however, the propagation of

---

* The famous "All the world's a stage" speech in *As You Like It* (1599) is in part based on an astrological reading of man's seven ages, each of the seven so-called planets guiding a particular period of life. The Sun oversees "young manhood"—the years twenty-three to forty-one.

Christianity, centered as it was upon an all-creating single deity, led the way to a scientific approach to the workings of the natural world through its rejection of the plethora of charms, omens, dreams, and suchlike foolery that went to make up astrology. This at first hardly helped astronomy. It is true that no medieval student could get his master's degree without understanding the rudiments of how to study the heavens; on the other hand, medieval Christian universities, taking their lead from the Church, never gave astronomy a prominent position: it may have been part of every introductory course, but was taught at an elementary level, with the main deficiencies of the Greek world-view persisting.

A 1635 Dutch drawing of "the astrolabe of Lansberg," based on Ptolemaic principles, according to the original caption— though it looks more like a sundial that the man is consulting

As the cultural climate warmed, and the first Christian universities in Europe were established in cities such as Oxford (1096), Paris (between 1150 and 1170), and Bologna (1158), the *Almagest* found its way into the curriculum of mathematical education in both Europe and the Near and Middle East, being taken as a natural sequel to Euclid's *Elements* and the treatises on spherical astronomy by Autolycus and Theodosius; even then, most students found it too difficult and resorted to interpretive texts that dealt with basic solar questions.

With the formation of universities there arose a figure who would come to dominate the intellectual landscape of Western Europe. The Franciscan friar Roger Bacon (c. 1214–1294) was within his lifetime hailed as "Doctor Mirabilis" ("the Wonderful Teacher") for his breadth and depth of understanding, and held by some to be the first true scientist. When Oxford was given its charter, he fought to have science made part of the curriculum, insisting that it was complementary, not antagonistic, to the Faith. "Everywhere," he wrote, there was "a show of knowledge concealing fundamental ignorance."[18] Pope Clement IV, who as Guy de Foulques had come to know Bacon while serving as papal legate to England, now wrote to his old friend asking him to undertake a book on how the sciences should be taught.

Within eighteen months Bacon had completed three large volumes, in which he offered not only his ideas for better scientific methods (for instance, that laboratory experiments be part of one's education), but also a vehement attack on the vices of clergy and monks. Over the next ten years, going far beyond his commission, he would anticipate an extraordinary range of discoveries and inventions: the magnetic compass needle, a method for constructing a telescope, reading glasses, the steamship, the airplane, the workings of radiology, the cause of rainbows, the principles of a camera obscura to project pictures—even television. He also wrote several essays on astronomy. But Pope Clement died before he could read any of this, and his successor, Nicholas IV, condemned Bacon's writings as heretical. By 1278 the Wonderful Teacher had been thrown into prison (most likely an unpleasant form of house arrest), which he was made to endure for the next fourteen years, dying at age eighty or thereabouts, in 1294.

Over the next two hundred years, astronomers remained preoccupied with the adaptation and manipulation of a bewildering profusion of tables, and for all the power of Bacon's reasoning, the universities continued to view science as a way to establish the mathematical relationships among the phenomena of nature rather than conducting an open-minded inquiry based on empirical research and experiment. As a result, a scholar proved his prowess through the mastery of the most difficult predictions (eclipses) and of the geometrical principles underlying the interminable tables.

The one exception to all this was a revival of fascination with Aristotle. Before the twelfth century his works were largely unknown in Christian Europe, but now they came back into fashion with a vengeance. In particular, people quoted *On the Heavens,* which proposed that science should seek out the *causes* of phenomena. According to Aristotle, a stone, for instance, falling toward the center of the Earth, increased its speed because, being of earthly nature, it was impatient to get home: all motion was a transition from "potency" into "act," the realization of what may exist in any object's basic nature. This might be physics back to front, but it won a wide following. (Arthur Koestler would remark that in moments when we curse an obstinate gadget or a temperamental machine, we all revert to Aristotle.)[19]

The end of the twelfth century saw two main schools of thought: the strict Aristotelians and the mathematical astronomers, who distrusted any theory that could not be supported by elaborate tables and calculations. Bacon had helped resolve the conflict between Aristotle's and Ptolemy's theories by introducing a new explanation for the ways the various celestial spheres might interact. Full of notions of outer convex and inner concave surfaces, his idea was no better than its predecessors, but it kept the peace—and kept both Greek thinkers in the classical canon. Aristotle and Ptolemy continued to be taught, and their incompatibilities, with each other and with observable phenomena, were tacitly accepted. Then in 1277 the bishop of Paris condemned 219 propositions of Aristotle's, declaring several of his doctrines (such as the eternity of the world) to be heresies. The Aristotelians retreated.

A century later, in 1377, another Parisian cleric, Nicole Oresme, wrote a commentary on Aristotle dismissing his arguments for the immobility of the Earth as purely relativistic. Although Oresme (later bishop of Lisieux) could not quite persuade himself that the Earth actually did perform a daily turn about its axis, he at least raised the possibility that it did so. This tentative probing of the ancient assumptions about an unmoving globe was, like the Arab criticisms, but one more dent in the bodywork of a trusted vehicle, and the Greek view of the heavens continued to trundle happily along, its engine carefully lubricated by both Church and Science. Heavenly bodies, the Sun included, still moved in uniform circles, and the Earth lay immobile at the center of a rotating universe.

But all that was about to change.

CHAPTER 9

THE EARTH MOVES

In the center of everything is the Sun. Nor could anyone have placed this luminary at any other, better point. . . . Not without grounds do some call it the World's Lamp.

—Nicolaus Copernicus[1]

People give ear to an upstart astrologer who strives to show that the Earth revolves, not the heavens or the firmament, the Sun and the Moon. . . . He's a fool who wants to turn the art of astronomy on its head.

—Martin Luther[2]

WHILE STILL A TEENAGER, GROUCHO MARX TOOK IT UPON HIMSELF to be his brothers' teacher, and one day he asked Harpo, two years his senior, how the world was shaped. Harpo confessed that he wasn't sure.

Groucho gave him a hint. "What is the shape of my cuff buttons?"

"Square."

"I mean the cuff buttons I wear on Sunday—not every day. Now: What's the shape of the world?"

"Round on Sunday, square on weekdays," replied Harpo, who shortly after took his vow of professional silence.[3]

The tale has been around for a long time, and the question for even longer. What shape is the world? In the first century A.D., Pliny the Elder concluded it was a globe, but he also had a hedging theory that it was "shaped like a pinecone." The encyclopedist Isadore of Seville would also declare the world circular but flat, like a wheel. The Venerable Bede thought it a sphere but could not imagine that people could live in the inverted antipodal ("feet-opposite") regions. These sages were mostly on their own, as from the fifth through the end of the tenth century, Europe returned to a belief in a rectangular world: mapping orthodoxy put Jerusalem at its center, but when in 999 the astronomer Gerbert became Pope Sylvester II, he adopted Pliny's globe as Church doctrine. In 1410, the appearance of Ptolemy's *Geography* in Latin confirmed the by then generally accepted notion that the world was round, and by century's end, as Columbus set out for the Indies by sailing west from Spain, sailors and scientists alike embraced the idea of a spherical Earth, even if they were unsure of its size. It took a far later generation to show that its rotation causes bulges at the equator, so it is not a perfect sphere after all.*

But where was our planet in relation to the rest of the universe? "Among the masses," writes William Manchester, "it continued to be an article of faith that the world was an immovable disk around which the sun revolved, and that the rest of the cosmos comprised heaven, which lay dreamily above the skies, inhabited by cherubs, and hell, flaming deep beneath the European soil. Everyone believed, indeed *knew*, that."[5] Well, not quite everyone: an increasing number of astronomers were finding that they could not make a geocentric universe fit all the facts. Muslims and Christians might endlessly revise the *Almagest,* but none of the variants they came up with could compute the planetary positions in a way that would square with the increasingly accurate observations of the sixteenth century. It was not that no one had figured it out: four hundred years before Aristarchus of Samos advanced his notion of a heliocentric cosmos, philosophers in northern India argued that if the Sun were the largest object in their universe, and if gravity was holding it together—they believed both to be true—then the Sun had to be at its center. There were others: early in the eleventh century, al-Biruni had come to roughly the same conclusions, and the German theologian Nicholas of Cusa (1401–1464) had speculated on the possibility. But before the advent of the printing press and with no one to record their thoughts, such musings mostly died in the dark.

When in 1450 Johannes Gutenberg finally invented movable type, not only could these theories be preserved in print, they could be widely disseminated. Scholars could now put together personal libraries, with access to the same printed text, which they could discuss by letter with colleagues, no matter how far distant. By 1500, printers' workshops could be found in every important Western European city, and there were some six to nine million copies in print of more than thirty-five thousand published titles.

---

* In 1956 Thomas Bailey's influential *The American Pageant* stated without a line of evidence that Columbus's "superstitious sailors . . . grew increasingly mutinous . . . because they were fearful of sailing over the edge of the world."[4] Both Tolkien (with Arda) and C. S. Lewis (with Narnia) envisaged a flat world at least once in their fiction, while about 1910 a schoolmaster in Mississippi was dismissed for teaching that the Earth was round. Otherwise one is left with a Monty Python sketch, "The Crimson Permanent Assurance," in which a pirated office building plunges off the edge of the world.

In 1465 a thirty-year-old astronomer from Königsberg, in Bavaria, Johann Müller (1436–1476), began to write and commission works on astronomy, along with almanacs and tables. These books and papers, given the need for tools to aid navigation and exploration and the ever-present enthusiasm for astrology, proved highly popular. Within five years, he had established himself in Nuremberg with his own observatory and printing shop, and by the end of the century nearly every astronomical manuscript of value had been disseminated throughout the West.

Enter Nicolaus Copernicus (1473–1543), a shy and retiring cleric in Polish-ruled East Prussia, who had became curious about the doctrines of heavenly spheres during ten years' study in Bologna and Padua, only to find himself, like so many before him, having to make adjustments to accommodate the contortions of Ptolemy.[6] How could Mercury and Venus never seem to stray far from the Sun, while Mars, Jupiter, and Saturn sometimes moved backward? He was compelled to the idea of a heliocentric system, which fitted the evidence far better than Ptolemy's solution—though still imperfectly, since like everyone else he continued to assume that the planets traveled in circular, not elliptical, orbits. He included Earth as another orbiting planet—a notion that he tried to dismiss as absurd, but that wouldn't go away.

Now thirty, he returned to the home of his bishop uncle (that "far corner of the Earth," as he sourly dubbed it), where he built a primitive observatory in one of the turrets in the wall surrounding the cathedral precincts. Following directions in the *Almagest,* he set about constructing the same crude wooden instruments that had been used for centuries, and spent his nights gazing at the stars. Around 1514, he began circulating copies of what later scholars would title *De hypothesibus motuum coelestium a se constitutis commentariolus,* or "Little Commentary," which laid out his vision of a Sun-centered system, and asked his friends and fellow scholars for their opinions. At this particular moment, the Catholic Church was imbued with a growing enthusiasm for learning and was happy to encourage original scientific research, as long as it did not openly challenge doctrine. When Pope Leo X sent an encouraging note, and liberal members of the Curia let it be known that they, too, approved, plans were made to print a more substantial version. One would have thought that the basic thesis was self-evidently controversial; yet at this stage there was no hint from official channels of the storm to come.

In 1532 Copernicus's new system was given its first presentation, by no less than the pope's private secretary, to an invited audience in the Vatican gardens. The secretary had been well briefed, and the guests were favorably impressed. Copernicus began to win a degree of fame without actually having committed anything to public print. Yet still he hesitated—partly, no doubt, because he was unable to find any direct evidence that the Earth *did* move, let alone around the Sun, partly because he must have suspected that Protestants, with their literal interpretation of the Bible, would be far less sympathetic than his friends in the Vatican. Almost thirty years would elapse before he allowed an extract from the work in progress to be included in a book on trigonometry.* The whole came to 212 folio pages and appeared in 1543 in an edition of a few hundred copies under the title *De revolutionibus orbium coelestium* (*On the Revolutions of the Heavenly Spheres*). By this time the ageing canon was seventy, afflicted with apoplexy, and paralyzed "from ear to heel"; he saw the first copy to come off press only hours before he died.

"The scorn which I had to fear," Copernicus writes in *De revolutionibus*, "on account of the novelty and absurdity of my opinion almost drove me to abandon a work already undertaken." That said, he was not bashful in claiming credit. It is true that in a dedicatory letter to Pope Paul III (the pontiff who, Dava Sobel reminds us, excommunicated Henry VIII and extended the remit of the Inquisition),[7] he acknowledges that he first learned about the possibility of a moving Earth from reading Cicero, and had discovered from Plutarch that others had shared such a view. But while he initially acknowledged the contribution of Aristarchus of Samos, in his last editing he struck all mention of him.[8] He outdid Aristarchus, however, in offering detailed models of the planets' passage around the Sun—Mercury in approximately eighty days (our current figure is 87.97), Venus in nine months (our figure is 224.7 days), the Earth-Moon system in a year, Mars in not quite two Earth years (1.88), Jupiter in twelve (11.86), and Saturn in thirty (29.4): all reasonable estimates. He also managed to determine within 5 percent of the correct values the planets' least and greatest distances from the Sun.

He proposed that the Earth orbited the Sun, rotating once every twenty-four hours (thus making the stars appear to revolve in the opposite direction). He placed the six known planets in their correct sequence from the Sun—Mercury, Venus, Earth, Mars, Jupiter, Saturn—but that is as far as he got; the spheres of the stars remained fixed, and the trajectories of the planets circular, not elliptical. Copernicus did not get everything right, but he delivered more than enough to stir up bitter argument. From the start, Protestants thought him blasphemous. Martin Luther (1483–1546), one of several leading church-

---

* The book was printed at Wittenberg, the same university to which Shakespeare sent that supreme procrastinator Hamlet; but Copernicus was hardly unique in his reluctance to publish. Darwin let *The Origin of Species* languish in a desk drawer for fifteen years.

men who viewed the new theory as not only shocking but gravely in error, asked plainly, "Who will venture to place the authority of Copernicus above that of the Holy Spirit?" After all, he continued, Joshua commanded the *Sun* to stand still, not the Earth.[9]

As long as Copernicus's ideas were represented as hypothetical, Rome herself remained silent, and the intellectuals in her flock followed suit. For fifty years after his death, most astronomers privately acknowledged the importance of what he had revealed while publicly maintaining their belief in a stationary Earth. "The acceptance of the heliocentric theory was far from general right up to the time of Newton," the astronomer Patrick Moore has written, "and in countries far away from the sphere of Mediterranean culture, the old ideas persisted for even longer."[10] In China and Japan, the Jesuits continued to teach the received cosmology. The notion of a huge and unknown universe of which the Earth was but one minute part would put much of the Christian apparatus of belief at risk, and that was not a thing to be undertaken lightly. The reaction of the baptized world recalls the wife of a Victorian canon of Worcester Cathedral who, informed that Darwin's theory implied the descent of man from apelike creatures, cried out, "Descended from the apes! My dear, we hope it is not true. But if it is, let us pray that it may not become generally known."[11]

Copernicus, ever the reluctant revolutionary, had aimed to revitalize the Ptolemaic tradition of mathematical astronomy, not to supplant it. Again and again his calculations fall into place alongside Ptolemy's—the very structure of *De revolutionibus* mirrors that of the *Almagest*, and its tables were no more accurate than those of the old orthodoxy. "Had it not been for Tycho Brahe and Kepler," concludes Otto Neugebauer acidly, "the Copernican system would have contributed to the perpetuation of the Ptolemaic system in a slightly more complicated form but more pleasing to philosophical minds."[12] Arthur Koestler is even more damning: in almost four hundred pages devoted to the period from Copernicus to Galileo, he buries the Polish canon's reputation—in character "a stuffy pedant, without the flair, the sleepwalking intuition of the original genius; who, having got hold of a good idea, expanded it into a bad system, patiently plodding on, piling more epicycles and deferents into the dreariest and most unreadable among the books that made history."[13] Yet Copernicus did strike the original blow in a genuine revolution. He was the first to take the idea of an Earth that rotates on its axis every twenty-four hours around a central Sun and to work it into a comprehensive theory. In 1546, three years after his death, there was born an astronomer who would model a cosmos that retained much of Copernicus's heliocentric theory but did not require the Earth to move. Tycho Brahe (his first name is Greek for "hitting the mark") came from a Danish noble family, and growing up Protestant inherited a worldview framed by a strict interpretation of the scriptures. He saw his first eclipse as a freckle-faced boy of fourteen, and was struck that it had been predicted to the day. How could imperfect man anticipate the majestic acts of the heavens? He took up astronomy to find out. At sixteen, he observed how the conjunction of Jupiter and Saturn, predicted by the still operative Julian calendar for August 25, 1563, was two days off if one went by Copernicus's tables, over a month if by Ptolemy's. (A conjunction occurs when two objects have the same celestial longitude and so seem to occupy the same space.*) The astronomical system, he decided, was in dire need of overhaul. That would be his mission.

The death of Tycho's foster father (his uncle Jørgen Brahe) in 1565 brought him into his inheritance, and he soon built an observatory at Herrevad Abbey, in Scania, ransacking Europe for the finest instruments, among them a celestial globe five feet in diameter that cost five thousand *dalers*, the equivalent of eighty years of a schoolmaster's salary.† A massive mural quadrant, a great quarter-circle made of brass, it had a radius of more than six feet and enabled Tycho to tell exactly when stars crossed the meridian. He added a huge sextant, which accurately measured the latitude and longitude of celestial objects. The complex also included a paper mill, a glassworks, a tubal intercom system, flush toilets, a private jail, an instrument factory, and a chemical laboratory—in effect, facilities sufficient to create an entire self-contained universe.

On the evening of November 11, 1572, Tycho was walking back from his observatory for supper when he was brought up short by a brilliant white light outshining Venus. "Amazed, and as if astonished and stupefied, I stood still, gazing with my eyes fixed intently," he wrote. "I was led into such perplexity by the unbelievability of the thing that I began to doubt the evidence of my own eyes." He called several servants, who confirmed what he saw, but fearing this might be a group delusion, he went on to stop peasants in the street to ask if

---

* Jupiter and Saturn are not only the largest of the Sun's planets, but also the farthest that are consistently visible to the naked eye. They align only every twenty years. One theory holds that their conjunction in 7 B.C. formed the "Star of Bethlehem" from which the Wise Men predicted the coming of the Messiah. They next meet in our skies on December 21, 2020.

† In other respects Tycho followed the pursuits more typical of a young man of his time. During Christmas 1566, while at the German university of Rostock, he attended a dance and, the worse for drink, got into an argument that led to a duel with broadswords in the dark. His opponent sliced off a sizable chunk of Tycho's nose, and thereafter the astronomer wore a prosthesis made of silver and gold blended into a flesh tone, with an adhesive balm to keep it from falling off. In 1901 his tomb was opened and his remains examined by medical experts, who claimed his nasal opening was rimmed with green, a sign that the false nose had been made of copper. Historians have speculated that he wore the lighter copper nose for everyday use and his more precious one on special occasions.

they could see what he was seeing. Over the next eighteen months the light, slightly northwest of the three stars that make up the right-hand stroke of the familiar W of Cassiopeia, remained at a point in the sky where no star had been observed before. Lying outside the zodiac, it could not be a planet, nor did it fit the descriptions of comets. Sometimes it blazed so intensely that it could be seen in daylight; but by December it had dimmed to the brightness of Jupiter, and by the following March was but a vivid point, modulating from white to red and finally to a leaden gray. Tycho monitored it until it disappeared around March 1574.

Were this a new object in the sky, that would require a dramatic revision to Aristotelian cosmology, for according to Aristotle change and decay were confined to the world below the Moon: everything above was immutable. But something in the heavens was all too evidently changing, for what Tycho had witnessed was a huge stellar explosion, a star in its death throes—a phenomenon that would one day be known as a supernova.

Tycho was the last astronomer of note to work without the aid of a telescope, but his was an age with a new respect for hard facts, and he became the most gifted observer on record, his naked-eye measurements accurate to the arc-minute—that is, one-sixtieth of one degree. (The human eye is a superb long-distance instrument, able to discern on a clear night a candle flame seven miles away.) Tycho rigorously catalogued the planets and stars, listing a thousand of the latter, and in 1573 published *On the New Star*, a pamphlet about the supernova that helped establish his reputation. The following year, he gave the Latin inaugural oration at Copenhagen University, with the entire academic community in attendance. "In our time," he began,

> Nicolaus Copernicus, who has justly been called a second Ptolemy, from his own observations found out something that was missing in Ptolemy. He judged that the hypotheses established by Ptolemy admitted something unsuitable and offending against mathematical axioms. . . . He therefore arranged, by the admirable skill of his genius, his own hypotheses in another manner and thus restored the science of the celestial motions in such a way that nobody before him has considered more accurately the course of the heavenly bodies. For although he devises certain features opposed to physical principles, e.g., that the Sun rests at the center of the universe, that the Earth, the elements associated with it, and the Moon move about the Sun in a triple motion, and that the eighth sphere remains unmoved, he does not for all that admit anything absurd as far as mathematical axioms are concerned.[14]

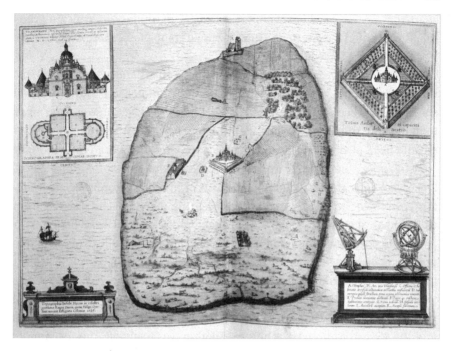

The castle and observatory of Uraniborg (Uranienborg in Danish) on the island of Hven

Tycho knew that the old system failed to make sense, but he was unwilling to let go of the Earth as the center of everything. He produced what became known as the Tychonic System, which resembled that of Copernicus except that it had two centers, the Earth and the Sun. The planets other than the Earth orbited the Sun, while the Sun orbited the Earth—a fine compromise between the Bible and science, if to our eyes untenable. But Tycho was an observer, not a theorist, and pressed on unperturbed.

His account of the supernova both intrigued King Frederick II and led him to worry that his leading astronomer might leave Denmark for more tempting pastures. The king decided to grant Tycho the remote island of Hven, situated midway between Zealand and Sweden, its cliffs a hundred feet high and in clear sight of the great fortress Frederick was raising at Elsinore. With it came a generous endowment consisting of the rents from the island's householders for the rest of his life: at one point in the 1580s, Tycho commanded one percent of Denmark's entire income.

He immediately set to work on two new observatories on Hven: the underground "Castle of the Stars," unshakable by any storm, and a larger complex aboveground, the "castle of Uranus," named after the Greek god of the skies. The latter, nicknamed his "Museum"—literally, "Temple of the Muses"—

boasted ramparts, pavilions, botanical gardens, fountains, an aviary, a print-shop, cellars, and towers. By the time it was all completed, "there was no other place quite like it anywhere in the world."[15]

In November 1577, Tycho was fishing one of the island's many ponds, his workers hauling in the nets, when he noticed what looked like another new star. As dusk faded into night, he saw that it had a reddish tail stretching away from the Sun, the unmistakable signature of a comet. (A word that originally meant "head of hair" in Greek; Aristotle appropriated it because he thought comets—as we now know, no more than conglomerates of ice and dirt—looked like hairy stars.) Tycho rushed to one of his observatories, where he determined the newcomer's position by measuring the angle and distance from its two nearest stellar neighbors. For four months, he tracked its fading dash across the skies, and found that it had no measurable "parallax" (a phenomenon exemplified by the seeming movement of objects in the foreground against those in the background when one moves one's head).

In a report to the king, he calculated that the object was at a distance of more than 230 times the radius of the Earth (four times that between the Earth and the Moon). Although Tycho's calculation was badly off, this, even more than the earlier supernova, convinced him that Aristotle was wrong: separate spheres could not be spinning above and below the Moon. He postulated instead a common atmosphere pervading the universe, in which novas and comets could burst out anywhere. The old doctrine of multiple spheres, he wrote in 1588, "a theory that authors have invented to save appearances, exists only in the imagination so that the motions of the planets in their courses may be understood by the mind."

From his mid-thirties on, Tycho became progressively more eccentric. He kept a dwarf named Jepp, who he believed had powers of clairvoyance and who sat under his table at dinner (unless required as a jester) feasting on tidbits fed him by his master, and also a tame elk, which one night wandered upstairs to an empty room and lapped up so much beer that it crashed back down the stairs, broke its leg, and died. His own fortunes would also take a turn for the worse: in 1588 Frederick II died (he, too, from a surfeit of drink) and, his son being only ten years old, for the next nine years there ruled a series of regents. All were Tycho's friends, so nothing changed; but when the new king, Christian IV, turned nineteen, he determined to shut off the favors this rattle-brained stargazer had enjoyed for over a quarter of a century. Tycho quit his island, taking with him his students, servants, instruments, and printing press, and settled in Catholic Prague. Here he built his last observatory—at Benátky Castle, about thirty miles outside the city, where, until his death in

1601, he spent his time writing poetry and summarizing his researches; but he published not a word more.*

The year before Tycho was forced out of Hven, he was sent the first publication by a twenty-five-year-old German schoolteacher of astronomy and mathematics. Johannes Kepler (1571–1630) was nearsighted, surly, reckless, erratic, and from a poor and severely dysfunctional family. (His mother had been accused of witchcraft, and her son fittingly had the devil's own job saving her from the stake.) He was also a genius. His *Mysterium Cosmographicum* (*The Cosmographic Mystery*) did not take one far in the way of explaining the universe, as it held to the notion of concentric spheres, but it did show its author's powers as an original thinker, and Kepler did place the Sun at the center. Tycho invited him to visit, but Hven proved to be too far from Kepler's home in Graz. Four years later, however, with Tycho now close to Prague, such a journey became practical; and when in 1598 Kepler, a Lutheran, saw his school closed by order of the Jesuit-educated Archduke Ferdinand, the die was cast.

On February 4, 1600, Tycho, fifty-three, and Kepler, twenty-eight, finally met, "face to face, silver nose to scabby cheek."[16] They spent three fractious and not particularly satisfying months in each other's company (the Dane suspicious and secretive, the young German obstinate and critical), during which Tycho insisted that before Kepler returned to Graz to bring order to his family's affairs, he should pledge in writing to reveal nothing about his host's work. But when, that same year, Tycho's principal assistant returned to Denmark, Kepler was invited to move in permanently. Although the two men seldom spoke to each other except at meals, when Tycho died the following October, Kepler found himself appointed to the post of imperial mathematician (a position that would have been Tycho's, had he not been too lofty an aristocrat for such a job). Kepler immediately set to work drawing on Tycho's vast and precise body of observations to advance his own theories, several of which (such as the motions of the planets) were the opposite of his benefactor's.

By 1605, Kepler's main researches were completed—he was to say that he had worked so hard on his calculations during these years that "he could have

---

* It was during this period—actually in 1592—that two of Tycho's cousins by marriage, Frederik Rosenkrantz and Knud Gyldenstierne (which means "Golden Star"), traveled to England on a diplomatic mission, in the course of which Shakespeare apparently learned about them and introduced them into *Hamlet* (completed around 1601) as two famously untrustworthy courtiers. It was probably not an unfair portrait, at least in the case of Rosenkrantz, who had years before impregnated a lady-in-waiting, for which outrage he had been sentenced to forfeit his title and lose two fingers, but escaped that fate by enlisting against the Turks. Tom Stoppard reimmortalized both courtiers in *Rosencrantz and Guildenstern Are Dead*.

died ten times." Interference from Tycho's heirs slowed down the publication of his findings, but at last, in 1609, Kepler's *The New Astronomy* appeared. In it, he established two cardinal principles: that the planets moved not in circles but in ellipses, and that their speed varied according to their distance from the Sun.

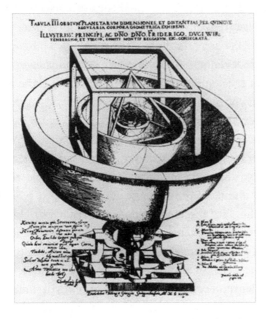

The frontispiece of Kepler's Cosmographic Mystery *(1596) shows the planetary spheres nested alternately with the five Platonic solids, so determining the sizes of the spheres and limiting their number to six. From this model, Kepler estimated the distance of each planet from the Sun.*

The Vatican declared that it considered only circular paths perfect; Kepler replied that celestial imperfection was there to enable God to make better music, since He had used a musical scheme to establish the positions and motions of the planets.* He was also able to strengthen the claims of both Copernicus and Tycho that the stars were immensely farther away than anyone had supposed. The universe for Kepler was vast beyond imagining.

That great philosopher Blaise Pascal (1623–1662), who would abandon science and devote himself to religion, was to respond to these revelations by having the atheistic "*libertin*" of his *Pensées* cry out, "The eternal silence of these everlasting spaces terrifies me!" It terrified Kepler too, who was heard to exclaim, "The infinite is unthinkable." But his findings strengthened rather than undermined his religious beliefs. "Whoever is too stupid to understand astro-

---

* In this he was returning to the Pythagorean belief in celestial harmony, according to which the heavenly spheres made music as they spun. Kepler realized that what previously had seemed the most harmonious scheme for the heavens—uniform circular motions—could generate only a monotone for each sphere; a heavenly body moving at variable speeds would generate more notes, which, combined with other bodies, would create complex polyphonic harmonies—a much richer system.

nomical science," he declared, "or too weak to believe Copernicus without affecting his faith, I will advise him that, having dismissed astronomical studies and having damned whatever philosophical opinions he pleases, he mind his own business and betake himself home to scratch his own dirt patch."[17]

Among them, Copernicus, Tycho, and Kepler laid the foundations of the scientific revolution of the seventeenth century. Although virtually no astronomer practicing at the time of Tycho's death believed in a heliocentric cosmology, all employed Copernican techniques; and as Owen Gingerich, Harvard's professor of astronomy, says, it was Kepler, "the world's first astrophysicist," who was "the man who really forged the Copernican system as we know it."[18] By the time of Kepler's death thirty years later, the entire profession was doctrinally Copernican, and accepted not only Tycho's evidence of how the stars moved but Kepler's theories of how the planets did, too. Cosmology had been transformed by these three men's discoveries—and by the appearance of the telescope. But that is a different story—the story of Galileo.

> In the year sixteen hundred and nine
> Science's light began to shine.
> At Padua City, in a modest house,
> Galileo Galilei set out to prove
> The Sun is still, the Earth is on the move.

So begins Bertholt Brecht's play *Galileo*, originally titled *The Earth Moves*. It opens with Ludovico, who is courting Galileo's daughter, calling on the scientist at his house in Padua, then part of the Venetian Republic. "I saw a brand-new instrument in Amsterdam," he tells his host, who is at this point a professor of mathematics. "A tube affair. . . . It had two lenses, one at each end, one lens bulged and the other was like that." (He gestures.) "Any normal person would think that different lenses cancel each other out. They didn't! I just stood and looked a fool."

Galileo asks if it is a recent invention.

"It must be," replies Ludovico. "They only started peddling it on the streets a few days before I left Holland."[19]

The lines Brecht gives the young suitor reflect what was actually happening in Holland, and Galileo's reaction in the play is also true to history. Ever the entrepreneur, he learned to grind and polish lenses and soon had put together his own, superior instrument. Venice was of necessity an unwalled city, which made early detection of enemy attack vital. On August 8, 1609, Galileo came before the doge and assembled court, whom he had invited to the tower of St.

Mark's Cathedral to demonstrate the applications of his device. As Brian Clegg writes, "The elderly senators had to be restrained from fighting over the chance to be the next to climb to the roof and scan the horizon for ships. They were like children with a new toy."[20] Having shrewdly presented his showpiece, Galileo had his academic stipend nearly doubled, from 520 florins a year to 1,000 (the equivalent of $300,000 now),[21] and his professorial tenure was confirmed for life.

The portraits we have of the early Galileo show him as beefy, ginger-haired, short-necked, and coarse-featured; they plainly suggest his stubborn nature and abrasive self-assertiveness. "A devoted careerist with a genius for public relations,"[22] he now claimed to be the original begetter of the instrument (which he preferred to call a *tubus*), and such was his skill as a lens maker that it was not until 1630 that anyone produced a higher magnification; but in fact there were several "begetters" before him.

Roger Bacon, in 1268, is probably the earliest person to have developed a far-seeing instrument in the West; Chinese astronomers had come up with something similar even before that. Leonardo has left intriguing hints in his diaries and notebooks that he might have discovered the telescope before the fifteenth century was out—"Make mirrors to make the Moon big" and "In order to observe the nature of the planets, open the roof and bring the image of a single planet onto the base of a concave mirror. The image of the planet reflected by the base will show the surface of the planet much magnified."[23] Over the years, however, a Dutch lens grinder, Hans Lipperhey, has been generally acclaimed as the main inventor. Late in 1608, he claimed to have discovered that the combination of a convex and a concave lens could make distant objects look nearer: in September of that year he presented a pair of cookie-sized lenses fixed inside a tube to the German prince Maurice of Nassau. By the following April, spyglasses could be bought in spectacle makers' shops all along the Pont Neuf in Paris. The Dutch prototype magnified by a factor of three; very soon there were instruments capable of magnifying more than twentyfold.*

It was Galileo who got the credit for using telescopes to examine the heavens. "It is hard to think of more surprising discoveries in the entire history of science," posits Noel Swerdlow. "In about two months, December and January [1609–10], he made more discoveries that changed the world than anyone has

---

* Magnifying glasses became common in the thirteenth century, but they were cumbersome, especially for writing. Craftsmen in Venice began making small disks of glass, convex on both sides, to be worn in a frame: spectacles. Because they were shaped like lentils, they became known as "lentils of glass," hence, from the Latin, lenses.

Upper left: Nicolaus Copernicus (1473–1543); upper right: Tycho Brahe (1546–1601), his false nose clearly visible; lower left: Johannes Kepler (1571–1630); lower right: Galileo Galilei (1564–1642)

ever made before or since."[24] Hyperbole, but understandable. Within seven months of his presentation to the doge, Galileo was describing his findings in terse, declarative style in *Sidereus Nuncius* (*The Starry Messenger*, although he himself seems to have intended the title to mean "a message from the stars"). "I have seen," he wrote, "stars in profusion, stars that have never been seen before, and which surpass the old, previously known by more than a factor of ten. But what will excite the greatest astonishment by far . . . is this, namely, that I have discovered four planets." The profusion of stars he refers to was the Milky Way. The telescope suggested, which the unaided eye could not, that the heavens had *depth*. The idea that the Milky Way was not a streak across the sky but a multitude of stars, the crowding together of uncountable brilliant spots of light, was astounding.

The "planets" that Galileo had seen were Jupiter's four major satellites,* now known as the "Galileans." Their discovery proved that planets other than Earth had objects in orbit around them, which strongly supported the Copernican view of the solar system: were Jupiter fixed to a crystal sphere, these moons would have shattered it. The telescope enabled Galileo to produce the first systematic drawings of the Moon, revealed that Venus had phases, and also showed the planets to be much larger than previously thought. There followed a greater discovery yet: the Sun had blemishes. It was "spotty and impure"—what Galileo had seen were sunspots. Moreover, he had watched them move along with the solar surface, which meant that the Sun itself was revolving. Preoccupied with determining the periods of Jupiter's moons (and protecting his position at court), he did not give as much attention to these markings as to his earlier finds, confessing he "did not know, and [could] not know, what the material of the solar spots may be," but they appeared to him most like clouds.[25]

Others before him may have seen objects cross the solar disk, but it was Galileo's account that astounded a wide audience and made him famous. In 1611 he made a triumphant visit to the Vatican,† where he was granted a private audience with Pope Paul V. It has ever been a fine art to be able to judge whether Rome would applaud or condemn, and Galileo seems to have reined himself in—perhaps mindful of a quip made by a curial cardinal that the Bible was "intended to teach us how to go to heaven, not how the heavens go."

Three other astronomers now came forward to claim that they, too, had

---

* The word "satellite" comes from the Latin *satelles* (an attendant at court) and was first used astronomically in a letter Kepler wrote to Galileo in the summer of 1610. Almost immediately afterward, for no discernible reason, Galileo broke off all contact, and refused throughout his life to acknowledge Kepler's law of elliptical orbits. For him, everything in the heavens went around in circles.

† At a banquet there a guest coined the word "telescope," Greek for "far-seer." Tom Stoppard's screenplay *Galileo* has the Jesuit scholar Robert Bellarmine neatly remark to him, "How cruel to make the Greeks give a name to the instrument you wield against their cosmos." Galileo's telescope survives in one popular visual aid: opera glasses. They began in their modern form in Vienna in 1823—two simple Galilean telescopes with a bridge in the center.

noted such marks. The English scientist Thomas Harriot (1560–1621) and two Germans, Christoph Scheiner (1573–1650) and Johann Goldsmid (his name Latinized as Fabricius, 1587–1616), each issued pamphlets, the first being Fabricius's, an account that he took proudly to the Frankfurt Book Fair in the autumn of 1611. Harriot, a steward in Sir Walter Raleigh's household,

Galileo as hero of science, on his triumphant visit to the Vatican in 1611

had made 199 observations of sunspots between December 3, 1610, and January 18, 1613; he just never published them. Scheiner, a Jesuit priest, had made his observations on October 21 to December 14, 1610, but did not publish his records until 1612. Galileo's first sunspot drawings were made between May 3 and 11, 1612. It genuinely seems that each of the four men separately detected the phenomenon and was not passing off another's findings as his own.

Whoever deserves the credit, it was Galileo's claim that touched off the ensuing debate. Scheiner's observation that the Sun was blemished was discounted by his fellow Jesuits, but he argued that the spots were not strictly solar, rather the silhouettes of previously undiscovered small planets revolving close to the Sun's surface—after all, he said, reverently echoing Aristotle, the Sun was perfect and could not be blemished. The following year Galileo replied with *Letters on Sunspots,* three public epistles in which he asserted that the marks did indeed lie on the Sun's surface. His evidence was that the spots displayed a peculiar speed-up, slow-down motion as they crossed the Sun, becoming elongated and foreshortened as they neared the solar edges—exactly the behavior of something fixed to the surface of a spinning globe.

In the last of his three letters, Galileo (possibly looking for a fight) made his first public endorsement of Copernicus's system. The letter enraged Rome, which had given clear notice of its intentions in the 1590s in the case of the Neapolitan philosopher Giordano Bruno (nicknamed "the Exasperated"). When Bruno declared that a rotating Earth in orbit around the Sun was unassailable fact, the Inquisition convicted him of being the worst kind of heretic, a pantheist who had demoted God from supreme creatorship, and had him transported on a mule to Rome's Campo de' Fiori, where he was hanged upside down, stripped naked, then burned at the stake, an iron spike driven through his tongue to prevent further blasphemies. For years, Catholics had been forbidden to read Copernicus unless the crucial nine sentences claiming his ideas were more than theory had been deleted. Even with these removed, the Congregation of the Index would finally ban the work (March 15, 1615) on the grounds that it defended "the false Pythagorean doctrine that the Earth moves and the Sun is motionless."

Galileo could no longer escape. He was summoned before the Holy Office, and on February 26, 1616, a report was filed stating that he had been commanded to give up Copernicanism and "to abstain altogether from teaching or defending this opinion, and even from discussing it." He was still not condemned, but told to direct himself to other studies. There matters rested for seven years, until Maffeo Barberini, once an astronomer himself, was elected pope as Urban VIII. A longtime friend of Galileo's, he now invited him to Rome. Over the course of six audiences the two men strolled the Vatican gardens discussing the heliocentric thesis. Urban is said to have told his old confrere that he could not revoke the 1616 condemnation, but urged him nonetheless to develop a formal comparison of the Copernican and Ptolemaic systems—on one condition: no conclusion was to be drawn as to the truth of either theory, since God alone understood the workings of the universe.

Galileo set to: it would take him nine years. Finally, with the approval of the local censor in Florence, he produced his *Dialogue on the Two Chief World Systems*, writing the first edition in vernacular Tuscan, the second in scholarly Latin.* "God pleased to make me the first observer of an admirable thing kept hidden all these ages." The weight of his argument was clear: the heliocentric version was indisputable; the Earth was in motion because mathematics de-

---

* One of Brecht's characters says that Galileo wrote "in the language of fishwives and merchants." Maybe; but he knew other tongues and when to employ them.

manded it must be. "The common use of Latin," writes H. L. Mencken, "which continued to be the language of the learned on to the beginning of the eighteenth century, made for an easy exchange of ideas, and what was discovered or suggested in one country was quickly taken up in all the others."[26]

Urban, already bruised by the battles of the Counter-Reformation, felt all "the fury of the betrayed lover,"[27] and let loose the Inquisition. In a formal trial the following year, Galileo was condemned "on vehement suspicion of heresy." Heliocentrism per se had never been declared heretical, either ex cathedra or by an ecumenical council; it was just, as one commentator puts it, that "Galileo was intent on ramming Copernicus down the throat of Christendom." Kepler, for one, was furious with his fellow scientist: "Some, through their imprudent behavior, have brought matters to such a pass that reading Copernicus's works, which was absolutely free for eighty years, is now prohibited."[28] In effect, Galileo forced the Church to silence him. It did, banning further sales of the *Dialogue* and ordering all extant copies confiscated. "It is not given to man to know the truth," Brecht's examining cardinal instructs Galileo. "It is granted to him to seek after the truth. Science is the legitimate and beloved daughter of the Church. She must have confidence in the Church."

Brecht claimed that the scientist was threatened with the rack and other instruments of torture, but of this there is absolutely no evidence (mark it down as an Urban legend). If anything, curial officials took special pains to avoid a collision. During his trial Galileo, by this time seventy, was housed in a five-room apartment overlooking the Vatican gardens, even accorded a valet and another servant to look after his food. Happily for the Church, on June 22, 1633, he fell to his knees in the great hall of the Dominican convent of Santa Maria sopra Minerva and recanted—perhaps because, in the end, he was a devout Catholic and saw the larger point that the Holy See wanted to make: science cannot be seen as the ultimate source of authority—or perhaps because, as he was forced to acknowledge as part of his retraction, he did not have incontrovertible evidence that he was right.

When his sentence was imposed, it was that he be confined to his villa in Alcetri, just outside Florence, where he spent the rest of his life at his studies, in particular on the science of dynamics. By 1637 he had lost his sight (though not from observation of the Sun), one eye at a time. The following year he received several visitors, including Hobbes and Milton, the latter of whom, six years on, in *Areopagitica*, recalled discussing cosmology with Galileo, "grown old, a prisoner to the Inquisition, for thinking in astronomy otherwise than the Franciscan and Dominican licensers thought." Milton returned to the subject in *Paradise Lost:*

Sollicit not thy thoughts with matters hid,
Leave them to God above, him serve and feare; . . .
Heav'n is for thee too high
To know what passes there; be lowlie wise.[29]

In the same poem, he describes Satan's landing on the Sun as creating a mark like a sunspot seen through a "Glaz'd Optick Tube" (i.e., a telescope). An endorsement? Anyway, his advice came far too late. At least Galileo, contrary to legend, did not spend a single day in a prison cell. He wrote two comic plays, lectured on Dante, and continued his academic studies before dying in 1642 at the age of seventy-eight. "In a generation which saw the Thirty Years' War," wrote Alfred North Whitehead, "the worst that happened to men of science was that Galileo suffered an honorable detention and a mild reproof, before dying peacefully in his bed."[30]

Almost a hundred years later, Benedict XIV would grant an imprimatur (Rome's formal permission to publish) to the first edition of Galileo's *Complete Works* (even though, paradoxically, the ban on Copernicus's work would linger on until 1828). Another 230 years would pass before John Paul II commissioned a study of Galileo's case, in 1979. After more than twelve years' further deliberation, in 1992 the Vatican finally conceded that Galileo had been justified in what he said. Then, in March 2008, Pope Benedict XVI announced that a statue of the great man would be erected within the Vatican walls. The head of the Pontifical Academy of Sciences (himself a nuclear physicist) declared: "The Church wants to close the Galileo affair and reach a definite understanding not only of his great legacy but also of the relationship between science and faith."[31] Even by the standard of this great institution, it was just a little slow.

THERE WERE REASONS why these arguments took so long to become accepted. Technologies profoundly affect belief systems, and only with the invention of the printing press could ideas spread widely. Still, where such notions involved a revolutionary new way of thinking, there was bound to be great resistance. Sigmund Freud (1856–1939) popularized the view that Copernicus had caused particular consternation by moving mankind from its preeminent position at the center of the universe. In fact, the distress was even greater in the wake of Galileo's revelations about sunspots. There was something horrible about the Sun's not being perfect—as if the ravages of the human face had been translated to the mighty disk and magnified. Before Galileo, the Sun had been an impeccable sphere. Suddenly it was grubby and mottled.

   With the advent of the telescope, seventeenth-century man had to frame things anew, not from imaginative interpretation—sun gods, chariots in the sky, dragons devouring the solar disk—but by repeatedly qualifying and sieving the latest evidence. It must have been profoundly deracinating to think that this great star was, despite being at the center of the solar system, only a small part of the Creator's purpose—and a spotted one at that.

CHAPTER 10

STRANGE SEAS OF THOUGHT

I do not know what I may appear to the world, but to myself I seem to have been only a boy playing on the seashore, and diverting myself in now and then finding a smoother pebble or a prettier shell than ordinary, whilst the great ocean of truth lay all undiscovered before me.

— Isaac Newton [1]

When an apple ripens and falls—what makes it fall? Is it that it is attracted to the ground, is it that the stem withers, is it that the Sun has dried it up, that it has grown heavier, that the wind shakes it, that the boy underneath wants to eat it?

— Leo Tolstoy, *War and Peace* [2]

I WILL MAKE AN END," WROTE THE FRIENDLESS AND ISOLATED SCHOOL-boy Isaac Newton (1643–1727) in his Latin exercise book. "I cannot but weep. I know not what to doe." His principal biographer describes him as "a tortured man . . . an extremely neurotic personality who teetered always, at least through middle age, on the verge of breakdown." [3] He never married—as a character in Tom Stoppard's *Arcadia* jokes, sex was "the attraction which Newton left out." One can hardly imagine him enjoying Galileo's description of wine as "light held together by moisture." He had no ear for music, dismissed great works of sculpture as "stone dolls," and viewed poetry as "ingenious nonsense." [4] Yet when Sir Isaac Newton died he was acknowledged to be the supreme genius of his age. Famous throughout the Western world, he had served for thirty years as president of the Royal Society (the world's greatest scientific institution, founded under the amateur physicist Charles II); twice as a member of Parliament; and as a surprisingly energetic warden of the Mint, in charge of all public coinage. He was laid to rest in Westminster Abbey beneath an ornate marble monument twenty-five feet tall, his effigy recumbent beneath a celestial globe. By his side, cherubs weigh the Sun and planets; the Latin inscription declares: "Mortals rejoice, that there has existed so great an ornament of the human race." No one in history had learned "what to doe" to greater effect.

Before he was twenty-four, Newton had started to formulate the principles of gravity—the basis of all modern predictive astronomy—and to prove that objects on Earth and in the sky move according to the same laws. He had already made important discoveries about the nature of optics, particularly the qualities of color and light, and would go on to develop a law of cooling, articulate the principles of the conservation of momentum, study the speed at which sound travels through the air, and theorize about the origins of the Sun. He was the first to explain the nature of tides, wrote innovatively about the construction of telescopes, and cofounded calculus—the mathematical discipline necessary for every scientific advance of the twentieth century. "Let no one suppose," declared Einstein in 1919, at the time his theories of relativity were bringing him world renown, "that the mighty work of Newton can really be superseded by this or any other theory. His great and lucid ideas will retain their unique significance for all time."[5]

ISAAC NEWTON WAS born into a marginally genteel farming family in Woolsthorpe-by-Colsterworth, in the eastern coastal county of Lincolnshire. His father's early death and his mother's hasty remarriage left him in the care of his grandmother—not unusual in those days, but probably not emotionally nourishing, either. As a child he built clocks and sundials and was known for his skill at telling the time by the Sun; by June 1661 he was studying at the College of the Holy and Undivided Trinity, then as now the most famous and largest of the colleges of Cambridge University.

Though Aristotle and the other great Greek philosophers still undergirded the prevailing orthodoxy, radical innovation had come in the person of René Descartes (1596–1650), who, while keeping much of his research secret for fear of the Church's deadly displeasure, in 1644 did publish *Principia Philosophiae,* in which he asserted that the Sun was just one of many stars and that each stood at the center of its own "vortex." For Newton, such a universe begged a host of questions. Why did any object tend to descend ever downward? Or move in a particular direction? Galileo had seen that the Moon had mountains and chasms not unlike Earth's; but if it were made of the same substance as our planet, what held it in the sky? Why did it circle the Earth rather than plummet down, or spiral away? Queen Elizabeth's personal physician William Gilbert may have provided a clue when he asserted in *De Magnete* (1600) that the Earth displayed what we now call a "dipolar magnetic field"—i.e., its poles have both a positive and a negative charge, making it "a great magnet." But what *was* a magnet?

All these conundrums Newton pondered in his room between Trinity's Great Gate and its chapel. But during the severe winter of 1665, plague spread from the Continent, leaping from parish to parish, killing thousands every week. In little more than a year, one out of every six Londoners was dead. Cam-

Sir Isaac Newton
(1643–1727)

bridge was not immune; the sixteen colleges shut down (they were still small communities: even by 1800 the university would have only four hundred students), and Newton returned to his grandmother's large stone farmhouse. His isolation lasted about nineteen months, during which he effectively created modern mathematics, mechanics, and optics. As he himself recalled, "I was in the prime of my age for invention and minded Mathematics and Philosophy more than at any time since."[6]

The power of his insight is revealed in the answers he found to two particular solar questions: the Sun's mass relative to that of the Earth, and its relative density: he reckoned it to be 28,700 times as massive (a considerable underestimate, but closer than anyone else had come), and slightly more than one-quarter as dense (within 2 percent of our current value). The law of gravity, however, proved more elusive: it would be nearly twenty years before he advanced the theory publicly, and by then he was at least in part inspired by others. Around 1639, another Cambridge-educated astronomer, Jeremiah Horrocks (1617–1641), had suggested that the Moon's motion, while affected by the Sun, was also influenced by some force emanating from the Earth. Thereafter, three members of the Royal Society, Robert Hooke (1635–1703), Edmond Halley

(1656–1742), and Sir Christopher Wren (1632–1723), as well as a French priest, Ismaël Bouillaud (1605–1694), all posited that this force diminished rapidly the farther the object was from the Earth's center. But it was Newton who both recognized that a general law was in play and first demonstrated how it worked.

He concluded that a body on the Earth's surface is drawn downward by a force some 350 times stronger than the tendency of the Earth's rotation to fling it outward—a force he named "gravity" (*gravitas,* meaning "heaviness"). The question was not whether such a force existed—Galileo had shown that it did—but whether it extended so far from Earth that it could be the same power that held the Moon in its orbit. Newton calculated that if the reciprocal attraction were proportionate to the size of the attracting masses, and diminished according to the square of the distance between them, then this explained not only how the Moon orbited, but how this same force shaped the orbits of all the planets—hence his term "universal gravitation." With this formula he had determined how the planets and the stars held their places and what caused the precession of the equinoxes, and explained the ebb and flow of the tides. Even now the majestic range of this reasoning ranks among the greatest of all advances in human understanding; but having made the calculation, he set it aside.

Many years later, he would recount to at least four people that he had been inspired by an apple in his garden. What if the power that made the fruit fall from the tree was not limited to a certain distance from the Earth but extended much farther? He never wrote of any single burst of insight, just that "I began to think of gravity extending to the orb of the Moon" as a force acting ever outward from its center.*

IN LESS THAN two years Cambridge reopened, and Newton, returning, quickly made several important discoveries about light. Plato had thought of vision as the result of particles shot out of the eyes; but was the essence of light outside or inside the observer? Aristotle had understood that light was necessary for color to exist, and Ptolemy had experimented with angles of refraction—but where did color come from? Was it a gift from the Sun?

---

* The story of the apple was created from a series of "contemporary accounts," the first being Voltaire's, in his *Essay on the Epick Poetry of the European Nations.* "[Newton] was walking in his garden," he wrote, "saw some fruit falling from a tree, and let himself drift into a profound meditation."[7] Similar reports appear in the memoirs of people who knew him, including Catherine Barton, Newton's niece (and Jonathan Swift's romantic fantasy), and in the work of his first biographer, William Stukeley, who unequivocally asserts that "the notion of gravitation came into his mind . . . occasion'd by the fall of an apple."[8] Whether or not there is any truth to the story—Newton's bedroom desk did overlook an orchard—one undeniable consequence was the naming of a new variety of apple: Newton's Pippin.

This time inspiration arrived in the form not of an apple but of an ornament. A few miles out of town, on the bank of the river Cam, lay Stourbridge Common, where once a year a large fair was held. Newton bought a prism from a local lens grinder, a simple triangular block of glass shaped like a Toblerone bar. Back in his rooms, he set down the prism: when the Sun fell on it, different colors emerged. Did the glass change the light, or did the sunlight contain the colors and the prism separate them out? Newton knew that a telescope produced a rainbow effect around whatever was being viewed, because every lens edge is a prism, but he was unconvinced by the explanation that white light, as it traveled through a lens, darkened at the thin areas, becoming red, and darkened even more at the thick areas, turning blue. He set up an ex-

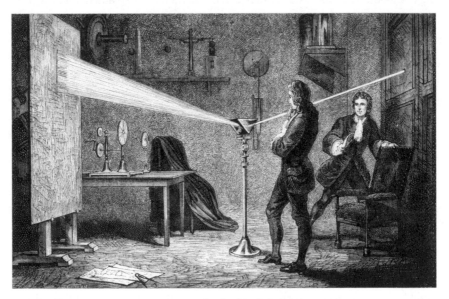

Newton using a prism to break white light into its spectrum,
as imagined by a nineteenth-century French artist

periment in which a thin sun ray struck the prism, which made the separation of light even clearer. "It was at first a very pleasing divertissement," he wrote,

> to view the vivid and intense colors produced thereby; but after a while applying myself to consider them more circumspectly, I became surprised to see them in an oblong form; which, according to the received laws of Refraction should have been circular. And I saw . . . that the light, tending to one end of the Image, did suffer a Refraction considerably greater than the light tending to the other.[9]

The different colors refracted differently, and in rainbow order, the prism bending them into different directions—which meant that the rays could be separated into components. Newton identified red, orange, yellow, green, blue, indigo, and violet. Indigo is not really a separate color, and orange is questionable, but Newton, prey to the charms of numerology, organized his observations under the magic number seven.[10]

His next task was to isolate one single color from white light and pass that through a second prism. He aligned two glass wedges and rotated one to direct first blue light, then red, through the second, to find that the blue rays were refracted by the first prism and slightly more by the next. He also observed that the second never created new colors or altered those received from the first. "*Light*," he noted, "consists of *Rays differently refrangible*"—that is, capable of being turned from one direction to another. So color was not a modification of light but a basic property; white light, which had been thought to be free of tint, actually contains all the colors the eye can see.

Descartes had thought color was induced by the rotation of the minute particles that made up light rays. Newton's thesis was that the colors we observe are separated by passing through objects, not by the objects themselves generating new colors. He dismissed Descartes's notion, establishing that—for example—daffodils are not intrinsically yellow, and rainbows are no more than the cumulative consequence of drops of water building up in the atmosphere during a shower and performing the same function as a prism. The colors of a rainbow—or of anything else—are a function of the way our eyes register individual wavelengths of light.

Culturally at least, Newton was on dangerous ground. Rainbows were icons of nature, closely associated with the Sun.[11] The Greeks had considered them the paths made by messengers between Earth and heaven. Incan armies carried banners of the rainbow, North American Indians believed that the dead dwelled in the "land of rainbows," the rebellious German peasant hosts of the sixteenth century marched beneath a rainbow ensign, emblem of apocalyptic hope. And now the great arc in the sky was suddenly contracted into no more than a side effect of passing raindrops.* More than a century later, the Romantics would be appalled at this demotion, and as late as 1817, Keats would denounce Newton

---

* The rainbow flag remains a symbol of progressive causes, among them the gay and lesbian rights movement, the different colors symbolizing diversity, while Rainbow Gatherings are conventions of hippies who come together on public lands to promote ideas of peace and community. The only supposedly malign rainbow is caused by the Moon—a "lunar bow" that appears in different shades of gray on showery nights and always opposite the Moon. Sailors used to believe it presaged a death among the crew.

for having virtually destroyed the poetry of the rainbow by reducing it to the spill of a prism. But he drank to the great man's health all the same.[12]

Newton could lose himself in an inquiry that knew few (if any) bounds. In one experiment, he stared for as long as he could bear at the Sun's reflection in a looking glass, repeatedly turning to a dark corner of his room to see what spots and colors floated in the darkness as an afterimage. He continued with this test over and over again until, fearing permanent damage, he shut himself up in a darkened room to wait for full sight to return. It took three days. In another experiment, intent on proving that color perception was affected by pressure on the optic nerve, he slid a darning needle around his eye socket until he could poke at its rear wall, noting dispassionately "white, dark and colored circles" for as long as he prodded away with "ye bodkin." He was never merely curious about a subject—obsession was his second nature.

Since light rays of different colors differ also in their refrangibility, Newton concluded that the indistinctness of an image formed by the main lens of a telescope derived from differently colored rays coming into focus at different distances. A single lens could not possibly deliver a distinct image, as the refracting telescope—like a prism—would disperse white light into colors, surrounding the images of stars and planets with false tints. So he invented the first practical reflecting telescope (known today as the "Newtonian" type). Grinding his own mirrors (not a pleasant job: in 1677 the great Dutch philosopher Spinoza would die at forty-four, his lungs rotted from years of inhaling the glass particles produced by his lens-making), he made a superior instrument by widening the mirror—though still only to an inch across. He then cast a two-inch mirror and ground it into spherical curvature, placing it at the bottom of a tube, which caught the reflected rays in a 45-degree secondary mirror, which in turn cast the image onto a convex ocular lens outside the tube—the lens he looked through. In 1671 he sent this little instrument, just over six inches in length, to the Royal Society, where the sensation it caused among the two hundred or so members encouraged him to publish *On Colour,* a work that he later expanded into the *Opticks* of 1704. This last laid out his theories in full and ended with what would become a famous set of rhetorical "queries," expounding his speculations on the nature of the physical world, which he predicted would be answered in generations to come.*

---

* Newton was not alone in his ponderings. Back in the fourth century A.D., the Chinese philosopher Ko Hung had written: "The human eye is colorblind, and the pupil shortsighted; this is why the heavens appear deeply blue. It is like seeing yellow mountains sideways at a great distance."[13] It would take until 1870 for the British physicist John Tyndall to discover that blue light scatters more easily than red, which accounts for the color of a sunny sky. In 1810, Goethe, a considerable scientist as well as a literary genius, published a 1,400-page treatise titled *Zür Farbenlehre* (*On the*

Meanwhile, despite Pope's famous epigram "Nature and Nature's laws lay hid in night: / God said, *Let Newton be!* and all was light," some in the academic world were less convinced. Many of Newton's peers were incredulous when they read in *On Colour* that light was made up of tiny particles that create movements in the ether. Deeply offended by this response (as by much else), Newton lapsed into characteristically venomous argument, constantly seeking redress for slights real or imagined, responding yet more fiercely to any criticism, often with the object of inflicting highly personal wounds, while refusing to make allowances for those he considered mere "splutterers in mathematics" (which, next to him, of course, they were).

The most enduring of his enemies was Robert Hooke, surveyor to the City of London, chief assistant to Christopher Wren, and curator of experiments to the elite cadre who formed the Royal Society. The sources of their rivalry date back to 1672, when Hooke first criticized Newton's theories on light, saying they had not been proved in enough detail. He was powerfully placed, as well as respected for his work on barometers, ultraviolet light, thermometers, wind gauges, and the nature of elasticity, and for demonstrating that air was essential to life. He had also written extensively on light in *Micrographia* (1665), an exposition on another of his inventions, the compound microscope.

Newton, furious, declared Hooke incapable of grasping what he had to say. Their reciprocal sniping continued unabated for years, not helped when Newton learned that the great Dutch mathematician Christopher Huygens (1629–1695) had proposed another challenge to his theory: that light consisted of waves, not particles. In fact, light exhibits the properties of both, but that discovery lay centuries ahead; so the dispute was further exacerbated by each side's being able to show that its opponents were not entirely correct, yet unable to prove its own version conclusively.

Events now move to a coffee shop in London's Strand, where in January 1684 Hooke, Halley, and Wren found themselves arguing about the attraction between the Sun and the planets. After much debate, Wren offered a handsome prize—a book of the winner's choice worth up to forty shillings, a good month's income for a laboring family—if any one of them could demonstrate within two months the kind of planetary orbit that would result were the Sun's gravitational attraction to obey an inverse-square law. The time expired, and

---

Theory of Colors), in which he reformulated Newton's definition. Newton had viewed color as a physical problem, which involved light striking objects and entering our eyes. Goethe realized that the sensations of color reaching our brain were also influenced by the mechanics of vision and by the way our minds process information. He argued that what we see depends upon the object in view, the lighting, *and* our perception.

none of the three had come up with an answer, so Halley was dispatched to Cambridge to put the issue to Newton.* As Halley later told it, the great man immediately replied, "An ellipse," adding that he had solved the problem of gravity some time before but not told anyone, although he would now set to work on putting the solution into publishable form. On hearing this, Hooke proclaimed that he had had the idea at least fifteen years before and had written to Newton back in 1679, arguing for just such a law. He may well have intuited the nature of gravity then, but he lacked the mathematical skills to prove it; and while he probably did feel cheated, history is littered with almost-discoveries.

Newton's mockingly false-modest response, in a letter to Hooke, put it succinctly: "If I have seen further, it is by standing on the shoulders of giants." Their fractious goading would end only with Hooke's death in 1703—after which, amid much political jockeying, his great rival was elected to succeed him as president of the Royal Society. But success could not alter Newton's nature. Despite the acclaim that met the *Principia* (from which he excised an acknowledgment to Hooke), he withheld the lion's share of his researches,[†] closeting himself in his room in Trinity for days at a time, careless of meals, toiling by candlelight, retreating almost completely into himself. He was a man ever at odds with his world.

---

* Newton had few friends, but Halley, an outstanding astronomer and physicist in his own right, was one. In 1705, having perceived that the characteristics of the comet of 1682 were nearly the same as those of the comets that had appeared in 1531 and 1607, Halley concluded that all three were the same object, returning not quite every seventy-six years, and forecast that it would reappear in 1758, which it did. While neither he nor Newton lived to see it, the fact that the comet maintained a steady orbit dramatically reinforced the case for Newton's formulation of gravity. According to Owen Gingerich, Halley was lucky in that only "his" comet was bright enough (among the 140 or so periodic ones now known) to be identified without aid of a telescope. At its closest, it shines with a light equal to a quarter of the Moon's. Halley's calculations enabled the comet's earlier appearances to be tracked back through history. Chinese astronomers observed it in 240 B.C. and possibly even as early as 2467 B.C. A Babylonian tablet records it in 164 B.C., and the Bayeux Tapestry in 1066. The most recent appearances were in 1910 and 1986. Halley (as the comet was titled immediately after its return in 1758) will next appear in 2061.

† Newton's unpublished writings amount to some 2.7 million words, including hundreds of pages about religion. These he sensibly kept secret: although submissive to the Church of England, he held that to regard Christ as God was a sin and that the Holy Trinity was a blasphemous doctrine—a deeply heretical position that could have imperiled his life. His papers remained in private hands until 1936, when they were sold piecemeal at Sotheby's. One of the buyers was the economist John Maynard Keynes, who acquired most of the writings on alchemy, chiefly to save them, and he reportedly read them in taxis going from one Treasury meeting to the next.[14] Keynes told a gathering at the Royal Society that he viewed Newton as "the last of the magicians, the last of the Babylonians and Sumerians."[15]

FOR ALL THE emphasis Newton placed on the notion that the fundamental laws of nature were universally the same, he didn't want a predictable or purely mechanical universe, but rather one that had a place for things of the spirit—thus alchemy, in its early days principally concerned with transmuting "inferior" bodies, especially metals, into "superior" ones by means of a "proper medicine." By the twelfth century this secretive activity had infiltrated European culture courtesy of the Arabs. At first, little distinction was made between it and chemistry, as both concerned themselves with differences and boundaries in material substances. But alchemy also embraced the exploration of nature in the form of procreation, fermentation, transmutation, and transfiguration.

Alchemists recognized seven main metals, each corresponding with a planet, the Sun being identified not only with gold as a substance but with "philosophical gold," the mysterious power hidden within it. All this was of supreme interest to Newton. "Noe heat is so pleasant & beamish as the suns,"

An Alchemist in His Workshop, *a nineteenth-century representation.*
Alchemy was full of nonsense, but it helped to develop chemistry as a science.

he wrote. The alchemists' holy grail was the so-called philosopher's stone, a perfect balance of elements and forces, which was believed to have the power to transmute any metal into gold and to confer on earthly man an all-seeing omniscience. When Milton's Satan lands on the overpoweringly radiant Sun, he finds it difficult to describe but reports that it might be akin to the philosopher's stone, which is "Imagined rather oft than elsewhere seen, / That stone, or like to that which here below / Philosophers in vain so long have sought."[16] This talisman went by many names, including "sun," and was deemed to take two main forms: a white stone, which could translate base metals into silver, and the red stone of the solar stage, which could translate them into gold.

In the garden shed outside his room, adjoining the wall of the college chapel and with a special chimney to carry away the fumes, Newton constructed a laboratory where a furnace burned day and night. When he passed a crimson-tinged alloy, cinnabar (red mercuric sulfide, known to painters as vermilion), through the fire, he extracted a liquid metal known as quicksilver, held by alchemists to be "philosophical mercury," or *Mercurius*, the material from which all things were made. Newton was so enamored of its power that he filled his room with crimson-upholstered furniture, crimson curtains, crimson cushions, even a crimson mohair bed. In the end, his constant handling of mercury built up the metal inside his body, inflicting tremors, sleeplessness, and some say paranoid delusions (one recent theory is that he came to suffer from Asperger's syndrome, a form of autism). Yet he was right in thinking alchemy possible; it just couldn't be effected by the chemistry of his time.

THE EPOCH OF discovery from Copernicus to Newton is generally known as the "Scientific Revolution." "Science was a minor partner in culture in 1600," observes James Gleick. "By 1800 it was *the* major part of culture"—to the extent that fashionable ladies had their portraits painted with sextants and telescopes at their feet.[17] Newton casts such a long shadow that it is easy to underestimate his successors. (Isaac Asimov once remarked that when scientists dispute among themselves who is the greatest scientist in history, what they are actually arguing about is who is the *second* greatest.) While the century after Newton's death saw far fewer discoveries about the Sun, important discoveries there were, including the first quantitative measurements of the speed of light, the notion of planets of other stars, the first reports of dark lines in the solar spectrum, and pioneering work on the relation of clouds to the Sun.

One question particularly fascinated scientists and public alike: How far was it from the Earth to the Sun? None of the existing calculations was even remotely convincing. This is where the passing of Venus between the Earth and

Sun—the visible rapid transit of a dot in front of the roiling solar surface—became vital.* Kepler's famous third law states that the cube of a planet's distance from the Sun is proportional to the square of the time that planet takes to complete its orbit—thereby giving us each planet's relative distance from the Sun; it does not, however, give its *absolute* distance.

Although Venus has made this passage for millions of years, it was the young Lancashire curate Jeremiah Horrocks, who, on November 24, 1639 (having rushed home from church, where he had been obliged to give two sermons), became the first to observe Venus crossing. He realized that if a transit were observed simultaneously from two locations set far enough apart, the data gained would be enough to work out the distance to Venus, the distance from Earth to the Sun, and finally the size of the whole solar system. Having sent a friend to witness the event from Manchester, Horrocks brought off his own observation, and contentedly wrote: "The object of my most sanguine wishes . . . just wholly entered upon the Sun's disk."[18] However, his larger ambition was not to be realized, for his friend's point of vantage proved too close to his own position to be useful, and two more transits would have to occur before anyone else effectively ventured such an observation.†

Horrocks's pioneering effort was a heroic beginner's attempt. In 1716 Halley published *A New Method of Determining the Parallax of the Sun, or His Distance from the Earth,* in which pamphlet he advocated a much more sophisticated approach: deploying as many observers as possible worldwide. By 1761 the scientific community was on notice for the next crossing. Joseph-Nicolas Delisle, a Parisian who had built an observatory and school of astronomy in St. Petersburg, dispatched astronomers to India, St. Helena, and other spots likely to provide a clear view of the June 6 transit. At least 120 observers were posted to more than sixty stations. But the event fell in the middle of the Seven Years' War, and two of the scientists, the astronomer Charles Mason and

---

* Venus crosses the Sun four times within a 243-year cycle. Transits come in pairs eight years apart, followed alternately by gaps of 121.5 years and 105.5 years. Occasionally, a "double transit" may not occur because Venus does not quite pass between Sun and Earth—as in the "missing" transits of A.D. 416, 659, 902, 1145, and 1388.

† See Lt. Cdr. Rupert Thomas Gould, *Jeremiah Horrox, Astronomer: A Paper Read Before Ye Sette of Odd Volumes on 28 November at Oddino's Imperial Restaurant* (London: Huggins and Co., 1923), pp. 23–24. Because the Sun's diameter is around 110 times that of Venus, the planet's silhouette in transit, if viewed from different places on Earth, is outlined against different points on the solar surface. If the transit is observed from places a known distance apart, simple geometry yields the distance to Venus: this is "triangulation," or "parallax" (from the Greek, for the value of an angle). How do people hundreds of miles apart compare their bearings on the silhouette? They must decide on a particular stage of the transit when everyone will observe; confirm the exact time they do so; and note the exact position of Venus on the Sun.

the surveyor Jeremiah Dixon (later to demarcate the Mason-Dixon Line), were aboard a ship attacked by a French frigate on their way to Sumatra, eleven of her company being killed. While Mason and Dixon's observation was aborted by war, clouds thwarted others.

Undismayed, the world's astronomers girded their lenses for the next passage, calculated to occur on June 3, 1769. This time seventy-six observational posts were spread around the globe. The Royal Society sent observers to northern Norway and Hudson Bay, while also funding Captain James Cook (1728–1779), the famed British explorer, to observe the transit from Tahiti.

Cook left England aboard the ninety-eight-foot *Endeavour* on August 26, 1768, taking with him a professional astronomer, Charles Green, and cratefuls of telescopes, clocks, and meteorological equipment. The French Ministry of Marine ordered all its commanders to refrain from interfering with Cook and immediately release him should they have detained him, since he "was doing important work for humanity." Cook duly arrived in Tahiti, without French interference.

With six weeks to the transit, he issued strict orders against the trading of metal objects with the island's females, who adorned their thighs with intricate tattoos of arrows and stars and who would happily exchange their favors for a nail or two (at first, it was "one ship's nail for one ordinary fuck,"[19] but hyperinflation soon set in). In their enthusiasm the crew of an earlier ship, the *Dolphin*, had pried so many nails from their craft as to render it almost unseaworthy. Despite Cook's best endeavors, every kind of metal object—cutlery, cleats, cooking utensils—kept vanishing.

The *Endeavour* returned to England on July 17, 1771. Cook's observations, along with everyone else's, were sent to Paris, where they were to be synthesized by the leading French astronomer of the day, Joseph Lalande. As he analyzed the results, it became clear that the host of observers had been unable to establish the exact moment when the edges of planet and star appeared to touch. As the black dot met the rim of the disk, it appeared to ooze into the Sun (whose edge has been theatrically designated "the Terminator"), a phenomenon that had caused the watchers to disagree by several seconds as to the moment of intersection.* Despite these variations (and the several problems caused by international time differences), Lalande managed to calculate the distance at 95 million miles, just two million miles off our present figure. In 1894 the American astronomer William Harkness would come up with the figure of 92,797,000 miles, at most only 158,000 miles below the modern value, which is variously estimated at between 92,955,887.6 and 92,750,600.02.

---

* This problem, known as the "Black Drop" effect, was probably due to refraction.

The most recent transit occurred on June 8, 2004. I witnessed its last hour, atop a friend's high-rise in mid-Manhattan. By then, as one observer was to put it, "the Sun's beauty spot had . . . shifted to the other cheek." The rim of Venus seemed uneven, due to the heat waves surging off the city. I recall having little sense of the planet, which seemed to be *within* the great star, an ant inching its way into a translucent orange. When it reached Terminator Point, it seemed as if a hole had been drilled in the Sun's bottom-right-hand quadrant, allowing its tiny visitor to escape. Today scientists can radar-range the planets, and deep-space probes have relegated transit campaigns to the history books. When the next transit takes place, on June 6, 2012, how many will realize the dramatic history of its tracking?

VENUS MAY HAVE preoccupied astronomers and governments, but the general public had moved on. New ways of playing with light—in camera obscura and camera lucida, magic box and trick lantern—were all the rage from Samuel Pepys's time on, while telescopes gained a particular hold on people's imagination. With the advent of yet more powerful models, the skies were becoming an open map. The American poet laureate Ted Kooser voices this new sense of advancing discovery:

> *This is the pipe that pierces the dam*
> *that holds back the universe*
> *that takes off some of the pressure,*
> *keeping the weight of the unknown*
> *from breaking through*[20]

Galileo's device had been used interchangeably for terrestrial and celestial purposes, a typical model being five or six feet in length. Magnification could be increased simply by extending the barrels; by the 1670s these were as long as 140 inches. A "telescope race" was on. However, this method of magnification had reached its limit; no further increases of power were possible until the 1730s, when James Hadley and others solved some of the practical difficulties that had plagued Newton's prototypes, and soon reflecting telescopes with primary mirrors up to six inches in diameter were being made. Sky searchers now had an arsenal of instruments: the achromatic lens (to counter defects in light dispersal), the optical glass, the micrometer (which measures small angles), and crosshairs (for fine alignment).

By 1780, in the hands of William Herschel (1738–1822), the reflecting telescope with parabolic ground mirrors enabled solar science to emerge as a specialized area of research. Building the most powerful telescopes of his time was only one of Herschel's accomplishments. The third child of an oboist in the Hanoverian foot guards, he had joined his father's band at fourteen and stayed with it until 1757, when his battalion became so heavily engaged in the Seven Years' War that music was put aside. On his father's advice, he absconded to England ("Nobody seemed to mind," he noted, although the elector of Hanover was also the British sovereign) and earned his living as a musical copyist, conductor, composer, music teacher, violinist, and organist in the fashionable spa of Bath. He soon became preoccupied with what he called "the construction of the heavens," fashioning hardwood telescope tubes "as elegant as cellos" that he topped off with magnifying eyepieces made of cocus, the wood used in oboes, all to maximize what he called "space-penetrating power."

In 1781 he became the first person to discover a planet when he identified Uranus, lying farther out than had ever been imagined: at a stroke its detection doubled the radius of the known solar system—and more than doubled Herschel's finances, allowing him to dedicate his life to astronomy. Summoned to an audience with George III, he wrote to his sister Caroline: "I will make such telescopes & see such things . . . that is, I will endeavour to do so." By 1783 this "middle-aged German musician residing in an English resort town,"[21] spending night after night on his observation platform, rubbing his face and hands with raw onions to keep out the cold, had established the general direction and speed of the Sun's movement in space, and analyzed the motions of seven bright stars to prove that part of their movement was due to the Sun's pull.

Two years later, with his sister as his long-suffering assistant, he used star counts to map the Milky Way, the results supporting Thomas Wright's proposition of 1750 that that galaxy was one vast spinning disk with the Sun situated within it. To ancient astronomers, the faint luminous band visible across the sky on clear nights was like a flow of milk yielded by some heavenly cow. The Greeks called it Kyklos Galaktikos, or "the Milky Circle," the road-building Romans Via Lactea. Although Democritus had been partially right in suggesting that the Milky Way consists of crowds of stars, Herschel showed that it was actually a huge aggregate of stars, nebulae, gas, and dust. "He has discovered fifteen hundred universes!" exclaimed the novelist Fanny Burney after visiting his Windsor observatory in 1786. "How many more he may find who can conjecture?"[22]

By 1800 Herschel had mapped some twenty-five hundred diffuse, cloudlike structures that he dubbed "nebulae" (the Latin word for "clouds"), classifying them and fixing their positions in the sky. He called the Orion Nebula, a knot of congealing gas sixteen hundred light-years* from Earth, "the chaotic material of future suns"—which is precisely right. The Sun's core he described as "a solid globe of un-ignited Matter."

In that same year—which also saw Beethoven's First Symphony and Wordsworth and Coleridge's *Lyrical Ballads*—he extended Newton's experiments on light by demonstrating that invisible "rays," as he was to name them, were to be found beyond the red end of the solar spectrum. One day he was writing up his experiments to quantify the relative heating power of different colors by letting a spectrum of light fall on a set of thermometers. To his surprise, he discovered that the greatest heat was generated beyond the red end of the spectrum, where nothing was visible. "Radiant heat," he reported, "will at least partly consist, if I may be permitted the expression, of invisible light." Building on Hooke's researches of a hundred years before, he had discovered infrared radiation—essentially the transmission of heat—which showed that the Sun's heat was mostly carried as invisible rays that behaved like light but could not be discerned by the eye.[23] Herschel's discovery would allow scientists to gauge from starlight how far off a star was, and how large.

Herschel also continued to be interested in the hardware of his science. While he and his colleagues had more than adequate tools to investigate the Sun's apparent motions, its distance from the Earth and the other planets, and the various oscillations of the Earth on its axis, they lacked the wherewithal for a really detailed study. When William's son John (1792–1871) followed him into astronomy, the two collaborated in creating a 20-foot-focal-length, 18.25-inch-aperture reflector, the first in a line of literally world-changing instruments.

One of William's biographers has characterized his achievement in transforming "the starry heavens from a static backdrop against which to measure planetary positions, into a vast dynamic region in which stars evolved from clouds of nebulous material."[24] Here Herschel was building upon the reflections of Immanuel Kant (1724–1804), whose *A General History of Nature and Theory of the Heavens* (1755) had proposed that the Sun and planets were formed by the condensation of a rotating disk of interstellar matter. Kant (himself a skillful lens grinder) added a "nebular hypothesis" of planetary formation, reasoning that diffuse nebulae—dim clouds of dust and gas that were only first being well observed in his lifetime—must collapse under the force of gravity. As they did so, they would spin out into a disk, which in time would

---

* A light-year, not a measure of time but a unit of length, is the distance that light can travel in one year—c. 5.9 trillion miles, or about 63,000 times the distance between Earth and Sun.

form itself into stars and planets. Kant, however, was no mathematician, and his ideas did not receive a convincing scientific basis until the great Pierre-Simon Laplace (1749–1827), who argued that the Sun and its system were formed by the gravitational collapse of an initially rotating diffuse gas cloud.

Throughout the nineteenth century, Laplace was cited as having provided the mathematical proof (not offered by Newton) that the solar system was built like a clock.[25] God had no place in such a universe. "Citizen First Consul, I have no need of that hypothesis," Laplace is said to have told Napoleon disdainfully, on being asked where the Creator fitted into his equations. By the time Laplace died in 1827, the battle lines were drawn for some of the greatest arguments of the nineteenth century. Astronomers had lost some of their awe of the Sun and no longer worried about its place in the Divine scheme. They had set about studying it as a star, and were wondering what it was composed of, how it acted on the Earth, what it could tell them about the rest of the universe. Until Newton's time it had been a great *pushing* object, imposing itself on mankind with its heat and blinding light. Newton had shown that the Sun actually *pulls* more—and has an impalpable power to organize everything around it.

CHAPTER II

ECLIPSES AND ENLIGHTENMENT

He . . . held out his right hand at arm's length towards the sun. Wanted to
try that often. Yes: completely. The tip of his little finger blotted out the sun's
disk. Must be the focus where the rays cross. If I had black glasses.
Interesting. There was a lot of talk about those sunspots when we were in
Lombard Street West. . . . Terrific explosions there were. There will be a total
eclipse this year; autumn some time.

—JAMES JOYCE, *Ulysses*[1]

WHEN I WAS A SCHOOLBOY OF PERHAPS TEN OR ELEVEN, ONE OF MY masters would end his class by making us keep silent for its final minute. It was his way of showing us how long time could feel. And so it did: those sixty seconds seemed to go on forever as we waited for the end-of-lesson bell, when we would burst out, like swimmers shooting to the surface.

Some forty years later, on November 23, 2003, I waited as the Moon moved across the Sun high in the heavens. I was standing with about sixty fellow eclipse-watchers on Antarctic ice three and a half miles deep, some way outside the Russian base of Novolazarevskaya—in Queen Maud Land, positioned at 70° 28′ W and 11° 30′ E. Ice surrounded us, its whole edge lit to a searing, fluorescent blue. The Sun's masking was to last one minute and forty-eight seconds, a brief enough period but one that would seem, again, to stretch for an eternity. Only this time I wouldn't want it to end.

My journey had started with a flight from New York City to Cape Town, South Africa, where a decommissioned Ilyushin-76 military cargo jet waited to take us on the six-and-a-half-hour flight to the ice strip near Novo. The night before takeoff, our group, ranging from an Arizona professor of astronomy to a pole dancer from Chicago realizing her life ambition, gathered for a briefing on conditions at base camp. There was the danger of being misled by mock suns, or sundogs, we were told—mirages created by light refracted through ice crystals; of "whiteouts," like being inside a Ping-Pong ball; and freakish "whirlies," compact, violent gusts that would scoop up any object and hurl it away. These, however, paled before the current weather conditions. Roaring winds of seventy-five miles an hour and temperatures of −18°F (−28°C) were making it too hazardous to travel. Continuous storms over the last twelve days had knocked down or blown away almost everything that had been set up for us: equipment left by a previous group was totally lost beneath the snow; even the two small transport planes were buried up to their wingtips. The site team had been unable to leave their tents for the last twenty-four hours, and the latest reports forecast a second, even fiercer front in two days' time. An eighteen- to thirty-hour weather break might give us just enough of a window to fly in, witness the eclipse, and lift off ahead of the second storm, but the official advice was plainly to abandon the venture. Our chain-smoking Russian pilot said we might have a chance. Next to me, someone whispered that no civil aviation plane had ever survived such an attempt. Our pilot stubbed out one more cigarette. "Well, I am ready if you are." And so we went.

We touched down in bright afternoon sunshine with a few hours to spare. Back in Cape Town, I had asked one of our group, Carol, a violinist with the Scottish Symphony Orchestra, if she would play for us if we smuggled in a violin. She had looked doubtful, but nodded. And that was how, within an hour of landing, the snowfields rang to Mozart's G Major Concerto, the Blue Danube Waltz, Elton John's "Something About the Way You Look Tonight," and

Solar eclipse, November 2003, Antarctica. The music being played was "Something About the Way You Look Tonight"—an appropriate choice, with just six hours to go before the eclipse.

Massenet's "Meditation" from *Thaïs*. A couple I had christened "the Lovers"— a New York–based British businessman and a svelte, dark-haired Parisian lawyer—suddenly emerged from our one tent, she in just a purple slip and ski boots, and tangoed in a tight embrace, their movements surprisingly sensuous in the cold. Behind them brooded the Ilyushin-76, and beyond that nothing but an endless expanse of frozen continent.

Although little wind was blowing by now, it was numbingly cold—about –8°F (–22°C). We assembled about 10:15 that evening and would remain in place, thickly muffled, until after 11:10 P.M. Through my protective shades, I watched the Moon devouring the Sun; this phase took about three-quarters of an hour. Then, about ten minutes before totality, great irregular alternating bands of shadow and vivid light began to pulse toward us from northwest of the star, like smoke rising from the snow: patterns made by rays from the last points of the solar crescent interacting with the Earth's upper atmosphere.

Just before total eclipse, we saw great red flares bursting off the Sun's surface, at about one, five, and eight o'clock. Our shadows grew darker and darker, until the Moon disk had finally put out the entire star, and all that remained was a black circle hanging in the sky, set against a glowing halo like a giant sunflower. The horizon turned a deep red, contrasting with an upper crest of brilliant green. I took off my protective glasses—the Sun was almost below the horizon now—and we watched its spectacular dance with the Moon and the Earth's edge. Slowly satellite parted from star, and a crescent reappeared, overwhelming the corona and forming a beautiful diamond ring. We were the first human beings in history to witness a total solar eclipse from the Antarctic.

WE WERE BUT the latest witnesses of what remains one of the great marvels of nature. And while we looked on in awe, we no longer did so possessed by superstition or fear. Science has not only enlightened us; it has learned to use eclipses for its advancement—a revolution indeed.

The tradition of pursuing eclipses in far-off lands is part of that revolution, one that began in the eighteenth century. Its first great practitioner was Captain Cook, of *Endeavour* fame. He had become intrigued by eclipses while reading Charles Leadbetter's *Compleat System of Astronomy*, which declared that a

Person being well skill'd in Astronomy . . . may, by the knowledge of Eclipses . . . determine the true Difference of Meridians between *London*, and the Meridian where the ship then is; which reduced to Degrees and Minutes of the Equator, is the true Longitude found at Sea.[2]

In August 1766, Cook experienced his first solar eclipse—off the Burgeo Islands along the Newfoundland coast. He was lucky to see it at all, as a fog lifted only at the last moment, but he seized the opportunity to work out the longitude of Newfoundland. His calculations were accurate enough that his reckoning was only three miles off across the whole width of the Atlantic; his paper to the Royal Society, delivered in April 1767, was so acclaimed that it secured him the official backing needed for his great voyages of discovery.

Over the next eleven years he would see the Sun go out four more times—each time totally: on May 10, 1771, off Ascension Island, in the South Atlantic; on September 6, 1774, off New Caledonia; on July 5, 1776, from Tonga (being twelve hours ahead, the occlusion coincided with the eclipse of British power in North America; the Sun must have darkened almost as the Declaration of Independence was ratified); and his last on December 30, 1777, from a place he christened Eclipse Island, off Christmas Island (the one in the central Pacific, about four thousand miles northeast of Australia). He knew from Tobias Meyer's *Nautical Almanac*, published in 1767, that wherever the Sun and Moon had identical coordinates, there he would find an eclipse. By the time he was murdered on Hawaii in February 1779, he had observed more solar eclipses than anyone in history.

The average eclipse lasts just under three minutes. The longest ever recorded was over the Indian Ocean in 1955 and lasted 7.08 minutes. Because Earth and Moon have to be precisely aligned for a total eclipse to occur over any given location, such events are relatively rare. A maximum of seven per year is possible (five solar and two lunar, or four and three); the minimum possible consists of two solar eclipses only. Within the whole solar system, Earth is the only planet to experience such pronounced shadowings. The likelihood of witnessing a total eclipse from any arbitrarily selected point is once every 375 years. To view an eclipse at all, one must be in a narrow band traced out by the Moon's shadow, at its widest no more than 110 miles. The shadow moves at 1,060 to 2,100 miles an hour, and completes its journey in three to four hours.*

As the scientific view of the universe deepened and grew in influence, observers began to speculate on a range of associated phenomena. On May 15, 1836, an annular eclipse passed over northern Britain. A retired stockbroker

---

* Every part of Britain and Ireland has witnessed a total eclipse at some time over the last two thousand years, except for a small tract at Dingle on Ireland's west coast. Yet the obscure island of Blanquilla, off northern Venezuela, was darkened by two occlusions within three and a half years, and a spot in the Java Sea saw four eclipses within fourteen years; a region near 29° E and 23° 30′ N in the desert of southern Egypt will see five such eclipses over the course of thirty-two years—February 14, 2325; June 20, 2327; February 5, 2334; July 31, 2353; and November 23, 2356.[3]

who had helped found the Royal Astronomical Society, one Francis Baily of Inch Bonney, near Jedburgh, described what have come to be known as "Baily's Beads." Just before the Moon started eating into the Sun, he wrote, a row of lucid points, like a bright necklace, irregular in size and spacing, formed around that part of the Moon that was about to enter the Sun's disk. What he was seeing were actually the last rays of sunlight streaming along those of the Moon's innumerable valleys that present themselves head-on to the Earth at the moment of eclipse; but it was the suddenness that impressed:

> Its formation, indeed, was so rapid that it presented the appearance of having been caused by the ignition of a fine train of gunpowder. Finally, as the Moon pursued her course, the dark intervening spaces . . . were stretched out into long, black, thick, parallel lines, joining the limbs of the Sun and Moon; when all at once they *suddenly* gave way, and left the circumference of the Sun and Moon in those points . . . comparatively smooth and circular, and the Moon perceptibly advanced on the face of the Sun.[4]

This same phenomenon had been described fifty-six years before, at a total eclipse over Penobscot, Maine, on October 27, 1780; but it was Baily's vivid description that secured his place in the poetry of astronomy.

Next to come under scrutiny were two happenings that we had seen in Antarctica: solar prominences, which look like reddish filaments, or blobs, in the lower corona but which are actually large accumulations of relatively cold gas suspended high in the Sun's atmosphere by its own magnetic field; and coronas themselves, the haloes of white light that appear around the Moon at totality—"Heaven's awful rainbow," in Keats's phrase. Coronas were described as early as A.D. 96, the term first being applied in 1563. At the eclipse of April 9, 1567, the great Jesuit astronomer Christopher Clavius reckoned them as the Sun's uncovered rim. Kepler disproved this, but only with the even less accurate explanation that the halolike effect emanated from the Moon itself; it would take another three hundred years to discover that the light came from the ionized gases that form the Sun's turbulent outer atmosphere. They have three components: polar rays, which extend into space for at least several times the Sun's diameter; and the inner and outer equatorial coronas, the long elliptical streamers that, concluded Mrs. D. P. Todd, the foremost amateur astronomer of her day, "delicately curving and interlacing, may tell the whole story of the Sun's radiant energy."[5] If only she had known how insightful her prediction would prove to be. The corona's pearly light is less than a millionth as intense as the central Sun's, and so can be seen only during total eclipses, but such observations tell us how other stars behave, whatever their distance from Earth.

OVER THE LAST five hundred years, scientifically prepared man has used eclipses to take advantage of the superstitions and fears of less informed peoples. Examples abound both in real life and in fiction. During his voyage of 1504, Christopher Columbus was stranded while reprovisioning in Jamaica. Some of his officers mutinied, and Columbus and the men loyal to him found themselves at the mercy of the island's Arawak inhabitants, who refused to bring food, leaving them to eat rats. From his astrological charts, he knew that a lunar eclipse was imminent, so he waited for the fateful day and sent a message to the Arawak that in his anger he would make the Moon rise "inflamed with wrath"; on cue, it began to disappear. The stunned islanders beseeched his mercy, whereupon he retreated to his cabin, used a half-hour sandglass to time how long the eclipse would last, then emerged to declare that God had granted his prayer. The Moon was restored, the ship restocked.

H. Rider Haggard (1856–1925) put the same idea to use in his 1885 classic, *King Solomon's Mines.* The first popular novel in English to be set in Africa, it recounts the adventures of three British gentlemen—the mighty hunter Allan Quatermain and his companions Sir Henry Curtis and Captain Good, R.N., together with their formidable attendant, Umbopa—searching for the fabled mines of the title in the imaginary kingdom of Kukuanaland, where they soon find themselves at the mercy of the great King Twala. Conferring with his companions, the blunt-spoken Captain Good pulls out an almanac: "Now, look here, you fellows, isn't tomorrow the fourth of June?"

Sure enough, it records that a total solar eclipse will begin at 11:15 Greenwich Mean Time, "Visible in these Islands—Africa, etc. . . . There's a sign for you. Tell them you will darken the sun tomorrow."

The intrepid quartet attempt to bluff their way through, but when Quatermain rhetorically inquires of King Twala whether any man can put out the Sun, the great chief laughs out loud. "That no man can do. The Sun is stronger than man who looks on him." Quatermain announces that this is precisely what his company will bring about, and next day at the appointed moment, the Sun starts to disappear. Haggard was more sympathetic than most of his contemporaries to African culture, and has the anciently evil witchdoctress, Gagool, cry to the trembling multitudes that all will be well: "I have seen the like before; no man can put out the sun; lose not heart; sit still—the shadow will pass." But the dying Sun sweeps her power away. "The dark ring crept on. Strange and unholy shadows encroached upon the sunlight, an ominous quiet

The eclipse trick is arguably one of the central tropes of the clash of civilizations. In 1949, Hergé published The Prisoners of the Sun, *in which Tintin, Captain Haddock, and Professor Calculus fall captive to an Inca kingdom that has survived into the modern world. The presiding chieftain commands that the trespassers be burned alive on a pyre lit by the Sun, but Tintin realizes that an eclipse is imminent, and at the crucial moment shouts out, "O God of the Sun, sublime Pachacamac, display thy power, I implore thee! If this sacrifice is not thy will, hide thy shining face from us!" The Sun goes out and the Indians run about in terror.*

filled the place, the birds chirped out frightened notes, and then were still; only the cocks began to crow." In the gathering stillness, a boy cries out, "The wizards have killed the sun. We shall all die in the dark." The Englishmen's power seems absolute, and the chastened king accedes to their demands.*

Haggard may well have inspired Mark Twain's *A Connecticut Yankee in King Arthur's Court,* in which a modern man, Hank Morgan, struck on the head by a crowbar, is precipitated into Arthurian England. It is June 19, 528, and the locals are again hostile, threatening to burn the hapless Morgan until he, too, learns that a solar eclipse is about to occur. Twain's satire appeared in 1889,

---

* H. Rider Haggard, *King Solomon's Mines* (Barre, Mass.: Imprint Society, 1970), pp. 172–74 and 185–86. This episode would become infamous, as it was fraught with scientific mistakes: the path of totality could never cover both "these [British] Islands" and southern Africa; there was a full Moon the night before the eclipse, although solar eclipses can only come about when the Moon is new, i.e., sunward of the Earth; and totality is described as lasting "an hour or more." In later editions, the chapter was rewritten, and the solar eclipse became lunar, with a consequent dimming of dramatic power.

less than four years after Haggard's novel was published in the United States, but it is also possible that Twain pinched the idea straight from history. He has his hero explain, "It came into my mind, in the nick of time, how Columbus, or Cortez, or one of those people, played an eclipse as a saving trump once, on some savages, and I saw my chance. I would play it myself, now; and it wouldn't be any plagiarism, either, because I should get it in nearly a thousand years ahead."[6]

During the nineteenth century, superstition was still strong enough that on September 7, 1820, the House of Lords suspended its divorce trial of Queen Caroline for the duration of an eclipse. But it was usually among the less educated that such atavistic fears persisted—giving rise to visions and religious inspiration. In the summer of 1831, a God-fearing Virginia slave acted on a vision vouchsafed him during the previous February's darkening: on the night of August 21, Nat Turner and seven companions massacred their master's household, then ravaged several other slave owners' houses nearby, killing more than fifty men, women, and children. After remaining at large for nine days, Turner was captured, tried, and hanged, but not before he had made his "Confession" about the effect the eclipse had had on him:

> By signs in the heavens that it would make known to me when I should commence the great work—and until the first sign appeared, I should conceal it from the knowledge of men—and on appearance of the sign I should rise and prepare myself, and slay my enemies with their own weapons. . . .[7]

William Styron imagined what Turner might have experienced in his 1967 novel, *The Confessions of Nat Turner*:[8] "I looked up then to see the sun wink slowly out as it devoured the black shape of the moon. There was no surprise in my heart, no fear, only revelation, a sense of final surrender."

The same year that Twain published *A Connecticut Yankee*, a total eclipse ignited a major new religion. It originated with an orphaned Paiute Indian, born in Nevada around 1858, adopted by a local farmer, and given the unassuming name of Jack Wilson. Like Turner, he grew up a devout child, and by his early twenties had become a local holy man, impressing his neighbors with the wisdom of his utterances—and taking a new name, Wovoka (which translates as "Cutter"—he had probably been a woodsman in his teens). In late 1888 he fell critically ill with scarlet fever. An eclipse fell at the New Year, and when the Sun reappeared, several Paiutes burst into the prophet's cabin to find him in a deathly coma. "The very hour of his translation," opines his principal biographer, Paul Bailey, "had been a climactic one. To be ushered out of mortality in the awful moments of the sun's death had all the portents of divine timing and celestial intervention. And the credulous Paiutes, after ascertaining with knife and fire on Wovoka's seemingly lifeless flesh that it was no ordinary sleep, were ready enough to ascribe the sun's escape from the black monster to the high assistance of their heaven-borne prophet."[9]

When Wovoka miraculously revived, he told his followers that he had been carried away to a green land where the dead lived again, and prescribed the special dance that would summon up their spirits. "Pretty soon now, the earth shall die; but Indians must not be afraid, because Earth will become live again, just like Sun died and came live again." He promised that the white race would be swept away by a torrent of mud and water; his people would become young and free of disease and pain; the land would be fertile, game would return in abundance, and an Indian paradise would repossess the country.

Such a vision was difficult to resist. In a matter of months the Washoe, the Bannock, and the Pit River tribes were treading the slow measures of what white men called the "ghost dance." Apostles from the Paiute carried his message to the Walapai, Cohonino, Mohave, and even the far-off Navajo. "No religious movement in history," asserts Bailey, "ever swept with greater speed or more dramatic impact across the face of the land." Soon Mormon settlers were shuffling alongside their native brethren as the new religion swept south, east, and north. The Bureau of Indian Affairs and the Interior Department became

Parisian women crowd together to view the eclipse of August 26, 1892. Some are using special filters to avoid eye injury.

A partial eclipse seen against a crescent on Islamabad's
grand Faisal Mosque, Pakistan, August 1, 2008

alarmed; but they need not have worried. Wovoka unwisely fixed a precise date—the year 1900—for this transformation. When nothing happened, the great movement that had been set alight by the eclipse of '89 quickly collapsed. By the time he died in 1932, Wovoka was little more than a faded footnote to history.

HOWEVER, NOTHING CAN get in the way of the sense of wonder we still feel when under the actual shadow of an eclipse. Novelists from Fenimore Cooper to Dorothy L. Sayers have recorded such scenes, either in their fiction or in their private notebooks. One of the finest descriptions, capturing that sense of momentary contact with the full mystery of created things, comes from Virginia Woolf, in Yorkshire to witness the 1927 eclipse, the first totality viewable in England for more than two hundred years. The next day, June 30, she wrote in her diary:

> Now I must sketch out the Eclipse. . . . Rapidly, very very quickly, all the colours faded; it became darker and darker as at the beginning of a violent storm; the light sank and sank; we kept saying this is the shadow; and we thought now it is over—this is the shadow; when suddenly the light went out. We had fallen. It was extinct. There was no colour. The Earth was dead. That was the astonishing moment; and the next when as if a ball had rebounded the cloud took colour on itself again, only a sparky ethereal colour and so the light came back. I had very strongly

the feeling as the light went out of some vast obeisance; something kneeling down and suddenly raised up when colours came. They came back astonishingly lightly and quickly beautifully in the valley and over the hills. . . . It was like recovery. We had been much worse than we had expected. We had seen the world dead. This was within the power of nature.[10]

I know of no other record so moving—yet the eclipse lasted only twenty-three seconds.* Length of time doesn't seem to matter—the experience is generally memorable. Isaac Asimov relates how, on June 30, 1973, he came to witness a total eclipse off the coast of West Africa aboard the cruise ship *Canberra.* "What impressed me most was its ending. A tiny dot of bright light appeared, and suddenly spread out in half a second to become too bright to look at without filters. It was the sun coming back with a roar—and that was the most magnificent astronomical spectacle I ever saw."[11] The paleontologist Stephen Jay Gould—like Woolf and Francis Baily—stresses the quality of *suddenness:* "The sky turns off as if a celestial janitor threw a switch. For the sun is powerful, and a fraction of one percent of sunlight is daytime, while totality is night time—and the transition is a moment, a twinkling of an eye." He adds, "In ordinary conditions, it takes several hours for a person's color discrimination to disappear. In an eclipse it happens in a flash—one moment there is color, then it dies out in a rush."[12]

For the eclipse of 2001, Gould, having ordered all his students to take time off to observe it, then journeyed from his Boston home (where the skies were shrouded by rain) to New York City to witness it in full. Later he describes "this remarkable event":

In this age of artificially induced full-body shake-me-ups . . . we hardly think that anything so subtle, albeit pervasive, as the character of surrounding sunlight could move our passions, or even invite our notice. . . . The sky did not darken precipitously over New York on May 10. But we are exquisitely sensitive to the usual character of light, even though we may not explicitly credit our awareness, and may not be able even to state what feels so odd. . . . The day turned eerily somber, while sunlight continued to reign—and people noticed, and trembled ever so slightly . . . noticed and stood still as a sky full of daylight darkened to

---

* Woolf would use her experience again in her essay "The Sun and the Fish," and once more, in 1931, when she re-created her moment of rapture in *The Waves,* in the visit to Hampton Court that marks the novel's climax.

the level of a clearly nonexistent thunderstorm. A woman said to her friend, "Holy shit, either the world is about to end, or it's going to rain—and it sure as hell ain't going to rain."

If these are unusual sentiments for a scientist, they do demonstrate that there is still something primal in our response. In John Updike's story "Eclipse," a man wanders into his backyard one afternoon with his two-year-old daughter to watch a near-total eclipse: "The day was half cloudy, and my impression had been of the Sun struggling, amid a furious knotted huddle of black-and-silver clouds, with an enemy too dreadful to be seen, with an eater as ghostly and hungry as time."

He calls out to his elderly neighbor sitting on her porch how she likes the eclipse. "I don't," she answers, adding, "They say you shouldn't go out in it. There's something in the rays." At last the darkness passes:

> *Superstition,* I thought, walking back through my yard, clutching my child's hand as tightly as a good-luck token. There was no question in her touch. Day, night, twilight, noon, were all wonders to her, unscheduled, free from all bondage of prediction. The Sun was being restored to itself and soon would radiate its influence as brazenly as ever—and in this sense my daughter's blind trust was vindicated. Nevertheless, I was glad the eclipse had passed, as it were, over her head; for in my own life I felt a certain assurance evaporate forever under the reality of the Sun's disgrace.[13]

MODERN TECHNOLOGY HAS helped us to understand parts of our own world through eclipses. In 1872, the great scientist Pierre-Jules-César Janssen (1824–1907) observed an eclipse at Guntur, in northeastern India, and just before and after totality discerned a yellow spectral line in the solar prominences, so bright as to indicate some previously unknown chemical element. The English astronomer Norman Lockyer independently arrived at the same conclusion, naming the new presence "helium" in the belief that it was to be found only in the Sun. By 1895, it had been isolated and identified as a substance found on Earth as well. Further, studies of past eclipses have revealed repeated slight changes in the Earth's rotation, and variations in its spin rate.[14] The preeminent scientific employment of eclipses, however, was made at the height of the First World War. Einstein had just published his full general theory of relativity, part of which stated that gravity would affect light, so that light emanating from a given source would not travel in a straight line.[15] Although this was

This photograph, titled Fifty Crows, *shows a total solar eclipse as seen from Chiapas, the southernmost state of Mexico, in 1991.*

a radical modification of Newtonian ideas, it explained a phenomenon that had long troubled astronomers: the discovery that the perihelion—the closest orbital point to the Sun—of the planet Mercury was precessing slightly faster eastward each year than the gravitational effects of the other planets could account for. Einstein's argument that gravity was a field, not a force, explained the discrepancy. On making his discovery, he wrote, "for three days I was beside myself with joyous excitement."[16] But it remained to be proved.

In 1917 a professor of astronomy in the neutral Netherlands sent a copy of Einstein's paper to the Royal Astronomical Society, where it was read by both the leading British astrophysicist of the day, Arthur Eddington (unpreoccupied by wartime duties because a strict Quaker), and by the astronomer royal, Sir Frank Dyson. The latter, a specialist in solar eclipses, realized that light's relationship to gravity could be tested by studying the stars visible in the immediate neighborhood of an eclipsed Sun. Were Einstein right, these would have their light bent by the Sun's gravitational field, and would show up on photographs as slightly displaced from their positions in photographs taken when the Sun was elsewhere in the sky.[17]

A solar eclipse was due on May 29, 1919, and Dyson, studiously ignoring the fact that he was trying to prove a theory that emanated from a scientist in perfidious Germany (feelings against which ran so high that less than two months after the eclipse, when the International Astronomical Union was formed, German and Austrian scientists were specifically excluded), won government funding for a two-pronged expedition, which sailed at the beginning of March, parting company at Lisbon. Eddington's party traveled to the Portuguese island of Principe, off West Africa; the other set off for Sobral, in northeastern Brazil, farther along the path of totality.

By how much would gravity bend light? Einstein's equation predicted that a ray just grazing the surface of the Sun would be pulled by 1.745 seconds of arc—twice the deflection assumed by classical Newtonian theory. (An arcsecond is 1/3,600 of a degree, so one is dealing with minute differences, hence the difficulty of measuring them.) Before he left London, Eddington, his principal assistant, E. T. Cottingham, and Dyson talked late into the night, and Cottingham asked what would happen if the eclipse photographs revealed not Newton's, not Einstein's, but *twice* Einstein's deflection. "Then Eddington will go mad," replied Dyson breezily, "and you will have to come home alone."

By mid-May, Eddington and his party were in place. When the pinpoints of five stars came into view, Eddington had eight minutes to take his pictures: "I did not see the eclipse, being too busy changing plates. . . . We took sixteen photographs." Over the next six nights he developed the plates, comparing them with photographs of the same stars taken when they had not been close to the Sun. In the first ten shots, thin cloud, while not enough to obscure the eclipse, was sufficient to mask the crucial stars; but the next two plates were usable, and quite enough to supply the proof he was looking for. On the evening of June 3, Eddington turned to Cottingham: "You won't have to go home alone." The two sets of photographs differed by almost the exact amount predicted by Einstein.

The expedition won headlines everywhere, the London *Times* declaring, REVOLUTION IN SCIENCE / NEW THEORY OF THE UNIVERSE / NEWTONIAN IDEAS OVERTHROWN, and *The New York Times* adding jocularly, STARS NOT WHERE THEY SEEM OR WERE CALCULATED TO BE, BUT NOBODY NEED WORRY, for Einstein knew where they were. It made the forty-year-old physicist a global hero overnight, and was also the first major instance of restored international scientific collaboration before a peace treaty had even been signed.[18]

Eddington (who years later would reconfirm general relativity with his own magisterial work on the behavior of light under the intense gravity of white dwarfs) celebrated by producing a pastiche of Fitzgerald's translation of the *Rubáiyát of Omar Khayyám*, two of whose verses run:

> And this I know; whether Einstein is right
> Or all his theories are exploded quite,
> One glimpse of stars amid the Darkness caught
> Better than hours of toil by candle-light. . . .

Oh leave the Wise our measures to collate.
One thing at least is certain, light has weight
One thing is certain, and the rest debate—
Light-rays, when near the Sun, do not go straight.[19]

The rest of the scientific world also rejoiced, if not so impishly. "It is doubtful," reflects Thomas Crump in his history of eclipses, "whether eclipse astronomy will ever again achieve a result of such cosmic importance."[20] It has yet to do so.

CHAPTER 12

THE SUN DETHRONED

GRAND MASTER: Oh, if you knew what our astrologers say of the coming age, and of our age, that has in it more history within one hundred years than all the world had in four thousand years before!

—TOMMASO CAMPANELLA (1568–1639), *The City of the Sun*[1]

I am greatly relieved that the universe is finally explainable. I was beginning to think it was me.

—WOODY ALLEN[2]

THE SUN DOES NOT STAND STILL, OR EVEN MOVE IN A SIMPLE PATTERN. At different times, and under different examinations, it may appear a solid orb, a ball of fire, or a constantly discharging source of winds, flares, spirals, and radioactive particles. Its latitudes rotate at different speeds, and its whole surface surges up and down about two and a half miles every 160 minutes—although even to talk about the Sun's "surface" is misleading since, being largely a collection of gases, it has none. Seen from other planets, it is larger or smaller than when viewed from Earth; and its atmospheric effects are different on Mercury, say, than on the torrid mists of Venus, and different again on our planet. By the end of the eighteenth century, astronomers had recognized that it was simply one star among many—just the one that happened to be nearest to us—and had estimated its distance, size, mass, rate of rotation, and movements in space to within 10 percent of today's values. But they continued to gnaw away at other problems.

What was going on inside it? What caused it to shine? How old was it, and how did it relate to other heavenly bodies? Only now were they finding many of the answers. Between 1800 and 1950, aided and abetted by advances in technology, the Industrial Revolution, and a new enthusiasm for scientific inquiry, hardly a year went by without astronomers, physicists, chemists, and geologists finding something that added to our understanding.

In just the first half of the nineteenth century, for example, absorption lines in the solar spectrum were mapped, and electromagnetic induction and electromagnetic balance were being discovered, both of which increased our knowledge of how the Sun draws us to it. During the same period, the Sun's energy output—the "solar constant"—was measured, leading to a greater appreciation of both its temperature and its influence on climate. In 1860, the Vatican's chief astronomer photographed the eclipse of July 18, showing that the corona and prominences were real features, not optical illusions or deceivingly illuminated mountains on the eclipsing Moon.

The list of specific discoveries runs close to two hundred. In just the single decade 1871–80, for instance, came electromagnetic radiation, the solar distillation of water (the Sun as purifier), and the first radical reestimate of the Sun's age at 20 million years and of the length of time it would have taken to contract to its current size—both of which were to touch off a major debate between scientists and literal interpreters of the Bible. Other discoveries were less contentious: The Sun's surface temperature was estimated at 9,806°F (5,430°C), while its core was reckoned to be gaseous, with temperatures steadily decreasing from its center to the surface. Several new instruments (the heliospectroscope, star spectroscope, telespectroscope) were invented, and all fixed stars were found to display only a few combinations of spectra, depending on their physical-chemical nature, "an achievement of as great significance as Newton's law of gravitation."[3]

In recent years writers such as Dava Sobel, Timothy Ferris, and Bill Bryson, as well as popularizing scientists including Carl Sagan and Stephen Hawking, have demystified such esoteric subjects as dark matter (invisible but still exercising attraction, whose presence we infer from its gravitational pull, though we don't know what it is) and black holes (familiar, but still difficult fully to imagine). But what about deuterium, the ionosphere, even the Coriolis effect? Such words and phrases are part of a discourse that can seem closed to most of us. But two or three of the main strands of research can be braided together to show how, between 1800 and 1953, our knowledge of—and our feelings about—the Sun were transformed.

TOWARD THE MIDDLE of the nineteenth century, the French philosopher Auguste Comte was asked what would be forever impossible. He looked up at the stars: "We may determine their forms, their distances, their bulk, their motions—but we can never know anything of their chemical or mineralogical structure; and, much less, that of organized beings living on their surface." Within a surprisingly few years mankind would have grasped those very truths. To put into context how quickly research expanded and focus shifted: in 1800 only one stellar "catalogue of precision" was available; in 1801, J. J. Lalande published one describing 47,390 stars; and in 1814 Giuseppe Piazzi added another 7,600. Others followed; between 1852 and 1859 alone some 324,000 were registered. Photographic charting began in 1885; and by 1900 the third and concluding volume of a work appeared that located a massive 450,000 stars, the fruit of collaboration between scientists in Groningen and John Herschel in Cape Town. It became abundantly clear that the Sun did not hang there as the supreme object but was just one more star, an unspectacular one at that. Scientists could now concern themselves with this multitude of other suns, many far larger and heavier than our own.*

Even this piece of knowledge was to be massively augmented: in the predawn hours of October 6, 1923, at the Mount Wilson Observatory, Edwin Hubble was examining a photograph of a fuzzy, spiral-shaped swarm of stars known as M31,[†] or Andromeda, which was assumed to be part of the Milky Way, when he noticed a star that waxed and waned with clockwork regularity, and the longer it took to vary, the greater its intrinsic brightness. Hubble sat down and calculated that the star was over nine hundred thousand light-years away—three times the then estimated diameter of the entire universe! As *National Geographic* reported the story: "Clearly this clump of stars resided far beyond the confines of the Milky Way. But if Andromeda were a separate galaxy, then maybe many of the other nebulae in the sky were galaxies as well. The known universe suddenly ballooned in size."[6]

Today we know that there are at least 100 billion galaxies, each one harboring a similarly enormous number of stars. The Earth had already been de-

---

* In the first quarter of the nineteenth century a German astronomer formulated Olbers' Paradox: were there to be an infinite number of stars in the universe there would be no night, the sky would be one huge glare, and we would be drowned in—indeed, could not have come into existence under—such overwhelming stellar energy.[4] A good explanation of why this is not so is worked out by fifteen-year-old Christopher Boone, the autistic narrator of Mark Haddon's 2003 novel, *The Curious Incident of the Dog in the Night-time:* "I thought about how, for a long time, scientists were puzzled by the fact that the sky is dark at night, even though there are billions of stars in the universe and there must be stars in every direction you look, so that the sky should be full of starlight because there is very little in the way to stop the light reaching Earth. Then they worked out that the universe was expanding, that the stars were all rushing away from one another after the Big Bang, and the further the stars were away from us the faster they were moving, some of them nearly as fast as the speed of light, which was why their light never reached us."[5] One should further add that no star shines for more than around 11 billion years.

† Certain objects have the letter *M* before their designated numbers in honor of Charles Messier, the French astronomer who in 1774 published a catalogue of 45 deep-sky sightings such as nebulae and star clusters. The final version of the catalogue was published in 1781, by which time the list had grown to 103.

moted by the realization that it circled the Sun, and not the other way around; William Herschel and his son, John, with their revolutionary telescopes, had revealed the riches beyond our solar system, and unwittingly demoted the Sun as well. Here was a third or even a fourth diminution: that not only were we orbiting what was just a minor star among a multitude in the Milky Way, but that the Milky Way itself was but one galaxy of an untold number. For scientists, stellar astronomy was the future; minor stars need not apply. As A. E. Housman, poet and classicist and also a keen student of astronomy, tartly observed, "We find ourselves in smaller patrimony."[7]

THIS DISPLACEMENT WAS occurring alongside the centuries-long and frequently vehement argument over the age of the Earth—and thus the age of both the Sun and the universe, an issue that provoked intense interest among astronomers, theologians, biologists, and geologists. The date of Creation had been made particularly exact by the elderly Irish Protestant archbishop James Ussher (1581–1656), who took all the "begats" in Genesis and the floating chronology of the Old Testament and—with a little helpful tweaking from Middle Eastern and Mediterranean histories, not least the Jewish calendar, which lays it down that Creation fell on Sunday, October 23, 3760 B.C.—arrived at the conveniently round figure of four thousand years between the Creation and the (most likely) birth date of Christ in 4 B.C.

Were it not for the London bookseller Thomas Guy, Ussher's dating might have sunk into obscurity, as had those of the hundred or so biblical chronologers before him. By law, only certain publishers, such as the Cambridge and Oxford university presses, were allowed to print the Bible, but Guy acquired a sublicense to do so, and in a moment of marketing inspiration printed Ussher's chronology in the margins of his books, along with engravings of barebreasted women loosely linked to Bible stories. The edition earned Guy a fortune, enough to endow the great London hospital that bears his name. If this were not enough, in 1701 the Church of England authorized Ussher's chronology to appear in all official versions of the King James Version. The "received chronology" soon became such an automatic presence that it was printed well into the twentieth century.

As the historian Martin Gorst puts it, "The influence of his date was enormous. For nearly two hundred years, it was widely accepted as the true age of the world. It was printed in Bibles, copied into various almanacs and spread by missionaries to the four corners of the world. For generations it formed the cornerstone of the Bible-centered view of the universe that dominated Western thought until the time of Darwin. And even then, it lingered on."[8] Gorst recalls coming across his grandmother's 1901 Bible and finding, set opposite the opening verse of Genesis, the date and time for when the world began: 6:00 P.M. on Saturday, October 22, 4004 B.C.*

Even in Ussher's time, however, freethinkers had begun to question the accepted chronology. Travelers returned from far-off lands with reports of histories that went back much further than 4004 B.C., and with the advent of new forms of inquiry based on scientific principles—*Nullius in Verba* (Take Nobody's Word for It) runs the motto of Britain's Royal Society—it became the turn of natural philosophers to debate the age of the globe.

In 1681, a fellow don of Newton's at Cambridge, Thomas Burnet, published his bestselling *Sacred Theory of the Earth,* which argued that the world's great mountain ranges and vast oceans had been shaped by the Flood of Genesis; he got around the Old Testament statement that Creation had taken only six days by quoting Saint Peter's pronouncement that "one day is with the Lord as a thousand years." A shelf of books appeared offering similar explanations. Then, later that same decade, two British naturalists, John Ray and Edward Lhuyd, after examining shells in a Welsh valley and ammonites on the northeastern English coast, concluded that both groups of fossils came from species that would have needed far more than 5,680-odd years in which to live and die out in such numbers. Their researches aroused a short-lived flurry of interest, followed by a period of neglect until the London physician John Woodward (1665–1728) posited that fossils were the "spoils of once-living animals" that had perished in the Flood. But this theory too was met with ridicule: faith in Ussher's chronology endured.

It was Edmond Halley who, in 1715, became the first person to suggest that careful observation of the natural world (such as measuring the salinity of the oceans) would provide clues to the Earth's age. His ideas were eagerly adopted by Georges-Louis de Buffon (1707–1788), who in his *Histoire naturelle* (1749) argued that the Earth was tens of thousands of years old, an idea that "dazzled the French public with a timescale so vast they could barely comprehend its enormity."[10] While Buffon did not sample the oceans, he did believe the Earth began not in the biblical single moment but as the result of a comet colliding with the Sun, the debris from this explosion coalescing to form the planets.[†]

---

* As recently as 1999, a Gallup poll reported that 47 percent of Americans believed God created the human race within the last ten thousand years. Stephen Hawking notes in *A Brief History of Time* that a date of ten thousand years fits curiously well with the end of the last Ice Age, "which is when archaeologists tell us that civilization really began."[9]

† Buffon simply ignored the biblical account of Creation, possibly encouraged by the appearance, in 1770, of the first printed blunt denial of any divine purpose, *Système de la nature,* by the forty-seven-year-old French philosopher Paul-Henri Dietrich d'Holbach.

Theologians at the Sorbonne erupted in fury, and Buffon was forced to publish a retraction—privately declaring, "Better to be humble than hanged." But he would not be kept quiet for long. When another Frenchman, the mathematician Jean-Jacques Dortous de Mairan, demonstrated that the Earth did not get its heat from the Sun alone but contained an inner source, Buffon was galvanized into writing a history of the world from its inception. If the Earth were still cooling, then by finding the rate at which it lost heat one could calculate its age. Over the next six years he conducted a series of experiments, ultimately concluding that the Earth was 74,832 years old (a figure he believed to be highly conservative—his unofficial estimate was a hundred thousand). His book was met by a chorus of disbelief. In April 1788 he died, and the following year, the Revolution erupted. Buffon's crypt was broken into and the lead from his coffin requisitioned to make bullets.

Undaunted, the intellectually curious continued to search for explanations for all they found on Earth, and by the nineteenth century the division between the Church and science was greater than ever. Between 1800 and 1840 the words "geology," "biology," and "scientist" were either coined or acquired specialized senses in the vernacular. Geology in particular was popular, and soon scientists and others were identifying and naming the successive layers of strata—in Byron's phrase, "thrown topsy-turvy, twisted, crisped, and curled"—that formed the Earth's surface. A shy young British barrister named Charles Lyell (1797–1875), building on John Woodward's example, used fossils to measure Earth's age, finding them in lava remains in Sicily that he estimated to be a hundred thousand years old.

By 1840, the evidence that the Earth was exceedingly "ancient" was overwhelming. Even if man had been around for only six thousand years, prehistoric time had expanded beyond comprehension. Then in 1859 came *The Origin of Species*, and the great argument that linked man's evolution to that of the apes, undermining a literal reading of the Bible story. Dr. Ussher's reckonings were at last no longer feasible. Cambridge University Press removed his chronology from its Bibles in 1900; Oxford followed suit in 1910. By then Mark Twain had already imagined the Eiffel Tower as representing the age of the world, and likened our share of that age to the skin of paint on the knob at its pinnacle.

Mighty doctrines, when overthrown, open up amazing perspectives. Once science had extended the age of the Earth to over a hundred thousand years, an array of new ideas emerged. What age to give now to the Sun? After Darwin, scientists had to consider that it had been releasing its energy over *millions* of years. Natural selection required that the solar system be of a previously inconceivable age. It is only because our Sun is not particularly powerful and burns relatively slowly, being tolerant of the sensitive carbon-based compounds that are the basis of earthly life, that evolution (not that Darwin ever liked that word) could occur.

Although the argument about Earth's age rumbled on through the latter decades of the nineteenth century, questions about its star were drawing more attention from scientists. As Timothy Ferris puts it, "The titans of physics chose to focus less on the Earth than on that suitably grander and more luminous body."[11] In the summer of 1899, a professor of geology at the University of Chicago, Thomas Chamberlin (1843–1928), published a paper on how the Sun was fueled that challenged one of the then-basic assumptions of astrophysics:

> Is present knowledge relative to the behavior of matter under such extraordinary conditions as obtain in the interior of the Sun sufficiently exhaustive to warrant the assertion that no unrecognized sources of heat reside there? What the internal constitution of the atoms may be is as yet open to question. It is not improbable that they are complex organizations and seats of enormous energies.[12]

Within five years the central principles of physics, and consequently the basic assumptions about the functioning of the Sun, had been remade.

IN THE SPRING of 1896, Henri Becquerel, a Paris physicist, inadvertently stored some unexposed photographic plates wrapped in black paper underneath a lump of uranium ore he had been using in an experiment. (Uranium, named after the planet in 1789, is the most complex atom regularly found in nature, composed primarily of uranium-238—as would be discovered, composed of 92 protons and 146 neutrons—plus a small percentage of the less stable uranium-235, three neutrons the fewer.) When after a few weeks he developed the plates, he found that it was as if they had been exposed to light, for they had been imprinted with an image of the silvery-white mineral, which, he thought, must be emitting a "type of invisible phosphorescence." Marie and Pierre Curie, following up on this lucky discovery, would later name the phosphorescence "radioactivity." But uranium was not easy to obtain, nor was the radiation it produced particularly impressive, so the Curies' findings were generally ignored. Then in 1898 they discovered that pitchblende, another uranium ore, emitted much greater amounts of radioactivity than other sources of the metal. Was it possible that there were elements on Earth that were releasing tremendous amounts of unseen energy, waiting to be tapped?

Soon the further, rarer element (termed "radium") that the Curies had extracted from pitchblende was being written up in mainstream papers as the miracle metal, the most valuable element known to man, capable of curing blindness, revealing the sex of a fetus, even turning the skin of a black man white. "A single gram would lift five hundred tons a mile in height, and an ounce could drive a car around the world."[13] Advertisements promoted its use

An advertisement for Tho-Radia, a cream containing radium and thorium, marketed in the 1930s as a breakthrough in beauty preparations

in "radioactive drinking water" for the treatment of gout, rheumatism, arthritis, diabetes, and a range of other afflictions. By 1904, scientists were also claiming radium to be a source of energy for the Sun.

The question was whether the energy was coming from within the atoms of radium or from outside. The great physicist Ernest Rutherford (1871–1937), a New Zealander who had worked on the nature of atoms in both England and Canada, began to explore atomic nuclei, combining forces with a Canada-based English chemist, Frederick Soddy (1877–1956). Radium, Rutherford had established, generates enough heat to melt its weight in ice every hour, and will continue to do so for a millennium or more: the Earth stays warm not least because it is heated by radioactive decay in the rocks and the molten core that lies at its center.*

---

* Most of the planet's internal heat is generated by four long-lived radioisotopes—potassium-40, thorium-232, and uranium-235 and -238—that release energy over billions of years as they decay into stable isotopes of other elements. The Earth's core is a hot, dense, spinning solid sphere, composed primarily of iron, with some nickel. Its diameter is about 1,500 miles, some 19 percent of the Earth's (7,750 miles). Surrounding it is a liquid outer layer around 1,370 miles

The two men were soon experimenting on thorium, a radioactive element not unlike radium, and found that it independently yielded a radioactive gas— one element transmuting into another. This result was so startling that when Soddy told Rutherford about his finding, the New Zealander shouted back across the lab floor, "Don't call it *transmutation*—they'll have our heads off as alchemists!" They proceeded to prove that the heavy atoms of thorium, radium, and other elements they discovered to be radioactive broke down into atoms of lighter elements (in the form of gases), and as they did so they threw off minute particles—which, dubbed alpha and beta rays, were actually the elements' main output of energy.

Further experiments by Rutherford showed that most of an atom's mass lay at its core, surrounded by a crackling web of electrons. He and Soddy speculated that radioactivity of the same kind could be powering the Sun, but while their work was considered worthy of further investigation, it was not seen as astrophysically revolutionary. Some forty years later, Robert Jungk would write in his groundbreaking account, *Brighter Than a Thousand Suns*, "Professor Rutherford's alpha particles ought really, at that time, to have upset not only atoms of nitrogen but also the peace of mind of humanity. They ought to have revived the dread of an end of the world, forgotten for many centuries. But in those days all such discoveries seemed to have little to do with the realities of everyday life."*

Soddy did his best to explain the discovery, first in *The Interpretation of Radium* (1912), then in *The Interpretation of the Atom* (1932), arguing that, before atomic decay was known, the only explanations for the Sun's energy output had been chemical, and thus short-term and puny; but once we had learned about radioactivity, the storm of subatomic reactions that revealed themselves was on something like the right scale to power the Sun.

Nearly a decade passed before anyone built on Soddy and Rutherford's discoveries. Eddington, fresh from his successes off the coast of West Africa, now conducted a wide-ranging study of the energy and pressure equilibrium of stars and went so far as to construct mathematical models of their tempera-

---

thick, 85 percent iron, whose heat-induced roiling creates the Earth's magnetic field. The two cores together comprise one-eighth of the Earth's volume, but a third of its mass. Over a period of 700 to 1,200 years, the inner core will make one full spin more than the rest of the planet, intensifying its magnetic field.

* Robert Jungk, *Brighter Than a Thousand Suns* (Harmondsworth: Penguin, 1960), p. 19. During the last year of the First World War, Rutherford failed to attend a meeting of experts who were to advise the high command on new systems of defense against enemy submarines. When censured, he retorted: "I have been engaged in experiments which suggest that the atom can be artificially disintegrated. If it is true, it is of far greater importance than a war." Ibid., p. 15.

ture and density ("What is possible in the Cavendish laboratory," he famously commented, "may not be too difficult in the Sun"). He estimated that the Sun had a central temperature of 40,000,000°F, and argued that there must be a simple relationship between the total rate of energy loss from a star (its "luminosity") and its mass. Once he knew the Sun's mass, he reckoned, he could predict its luminosity.

To the ancients it had seemed obvious that the Sun was on fire, but to late-nineteenth- and early-twentieth-century physicists, that was unacceptable: it was just too hot to be chemically burning. So the question remained. As John Herschel put it, the "great mystery" was

> to conceive how so enormous a conflagration (if such it be) can be kept up. Every discovery in chemical science here leaves us completely at a loss, or rather, seems to remove farther the prospect of probable explanation. If conjecture might be hazarded, we should look rather to the known possibility of an indefinite generation of heat by friction, or to its excitement by the electric discharge . . . for the origin of the solar radiation.[14]

Since research into the Earth's age was now suggesting it was more than two million years old, the Sun must have been shining for at least that length of time. What process could possibly account for such a startling output of energy? As the Ukrainian-born American physicist George Gamow (1904–1968) put it, "If the Sun were made of pure coal and had been set afire at the time of the first Pharaohs of Egypt, it would by now have completely burned to ashes. The same inadequacy applies to any other kind of chemical transformation that might be offered in explanation . . . none of them could account for even a hundred-thousandth part of the Sun's life."[15]

Cosmologists looked to Gamow's fellow astrophysicists for an answer, and Eddington offered two: the first was that electrons and protons, having opposite electric charges, annihilated each other within the Sun's core, with the concomitant conversion of matter to energy. About a year later he advanced his second, and correct, solution: the Sun, by fusing protons, created heavier atoms, in the process converting mass to energy. But how could such fusion take place in the Sun's consuming heat?

A feature of these years was that many vital discoveries were made by outsiders—scientists whom no one had regarded as solar physicists before their contribution to the field.[16] Eddington had a young disciple: Cecilia Payne (1900–1980) had been five years old when she saw a meteor and decided to become an astronomer. After college, she was introduced to Eddington, who advised her to continue her studies in America; and she became the first student, male or female, to earn a doctorate from the Harvard College Observatory. The examiners judged her 1925 thesis, a photographic study of variable stars, the best ever written on astronomy.

Her solution to the temperature problem was to use Rutherford's discoveries of atomic structure to show that all stars had the same chemical makeup: where their spectra differed, it was from physical conditions, not from their innate composition. Hydrogen and helium were by far the most abundant of the Sun's fifty-seven known elements, as they were in other stars.[17] Despite this conclusion, in her final list of chemical elements in the Sun she omitted hydrogen and helium, branding her own argument as "spurious."

It later transpired that her supervisor, the renowned Princeton astronomer Henry Norris Russell, had tried to talk her out of her theory. "It is clearly impossible that hydrogen should be a million times more abundant than the metals," he wrote to her, reiterating the conventional wisdom.[18] But Payne's arguments nagged at him; he reanalyzed the Sun's absorption spectrum and finally accepted that she was right: giant stars indeed had an outer atmosphere of nearly pure hydrogen, "with hardly more than a smell of metallic vapors" in them. Stars fuse hydrogen into helium, releasing a continuous blast of energy. And when the Sun's great energy was generated by the transformation of chemical elements taking place in its interior—as Gamow impishly noted— what was happening was precisely that "transformation of elements" that had been so unsuccessfully pursued by the alchemists of old.[19]

The next step was to understand nuclear fusion. The late 1920s and early 1930s witnessed a shift to research on the atomic nucleus,[20] one particular center being the Institute for Theoretical Physics at Copenhagen University, directed by the estimable Niels Bohr (1885–1962), "who dressed like a banker and mumbled like an oracle."[21] By the 1920s Bohr had attained international stature and was able to attract many of the greatest physicists of the day, among them George Gamow. This remarkable Ukrainian had a reputation both for scientific innovation and for playfulness (he would illustrate his scientific papers with drawings of a skull and crossbones to indicate the danger of accepting hypotheses about fundamental particles at face value). In 1928 he showed how a positively charged helium nucleus (an "alpha particle," of the kind emitted in such great numbers by the Sun) could escape from a nucleus of a particular metal found on Earth— uranium—despite the binding electrical forces within the metal.*

---

* The general spirit of the Copenhagen hothouse can be gleaned from a trip several of the scientists made to the cinema in 1928. After seeing a Western, Bohr contended that he knew why the hero always won the gunfights provoked by the villain. Reaching a decision by free will always takes longer than reacting instinctively, so the villain who sought to kill in cold blood acted more slowly than the hero, who reacted spontaneously. To test this "in a scientific manner," Bohr and

Gamow not only showed how alpha particles broke *out* of nuclei, he went on to show how they could break *in*. At Cambridge, two of his colleagues, John Cockcroft and Ernest Walton, set about applying Gamow's theory, testing whether very high voltages would propel the particles through the outer walls of the nuclei. In 1932 they succeeded: for the first time a nucleus from one element had been broken down into another element by artificial means, a feat that became known as "splitting the atom." In that same "year of miracles," another Cambridge man, James Chadwick, discovered the neutron, the most common form of particle lurking inside virtually all nuclei. Suddenly, a whole variety of powerful reactions could be detected and even induced in the subatomic world. What was now obvious was that such discoveries bore out Payne's view that these were the reactions powering the Sun.[23]

The high tide of crucial new work roared on. In 1934, the French physicist Frédéric Joliot and his wife, Irène Curie (daughter of Pierre and Marie), proved that by bombarding stable elements with alpha particles, "a new kind of radioactivity" could be affected. A few weeks later, the Italian physicist Enrico Fermi bombarded uranium with neutrons and reported similar results.

In 1938–39, Hans Bethe (1906–2005), the great German American atomic physicist, originally from Strasbourg but by this time at Cornell, wrote a series of articles culminating in "Energy Production in Stars," which explained how stars—including the Sun—managed to burn for billions of years. He had been cataloguing the subatomic reactions that had been identified to date, but up until 1932 these had been only a fistful. Suddenly there came a landslide, and he could now see which particular ones explained the Sun's workings. He posited that its abundant energy was the result of a sequence of six nuclear responses, and it was this process that lit all the stars. Simply put: the Sun was what would become known as a nuclear reactor.[24]

That same year, the German physicists Otto Hahn and Fritz Strassmann demonstrated that what Fermi had witnessed in 1934 was actually the bursting of the uranium nucleus. Two of their colleagues, Lise Meitner and Otto Robert Frisch, went even further, showing that when a uranium atom is split, enormous amounts of energy are released. (Frisch asked a biologist colleague for the word that described one bacterium dividing into two, and was told "fission," and thus the splitting of atoms acquired its name.)

The Hungarian Leo Szilard was the next to make his mark. Although even Einstein believed that an atomic bomb was "impossible," as the energy released

---

his band of fellow researchers sought out the nearest toy shop and bought two guns, with which a showdown was duly enacted, Bohr taking on the role of hero and the six-foot-four Gamow that of villain. Bohr's theory was adjudged correct.[22]

by a single nucleus was so insignificant, Szilard, spurred on by memories of reading H. G. Wells's 1914 novel *The World Set Free*, which predicted just such a weapon, confirmed that neutrons were also given off in each "splitting," creating the possibility of chain reactions whereby each fission set off further fissions (an idea Szilard conceived while waiting at a stoplight one day in London), so that the energy of splitting a single uranium nucleus could be multiplied by many trillions, releasing energy exponentially.[25] And not only was fission possible; it could be produced at will. Opined the Cambridge scientist C. P. Snow, "With the discovery of fission ... physicists became, almost overnight, the most important military resource a nation-state could call upon."[26]

By the advent of the Second World War, both Allied and Axis scientists were keenly aware that nuclear fission might be converted into some kind of weapon, although no one was sure how.* Still highly skeptical, Einstein de-

---

* With the war under way, one of the routine duties of R. V. Jones, the scientific adviser to MI6, was to read through a monthly summary of German scientific publications. Early in 1942 he perused the latest report, and the next minute was running like a lamplighter down the corridor. "The German nuclear physicists have stopped publishing!" The War Cabinet met within the hour.[27]

Albert Einstein (1879–1955) with the American theoretical physicist
J. Robert Oppenheimer (1904–1967) during the time they worked together
before the Manhattan Project

clared that the challenge of inducing a chain reaction and making a fission bomb was akin to going out to shoot birds at night in a land where there were very few birds; but secretly, in a letter to President Roosevelt dated August 2, 1939 (drafted along with his friend Leo Szilard), he urged that funds be devoted to developing fission-based weapons: "A single bomb of this type, carried by boat and exploded in a port, might very well destroy the whole port together with some of the surrounding territory."[28]

Much alarmed by this letter, that autumn Roosevelt earmarked a small amount of money for the study of fission: small because it was assumed that any bomb would require so many tons of uranium that it was theoretically possible but probably not a practical option. However, in early 1940 two German refugees in Britain, Otto Frisch and Rudolf Peierls, calculated that only a few pounds of the 235 isotope were required. Other scientists in Britain conceived a gaseous diffusion technique, and put together, these two advances galvanized the government to lobby for the U.S.-based research to be shifted into more professional hands and be more effectively financed.

The Americans listened. In 1942, Brig. Gen. Leslie R. Groves, Jr., took over what became known as the Manhattan Project. Guided by the solar physicist J. Robert Oppenheimer (1904–1967), Groves brought together the top nuclear scientists of the day, supported by unprecedented funding and manpower.[29] The project took over some thirty sites across the United States and Canada— Oak Ridge, Tennessee; several in central Manhattan; Chalk River, Ontario; Richland, Washington; and its headquarters at Los Alamos ("The Cottonwoods"), a small school on a ranch near Santa Fe, New Mexico. At its peak the project employed more than 130,000 people (most of whom had no idea what they were working on).

When a nucleus of uranium-235 absorbs a neutron, it splits into atoms of strontium and xenon, releasing energy plus twenty-five neutrons for every ten atoms. Uranium-238, on the other hand, absorbs neutrons and does not divide, so there is no such reaction. A bomb needs to be 80 percent pure uranium-235; otherwise the uranium-238 blocks the chain reaction. For the scientists on the Manhattan Project, the problem was how to separate the two. The breakthrough came at Oak Ridge: a series of "racetracks"—really huge belts of silver magnets—were devised to pull gaseous uranium through vacuum chambers, separating the weapons-grade U-235 from its heavier, benign sibling. "You used to walk on the wooden walkways and feel the magnets pulling at the nails in your boots," a scientist there recalled. On one occasion, a man carrying a sheet of metal walked too close to the racetrack and found himself stuck to the wall. Everyone yelled for the machines to be shut down, but since it would have taken a day or more to restart everything, the engineer in charge refused. Instead, the unfortunate had to be pried off with two-by-fours.*

When the Third Reich surrendered on May 8, 1945, the Manhattan Project was months away from a weapon. To hasten victory in the Pacific, Oppenheimer decided to conduct a test, and on July 16, in the desert north of Alamogordo, New Mexico, released the equivalent of nineteen kilotons of TNT, far mightier than any previous man-made explosion. The news was rushed to Truman, who tried to use it as leverage on Stalin at the Potsdam Conference, to no avail. After listening to the arguments from his scientific and military advisers, and hoping to avoid an invasion calculated to exact a million and a quarter Allied dead—nearly double the British and United States fatalities in the whole war up to that point—Truman ordered that the weapons be used against Japanese cities, and on August 6, a uranium-based bomb, "Little Boy," was dropped on Hiroshima.[†] Three days later, "Fat Man" was loosed on Nagasaki. The pilot on the latter mission recalled, "This bright light hit us and the top of that mushroom cloud was the most terrifying but also the most beautiful thing you've ever seen in your life. Every color in the rainbow seemed to be coming out of it."[31] The bombs killed at least 100,000 people outright, another 180,000 dying later of burns, radiation sickness, and related cancers. The Atomic Age had begun. Hans Bethe, aghast, devoted the rest of his life to checking the weapon's "own impulse," as he put it: "Like others who had worked on the atomic bomb, I was exhilarated by our success—and terrified by the event."[32] Winston Churchill told the House of Commons that he wondered whether giving such power to man was a sign that God had wearied of his creation. "I feel we have blood on our hands," Oppenheimer told Truman, who replied, "Never mind. It'll all come out in the wash."

On September 23, 1949, the Soviet Union tested its first nuclear device, also a fission bomb. The amount of energy practically releasable by fission bombs (that is, atomic bombs—or A-bombs—which produce their destructive force through fission alone) ranges between the equivalent of less than a ton of TNT to around 500,000 tons (500 kilotons). The other, vastly more powerful category of bomb obtains its energy by fusing light elements (not necessarily hy-

---

* See Sam Knight, "How We Made the Bomb," London *Times*, July 8, 2004, T2, p. 14. The early labs were hardly sophisticated places: stray radiation induced radioactivity in everything from gold teeth to the zippers on trouser flies. A large proportion of those who worked on the Manhattan Project died at suspiciously early ages.

† In May 1946 *The New Yorker* sent the Japanese-speaking journalist John Hersey to write an extended account of what had befallen. His thirty-thousand-word report took up the whole of the August 31 issue (no cartoons, verse, or shopping notes), which sold out within hours. One haunted subscriber had preordered a thousand copies: Albert Einstein.[30]

XX-28 George, a 225-kiloton thermonuclear bomb, exploding on May 8, 1951

drogen) into slightly heavier ones—the same process that fires the stars. Duplicate this, and one can make weapons of almost limitless power. Known variously as the hydrogen bomb, H-bomb, thermonuclear bomb, or fusion bomb,
these work by detonating a fission bomb in a specially manufactured compart-
ment adjacent to a fusion fuel. The gamma rays and X-rays emitted by the explosion compress and heat a capsule of tritium, deuterium, or lithium
deuteride, initiating a fusion reaction.

In 1952 "a blinding flash of light" from an explosion that the United States
set off on a small island in the South Pacific signaled the detonation of the first
hydrogen bomb, and for a split second an energy that had existed only at the
center of the Sun was unleashed by man on Earth; the explosion made a crater
in the ocean floor a mile wide. In the intense debate that followed, some scientists, such as the ferociously anti-Communist Edward Teller, who had originally posited such a bomb, argued that nuclear power was a good thing and
that research should be continued; others, Oppenheimer, Einstein, and Bethe
among them, felt that their dreams for what physics might accomplish had
turned into darkness and blood. What went almost unnoticed was that splitting the atom heralded yet one more demotion for the Sun: its otherworldly energy was no longer unique. As Bohr exclaims to Werner Heisenberg in Michael
Frayn's play *Copenhagen,* "You see what we did? We put man back at the center
of the universe."

PART THREE

The SUN on EARTH

In the 1950s, the English author Stephen Potter made fun of the sunbathing craze in his bestselling book Lifemanship, laying out three tanning options.

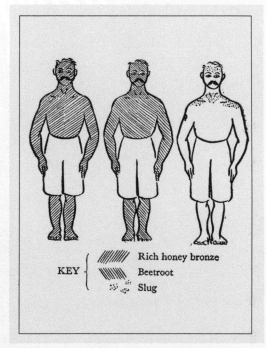

KEY {
Rich honey bronze
Beetroot
Slug

CHAPTER 13

SUNSPOTS

Some have thought that they were great masses of scoria, *or cinders, or dross, floating on a sea of molten rocks; others have thought that they were the tops of mountains appearing above a great sea of fire; others still have thought that they were clouds of black smoke, floating over the face of the Sun; but now it is quite generally agreed, among men of science, that they are vast openings through the envelope of the Sun, allowing us to see away down, down, down into an immense abyss.*

—WILLIAM URMY, 1874[1]

CASTEL GANDOLFO, THE PAPAL SUMMER PALACE, LIES SOME THIRTEEN miles southeast of Rome, overlooking Lake Albano. One sun-drenched afternoon in late October 2003, I climbed the steep hill to what is as much stronghold as spiritual residence. The estate extends all the way down to the lake—136 acres embracing a working farm, an array of monuments, a garden designed by Bernini, an imposing baroque fountain, and the ruins of a villa of Domitian's, acquired by Clement VIII at the beginning of the seventeenth century. The whole site is spectacular.

I was there because the papal residence also houses a fully operational observatory and an extensive astronomical library. One of Clement's sixteenth-century predecessors, Gregory XIII, commissioned a study of the calendar, an initiative that inspired what became three observatories, two just outside Rome and the third within the Vatican itself. In 1891, Leo XIII moved the Vatican observatory to a spur of the Seven Hills behind St. Peter's, where for more than four decades the Church's scientists worked; but eventually the growth of Rome so brightened the night sky that observation became almost impossible, and in 1933 the center relocated to Castel Gandolfo. Two new telescopes were built and an astrophysical laboratory installed. The library burgeoned: *The Da Vinci Code* avers that more than twenty-five thousand books on astronomy are housed there, although its longtime librarian, Juan Casanovas of the Society of Jesus, doubts that number.

Father Casanovas was at the castle entrance to welcome me. An imposing figure, over six feet tall and with completely white hair, he seemed happy to help a fellow enthusiast. I had written asking if I could see some of his prize volumes, look over the rooftop observatory, and talk to him about sunspots, on which he is an authority. We walked together up a narrow side staircase. The whole place seemed deserted and full of echoes, the pope along with his retinue having long since returned to Rome.

Galileo published these drawings in 1613 in his History and Demonstrations Concerning Sunspots and Their Properties. *When his thirty-five suns are made into a flip book, the motions of the spots are easy to track, as the drawings were made close together in time.*

As we reached the library, Father Casanovas told me of its first editions of Copernicus, Newton, Kepler, and Tycho, but what he led me to was an outsize volume bound in weathered brown leather: Galileo's personal register, in which he recorded his first sightings of blemishes on the Sun. I turned the leaves gingerly, taking in his original diagrams, evidence that the great daytime star was not after all immaculate: a revelation sufficiently dangerous in its theological implications that for over two years Galileo told no one. Then came Fabricius, and Harriot, and finally Scheiner, and the Pisan's competitive instincts took over.

Father Casanovas moved away from my desk, leaving me to my musings. What had it been like for those even before Galileo, such as Aristotle's best pupil, Theophrastus, who saw spots around 325 B.C.? Why hadn't his eyes been burned out of his head? Likely enough, instead of looking at the Sun directly, he had examined its reflection, or gazed at it through some translucent mineral, as had the Chinese, observing it through wafers of jade. The sun-watchers of Xi'an and elsewhere evidently had no idea what sunspots were; neither had Galileo, but as he examined these impurities stealing across the solar surface he at least deduced that the Sun rotated, and used the very drawings in front of me to estimate how fast it did so.

My thoughts were interrupted by Father Casanovas, who appeared suddenly at my shoulder to indicate that I should follow him, and strode off down one of the long corridors, so that I had to hurry to keep up. Several turnings later we arrived at his office: everything in place, files neatly stacked, working papers carefully squared off. A computer was humming, its screen a kaleidoscope of red, orange, and yellow, the latest images from SOLO, the Solar and Heliospheric Observatory, the joint project of NASA and the European Space Agency to photograph the Sun. I had never seen pictures so dramatic. "This month is a high point for sunspots," Father Casanovas explained. "Huge storms are coming because of them." He rummaged through a shelf of books and, with a grunt of satisfaction, handed me a slim volume, in English, on the history of sunspots. The cover proclaimed its author: J. Casanovas of the Specola Vaticana.

That evening, back in my hotel room near Rome's great central station, I settled down to read. Bit by bit I came to see that the discovery and understanding of sunspots are one of the most exciting stories in science, and also hugely intricate, stretching over continents and centuries. The two earliest reports of sightings occur in that classic Chinese text the I Ching (The Book of Changes, which dates back anywhere from five thousand to eight thousand years), and tell of a *dou* and a *mei* on the Sun (both meaning a darkening or obscuration).[2] Over later ages, Chinese and Korean astronomers noted about 150 spots, likening them to hen's eggs, swallows, ravens, and other birds. Virgil wrote about the Sun's being "checkered with spots" in one of his pastorals; Gregory of Tours, in the late sixth century, described "blood-red" clouds on the Sun; and Einhard in his *Life of Charlemagne* (c. 807) recounted a "black-colored spot [that] was to be seen for seven days at a stretch." On December 8, 1128, John of Worcester, a monk who contributed to the Anglo-Saxon Chronicle, which detailed English life (including astronomical events) from the birth of Christ to the accession of Henry II in 1154, drew the Sun marked by two large dark spots—a drawing that modern astronomers believe to be accurate. He lacked a telescope, so for him to detect not only the spots themselves but also their penumbras (outer shadows) implied that these maculations were large indeed. Even so, not even this discovery excited comment. Centuries before, Aristotle had declared the heavens incorruptible, and the Church in due course concurred:

so in Europe at least, any such sightings were either ignored or ascribed to transits of Mercury or Venus.

The introduction of the telescope would change all this. In his history, Father Casanovas equates the impact of Galileo's description of his sightings, as published in *The Starry Messenger* in 1610, with that of man's first landing on the Moon.[3] However, early telescopes were of poor quality, so most observers continued to believe that sunspots were somehow not actual parts of the Sun, and whatever interest Galileo had stirred soon faded away (Cyrano de Bergerac, a close friend of one of Galileo's pupils, being one of the few who continued to be fascinated by them).*

The next advance came about by accident. As telescopes increased in number, generations of amateur astronomers strove to discover the planet that was believed to orbit between the Sun and Mercury. Samuel Heinrich Schwabe (1789–1875), a Dessau pharmacist-turned-astronomer, knew his best chance of detecting this hypothetical body was during its passage in front of the Sun; but he also realized that he might confuse such an object with a sunspot. So, from October 30, 1825, on, he meticulously recorded *everything* he saw in the heavens. After nearly two decades, he had still not found his planet; but he had tripped over something far more important.

"From my earlier observations," he wrote, in an article that appeared in 1843, "it appears that there is a certain periodicity in the appearance of sunspots."[5] His accompanying table gave powerful evidence of a cycle, with spots appearing in clusters. Once a circuit was completed, the solar disk might be clear for weeks, but the existence of the cycle itself seemed indisputable. A summary of his observations shows, for example, that the low point of one cycle was in 1833, with the fewest clusters and the highest number of days without visible spots; the next low point fell about ten years later. The high points can be seen in 1828, and again in 1837:

---

* Curiously, one possible exception to this lack of interest lay in France. In *Sunspots and the Sun King*, Ellen McClure, a professor of French history, points out that Henri IV was assassinated in the same year that Galileo made his first observations, and suggests that both events represented a challenge to order and hierarchy. Henri's murder renewed questions about his legitimacy; Galileo's discovery brought the Sun into the realm of terrestrial corruption. It became urgent to restore faith in an order grounded in asserted permanence and transcendence. Efforts to achieve this, she writes, "resulted in the creation of the Sun King, whose authority and monarchical identity [were] constructed at least in part to counter the destructive implications of the sunspots."[4] Two hundred years on, during the Congress of Vienna, Napoleon was lampooned in the British press in *Napoleon and the Spots in the Sun*, which suggested that the defeated Boney had been packed off to the Sun because he had nowhere else to go and, further, that he, being himself but a cluster of sunspots, blocked sunshine from the Earth, causing that year's dismal weather.

Year	No. of clusters	Days when no spots were observed	Observation days
1826	118	22	277
1827	161	2	273
1828	225	0	282
1829	199	0	244
1830	190	1	217
1831	149	3	239
1832	84	49	270
1833	33	139	267
1834	51	120	273
1835	173	18	244
1836	272	0	200
1837	333	0	168
1838	282	0	202
1839	162	0	205
1840	152	3	263
1841	102	15	283
1842	68	64	307
1843	34	149	324

At first Schwabe's analysis attracted little attention, but when the Irish astronomer and explorer Edward Sabine (1788–1883) saw the article, he realized that the cycle Schwabe had uncovered correlated with the fluctuations he had observed in the Earth's magnetic field. Then a Swiss observer, Rudolf Wolf (1816–1893), came up with a way to compute the average number of spots over even longer periods, refining Schwabe's estimate to 11.1 years.

Over the next twenty years, Wolf put together a census reaching as far back as 1745. As he continued to reconstruct yet earlier activity, he realized how few spots had been observed between 1645 and 1715. This hiatus coincided with the coldest part of the so-called Little Ice Age in Europe and North America alike, when tidal waters like the Thames and even the canals of Venice iced over. But because of the lack of interest in sunspots, it would be more than two centuries before anyone made a connection.

However, it would take the scientific community a while to accept both Schwabe's correlations and Wolf's "measure of solar acne."[6] Enter that energetic Prussian Baron Alexander von Humboldt (1769–1859), who has been described as "a combination of the measured studiousness of astronomer Carl Sagan and the down-and-dirty enthusiasm of *Titanic* discoverer Jim Ballard."[7] Humboldt was intrigued by Schwabe's researches, having in his youth spent five years traveling throughout South and Central America (European newspapers reported him dead on three separate occasions), where besides studying its plants, animals, rivers, and volcanoes he had taken magnetic measurements wherever he went, and had found that field strength varied widely. Sunspots seemed the likely cause. His enthusiasm would make the study of the relationship between sunspots and magnetic fields an acceptable scientific activity (so much so that now more than two hundred terrestrial magnetic observatories are dispersed over the globe), and in 1851 he included Schwabe's updated table in his encyclopedic five-volume compilation of the natural sciences, *Kosmos.* His endorsement made scientists all over the world take Schwabe seriously.[8]

The German astronomer Christoph Scheiner (1573–1650)
observing sunspots with the help of an assistant

Schwabe also influenced the Englishman Richard Carrington (1826–1875), a gentleman brewer and experienced astronomer who distilled over a decade of research into *Observations of the Spots of the Sun* (1863). On September 1, 1859, Carrington had been tracking a group of spots from his observatory when all at once, as he wrote, "two patches of intensely bright and white light broke out." As he stared, incredulous, the two spots intensified and became kidney-shaped. He dashed outside, hoping to find another witness, but the flare was brief, just five minutes, the result—as we now know—of magnetic currents crisscrossing and short-circuiting one another at 420,000 miles an hour. Less than seventeen hours later (the light and X-rays from a flare take only eight minutes to reach Earth, but the heavier particles from each eruption need between eighteen and forty-eight hours), a great magnetic storm erupted over the globe, its auroras staining the skies as far south as Cuba. Was there a causal relationship between flare and storm? Carrington thought so, but cautioned, "One swallow does not make a summer," and never took his hunch further.

Yet "Carrington's flare," writes Stuart Clark, "was a tipping point in astronomy. The sudden demonstration of the Sun's ability to disrupt life on Earth catapulted astronomers into a headlong race to understand the nature of the Sun."[9] By the 1890s, the prevailing wisdom was that sunspots were made up of strong cyclones, which prompted the great American astronomer George Ellery Hale (1868–1938) to examine them for magnetic activity. Employing a spectroheliograph (a combination of spectroscope and something similar to a movie camera), he took photographs of the Sun that showed large blotches of hydrogen being swept into the center of a sunspot as if caught in a whirlpool.* He observed two large flares, each of which was followed by a huge magnetic storm on Earth, 19.5 and 30 hours later, respectively. By 1908 he had demonstrated that the spots were indeed gigantic cyclones in the Sun's atmosphere, similar in formation to the hurricanes and whirlwinds that build up in the West Indies and ravage the Gulf Coast. Sunspot activity and the Earth's climate seemed not only connected; they seemed to be so *magnetically.*

Hale's conclusion that the spots were intensely magnetic was consistent with observations of the streaks visible above sunspot areas during total eclipses, which resembled the field lines of a bar magnet. Looping out from the Sun's surface, these streaks speckle it with positive and negative lines of magnetic force, and channel the storms into eruptions that extend thousands of miles beyond the surface before being dragged back.[10] Hale's reasoning let observers gauge sunspot radiation, which originates in layers deep within the Sun: it also suggested what the spots were and how they affected the climate of Earth.

Sunspots could be said to be a bit like snowflakes, in that each one is unique but of similar structure. All are roughly circular, averaging from 1,865 to 18,650 miles in diameter—although Hale saw one of 81,000 miles, ten times Earth's diameter. At each spot's roiling center lies what is called an umbra ("shadow"), which appears dark only because it contrasts with the more luminous surface around it, but if isolated would be as bright as a full Moon set against a black sky. An umbra typically has a temperature of about 4,300 kelvin (a centigrade measure of heat, where −273°C is absolute zero), roughly

---

* The first photograph of the Sun, a daguerreotype, was taken in 1845 by the French physicists Armand-Hippolyte-Louis Fizeau (1819–1896) and Léon Foucault (1819–1868)—the latter best known for the Foucault pendulum, which demonstrates the Earth's rotation.

2,100 K cooler than the surrounding photosphere and embedded about 450 miles deeper. The umbra, comprising nearly a fifth of the area of an average spot, is surrounded by a fibrous gray filamentary penumbra, which resembles the petals of a flower. Taking up the remaining 80 percent, it is about three-quarters as hot as the photosphere (which constitutes the surface of the spot). In summary, a sunspot appears in three different forms, depending on depth: the photosphere, the penumbra, and, at its deepest, the umbra.[11]

It is still a mystery why a spot's core should be less hot than either its penumbra or the solar surface; all we know for sure is that within each of these great funnel-shaped vortices in the outer layers of the Sun a giant cooling process is at work, and that the most powerful focus of its magnetic field lies at its darkest, or coldest, point, the umbra.[12]

Some years before Hale's researches, the superintendent of the Royal Observatory at Greenwich, Edward Walter Maunder (1851–1928), had been making some important discoveries of his own. Keen on sunspots since the age of fourteen, he was measuring how large they could get by the time he was twenty-six. Over the next thirty years, he embarked on a massive photographic project, accumulating several thousand pictures of five thousand groups of spots, coming to think romantically of this task as capturing the Sun's portrait, and once writing that taking the spectrum of a sunspot was like seeing into its soul.

He found that the magnetic flux increased rapidly during a spot's early life before sinking into a slow decline over a matter of days. His photographs showed that when the latitude of sunspots was plotted against a series of eleven-year cycles, their positions formed a pattern that (with a little imagination) resembled three butterflies traveling west: the spot farthest ahead in its migration was designated the "leader," and had one magnetic pole, while its followers showed an opposite polarity. Just as Galileo had demonstrated, all spots crossed the solar disk in straight lines, and hunted in pairs. They would start out close together, then move apart as the group traveled, sometimes at distances that were up to 20 degrees of the Sun's circumference, while always remaining parallel to the equator. At the end of every cycle the polarity reversed, so that the Northern Hemisphere "leaders" took on negative polarity and the Southern, positive. A cycle has to run through two successive eleven-year sequences—twenty-two years in all, give or take a few months—to complete itself. Surprisingly, at the height of the cycle, when sunspots were most numerous on the solar disk, the Sun shone with a greater intensity than it did when its spots were fewer. A cycle is part of an overall pattern of solar activity involving not just the movement of spots but also of prominences, flares, showers of solar wind particles, cosmic rays, and energetic protons, a minute but potent proportion of which plunge to Earth, some within fifteen minutes of leaving the solar surface.

To put it another way: as the Sun throws its tantrums, the magnetic fields erupting from its interior discharge high-energy cosmic rays. Maunder speculated that sometimes these rays are deflected past Earth, which cools as a result, so that, for example, in 1536 Henry VIII and his court were able to sleigh over the Thames all the way from central London to Greenwich, while during the terrible winter of 1709 wine froze in the glasses before the Sun King at his banqueting table.

THE NEXT STEP forward came from an unlikely place. Shortly after the First World War, Andrew Douglass (1867–1962), an academic at the University of Arizona, pioneered a new science called dendrochronology—the study of tree rings, whose successive widths (broad in years that were good for growing, narrow in bad years) provide a record of climatic variation over a tree's lifetime. He assumed that trees, the longest-living organisms on Earth, were the only objects of the plant kingdom that could offer reliable records, since everything else died into the soil, and noted that years of rapid growth were followed by periods of retarded development, then rapid growth resumed. On average, ten to twelve rings separated any two intervals of development.

One morning in 1922, out of the blue, he received a letter from Maunder outlining his theory about the effects of the absence of sunspots between 1645 and

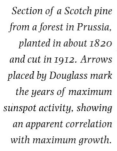

Section of a Scotch pine from a forest in Prussia, planted in about 1820 and cut in 1912. Arrows placed by Douglass mark the years of maximum sunspot activity, showing an apparent correlation with maximum growth.

1715 and suggesting, "You should see this in the tree rings."[13] His curiosity piqued, Douglass began studying beams from old buildings as well as very old trees, for instance, Arizona pines and California redwoods; and sure enough, their rings revealed a pattern of slow growth for the same period that sunspots vanished and Earth suffered its long cold snap. This was still not enough for general acceptance, however, and Maunder died in 1928, his theories unrecognized.*

Maunder had not been alone in his theories about the effects of sunspots—or in having to endure the skepticism with which such theories were met. From the time spots were first observed, some people have speculated about how they affect life on Earth, while others have mocked them: as early as December 1795 William Herschel delivered the first of what was intended to be a series of lectures on the Sun and its effects on Earth to the elite members of the Royal Society, in which he said that he had uncovered five periods of low sunspot activity during which the price of wheat had been high, a fact he connected to unusually long spells of drought. Most of his audience laughed at him (so much so that he felt forced to cancel the remaining talks)—but then he was also airing the theory that the Sun was cool at its center and inhabited, a belief that in Herschel's day was enough to have one declared a lunatic (or perhaps a solatic).

To ascribe happenings on Earth to spots on the Sun seemed destined to remain the stuff of fiction, and sure enough, in 1892, Mark Twain published a novel, *The American Claimant*, in which the idea of solar blemishes controlling our climate is carried to its logical extreme—sunspots as big business. By the novel's end, the scheming Colonel Mulberry Sellers comes up with his grandest plan yet for making a fortune: he will reorganize the Earth's climates, furnishing them to order and taking old climates in part payment. How? By utilizing "the spots on the Sun—get control of them, you understand, and apply the stupendous energies which they wield to beneficent purposes in the reorganizing of our climates."†

Twain, of course, was making fun of crazy business schemes as much as he

---

* In 1937, the memorably named scientist H. True Stetson, of MIT and Harvard, came up with new evidence: rabbit skins. The pelt records of the Hudson's Bay Company showed remarkable variations in the number of fox, lynx, and rabbit taken over periods of ten to eleven years. Nearly every peak corresponded "quite closely" with the scarcity of spots. "If sunspots have anything to do with rabbit population," Dr. Stetson continued, "and sunspot years have been favorable for tree growth, one might wonder why rabbits appear most numerous near sunspot minima and most scarce near sunspot maxima. Perhaps the trappers, stimulated by sunspots, have been more energetic in depopulating the rabbit world during years of sunspot maxima, or perchance other animals who are natural foes of the little four-footed creatures thrive best at sunspot maxima."[14] This argument, too, failed to sway the doubters.

† This recalls an even earlier satire, Samuel Johnson's 1759 philosophical novella, *Rasselas*. In the chapter "The Astronomer Discovers the Cause of His Unease," the savant in question tells Imlac, the story's narrator, "I have possessed for five years the regulation of weather, and the distribution of the seasons. . . . The Sun has listened to my dictates, and passed from tropic to tropic by my direction; the clouds, at my call, have poured their waters, and the Nile has overflowed at my command."[15]

was satirizing the wayward theories of scientists. But it would take a while for Maunder's conviction that a connection *did* exist between solar activity and fluctuations in the Earth's magnetic field to find widespread acceptance. After World War II, physicists intensified their monitoring of solar events. Yet for all their prodding and probing, evidence of the effects of sunspots remained sporadic: as late as the 1960s, a young researcher entertaining any correlation between solar phenomena and patterns in the terrestrial climate branded himself a crank. Then came Eddy.

John A. "Jack" Eddy (1931–2009) was an astronomer based at the University of Colorado, which has a High Altitude Observatory where this versatile scientist was able to study Jupiter's atmosphere, the solar corona, the history of solar physics, even American Indian astronomy. Intrigued by Maunder's theories, which he later likened to "deciphering the Dead Sea scrolls of solar physics," he couldn't understand why they had been written off. In the early 1970s, he went to the Laboratory of Tree-Ring Research in Tucson, but was unable to replicate Douglass's match between tree rings and climatic variations. Undaunted, he next tracked the history of auroras—which, after all, owe their life to sunspots—and established how few there were during the era that he titled the "Maunder Minimum." When the Sun was magnetically active, he discovered, it grew many spots, whose increased aggregate magnetism reduced the Earth's exposure to radiation. What does not get through collides with other particles in the atmosphere, amalgamating into carbon-14 (the element's main radioisotope), which shows up in tree rings. And, said Eddy, coming to a triumphant point, the rings displayed a prolonged increase in carbon-14 between 1650 and 1715.*

Judging that he might have found the final piece of the puzzle, Eddy now turned to the ideas of the Serbian engineer Milutin Milanković (1879–1958), who believed that ice ages on Earth were caused by small variations in the amount of sunlight, the result of gradual cyclic changes in the shape of the Earth's orbit.[17] He argued that three types of such orbital variations (sometimes referred to as stretch, wobble, and roll) manifest in cycles operating over periods

---

* Since Eddy, scientists have looked again at the annual layers of silt brought down by rivers and found in lake sediment cores, and also studied cave minerals and deposits, pollen counts, geological boreholes, coral, and mountain glacier deposits—all of which bear witness to the chill presence of the Maunder Minimum. Recent research has also helped explain one of the great puzzles of the Little Ice Age—that it appears to have been a peculiarly European phenomenon. Peaks in ultraviolet radiation emitted by the Sun have been shown to boost ozone formation in the stratosphere—the layer of the atmosphere that lies twelve to thirty miles above our heads—which in turn absorbs more ultraviolet and heats up. Stratospheric winds affect the weather we experience, and Europe is particularly susceptible because it sits under the northern jet stream at a longitude that makes it vulnerable.[16]

of approximately 100,000, 22,000 and 40,000 years, respectively, and that these changes affect the amount and incidence of solar radiation reaching the Earth.[18]

The first cycle, he argued, is dictated by the shape of the Earth's path around the Sun, and the degree to which it departs from a circle (its "eccentricity") to become more elliptical. As the circle "stretches," the distance between Earth and star alters, thus affecting the amount of radiation we receive.[19] The second cycle arises from the "wobble" in the Earth's axis of rotation, which affects when the seasons fall, slowly swinging the Northern Hemisphere—and reciprocally the Southern—closer to, then farther from, the Sun. The Northern Hemisphere contains much more land than the Southern, and land reacts more rapidly to temperature changes than do oceans. The resulting swings in heating modify weather patterns. The third cycle is created by a small oscillation in the tilt of the Earth's axis of rotation. Over the course of about forty thousand years, this tilt varies between 21.5 and 24.5 degrees, and when it is at its smallest (as at present, remaining so for about 9,800 years from now), differences between summer and winter are reduced.*

By employing Milanković's theories about cycles, Eddy was able to give more credibility to Maunder's claims. He for now buttressed his case with evidence of temperature changes stretching back nearly three hundred thousand years, using data from mud and clay dredged up from the seabed: such temperatures rose and fell over the centuries with the amounts of sunlight received—and fit in well with Milanković's calculations. In a landmark paper published in *Science* in 1976, Eddy concluded that Earth has undergone a total of eighteen periods of minimal sunspot activity in the last eight thousand years—one of which was the Little Ice Age. Maunder was finally vindicated, and his conclusions about what—thanks to Eddy—became known as the Maunder Minimum would be acclaimed as "the most significant event in the history of solar observation."

WITHIN OUR SOLAR system, only Venus, Earth, and Mars have solar-influenced atmospheres. Venus is a boiling mess, Mars a frozen desert, while the Earth, though subject to far less extreme temperatures, is under a constant flux, its climate overwhelmingly driven by the Sun. But how much of what happens to us derives specifically from sunspots, which are, after all, just the visible manifestations of magnetically active regions on the Sun?

---

* The traditional objection to Milanković's theory about ice ages is that if their timing were set by variations in the sunlight falling on a given hemisphere, why didn't the Southern Hemisphere get warmer as the Northern Hemisphere cooled, and vice versa? The answer is that changes in atmospheric CO_2 and methane link the two hemispheres, warming or cooling the planet as a whole; thus global warming and the natural emission of greenhouse gases have reinforced each other.[20]

Generally, spots are blamed (rarely thanked) for many events, from the drop in cast iron production to the incidence of suicide in northern climates; time taken to complete sea voyages; shipwrecks, disappearing radio programs (especially in the short-wave, high-frequency bands), carrier or racing pigeons losing their way, enormous increases in car production, myocardial infarctions, convulsive seizures and hallucinations, even the outbreak of epidemics, wars, and revolutions—it being argued that the French Revolution began in 1789 rather than four or five years later because the winter of 1788 was extremely cold.*

Most physicists take it as given that sunspots can disrupt compasses and various electrical transmissions,† yet it took a while for word to reach beyond the scientific community. In 1953, for instance, during the proceedings of the House Un-American Activities Committee, Roy Cohn was interrogating Raymond Kaplan, a supposed Communist sympathizer and chief radio engineer for the Voice of America. The VOA had been set up in February 1942 to promote favorable views of the United States abroad, and Cohn had tagged it as one of seventy-eight suspect organizations intent on subverting American interests. He pointed out that some programs were not being broadcast as widely as they should. Disloyal people were responsible: could Kaplan name names?

COHN: There are VOA towers where radio messages could not get out
    to countries that we wanted to reach—
KAPLAN: The transmission towers couldn't reach certain countries—
COHN (raising voice): Disloyal people working for—
KAPLAN (interrupting): It's more complicated than that. For example,
    sunspots: they affect the transmission, it's not entirely clear how—
COHN (interrupting): Sunspots? [General laughter] (Softly, but derisively)
    Sunspots . . . or perhaps disloyal Americans selling out our country.

Within three weeks of that hearing, Kaplan threw himself in front of an oncoming truck, dying at the age of forty-two.[22]

---

* Both the Russian scientist Valerii Orloff and the American historian William James Sidis have connected revolutions and sunspots.[21] The American Revolution, the French Revolution, the Paris Commune, and both Russian revolutions (1905 and 1917) all fell fairly close to the years of maximum solar activity.

† Long-distance radio signal reception can be seriously compromised by varying sunspot activity; see Brody, *The Enigma of Sunspots*, pp. 162–63. During the preparations for D Day, precautions were taken to ensure that the Allied landings did not coincide with a shortwave blackout due to an errant solar flare. Daily "top secret" reports on the Sun's state would be filed from an observatory high above the Fremont Pass in Colorado, which strategists thousands of miles away studied carefully before planning the attack.

There is now no doubt that sunspots can violently disrupt events on Earth. Following the invention of electric telegraphy and the telephone, telegraph and telephone lines were strung across the world. During solar storms, some operators were almost electrocuted by sunspot-induced surges. "At one place the operator received seven shocks," reads one account. "At another place the telegraphic apparatus was set on fire; and in the city of Boston a flame of fire followed the pen of a telegraphic instrument."[23] On March 12, 1989, electric currents streaming up to sixty miles above the ground hit Québec's power grid and in less than sixty seconds incapacitated half its generating system; the shutdown lasted more than nine hours, left 7 million people without electricity, and cost the province between 3 and 6 billion dollars.[24] As the world becomes more reliant on electronics, satellites, and similar technologies, we can expect even greater dislocations, especially since the erratic nature of sunspot flares can leave scientists as little as thirty minutes to give warning. (During the Vietnam war, a large number of mines dropped into Haiphong Harbor blew up simultaneously in response to a large solar flare.) And even the immense magnetic storm of 1989 was only *one-third* as intense as that which Richard Carrington had witnessed 130 years before, when some 20 billion protons sandblasted every square inch of our planet's atmosphere. The next solar maximum is going to be the most active for the last fifty, or possibly hundred, years, and is expected to start sometime between 2010 and 2012.

So it seemed like a good idea to return to Eddy and his tree-ring evidence and check out the current thinking. In January 2005 I visited Mike Baillie, professor emeritus in the School of Geography, Archaeology and Palaeoecology at Queen's University, Belfast, who specializes in dendrochronology. Despite the fact that he has constructed a table of tree rings going back seven thousand years, he turned out to display a bracing skepticism (and tart humor) toward his chosen field. "We really don't know what the relationship is between sunspots and tree rings," he admitted as we pored over various papers in his overflowing office, about the same size as Father Casanovas's but at the opposite end of the tidiness scale. "Eddy supposes that there were no 'global events' before tree chronologies began; then one discovers that there were. And one must be careful: a 22.2-year sunspot cycle doesn't just translate into two eleven-year cycles."*

---

* By the end of the twentieth century, among economists "sunspots" became a shorthand for any extrinsic uncertainty that sets a crisis in motion—an accidental random shock that, by altering expectations, leads to an economic downturn. Then came a change of perception. Roger Guesnerie is an authority on sunspots as a branch of chaos theory: "The idea of sunspot equilibrium," he writes, "has been one of the most important ones . . . in economic theory in general over the last twenty years. It has made us reassess our understanding of economic fluctuations, of dynamic economies, and of the rational expectations hypothesis." Thus "sunspot theory" began as an insult and has developed into a useful instrument of descriptive economics.[25]

At this point he got up from his chair—dislodging two small bits of stone and a stack of student papers—to hunt down a thin magazine: *The Tree-Ring Bulletin*, containing a long article by two researchers from the University of Arizona.[26] The authors argue that while it is conceivable that general trends in sunspot activity might be reflected in tree-ring growth, they couldn't find any convincing evidence of a consistent relationship. "This came out in 1972," said Baillie, standing over me as I read. "There's been nothing better since." Eddy produced his groundbreaking essay a full four years later—without ever mentioning this awkward argument. My host turned back to his chair. "Of course, sunspots *do* probably have an effect on climate, but so do other factors. . . ." For a moment he seemed lost in his own thoughts. "We don't know what effect volcanoes have. We don't have good returns for past tsunamis, or how often the Earth has been struck from space; but people think I'm a little crazy worrying about all this."[27]

He didn't seem crazy at all. Baillie was talking several years before the eruption of Eyjafjallajökull in April 2010, but in April 1815 Indonesia's Mount Tambora exploded with even greater force—so much so that a third of it vanished in rocks and dust, in possibly the biggest eruption in recorded history: ten thousand people were killed outright, with ash and sulfuric acid blasted twenty-seven miles into the atmosphere. Global temperatures dropped, food prices soared, there were riots across Europe, famine, and epidemics.[28] Byron was so depressed that he wrote a poem about the Sun itself being destroyed: "I had a dream. Which was not all dream / The bright sun was extinguish'd, and the stars / Did wander darling in the eternal space / Rayless, and pathless, and the icy earth / Swung blind and blackening in the moonless air."

For all that, it does not help when other scientists propose so many cycles, and correlate them with so many different events: besides the 11-year one, the Sun is supposed to exhibit 27- and 154-day cycles, the 22-year cycle of magnetic reversal found by Hale, a Gleissberg cycle of about 80 years, and a 200-year cycle called the Seuss. Clearly, there are still questions to be answered. Meanwhile, new branches of research are springing up all the time. Data from SOHO have recently shown that gas flow affecting magnetic fields near the Sun's poles may have been responsible for the lack of sunspots, flares, and other storms from 2008 through the first half of 2009, which have extended the usual lull at the end of the eleven-year solar cycle for an extra fifteen months, with sunspots virtually disappearing.[29] But as of mid-April 2010, the sun appears to be waking up again, and a huge coronal mass ejection shot out of the Sun at about 1.1 million miles per hour, smashing into the Earth's upper atmosphere and setting off dazzling auroras around the Arctic and Antarctic—the most spectacular solar storm to hit Earth for three years.[30] We are back almost where we began: The Sun clearly creates climate; but just how much does it *vary* it? A vital question, tied as it is to the great debate on global warming.

CHAPTER 14

THE QUALITIES OF LIGHT

For the rest of my life I will reflect on what light is.
                    —ALBERT EINSTEIN, 1917[1]

We say "light," but we are thinking only of the Sun.
                    —GRAHAM GREENE,
                    The Power and the Glory[2]

NOT SURPRISINGLY, LIGHT SUFFUSES MUCH OF MY ACCOUNT OF THE
Sun. This chapter begins with mankind's attempts to track its speed and examines how light releases color, then shifts to cultural history, following it from where it is brightest through how it affects those on playing field and battlefield, through its diminutions, until it finally gives way to night.

What *is* light? Pythagoras' contemporary, the philosopher and poet Empedocles, brilliantly intuited that it was a streaming substance, adding that, because it travels so fast, we are not conscious of its motion. But he also held, along with Plato, that there was a "fire within the eye," as if we see through some kind of lantern. The more traditional belief was that an object emitted particles of light that bombarded the eye of the beholder, and that light was transmitted instantaneously. Not until 1632 was the idea advanced that light might have a measurable speed—the year that Galileo, in his polemic *Dialogue Concerning the Two Chief World Systems,* put the notion into the mouth of the credulous Simplicio:

> Everyday experience shows that the propagation of light is instantaneous; for when we see a piece of artillery fired at great distance, the flash reaches our eyes without lapse of time; but the sound reaches the ear only after a noticeable interval.[3]

Galileo was certain that light had a finite velocity, but his experiments were crude (he simply placed two lanterns on hilltops less than a mile apart), and he never came close to appreciating the tremendous speeds involved. The first serious estimate—when the Danish astronomer Ole Rømer (1644–1710), working from the Paris Observatory, calculated it to be 138,000 miles per second—was overshadowed by the debate between Huygens and Newton over whether light was made up of waves or particles. In 1728 the English scientist James Bradley, working on the premise that light travels at ten thousand times the Earth's speed in orbit, came up with the remarkably accurate figure of 185,000 miles per second (mps), which means that sunlight reaches us a little over seven and a half minutes after leaving the Sun.

Nearly a century later, further experiment by the French rivals Hippolyte Fizeau and Léon Foucault, one employing revolving mirrors and the other rotating toothed wheels, yielded results of 190,000 mps.* Then in 1859, Gustav Kirchhoff (1824–1887), a physicist from East Prussia, demonstrated that one could derive detailed information about celestial objects by studying the light they emitted: a transforming discovery that would lay the foundation for the new discipline of astrophysics. This was joined to a further breakthrough just a few years later—the equations that Einstein was to call "the most profound and the most fruitful that physics has experienced since the time of Newton": the great Scots scientist James Clerk Maxwell (1831–1879) discovered that a particle or photon of light was actually a self-sustaining bundle of constantly oscillating electric and magnetic fields. As the physicist Michio Kaku observes, "Maxwell suddenly realized that everything from the brilliance of the sunrise, the blaze of the setting sun, the dazzling colors of the rainbow, to the firmament of the stars in the heavens could be described by the waves he was scribbling on a sheet of paper."[5] Maxwell was even able to calculate the speed of light from the basic parameters of electricity and magnetism: almost a lifetime later, quantum physics would show that while light moves through space like a wave, it behaves like a particle when it encounters matter.

In 1887, Albert Michelson (1852–1931), a Polish Jewish immigrant to the United States, redesigned Foucault's experiment, and came up with a speed for light of 186,355 mps—twenty times the Frenchmen's accuracy, and enough to make Michelson famous.[6] In the 1880s, the general opinion was that light

---

* Fizeau's rotating wheel had regular small gaps in it. The light, passing through from a source on the farther side of the wheel, was reflected by a mirror and, if the wheel was rotating fast enough, returned through a different gap. Its speed could be calculated using the distance from mirror to wheel, speed of rotation, and spacing between the gaps. Foucault altered Fizeau's method to include a rotating mirror, rather than a spinning wheel. A light source was shone on the mirror, where it was reflected onto a distant second mirror, concave and fixed, then bounced off the fixed mirror back to the rotating one. If the mirror was rotating at high speed, the returning light hit it at a slightly different place, shifting the beam from its original path. The speed of light could then be measured by taking into account the speed of mirror rotation, the amount of shift, and the distance between the two mirrors.[4]

waves had to be transmitted through a "luminiferous"—light-bearing—aether (as it used to be spelled), assumed to be invisible and all-pervasive. Working with the chemist Edward W. Morley (1838–1923), Michelson argued that light required no such medium, and that its speed is constant whatever its situation: thus a beam from the headlight of a speeding train will not move faster than that from a stationary beacon. In this he overreached, and Einstein would prove that its speed does vary: in deep space in a hard vacuum, far removed from solar and planetary masses, it will travel more slowly than when near the Earth or similar large masses (though not just size matters—through diamonds, it travels less than half as fast).

Until recently, the slowest speed recorded was just over thirty-eight miles per hour—slower than a racing bicyclist—through sodium at –272°C. In 2000, a team at Harvard brought light to a standstill by shining it into a "bec" (which stands for "Bose-Einstein condensate," a medium for cooling atoms to extremely low temperatures).[7] So light not only travels at different speeds; it can, amazingly, be halted altogether.

In 1802, two years after William Herschel discovered infrared radiation, the English physicist William Wollaston (1766–1828) found that if he shone light through a prism in front of which he had placed a thin slit, the resulting spectrum displayed a series of parallel dark lines, like the cracks between piano keys; not until twelve years later did Josef Fraunhofer show that different wavelengths produce different colors, assigning a quantity to each color, thus making it possible to match the range of wavelengths to the visible "spectrum" (Newton's word). The color resulting from the shortest wavelengths was violet, red from the longest: anything beyond these two extremities we cannot see.*

Fraunhofer next set about passing a sunbeam through a prism. Unlike his predecessors, he saw something unexpected—"an almost countless number of strong or weak vertical lines which are darker than the rest of the colored image." Surprised by this discovery—some lines appeared almost a perfect black—he knew this was no optical illusion; rather, these were "gaps" in the solar spectrum, registering the absence of certain chemical elements in the

---

* Our eyes, each with some 125 million sensors, are good detectors but not great ones. The retina, or screen at the back of the eye on which the lens casts its image, contains two different elements. Rods register form; cones (so named because of the shape of the receptive part of the cells), color. Animals that are active mostly during daylight use a mixture of rods and cones, but the retinas of nocturnal creatures are packed almost entirely with rods. Owls, for instance, perceive the world largely in monochrome, but are able to do so in light so dim that other creatures cannot see at all.[8] Most vertebrates have at least two classes of cones; many birds, turtles, and fish, four or five. Humans get their sense of color from three different types of receptors (or cones) in the eye's retina. As they are stimulated in different proportions, our visual system constructs the colors we see.

Sun itself. As no two elements share the same lines, he could examine each set of colors and identify the element that caused it.

In 1854, an American, David Alter, ratcheted the argument up a notch by suggesting that each element had its own pattern of colored lines—a unique signature. The science writer Stuart Clark explains the significance of this finding: "If astronomers could determine which lines were produced by which vapors, they would gain a power beyond belief: the ability to deduce the chemical composition of celestial objects."[9] A few years on from Alter, Gustav Kirchhoff and Robert Bunsen (forever celebrated for the Bunsen burner—not his invention, but named after him)* combined prism and telescope into something they called a spectroscope, which one evening, on a whim, they turned on a fire raging in the nearby city of Mannheim and detected the presence of barium and strontium. If they could analyze the makeup of a local conflagration, why not the surface of the Sun? And so, several experiments later, they did. They argued that Fraunhofer's dark lines derived from the absorption of wavelengths in the Sun's atmosphere, and that the Sun's surface consisted of a hot radiant liquid, the studying of which would reveal all its parts.[10] They had done what Comte believed impossible: investigated the chemical composition of an object without actually having chipped off a piece for analysis.[11]

At the same time that Kirchhoff and Bunsen were building on Fraunhofer's ideas, another major figure of nineteenth-century physics, John Tyndall (1820–1893), discovered that when a beam passes through a clear fluid holding small particles in suspension, the shorter blue wavelengths scatter more than the longer red ones: a clear, cloudless daytime sky shows as blue because particles in the air scatter blue light more than red.[†]

---

* Bunsen was one of history's great experimenters. While still a young man, he was working on a sample of poisonous cocodyl cyanide when it exploded, shattering the protective mask he was wearing, destroying the sight in his right eye and almost destroying him. Undeterred, by 1861 he had identified sodium, calcium, magnesium, iron, chromium, nickel, barium, copper, and zinc in the Sun. He claimed that he never found time to marry, although a better reason may be that the chemicals he worked with made him stink. The wife of one of his colleagues said of him, "First I would like to wash Bunsen and then I would like to kiss him, because he is such a charming man."

[†] It may be no coincidence that our vision is adjusted to see the sky as a pure hue: we have evolved to fit in with our environment, and our ability to separate natural colors clearly is likely a survival advantage. Yet "blue" went unremarked until quite recent times: in the many hundred allusions to the sky in the Rig Veda, the Greek epics, even the Bible, there is no mention of the color.[12] In a 1968 study, Elizabeth Wood writes: "Physicists have found that light-scattering by the air molecules themselves accounts for the hue and density of the blue of the sky, whereas scattering by particles larger than molecules detracts from the blue color and gives the sky a milky appearance."[13] Light under the sea's surface is blue because water absorbs its longer wavelengths through depths up to about sixty-five feet; if you cut yourself below a certain level, your blood will not appear red.[14]

———

THE SUN IS reckoned to be brighter than 85 percent of the stars in the Milky Way (most of which are red dwarfs), but the brightest sustained light on the planet is not the Sun but the Sky Beam at the Luxor Resort and Casino in Las Vegas: aimed straight into space, it derives from thirty-nine xenon lamps, each of seven thousand watts, every one about the size of a washing machine. The resort's technical manager has explained that every night, before the beam is switched on, strobe lights flash for thirty seconds: "We don't want to surprise any pilots."*

Sunlight still has its special power, and throughout history soldiers in particular have learned both how to use that power and to guard against it. When warfare still allowed for some mutual consideration, soldiers would put a reflecting object next to the wounded, so the enemy would hold its fire. But more often than not, sunlight has been employed offensively. In 1805, during the Battle of Austerlitz, Napoleon ordered his men to leave their commanding position on top of a small hill, ceding the ground to the Austro-Russian army. The next day, the French troops were hidden by fog, but the enemy was exposed by the morning sunshine as a perfect target. Arriving at the battlefield of Borodino seven years later, Napoleon triumphantly alluded to his *soleil d'Austerlitz*. On occasion he would instruct his cavalry to put stockings of white fabric over their helmets; otherwise, reflected sunlight would enable enemy artillery to find its range at a substantial distance. (The straightness of Roman roads is probably due to legionaries' polishing their shields to provide bearing reflections to keep those roadways straight.)[17]

Reflected light brought real danger in the field, particularly in the days of colorful uniforms, which would disappear from the battlefield as field glasses became pervasive and the range, accuracy, and capacity of weapons improved. In *The Wouldbegoods*, E. Nesbit has her child narrators chat with an artillery

---

* *National Geographic*, October 2001, p. 33. Las Vegas and the Sun have an unusual quality in common: an abundance of neon glow. In 2005, scientists discovered that this element, the fifth most common in the cosmos, is some three times more abundant in the Sun than previously thought.[15] Las Vegas is perhaps uniquely a place where neither light nor time is seen as a welcome necessity. "The absorptions of casino life make mock of the futile cycles of sun and moon," writes Anthony Holden in his poker classic *Big Deal*. "Daylight, in Vegas, is the pushy, intrusive irritant delaying the beauties of neon-clad light—brighter, more beautiful and far more exciting. There are no clocks on view on any Las Vegas casino floor, another deliberate ploy by the management to speed your passage further and further from reality. The only time you keep in Glitter Gulch is that set by your body clock. . . . For reasons no one can satisfactorily explain, poker is essentially an act of darkness. By which I mean that it feels odd to play poker in daylight."[16] Las Vegas is not so much a city of darkness, however, as one of sinister light.

battery about to embark for the Boer War, whose friendly commander tells them how, with the introduction of the rifle, every kind of visibility is an invitation to die: "The guns will be painted mud-color, and the men will wear mud-color too."[18] The term "mud-color" would become *khaki* (Hindi for "dirty" and soon to be synonymous with war's new loss of glamor). That great unorthodox Catholic Charles Péguy, at the Battle of the Marne in September 1914, took out his field glasses to scan an enemy position and was instantly shot through the head. For much of that war, the French continued to wear their blue coats and scarlet trousers—a machine gunner's dream. But that conflagration was a hinge point: by 1914, countries like Britain already had their men in khaki; the rest would soon follow.

Sunlight can reveal, but it can hide too. In air combat, pilots seek to position themselves "up-sun"—that is, with its glare behind them—before they swoop down on the enemy.[19] The same principle applies at sea: on November 1, 1914, off the coast of central Chile, a Royal Navy squadron of four ships commanded by Rear Admiral Cradock took on eight German ships led by his friend Vice Admiral von Spee, each maneuvering to get the afternoon sun into the other's eyes, a serious disadvantage in the days before radar, when the naked-eye calculations of the gunlayers and fire controllers were crucial. A historian of the campaign describes what followed: "Cradock attempted to close the fire battle quickly, to find the sun behind him blinding the German gunners, but von Spee, with speedier ships, kept out of the range of the British guns and delayed the battle until the sun had set."[20]

"We were silhouetted against the afterglow," recalled an officer on one of the surviving British ships, "with a clear horizon behind us to show up the splashes from falling shells while the [enemy] ships were smudged into low black shapes scarcely discernible against the background of gathering darkness." The Germans closed in for the kill: in all, 1,654 officers and men lost their lives, while just three German sailors were wounded in the first British naval defeat since 1812.*

Sunlight can also dictate fortunes in the sporting arena. In NASCAR racing, a strong Sun helps the cars' tires to grip better. But the cheaper stands at American football and baseball games are called "bleachers" be-

---

* The Sun's rays can also take their toll in times of peace. A young Bob Dylan crashed his Triumph 500 motorcycle at the crest of a hill near Woodstock. As he told the playwright Sam Shepard, "I was driving right straight into the sun, and I looked up into it even though I remember someone telling me a long time ago when I was a kid never to look straight at the sun 'cause you'd get blinded." Again, in August 1947, the novelist John Dos Passos was driving with his wife through Connecticut when he was similarly blinded and crashed into a roadside truck, Dos Passos losing the sight in his right eye and his wife her life (a tragic episode that he later turned to fictional account).[21]

cause they are whitened by exposure to the Sun, and baseball and cricket are filled with examples of catches missed under its glare.[22] Not only catches: Visiting Lord's, the game's headquarters in England, the cricket-loving G. H. Hardy, a world authority on number theory, saw a batsman unsighted by a reflection from an unknown source. It was eventually tracked down to "a large pectoral cross reposing on the middle of an enormous clergyman. Politely the umpire asked him to take it off"—to the delight of the gleefully anticlerical Hardy.[23]

Solar interference can also be used to rationalize bad decisions, something I have experienced firsthand. In 1994 the Commonwealth Fencing Championships were held in a large ski chalet perched above a resort seventy-eight miles north of Vancouver. It was an exhilarating venue, with snow-covered mountains complementing the all-white-clad fencers. I was a member of the Northern Irish team, and by late afternoon the saber event was down to its last eight competitors. As I squared off against my Canadian opponent, I knew that whichever of us won would receive at least a bronze medal. The bout swung first one way, then the other, until we reached 14–all, with one hit to go. My opponent attacked, I parried and riposted. As he made no attempt to defend himself, but continued with a second attack, I knew the hit was properly mine. However, I hadn't reckoned on the referee. In international fencing, an official's nationality is not always scrutinized, and here ours was—Canadian. He said he couldn't award the hit, as the Sun had made it impossible to see whether I had parried the first attack effectively. As the setting Sun was *behind* him, and the chalet lighting was good, this was a difficult decision to accept, and some of my teammates gave loud vent to their feelings. The fight eventually restarted, and this time my opponent made a successful lunge, to win the bout and eventually the gold.

"LOVE IS SUNSHINE, hate is shadow," wrote Longfellow. The power of light is one thing, but its gradual diminution over the latter day, and indeed its momentary absence, are both part of the Sun's story. Shadows at their simplest are where light cannot reach, sometimes called "light's negative traces," daytime darkness. Many ancient societies held men's shadows to be their souls. From their point of view, a shadow, having no substance, is a nothing, an existential void. The East African Wanika people are even afraid of their shadows, perhaps believing that these inseparable emanations can watch all their actions and bear witness against them. Zulus say that a corpse throws no shadow. Peter Pan's snaps off when he leaps out of the Darlings' window (to be put in a drawer and later sewn back on by Wendy), Barrie perhaps subconsciously responding to the idea of a shadow being the externalization of one's self, and Wendy and her siblings are being asked to leave their real selves behind. "Who is it that can tell me who I am?" pleads King Lear, to which loyal Kent replies, "Lear's shadow."

At the other end of the cultural scale comes the ritual of Simloki, or Soldier Mountain, a five-thousand-foot extinct volcano near the California-Oregon border. As the Sun retreats in late afternoon, Simloki's shadow stretches until it falls on the far side of a valley some twelve miles away about an hour and a

The great ball of Krishna, Mahabalipuram,
on the southeast coast of India, 1971

half later. Simloki is a holy place to the nearby Ajumawi people, its shadow venerated as a spirit of great power. According to legend, two of the tribe's outstanding mythic forebears agreed to race against the shadows of large trees or mountains as these swept across the valley. From this tradition came an earthly contest. Anyone who outraces Simloki—a possible outcome, though rare—is believed to receive supernatural graces. Since the shadow is itself a spirit being, with commensurate powers, it is a unique opponent, and must be treated with proper respect. Thus, should a runner glance over his shoulder in midrace, he will be instantly struck down. Ajumawi braves make sure never to look back.

Webster's dictionary lists twenty-three meanings for "shadow," including a parasite, a spy/detective, and a stand-in for the devil, while "to cast a shadow" is to threaten happiness, friendship, fame. In the Middle East, those who bring bad luck "cast a yellow shadow"; and in *The Lord of the Rings* Tolkien creates the forbidding land of Mordor, "where the shadows lie." In ninja fiction, the cult stories based on the fourteenth-century Japanese assassins, these deadly figures can morph into "living shadows." More innocently, Japanese culture arranges flowers into *In* (shade) and *Yō* (sun). Shadows have even played their part in murder trials. In the 1920s a famous advocate nailed his man by showing a photograph of the defendant coming out of a church, where the shadows indicated a different time of day than the man claimed.[24]

A peasant shades himself from the Sun, Cambodia, 1952.

    The equation of darkness with evil acts is natural enough; yet one category of people specifically work at night to avoid the day: the sub-untouchables of India. To set eyes on such a person, even at a great distance, was deemed to pollute the viewer, and these wretches, if they ventured out in daytime, were usually murdered.* India was, and is, full of such dark laws. For instance, through the centuries it has been regarded as a profound offense to stand in a king's shadow or, worse, to let one's own fall across one's sovereign. Pollution by shadow—getting between the Sun and a person of higher caste—was a deep

* *Encyclopaedia Britannica*, 1973 edition, vol. 5, p. 25. On the other hand, in many societies it was at night that servants frolicked, their masters long asleep. Slaves' spirits rose as the Sun fell. A book on the subject of night quoted a North Carolina planter's judgement that for slaves, "'Night' was 'their day.'"[25]

outrage, while even a man of much lower birth might go out of his way to avoid having his shadow trodden on: a French politician, visiting in 1957, recalls villagers spending their daylight hours in trees to avoid this degradation.[26]

Darkness was not always to be shunned, and in hot climates shade was often coveted, even seen as a symbol of majesty. As potentates were frequently on the move, the twin demands of comfort and symbolism led to the invention of the umbrella (named from the Latin *umbra*, "shade" or "shadow") as a portable canopy. The sixth-century Sanskrit epic *S'akuntala* likens a ruler thus protected to a sunshade for his subjects, in that "the sovereign, like a branching tree, bears on his head the scorching sunbeams, while the broad shade allays the fever of those who seek shelter under him."[27]

In Hindu mythology, the Lord Vishnu visits the underworld carrying an umbrella, while in ancient Egypt and other Arab lands, umbrellas (often made from palm fronds) were carried over princes of the blood as one of their privileges. But soon that amenity was democratized, and an 1871 health manual for Anglo-Indians states authoritatively that "during the hot season it is unsafe for any European to walk out in the sun, on the plains, between eight o'clock in the morning, and four in the afternoon . . . without at least the protection of an umbrella."[28] Robinson Crusoe constructed one for himself in imitation of those he had seen in Brazil, "where they were very useful in the great heats there. . . . It . . . kept off the sun so effectually, that I could walk out in the hottest of the weather." Early versions in both Britain and France were consequently known as "Robinsons."*

Light modulates continuously over the day, each stage with its own character: dawn, sunrise, noon, twilight, dusk, gloaming, sunset. In Robert Louis Stevenson's *The Strange Case of Dr. Jekyll and Mr. Hyde* (1886), Mr. Utterson beholds "a marvelous number of degrees and hues of twilight."[29] The battle-weary subaltern in R. C. Sherriff's *Journey's End*, as dawn rises over the wire and trenches of the ravaged Western Front, exclaims, "I never knew the Sun could rise in so many ways till I came out here. Green, and pink, and red, and blue, and grey. Extraordinary, isn't it?" When the Sun is rising or setting—shining flat across the ground—every feature on Earth is caught up and dis-

---

* Quoted in S. Baring-Gould, *Historic Oddities and Strange Events* (London: Geoffrey Bles, 1945), pp. 132–38. In Europe, carrying an umbrella was long regarded as effeminate. The first Englishman who, from 1750 on, regularly sported one (made from oiled silk) is said to have been the sickly but dandified Jonas Hanway, founder of the Magdalen Hospital. Women may have carried them at an earlier date, and Beaumont and Fletcher allude to the new machine in their 1624 comedy *Rule a Wife and Have a Wife:* "Now are you glad, now is your mind at ease; / Now you have got a shadow, an umbrella, / To keep the scorching world's opinion / From your fair credit."

We now use almost horizontal light to discover the site of lost villages. The means is low-level oblique photography, a child of necessity born of the First World War. Air photographs taken at an angle had been produced even before the invention of the airplane, but after 1918 the skills acquired by air observers in the Great War was developed for peacetime surveys, and the Second World War only refined photographers' ability to discover what was lying beneath what was obvious on the ground. Once a land surface has been disturbed by human or natural agency, the imprint of that disturbance remains indefinitely, and where features exist in relief, they can be effectively photographed from above by taking advantage of shadows. Sometimes such photos not only reveal new features but suggest an order of succession—an introduction to not one but several older worlds. Skillful attention to such "shadow-sites" emphasizes outstanding features of form and shape better than ground surveys, and can convey the variations in tone of soil or surface growth that reveal the existence of buried structures: the discovery in the 1920s of Woodhenge in Wiltshire, the timber counterpart of Stonehenge, is one example: literally thousands of lost villages and Roman camps have been revealed by this method.[30]

played. Twilight inspired Hegel famously to write: "Only when the dusk starts to fall does the owl of Minerva [the Roman symbol of wisdom] spread its wings and fly"—his way of saying that philosophy comes to understand a historical condition just as it is on the point of disappearing.[31]

Just before the Sun sets there comes "that brief time between day and night when great plans are hatched and all things are possible."[32] Vladimir Nabokov in his memoirs makes a point of mentioning "the lovely Russian word for dusk"—*soomerki.*[33] It is at twilight, but only then, that an Orthodox Jew may study secular matters.[34]

Writers have risen to this moment. To Rat, at dusk in *The Wind in the Willows*, "the light seemed to be draining away like flood-water." "Real dusk," writes Henry James in *The Portrait of a Lady*, resorting to a parallel image, "would not arrive for many hours; but the flood of summer light had begun to ebb, the air had grown mellow, the shadows were long upon the smooth, dense turf." Others have called twilight "blind man's holiday," *l'heure verte*, even *entre le chien et le loup*, the time when the light is such that it is impossible to distinguish between dog and wolf.[35]

The fullest moonlight is just too weak to awaken color discrimination in the human eye, but because the Earth is round, most clouds catch the Sun's rays long after it has set beneath them, forming at such a height that they reflect the sunlight just over the horizon and glow with a ghostly hue—a phenomenon called "noctilucence." The distinctive light cast by the dying Sun on the Earth, seen as a blue-gray front of light that rises in the east after a clear sunset and often tinged with pink, is "twilight edge." Moments later comes "civil twilight," the minutes after sunset until the Sun is 6 degrees below the horizon; "nautical twilight" covers between 6 degrees and 12 degrees below, and between 12 degrees and 18 degrees is "astronomical twilight," an acknowledgment that it is finally getting dark enough for sky watchers to begin work. Once the horizon has lifted 18 degrees above the Sun, true night begins.

Reflecting on his travels in Africa, Carl Jung wrote, "When the great night comes, everything takes on a note of deep dejection, and every soul is seized by an inexpressible longing for light. . . . It is a maternal mystery, this primordial darkness. That is why the sun's birth in the morning strikes the natives as so overwhelmingly meaningful. The moment in which light comes is God. That moment brings redemption, release. . . . The longing for light is the longing for consciousness."[36]

CHAPTER 15

BENEATH THE BEATING SUN

The heavens are red-hot iron and the earth is burning brass,
And the river glares in the sun like a torrent of molten glass,
And the quivering heat haze rises, the pitiless sunlight glows
Till my cart reins blister my fingers as my spectacles blisters my nose.
Heat, like a baker's oven that sweats one down to the bone,
Never such heat, and such health, has your parboiled nephew known.

—RUDYARD KIPLING,
"Dear Auntie, Your Parboiled Nephew"[1]

I AM A TYPICALLY FAIR-SKINNED ENGLISHMAN, GINGER-HAIRED AND freckled: even a few minutes under the Sun and any areas I have left uncovered will turn an unsightly red, painful reminder that its rays are not only a source of vitamins: they also scorch. From wretched childhood afternoons on, I have been aware of what the Sun can do to the human body, even though my experiences have generally been of British summers, the mildest broiling possible. In recent years, I have stood in the middle of the Arizona desert, where temperatures climb over 120°F (49°C), and pondered what it must be like to be burned alive. The force of the southwestern sun bleeds color out of fabrics and wood faster than anywhere else on the continent, but even that heat does not come close to matching the hottest surface temperature ever recorded—136°F (58°C) on September 13, 1922, in the Libyan Sahara. In the United States, Greenland Ranch, California, seared at 134°F on July 10, 1913, holds the official record, although the true all-time high was most likely endured during forty-three consecutive days between July 6 and August 17, 1917, in Death Valley, California—so named in 1849 by one of eighteen survivors of a party of thirty that attempted a shortcut to the goldfields.

In his book *Equator*, Thurston Clarke describes a journey he made in 1986, stopping off at points along the zero-degree latitude, easterly from French Martinique, in the Caribbean, to Mount Cayambe, in Ecuador, whose name is Spanish for "equator." "I have become a connoisseur of heat," he writes at the outset, and indeed he creates a convincing taxonomy of his subject: "There is the heat that reflects off coral and scorches and softens the face like a tomato held over a fire. There is the greasy heat of a tropical city, a milky heat that steams a jungle river like a pan of nearly boiled water, a blinding heat that explodes off tin roofs like *paparazzi*'s flashbulbs, and a heat so lazy and intoxicating that all day you feel as though you are waking from a wine-drugged nap."[2]

When he reaches Lambaréné, the island where Albert Schweitzer established his mission amid the swamps of the Ogooué, the principal river of Gabon, in west-central Africa, he says he feels as if "wrapped in hot towels," and recounts the time that Le Grand Docteur, making his sweat-soaked rounds, was asked whether such heat bothered him. He never permitted himself to think about it, he said, then added, "You must have great spiritual activity to fight the heat." A handful of great men throughout history—George Washington and General Douglas MacArthur among them—have added to their aura of majesty by apparently not sweating at all. Barack Obama is said to be of their number.

Heat has long been the enemy of the unprepared traveler, particularly of soldiers on the move. In June 1578, Sebastian I of Portugal took a large force to northwest Africa, where his "armor grew so hot that he had to have water poured over his body under the metal plates, and the suffering of his troops, who could not afford this luxury, must have been extraordinary."[3] The army was annihilated, the prelude to a run of disasters: within three years Portugal had fallen to the Spanish Hapsburgs. In China at around the same period, a different kind of army under a hot sun—coolies digging canals—ended each day with their backs "chapped like the scales of a fish." Some historians still perpetuate the legend that when, on June 24, 1812, Napoleon and his Grande Armée of 691,501 (a suspiciously exact figure, but still an acceptable approximation for the largest force in European history since Xerxes invaded Greece) crossed the river Niemen and headed toward Moscow, they were destroyed by the cold. By the time the army reached Smolensk in mid-August, one-third had indeed been lost—but to a weeks-long heat wave. As Pierre Bezukhov comments in *War and Peace*, "Russia and hot weather don't go together."[4]

The Sun imposes a steep threshold of effort in a very short time, and the wise general seeks to protect his troops from the heat of day; but sometimes military tradition or plain ignorance trumps common sense—Scottish highland regiments, for instance, got up in kilts, found the backs of their legs got charred (especially after being in skirmishing line on their bellies), compromising their ability to march. For centuries, British troops died routinely of heatstroke during battle, because there was no drill for taking off one's coat—a deadly tradition that continued well into the twentieth century. In the Western Desert of Egypt and Libya during World War II, British soldiers in the Eighth

Army wore open-necked shirts and shorts, their quartermasters ignoring the fact that the best way to keep off the Sun would have been with loose-fitting, lightweight clothes that covered more skin.

Certainly Indians have long understood this and dressed accordingly—with good reason. Their country seems to have been the most systematically awful place for heat, a place where things *teem* most, where unfriendly micro-organisms and long-lasting infections conspire against you. Harry, Olivia's lover in Ruth Prawer Jhabvala's *Heat and Dust,* exclaims: "I stayed in my room all day yesterday and this morning. What else can you do in this hideous terrible heat? Have you looked outside? Have you seen what it's like? . . . No wonder everyone goes mad."[5] A junior officer under the Raj who failed to wear his topee (sun helmet) outdoors would be confined to barracks for fourteen days. For the British, going out at midday became almost an act of impiety—certainly of self-destruction: Carnahan, the broken protagonist of Kipling's *The Man Who Would Be King,* dies after wandering about bareheaded at noon for a mere half-hour.

Mad dogs and Englishmen—Noël Coward's famous ditty begins:

> *In tropical climes there are certain times of day*
> *When all the citizens retire*
> *To tear their clothes off and perspire.*
> *It's one of those rules that the greatest fools obey,*
> *Because the Sun is much too sultry*
> *And one must avoid its ultry-violet ray.*[6]

The song had its origins in Coward's disdain for British colonial society, whose refusal to adapt to foreign habits he observed in Malaya, which he visited in 1930. He may have been inspired by Kipling, who wrote in *Kim* that "only the devils and the English walk to and fro without reason," and later, "and we walk as though we were mad—or English"; but the phrase is Coward's.[7]

Outposts of the British Empire seldom provided a welcoming climate, and the lack of knowledge about how to deal with weather extremes accounted for much suffering, even loss of life. Charles Sturt, leader of the Royal Geographical Society's 1844 expedition into the Australian desert in search of a legendary inland sea, recorded: "So great is the heat that . . . our hair has ceased to grow, our nails have become as brittle as glass. The scurvy shows itself upon us all. We are attacked by violent headaches, pains in the limbs, swollen and ulcerated gums."[8] In British East Africa (modern Kenya) in the early years of the twentieth century, everyone buttoned on spine pads of quilted flannel, as it was believed that the Sun affected the backbone. Men wore cummerbunds under their revolvers to protect their spleens, while people with metal-roofed houses at high altitude wore hats indoors, as they thought the Sun penetrated iron. The first permanent doctor in Nairobi, setting up practice in 1913, held that people with blue eyes should wear dark glasses, and everything should be lined with red—a color believed to modify solar rays.[9]

Try as they might, the colonial British were trapped. In his fine study of the Indian Mutiny, Richard Collier writes:

> Dawn, 14 June 1857. Implacably the sun rose above Cawnpore. . . . Even in other years, granted the gift of shade, such heat had driven men to lock themselves in darkened rooms and blow out their brains. On this day five, perhaps six, people within Sir Hugh Wheeler's entrenchment would die of sunstroke. A band of steel seemed to encase their temples followed by a great drowsiness; they sank vomiting to their knees, faces blackening as they died. Today some muskets would split apart like fire-crackers, touched off by the sun. From the mud walls, men would see, across the arid sandy plain, a strange chimera of forest glades and blue sparkling water. Against their wills they would suck in the unbelievable, unbreathable smell of the entrenchment, like the contents of a thousand privies rotting beneath the sun. Small children pleading for a drink of water would be told to suck on a leather strap to quench their thirst.[10]

Even before the uprising, life had been barely tolerable. Families sat in semi-darkness, officers in undress summer uniforms, their ladies in loose white muslin. Church services, so much a part of Empire, were kept brief, but because of the heat few would kneel in worship. In one Calcutta station,

> the British did not know it but they were really fighting a losing battle against the sun, a sun that caused tempers to flare murderously, that made social standards at times of rigid importance, at other times a burden not worth the shouldering. For eight months of the year the sun was a monstrous weight pressing down on all their lives, curtailing movement, hedging them in darkened rooms. The sun sent pianos out of tune, dried ink as it touched paper, melted peaks of officers' bell-shaped shakos [tall, cylindrical caps] to a sticky black glue. . . . By common consent the greatest beauty had the palest cheeks in the station. . . . Houses were designed with one purpose: to cheat the sun of victory.[11]

Intense heat of this kind can bring equatorial illnesses peculiar to countries, such as miliaria, where sweat collects beneath the skin—one of the commonest diseases for soldiers fighting in the tropics during World War II; tropical acne; cataracts; cholinergic urticaria, an eruption of 1-to-3-mm wheals, mainly on the torso; phototoxicity and photoallergy—eruptions caused by the interactions between drugs and sunlight; solar eczema; and homopolar ringworm, in which the worm raises bumps on the skin where hair doesn't grow. Such diseases accounted for 12 percent of all outpatient visits in the Vietnam War.[12]

Great heat has ever been wrapped in fear and superstition. In ancient Greece, high noon was when demons came out, and it is no accident that Pan, a distinctly sinister god, is Lord of Noon. Of course, all regions are not created equal. The incidence of sunshine over the Earth varies from more than four thousand hours per year (more than 90 percent of the longest period possible) to fewer than two thousand. The Sahara gets the most, areas around Iceland and Scotland the least.

Until the current era of extreme population mobility, a person's skin color was likely to be influenced by where he or she lived. Around the equator, dark-skinned people would prevail, because dark skin contains above-average amounts of melanin—a pigment that is also a sunblock—which helps them adapt in areas saturated with light; light-skinned people would be more common toward the poles.

Skin color has come to have symbolic and even moral values divorced from its origins, so that throughout human history it has been used to assess the relative superiority of races as well as professions and classes. As the science writer Jonathan Weiner has noted, "Skin color, the biggest visible difference between the populations of human beings on this planet, around which so many bitter and divisive myths have been spun, is really nothing more than a schematic diagram of planetary levels of ultraviolet radiation."[13] Some dark-skinned people hold that as their color shows that they are nearer to the Sun, it also implies greater closeness to God; but in general there has been and continues to be a prejudice against darker skins in much of the world. Common definitions for blackness listed in the *Oxford English Dictionary*, some dating back to the 1500s, include "foul," "dirty," "wicked," and "horrible." Barack Obama tells in *Dreams from My Father* of flicking through an old copy of *Life* magazine when he was a child and coming across a picture of a man who had been chemically treated to lighten his complexion. His skin, writes the president, had a "ghostly hue," as if "blood had been drawn from the flesh."[14] The man "expressed some regret about trying to pass himself off as a white man, was sorry about how badly things had turned out. But the results were irreversible. There were thousands of people like him, black men and women back in America who'd undergone the same treatment in response to advertisements that promised happiness as a white person."

Even when it didn't have racial—and racist—implications, the prejudice favoring lighter skin has been consistent through the ages, partly because skin color has served as a signifier of social, economic, and religious standing. Those who performed manual labor outdoors were much more likely to be tanned than those whose professions were better paid and more highly esteemed (or those who were wealthy enough that they did not have to work at all).* In Minoan Crete (2600 B.C. on), women kept out of the light to maintain an ethereal look. From the mid-tenth century, European women were whitening their skins with potions every bit as destructive as the one used by the man in *Life*. Arsenic was the skin whitener of choice, even though its deadly properties were well known—as shown during the Italian Renaissance, when one Giulia Toffana created an arsenic-based powder designed for wealthy women to kill their husbands. The women applied the powder to both their pudenda and their cheeks; husbands who kissed too passionately rarely survived. Some six hundred widowhoods later, Toffana was detected and put to death, but still women whitened their faces with arsenic. Not only their faces: in 1772, British ladies started bleaching their hands with the poison to give them a porcelain tone, prompting the great potter Josiah Wedgwood to advertise black teapots, against which background the hands of the hostess looked even whiter.

Shakespeare seems not to have shared this preference. Three of his most delightful heroines are described as sun-browned—Olivia in *Twelfth Night*, Julia in *The Two Gentlemen of Verona*, and Beatrice in *Much Ado About Nothing*. He himself, according to Sonnet 62, was "Beated and chopp'd with tanned antiquity,"[16] as he realizes when looking in a mirror. He was certainly well aware of the prejudice of the day. Sonnet 127 asserts, "In the old age black was not counted fair / Or if it were, it bore not beauty's name."

During the reign of Queen Elizabeth I some women were so keen to appear pale-skinned that they used whitening agents that contained lead oxide, which, accumulating in the body, brought on grave pathologies, even paralysis and death. Others spread egg white over their faces to obtain a glazed look, and

---

* Ancient Greek statuary is an odd exception: Pliny the Elder (c. A.D. 23–79) informs us that the statues per city—Rhodes, Athens, Olympia—numbered in their thousands, and that one could go nowhere in the Greek world without encountering life-size bronze simulacra. While many statues were of painted stone, or polychrome, or gilded wood, more suggested that to be bronzed (although we do not know exactly what kind of color this would have been) was to be perfect, or at least envied. What must it have been like to have lived among such figures? Would it have suggested that to be suntanned was to emulate the gods?[15]

painted thin blue lines on their foreheads to give off a translucent glow. Elizabeth herself, traversing England on horseback to urge loyalty on her people, did not have the luxury of remaining indoors; but back at court she hid her weatherbeaten features behind a lotion called "ceruse," a lead-based skin whitener used in ancient Rome and revived in the Renaissance. In seventeenth-century journeyman portraits, women's faces are customarily rendered as corpselike white ovals. Later, the development of coach travel reinforced the fashion: being pale showed you could afford covered transport.

Cleopatra Dissolving the Pearl, 1759—Joshua Reynolds's portrait of the courtesan and actress Kitty Fisher. Lead-based potions she took to lighten her complexion killed her in 1767.

Under Charles II, heavy makeup was universal at court. The most perilous of the beauty aids employed were white lead, arsenic, and mercury, all of which not only ruined the skin over time but also made the hair fall out, corroded the stomach, brought on the shakes, and often killed the aspirant beauty outright. The danger in such cosmetics became widely known through the dramatic deaths of the courtesan Kitty Fisher and of Maria, countess of Coventry, the elder of the two legendary Gunning sisters of George II's day. Kitty and Maria maintained a famous rivalry as a result of Kitty's affair with Maria's husband, George William. As for Maria (1733–1760), she caught the eye of the notorious playboy Augustus Henry Fitzroy, third duke of Grafton and from 1767 to 1770 functionally Britain's prime minister, but the effort of retaining his affections came at great cost: she died at twenty-seven, after too many years of lead-based improvements. Catherine Maria "Kitty" Fisher died in 1767, apparently from lead-based preparations, seven years after her rival.

As the eighteenth century gave way to the nineteenth, whiteness contin-ued to be the primary prerequisite of beauty. In America, no southern belle or northern debutante dared go out without her parasol for fear of losing her lily-white coloring. Women in the South at the time of the Civil War took to chew-ing newspaper, as they had discovered that something in the ink whitened their faces. Freckles (the result of one's melanin granules being distributed un-evenly: a freckle is just an activated accumulation) were also to be avoided. Scarlett O'Hara's younger sister Suellen, finding that the family carriage has momentarily stopped in the Sun, cries out in dismay: "Oh, Papa, can't we get going? I can feel the freckles popping out on my face."

Even as late as the 1880s, ladies in society were prepared to resort to extreme measures. The Parisian beauty Amélie Gautreau, the notorious "Madame X" of Sargent's portrait, worked hard on her appearance: her "skin was suspected by some of her rivals to be the work of an enameller and by oth-ers to have been achieved by judicious doses of arsenic."[17] But by Madame X's time, fashions were changing. In Victorian Britain, people had started to shun makeup, associating it with prostitutes and actresses (considered by many to be one and the same). Any hint of tampering with one's natural color was frowned upon as vulgar—or worse. White skin as a mark of class may also have lost its claims as a result of urbanization, and because of the changing labor patterns brought on by the factories of the Industrial Revolution. With indoor work the norm for most working-class people, a pale skin was no longer a signifier of status. And with few officers returning from the trenches of World War I with pale skin, a tan or ruddy color was a sign of patriotism.

Besides, a new enthusiasm was in the wings. Seaside vacations, which had originally been recommended for reasons of health, were becoming a mark of the smart set. Those who had the money and the time to enjoy themselves wherever they wanted were choosing to go to the beach, and tanned skin came to be seen as a sign of membership in the propertied classes. As Paul Fussell recognized:

> For the Mediterranean to be re-appropriated after the [First World] War, its most ubiquitous natural asset, the sun, had to be redeemed from the social stigma it had borne in the nineteenth century. Then, the better sort of people had tended not to sit in it, believing that if its ef-fects were indispensable to the welfare of flora, they were of very dubi-ous value to persons.[18]

When in 1920 the couturier and society trendsetter Coco Chanel accidentally developed a tan while cruising the Mediterranean aboard a nobleman's yacht, her bronzed skin changed the course of fashion. Setting her imprimatur on the new look, in 1929 she declared, "A girl simply has to be tanned." The social prestige, or otherwise, of the color achieved was keenly debated: How intense should one's tan be? A golden quality was best, said those in the know.

Chief among the fashionable seaside venues was the French Riviera. Such consumptive luminaries as Robert Louis Stevenson and Aubrey Beardsley were early visitors; then came Greta Garbo, Cole Porter, Dorothy Parker, Anita Loos, Noël Coward, Zelda and Scott Fitzgerald ("[It] was a way of life, lived outdoors on the beach and under the sweltering sun").[19] A year-round tan was especially envied.

But even as people were discovering the delights of beach life, there remained some ambivalence. In 1916, the first British *Vogue* advertised a Helena Rubinstein cream that allegedly removed "impurities" including "sunburn, freckles and discoloration." ("Burn" and "tan" were interchangeable, with "tanning" having the unpleasant association of prepared leather.) And in the 1920s, the open touring car, which allowed one's face to brown, went quickly out of fashion.

Hollywood was developing its own standards: black-and-white film tended to make tans look an unhealthy gray, so stars who began their careers in this period, such as Garbo, Ingrid Bergman, and the two Hepburns, either kept themselves out of the Sun or used enough makeup so that their skin appeared luminously white onscreen. Another potential problem was identified quite clearly by Rudolph Valentino, who was careful to stay out of the Sun because he tanned so deeply that "I become like a Negro." He lightened his skin with a makeup prepared for him by Max Factor, while himself grinding up additional pigments to hasten the process so that he might escape having to play bit parts as a swarthy villain.[20]

Conversely, the legendary Josephine Baker, having escaped black St. Louis (where she was judged too light-skinned), became the toast of Paris as the star of "La Revue Nègre," the black stripper in a banana skirt. Suddenly, Parisian women wanted to imitate the Baker look, and she was soon promoting Bakerfix hair gel and Bakerskin darkening lotion—even though her own routine was devoted to skin-lightening milk baths and lemon rubs.[21]

While most white people certainly did not aspire to look black, the fashion in skin color among Western Europeans and trend-conscious Americans was beginning to change. Cary Grant would work on his tan almost daily, spending his vacations from filmmaking on the nearest sunny beach, holding a mirror under his chin to reach nooks hidden from the light. Over the same period there was a corresponding change in lifestyles. Women left their houses to enjoy outdoor life—going on hikes, picnicking, playing lawn tennis, and engaging in other acceptably "feminine" activities. Throwing away centuries of tradition, on beaches throughout Europe and the United States, they sunbathed, wore decorative sun hats and shawls—not for protection but as fashion statements—and went swimming. Afterward, they brushed brown and beige-tinted powders and creams on the areas of the skin that the Sun had missed.

To cater to the new enthusiasm, Ambre Solaire sun cream was invented in 1936. Across the classes, the suntan had arrived. By the end of the 1930s the fashion world was featuring clothes for women to show off their new tans: shoes were worn without stockings, and sleeveless dresses became stylish; bathing suits that had covered women's legs now left them bare. For those without the opportunity to be in the Sun or who didn't tan easily, cosmetics based on the leg makeup that was women's answer to the scarcity of silk and nylon stockings during the Second World War appeared on the market.

In France, access to the Sun—for both men and women—was finally democratized in June 1936, when the National Assembly introduced the two-week paid vacation, *les congés payés,* which "brought a glow to workers' pale cheeks."[22] Now at last working-class townspeople were learning how to picnic and to bicycle and getting their first amazed glimpse of the sea—but leisure time was so new to them that they had no vacation clothes, and were photographed at play in suits and hats. Stephen Potter, author of a series of humorous books on "Lifemanship," brilliantly caught the pallid Englishman's attitude to sunbathing at the start of the 1950s. The first action at a summer weekend, he writes, of "that confirmed hostess-nobbler, P. de Sint" is to strip to the waist and sunbathe, and his skin soon turns "a rich honey bronze." This is when his nemesis, Cogg-Willoughby, strolls by.

COGG-WILLOUGHBY: By Jove, you brown easily.
DE SINT: Do I?
COGG-WILLOUGHBY: Yes. You're one of the lucky ones.
DE SINT: Oh, I don't know.
COGG-WILLOUGHBY: They always say that the Southern types brown
    more easily.
DE SINT: Well, I don't know, I'm not particularly . . .
COGG-WILLOUGHBY: Oh, I don't know . . . Mediterranean . . .

Adds Potter: "Cogg-Willoughby was able to speak this phrase with an intonation which suggested that de Sint was of Italian blood at least, with, quite probably, a touch of the tarbrush in his ancestry as well." Thereafter, de Sint spends the rest of the weekend "trying to cover up every exposed inch of his body."[23]

But Potter, if he were ever a serious combatant, was fighting against the tide. Within a decade of the publication of *Lifemanship*, the reversal in skin-tone preferences was complete: bronzed skin on Caucasians signified both health and social standing, for reasons that were a reversal of those that had previously favored paleness. President John F. Kennedy was at least as obsessive as Cary Grant in his commitment to the perfect tan. During the week that he was preparing his celebrated inaugural address, he spent as much time tanning as he did working on his speech.*

The downside was that Kennedy's skin deteriorated rapidly. He might have looked youthful on television—in contrast to the sweating, sallow Nixon—but Gore Vidal, who shared a stepfather with Jackie, wrote that his stepbrother-in-

John F. Kennedy deepens his tan at Palm Beach during the winter of 1944.

law looked older than his photographs: "The outline is slender and youthful, but the face is heavily lined for his age."[25] By 1961 the Sun was already taking its toll.

By the 1970s, two entire generations had baked their bodies, and two decades later, Disney was having trouble getting its female staff at Euro Disneyland to wear stockings—a good tan was worth flaunting. Conversely, skin

---

* According to a recent biographer, "In 1961 a tan was considered evidence of well-being and a way to disprove the rumors about Kennedy's poor health. Hugh Sidey, who covered him for *Time*, remembers the whole family being obsessed with tanning. Jackie used a reflector, Bobby pursued the grail of a full body tan, and Kennedy himself had been traveling with a sun lamp since his 1946 congressional campaign. He worked tirelessly at keeping his tan dark enough to show up on black-and-white television."[24]

whiteners and lighteners (such as "Fair and White," from France) are still popular among many dark-skinned people worldwide. A label on a cosmetic manufactured in Paris but sold in Southeast Asia asserts that its product is "the first technology which regulates the different steps in the skin pigmentation process, to perfect the whitening efficiency." In 2007, plastic surgeons in London announced that they had developed creams using plant extracts and concentrated vitamin C to lighten skin—adding that they hoped to drive backstreet suppliers of potentially dangerous preparations (often containing bleach) out of business.

From the Hawaiian prostitutes in James Jones's 1951 novel *From Here to Eternity* (who were kept out of the Sun, as their servicemen clients preferred their women as white as possible) to Jennifer Lopez, Michael Jackson, and the heroines of modern Japanese novels, it still holds true: light-skinned people generally want darker skins, and vice versa.

THE WHOLE FASHION for the seaside holidays that started the trend for tans had begun with concerns about health—and eventually took on a distinctly erotic undertone. In a 1990s essay, the novelist John Fowles writes about the transformation of his hometown of Lyme Regis, in Dorset, as early as the eighteenth century, specifically because of the belief in the health benefits of sea and sun:

> How can something so nice have been ignored for so long? The change came, like most major human changes, from a conjunction of two factors. . . . In this case the medical profession and the first Romantics spoke as one.[26]

By 1780, Fowles asserts, the sea—its water, its air, the light and relief it gave landscapes—was all the rage. Between 1803 and 1804 Lyme played host to its most famous habituée, Jane Austen, and her family, and *Persuasion* shows her marked approval of what the town had to offer. As one of her characters comments: "I am quite convinced that, with very few exceptions, the sea-air always does good. There can be no doubt of its having been of the greatest service to Dr. Shirley, after his illness. . . . He declares himself, that coming to Lyme for a month, did him more good than all the medicine he took."[27]

The appeal of the French Riviera had similar origins. By the late eighteenth century it had become a fashionable destination for the British upper classes. Its popularity was solidified and much expanded in the 1830s, when Henry Peter Brougham (1778–1868), lord chancellor of England, traveling with an ailing sister, began visiting Nice at a time when it was little more than a fishing village on a picturesque coast, and built a house there. By the time of his death, Nice and its neighboring towns had become the sanatorium of Europe, and by 1874, shortly after the railway reached the Riviera, one observer calculated that "between seven and eight thousand English invalids . . . annually spent the winter in the south."[28]

The health benefits of being in the Sun got another boost in 1890, when the German bacteriologist Robert Koch showed that sunlight was lethal to the tuberculosis bacillus, and sanatorium designers were encouraged to exploit this discovery. However, most doctors were unconvinced, and the Sun's use as a medical aid in modern times is primarily due to Oskar Bernhard (1861–1939), who was surgeon-in-chief at a hospital in the Swiss valley of Upper Engadine when, on February 2, 1902, an Italian was admitted with a severe knife injury. Eight days after surgery the wound still gaped widely, and attempts to restitch it failed. The surrounding tissue was soft and spongy, and the whole discharged

English children were subjected to sunlamp treatment in an effort to safeguard them against tuberculosis, which spread alarmingly in both Europe and the United States after the First World War.

profusely, with no drying treatment having any effect. Dr. Bernhard then took the unusual step of exposing his patient to the Sun:

> By the end of the first hour and a half there was a marked improvement noticeable, and the wound presented quite a different appearance. The granulations became visibly more normal and healthy, and the enormous wound skinned over quickly.[29]

Bernhard began to treat other patients with sunlight, and found further healing effects: the discharge from foul-smelling wounds rapidly became odorless, and in addition to cleansing the lesions, sunlight had an analgesic effect. He decided to treat open tuberculous cavities the same way. At that time, the disease was rampant, claiming a million lives in Europe every year. Bernhard's approach was soon widely copied, one expert declaring:

> Every room occupied or visited by patients should be flooded by sunlight whenever possible, because of all disinfectants, sunlight has been shown to be the most powerful.[30]

In 1903 the Dane Niels Finsen was awarded the Nobel Prize for his use of artificial sunlight to cure tuberculosis of the skin. He was the theorist; others, like Bernhard, were the main practitioners. That same year, another Swiss physician, Auguste Rollier (1874–1954), began moving his patients to five thousand feet above sea level, the better to catch the ultraviolet light: the first patient he cured was his fiancée. Rollier pioneered a treatment in which very slow tanning in cool conditions was combined with rest and fresh air: a patient's feet would be uncovered first, followed gradually, over the course of fifteen days, by the lower legs, the thighs, the abdomen, and the chest, until finally the entire body could be exposed. It seemed to work.

But treating tuberculosis with sunlight remained on the borderline of accepted medical practice. When in 1905 Rollier presented his first results to a Paris audience, they walked out. Yet at his peak he was operating thirty-six clinics, with more than a thousand beds. (The leading practitioner of heliotherapy in Britain, Sir Henry Gauvain, did not believe that sunlight alone could cure advanced cases, and combined his patients' exposure to sunlight with sea bathing.)

Rollier's work attracted the notice of architects, and the number of designs for sunlit buildings increased. Le Corbusier may have been the standard-bearer in France, but between the wars, numerous beach resorts were built in southern Britain to draw people out of their bad housing and—for a while at least—

into healthy, "life-enhancing sunshine." By 1933, sunlight was said to be a beneficial treatment for more than 165 diseases. In the years after the Second World War, it was a standard medical precept that the more Sun we got the better. Schoolroom windows were made so that ultraviolet rays could penetrate easily, and boarding-school matrons would line up their charges at bath time to stand under sunlamps.[31]

The Sun's health benefits were first taken up in earnest in Germany with the treatment of the wounded during the First World War, after which sunbathing was recommended for those recovering from the flu epidemic of 1918–19 (which killed 21 million worldwide) and for children with vitamin deficiency ailments inflicted by the Allied blockade. Almost immediately, the "naturalism" movement began to advocate nudity for the sake of one's general well-being: indeed, nudists were the first modern sun worshippers.

After the First World War, a generation of writers grew up basing their philosophies on various aspects of "solar" thinking. The critic Martin Green has labeled them *Sonnenkinder*—Children of the Sun—dividing them into the dandy, the rogue, and the naïf, all set on bucking the traditional culture.[32] Many artists and writers were among these early disciples. As one historian wrote of Rupert Brooke and his intimates, "These Children of the Sun were enthusiastic nudists."[33] Before him, Walt Whitman had made naked sunbathing part of his nature cult, and in the early twentieth century Hermann Hesse followed suit, traveling to Italy to cure his headaches and gout.

For many years, "sunbather" was a euphemism for "nudist," while one of the first associations in Britain devoted to nudity was called the "Sun Bathing Society," and most naturalists belonged to "sun clubs." Giving naked sunbathing (and by clear implication its erotic potential) the royal nod while reveling in the eastern Mediterranean sun in June 1936, Edward VIII stripped down to demonstrate his longing for "freedom" on the yacht *Nahlin* with Mrs. Simpson. Other European nations at first responded fitfully to the craze, but from around 1925 there set in "the Solar Revolution," one of the most startling reversals in modern social history.[34] By the end of the decade the Sun had become the design motif of the era, appearing on everything from cufflinks to stained-glass suburban windows, garden gates to wireless sets.[35] But the enthusiasm remained greatest in Germany. Stephen Spender witnessed it firsthand:

> Thousands of people went to the open-air swimming baths or lay down on the shores of the rivers and lakes, almost nude, and sometimes quite nude, and the boys who had turned the deepest mahogany walked among those people with paler skins, like kings among their courtiers.

The sun healed their bodies of the years of war, and made them conscious of the quivering, fluttering life of blood and muscles covering
their exhausted spirits like the pelt of an animal: and their minds were
filled with an abstraction of the sun, a huge circle of fire, an intense
whiteness blotting out the sharp outlines of all other forms of consciousness.[36]

Spender's description also hints at the erotic charge of baring oneself to
these life-giving rays. As Fowles wrote of Lyme Regis:

For many decades sea-bathing remained what it had been to Jane
Austen, a medicinal activity . . . and those who still braved Neptune direct did it from wheeled cabins. But the Victorian spirit . . . began to see
the siren plainly—that is, to sense the always implicit eroticism and
sexuality of the beach.[37]

It was not just the bodily display: the Sun has always been an aphrodisiac—
"hot" had the meaning of "lustful" well before 1500. Only under the Sun,
wrote Lawrence Durrell, could "the essential male and female relationship"
flourish "uncomplicated by mirages and falsities." His fellow novelist Henry
Green went further: "The weather lies at the root of the way women and men
behave," with the result that "the English in their relations with each other are
less frank than other nationalities to the extent to which their skies are less
clear and so by the less amount of sun they have."[38] And indeed, many English
people looked with dismay at the new fashion, and the skimpiness of attire that
was part of its appeal. In Alan Herbert's *Misleading Cases*, he creates the story
of a man arrested around 1920 for wearing bathing drawers and not a full
bathing costume. A newspaper of 1925 records how in the south coast town of
Bournemouth attendants were empowered to prevent visitors from sitting on
the beach in their swimming suits, and bathers were commanded to "walk
straight into the sea and straight back to their bathing tents."[39] Sunbathing
continued to be banned in Bournemouth until the early 1930s. And as late as
1941, in Agatha Christie's *Evil Under the Sun*, sunbathing is presented as close
to exhibitionism.

Albert Camus, that great French-Algerian chronicler of philosophical
dread and human absurdities, has provided the nearest to what one might call
a philosophy of sunbathing. In 1939, in "Summer in Algiers," a long essay on
its pleasures, he wrote, "Algiers and a few other privileged coastal towns open
into the sky like a mouth or a wound."[40] One fell in love first and foremost with
the sea, but also "a certain heaviness of the sunlight." Following from that was

not the sunbathing hedonism of the rich, but the delights of the poor and the dispossessed, whose "pleasures have no remedies" and whose "joys remain without hope." In the summer, the young men go down to the sea to enjoy "the water's gentle warmth and the women's brown bodies":

> Not that they have read the boring sermons of our nudists, those protestants of the body. . . . They just "like being in the sun." It would be hard to exaggerate the significance of this custom in our day. For the first time in two thousand years the body has been shown naked on the beaches. . . . Swimming in the harbor in the summertime, you notice that everybody's skin changes at the same time from white to gold, then to brown, and at last to a tobacco hue, the final stage the body can attain in its quest for transformation.

More prosaically, in 1986 a leading dermatologist commented: "Why does anyone insist on tanning their hide? Because it's sexy. It's a cultural characteristic; a sign of health."[41]

CHAPTER 16

SKIN DEEP

Some people get too much sun down here. You must be careful.

—U.S. Head of Special Operations,
Rio de Janeiro, to Ingrid Bergman in
Alfred Hitchcock's *Notorious* (1946)

*Sunlight stimulates and enlivens; it is of help in almost all conditions. It is
the greatest of all natural tonics—like good champagne, it invigorates and
stimulates; indulged in to excess it intoxicates and poisons.*

—SIR HENRY GAUVAIN,
The Times, May 11, 1922

SUNSHINE CAUSES TWO COMMON SKIN CANCERS, BASAL CELL CAR-
cinoma and squamous cell carcinoma, and possibly a third: melanoma. More
than a million cases of the first two are diagnosed annually in the United
States, the mortality rate rising at about 3 percent annually since 1980.[1] One
American in six will develop skin cancer. In Britain, the affliction is twice as
common in teenagers and children as twenty years ago, and in the population
as a whole has risen by 25 percent in five years. The annual number of non-
melanoma skin cancers diagnosed in the United Kingdom is one hundred
thousand, a figure that could triple by 2035.[2] But take any country with hot
summers and a passion for sunbathing, and the statistics point in the same direction.

Skin is the body's largest organ (laid out flat, the average adult man's
would cover about twenty-one square feet), and in many ways the most vulnerable. Less than a millimeter thick in places, it is made up of several layers,
principally the epidermis (the entirety of which our bodies replace every
thirty-five to forty-five days), the dermis (containing our nerve endings, as well
as our sweat glands and hair follicles, which together control our temperature), and a layer of subcutaneous fat (containing the tissue that acts as energy source, cushion, and insulator). The epidermis has five sublayers, the

innermost of which, known as the Malpighian, is what is affected by tanning. Within this layer are the cells that produce melanin, the pigment that protects other tissue and increases in response to sunlight and causes us to darken.

Lack of melanin causes the serious condition known as albinism, a term coined by Balthazar Tellez (1595–1675), a Portuguese explorer who was sur-

A mother with her albino child, photographed in 1905 during a famine in southern Sudan. Most forms of the condition result from recessive genes inherited from both parents.

prised by how many of the West Africans he encountered had light skin and blond hair. Albinism is set off by a genetic disorder characterized by a lack of pigment in the eyes, hair, and skin (more rarely in the eyes alone). The principal gene causing albinism prevents the system from creating the usual amounts of melanin. It affects males and females equally, and most species of animals—mammals, fish, birds, reptiles, and amphibians. Nor are plants immune: the albino cactus, *Mammillaria albilanata*, protects itself from the Sun by an extremely dense profusion of stark white thorns. Because their lack of melanin means that they will burn under the slightest exposure, albino organisms must spend their lives defending themselves against the Sun. Up to one in seventy-five human beings carries the albino gene. It is a miserable inheritance: African albinos, for example, have a rate of skin cancer a thousand times greater than their pigmented brethren. None over the age of twenty is free of the disease, and in the sunniest parts of the continent only about one in ten lives past thirty.[3] They are often legally blind or afflicted by severe vision problems, because their retinas do not develop and are unable to absorb enough incoming light. Since melanin also affects parts of the body that never see daylight, their hearing and nervous systems are also compromised.

Unsurprisingly, as with so much related to the Sun, albinism's outward stigmata have aroused myths and superstitions around the world, particularly in poorer countries. In Jamaica, albinos are viewed as lower forms of being, to be held apart as cursed. In Zimbabwe, a folk myth has it that having sex with an albino woman will cure a man of HIV; many such women have been raped (and infected) by HIV-positive men. Elsewhere in sub-Saharan Africa, albinos are deemed to possess magical powers, a belief that has led to the ghastly practice of their being hunted down—forty-five of Tanzania's seven-thousand-strong albino population have been killed for body parts since 2007, fetching two thousand dollars a corpse. Fishermen on the shores of Lake Victoria weave albino hairs into their nets to increase their catch.[4]

OUR BODIES PRODUCE two different types of melanin: eumelanin (brown) and phaeomelanin (yellow and red). Red hair is an index of the body's creating more of the latter, less of the former, which induces poor tanning. Skin changes color because of the increased production of melanin under ultraviolet light. Laboratory tests on people of different skin types at the equivalent of noon on a summer day have showed that the most sensitive skins—those with the least amount of melanin—burn after fourteen minutes of exposure; the most resistant dark skin takes more than seven times as long. Ultraviolet light acts in three forms, A, B, and C. It is UVA, "black light," that tans the skin; at sea level, 99 percent of UV radiation is UVA. UVB is what reddens the skin in sunburn, and is considered the main cause of basal and squamous cell carcinomas, and possibly a significant one of melanoma. Although less likely than UVB to induce sunburn, UVA penetrates the skin more deeply and is considered the chief culprit behind wrinkling, leathering, and other aspects of "photoaging"—it not only increases UVB's cancer-causing effects, but also may directly cause some cancers, including melanomas. Since UVC is filtered out by the atmosphere and hardly ever reaches us, we don't have to worry about it for now, but the depletion of the protective ozone layer may change that in the future.*

In 1985, the American Academy of Dermatology (AAD), alarmed at the

---

* A recent *New Yorker* cartoon shows a man on a beach swathed from head to foot in protective clothing. "I'm not a Buddhist priest," he explains. "I'm a dermatologist." Meanwhile, sportswear companies have built extra sun protection into their clothes by washing them in sunblock, applying a coating to absorb rays, or choosing tightly woven fabrics that keep out ultraviolet light. Most contact lenses now contain ultraviolet protection.

growing incidence of these cancers, became the first medical organization to launch a public health campaign. Three years later, an AAD conference concluded that "there is no safe way to tan." By the end of that year, a leading modeling agency director (Eileen Ford) added her voice: "The tanned look is dead." Widely publicized as they were, these views had minimal effect. In a 1997 *Seventeen* survey, two-thirds of the teenagers polled said they "look better with a tan and feel healthier, more sophisticated." As *Women's Wear Daily* noted in 2000, "It will be a cold day in hell before there's a shortage of bodies sun tanning on the beach, and this summer, a sun-roasted hide is more fashionable than ever."

Today between 20,000 and 24,000 tanning salons are listed in the U.S. Yellow Pages, claiming 22-million-plus regular clients each year. The light emitted by the sunlamps in these establishments is about 95 percent UVA, 5 percent UVB. The industry turns over $5 billion a year—Sarah Palin, while she was in office, even installed a tanning bed in the Alaska governor's mansion—despite the fact that in recent years the World Health Organization, the American Medical Association, and the AAD have all labeled such beds the health-peril equivalent of cigarettes. The United Kingdom—which has eight hundred tanning salons—is averaging a hundred deaths a year from sun beds alone, and studies over the last four years have suggested that both outdoor and indoor tanning can be addictive. In November 2009 the Scottish Parliament forbade under-eighteens to use sun beds, and similar legislation is being considered in other countries.[5]

An oddity on any list of Sun-related ailments is the long-held fascination with staring at the Sun. For normal unprotected eyes, it takes about thirty seconds of direct exposure to inflict damage (in most people, natural reflexes make them look away, contracting their pupils). When injury does result, it takes the form of a blind spot in the center of one's field of vision ("solar retinopathy"): this can be temporary but more often proves permanent. Yet it is apparently so widespread an enthusiasm that the *Guinness Book of Records* now specifically refuses to consider entries for the longest time spent staring at the Sun.[6]

In 1840, Gustav Fechner (1801–1887), the German philosopher who founded experimental psychology, was blinded for three years as a result of a Newtonesque experiment, in his case peering at the Sun through colored glasses to investigate the effects on the retina. His injury was the more serious because he was looking through a hole in the shutter of a darkened room, which meant his pupil was expanded. Thomas Harriot, the co-rediscoverer of sunspots, made protracted observations of the noonday Sun and reported, "My sight was after dim for an houre"; while G. K. Chesterton (1874–1936) wrote a whole murder story, "The Eye of Apollo," based on the deceiving notion that one could, with training, look the Sun in the face.[7]

Even more dangerous than high noon is sunset, because it makes sun-gazing easier, even though the strength of UV rays is barely diminished. A similar danger lurks in the dim light of a solar eclipse. In 1999, after the occlusion over England, one London eye hospital reported that 10 percent of its admissions that day experienced some permanent loss of vision. Upon examination, doctors could pinpoint the actual phase of the eclipse that a patient had watched by noting the "sickle" on the patient's retina—the arc of swelling that corresponded to the crescent-shaped solar portion left uncovered at that moment by the Moon. Besides eclipse watchers, the people most susceptible to solar injuries are those who gaze up at the sun in a drug-induced stupor.*

TOO LITTLE SUNLIGHT can be as much of a problem as too much: for example, the very same rays that cause cancers also provide essential vitamins. Before urbanization, most people spent hours each day in the Sun, so lack of exposure was not an issue; but for the last two hundred years ever more people in the Northern and Western hemispheres have lived mostly indoors, so that many have grown up with insufficient sunlight and thus insufficient vitamin D—the substance that is produced in our bodies by exposure to the Sun. Besides strengthening our immune systems, this vitamin is vital for the absorption of calcium, which protects against rickets (a condition that softens bones and consequently deforms skeletons), and also combats osteoporosis, multiple sclerosis (people born in October—whose mothers have had a summer of sun—rarely get the disease; although this may be because UV light is doing something beyond making Vitamin D), rheumatoid arthritis, hypertension, premenstrual tension, diabetes, influenza, and several cancers.[9] It has been documented that deaths from cardiovascular disease are higher in the winter, when less Sun reduces levels of the vitamin.

The vitamin comes in two forms: D_2, derived from plants; and D_3, from ultraviolet B (UVB) rays and from animals (cod liver oil, salmon, mackerel, sardines, and fortified dairy foods are particularly rich sources). According to Dr. Michael Holick, the distinguished Boston University professor of medicine, between 90 and 95 percent of our intake comes from casual exposure to sunlight.[10] Both types are converted into an active form in the liver and kidneys.

---

* It also appears that about 25 percent of people sneeze when exposed to sunlight. Francis Bacon commented at length on the subject in *Sylva Sylvarum* (1635), so the malady has been known for centuries. Dr. Tom Wilson, a leading pathologist, writes, "We do not know exactly why this happens, but it might reflect a 'crossing' of pathways in the brain, between the normal reflex of the eye in response to light and the sneezing reflex. There is no apparent benefit from 'sun-sneezing,' and it probably is nothing more than an unimportant (but annoying) holdover of evolution."[8]

The young girl in the center of the photo, who suffers from the rare disease xeroderma pigmentation (or XP), which affects one person in a million, must wear protective clothing for even the briefest exposure to daylight, and goes through three bottles of sunscreen a week during the summer.

Between twenty and thirty minutes provides 10,000 units of the vitamin: an infant needs about 200 units every twenty-four hours, an elderly person 600, an adult about 1,000. The elderly are most at risk, in part because they tend to stay indoors—in Europe as a whole, one old person in three lacks adequate vitamin D.[11] In Saudi Arabia, despite the ample sunshine, women get rickets and other conditions caused by a deficiency of vitamin D because their traditional clothing covers nearly all their skin.*

Some people have no choice but to stay indoors: for them, going out in the Sun is at best a time of discomfort, at worst a death sentence. In the case of one disease, its small band of sufferers endure a nighttime existence of social isolation perhaps relieved occasionally by going to children's camps where they can

---

* This prompts the question, How long can one live without the Sun? Some people seem to exist almost entirely without sunshine. There have been communities that dwell in deep caves—the Trogloditae, found in desert communities in Tunisia, Morocco, and Libya, and the Miao, who have dwelled for centuries in the mountain caverns of southwestern China's Guizhou Province. In 1989 a young Italian, Stefania Follini, spent 130 days in a New Mexico cave, sleeping ten hours at a time but often staying awake for twenty-four. She so lost track of time that when Maurizio Montalbini, the sociologist conducting the experiment, informed her that her allotted four months' isolation was up, she thought she had been underground for only half that time. It would appear that the biological clock is not a perfect instrument, and will tend to run fast or slow if deprived of synchronizing clues.[12] But so far none of those who have experimented with periods underground have suffered long-term ill effects.

explore caves and build fires by moonlight, or by venturing forth in the day clad like a nuclear-reactor worker. They have xeroderma pigmentosum (or XP), which affects one person in a million—six thousand or so on the planet— and are incapable of repairing the damage ultraviolet light does to cells, leaving them a thousand times more vulnerable to skin cancers.* On average, the rest of us can withstand sixty years of regular exposure to sun before skin cancers may develop; XP sufferers develop the same tumors by the age of ten.[13]

I WILL LIMIT the list of disorders brought on or aggravated by the Sun: they make for distressing reading. But one ailment with wide-ranging implications is porphyria cutanea tarda (from the Greek for "red/purple"), a flaw in the body's ability to metabolize the red pigment in blood. While sunshine does not cause the disease, it exerts a profound effect on anyone who has it, which explains the fear of the Sun among sufferers.

Briefly: a component of hemoglobin, which helps transport oxygen around organs and tissues, is produced in eight stages, each catalyzed by a separate enzyme. In porphyria, one step fails to take place, causing a backup. The body compensates by dumping the compounds, often into the skin, which forces pigments to accumulate there and in bones and teeth. Even at its mildest, the disease causes abdominal pain, nausea, vomiting, weakness, confusion, rapid heart rate, and urinary problems.[14] At its worst, porphyrins, unthreatening in the dark, are transformed by sunlight into caustic, flesh-eating toxins, and if left untreated can eat away a victim's ears and nose and erode lips and gums to reveal red, disease-bared teeth.

The disease has become inextricably entwined with the vampire myth in general and Dracula in particular. Such myths go back thousands of years and appear in almost every culture (India boasts vampirelike supernatural beings who dissolve at the Sun's rising). The word "vampire" came into the English language in 1732, possibly from the Slavic verb "to drink," via bestselling books about a Serbian farmer who had reportedly been killed by a vampire and returned from the dead to feast on his neighbors. In 1897 the legend was reworked by an Irish writer named Abraham "Bram" Stoker, who combined the unlikely callings of theater manager and legal scholar. His inspiration was Henry Irving, the overbearing actor-manager of London's Lyceum Theatre, who he hoped would play the count in a stage version; and he gave his famous figure many of the marks of porphyria—including a dread of sunlight.[15] Ac-

---

* One notable character in fiction who suffers from XP is Christopher Snow in *Fear Nothing* and *Seize the Night* by Dean R. Koontz, while in *The Hobbit*, trolls turn to stone when exposed for even a second.

Max Schreck in the greatest of all vampire films— Nosferatu: A Symphony of Horror, *directed by F. W. Murnau in 1922. The name in the title is Slavic for "plague carrier."*

cording to Stoker, vampires can attack only between sunset and sunrise, so while Dracula may walk about even under the brightest day, he cannot exercise most of his powers except when the Sun is down.

The portrayal of vampires continued to flourish without much change or critical commentary until in 1985 the biochemist David Dolphin suggested that it was porphyria that underlay the original legends, pointing to the number of similarities between it and vampirism: porphyria victims are intensely sensitive to sunlight, and even mild exposure can cause disfigurement (hence the myth of the vampire as werewolf); to avoid sunlight, people with porphyria go out mainly at night; today porphyria can be treated with injections of blood products (centuries ago, sufferers might have treated themselves by drinking blood);* por-

---

* At death, a great deal of blood generally collects in the lungs. As the body decomposes, internal gases build up. After about four days, the pressure causes a "bloody purge," as a double lungful of blood is expelled through the mouth and nose, both of which remain full of blood, the face probably covered in it, and a large puddle collecting around the body. This may be the origin of the belief that some corpses were not "properly dead" but continued to live and began to drink the blood of fellow humans. Ancient methods of dealing with the "undead" may also have lent even more credence to the belief that they were still alive: when the bodies of suspected vampires were exhumed and a wooden stake driven through the chest, any pooled blood would spurt out, while the escaping gases would generate what sounded like a sigh. This was taken to be the release of the spirit of the undead from the curse of vampirism.[16]

phyria is inherited, but the symptoms may remain dormant until triggered by stress; and garlic contains a chemical that makes the symptoms worse, turning a mild attack into an agonizing reaction.[17]

Professor Dolphin's argument was never widely accepted, and indeed any discussion of vampires and porphyria may seem an exercise in melodrama; but the long-standing connection does underscore our deep-seated belief that sunlight ensures health, along with its converse—our fear of those who, whether porphyric or albino, seem to shun the light.

PORPHYRIA HAS A distant cousin: psoriasis. More than any other skin ailment, it is amenable to the healing effects of sunshine. Although the actual name (from the Greek *psora*, "to itch") was not introduced until 1841, the condition was first discussed by Hippocrates (460–377 B.C.): some scholars believe it to have been included among the skin conditions called *tzaraat* in the Bible, and it has frequently been described as a variety of leprosy.* Unpredictable and infuriating, it is also the most baffling of disorders, characterized by skin cells that multiply up to ten times faster than normal: as they reach the outer skin and die, their sheer volume leaves dry red patches covered with white scale. Psoriasis typically breaks out on the knees, elbows, and scalp but can affect the torso, palms, and soles. Its prevalence in Western populations is around 2 to 3 percent.[18]

Sufferers range from the late playwright Dennis Potter to the musicians Art Garfunkel and Elton John and include three revolutionaries, Joseph Stalin, Abimael Guzmán (the leader of the Shining Path movement in Peru), and Jean-Paul Marat (1743–1793), the radical journalist of the French Revolution, famously murdered while in his bath. Marat spent most of his days in that tub, in water mixed with medications and with a vinegar-soaked cloth around his head like a turban, because he had developed psoriasis while hiding from the police in the Paris sewers. Had he ventured outdoors, under the healing rays of the Sun, he might twice over have saved his skin.

Another victim was John Updike, who wrote that the disease "favors the fair, the dry-skinned, the pallid progeny of cloud-swaddled Holland [of whom Updike was one] and Ireland and Germany." In his memoirs he recounted that he had been under its thrall from the age of six. "Only the Sun, that living god,

---

* The history of psoriasis is littered with treatments of dubious effective but high toxicity—the application of cat feces, for example, in ancient Egypt; onions, sea salt and urine, goose oil and semen, wasp droppings in sycamore milk, and soup made from vipers have all been applied. In Turkey, "doctor fish" in the outdoor pools of spas are encouraged to feed on the skin of psoriatics, and cleverly consume only the affected areas. There is still no real cure.

had real power over psoriasis," he writes. "A few weeks of summer erased the spots from all of my responsive young skin that could be exposed—chest, legs, and face." As an adult, he attended Massachusetts General Hospital for ultraviolet light treatment "in a kind of glowing telephone booth." For Updike, the relationship with the Sun became intense, almost personal:

> From April to November, my life was structured around giving my skin a dose of sun. In the spring, at Crane's Beach, though the wind off the sea was cold, the sun was bright and the hollows in the dunes were hot. I would go there alone, with a radio and a book, and without benefit of lotion inflict upon myself two hours of midday ultraviolet. I wanted to burn; my skin was my enemy, and the pain of sunburn meant that I had given it a blow. Overnight, the psoriasis would turn from raised pink spots on white skin to whitish spots on red skin; this signaled its retreat, and by June I could prance back and forth on the beach in a bathing suit without shame.

But then something odd happened. Though sunlight healed Updike's diseased skin, it also gave him headaches and made him nauseous, dislocating any attempts he made at consecutive thought—"Writing," he lamented, "is a thoroughly shady affair." Then, in his early forties, he found the psoriasis was fighting back: "My skin, once so thirsty and grateful for solar rays, and so sensitive to them, had grown toughened and blasé. . . . At forty-two, I had worn out the sun."[19]

THE SUN (or rather its absence) is associated with one other very different illness: SAD, or seasonal affective disorder, so named in 1982 and defined as depression caused by the absence of sunlight during winter months. The official definition requires that a patient exhibit a lifetime of mostly winter depressions, including at least two straight years without any episodes out of season. Sufferers become anxious, irritable, and incapable of concentrating or acting decisively, and avoid social activities. They tend to eat and sleep excessively, their sexual drive decreases, and they are beset by fatigue, feelings of worthlessness, and guilt.[20] Women are three times as likely as men to be affected. SAD diagnoses account for about one case in three of clinically recognized depression, and only about 10 percent of its major manifestations. (A milder form, "winter blues," is said to be at its worst from December through February, mainly afflicting those aged eighteen to thirty.)

SAD is believed to involve a disruption of the body's "circadian rhythm" (from the Latin for "around a day")—the roughly twenty-four-hour cycle that governs biological processes in both plants and animals and acts as a body clock, affecting sleep, alertness, and hunger. The circadian "clock" is located in the brain's suprachiasmatic nucleus (SCN), which is in the hypothalamus; it takes in information about the length of day and night from the retina and passes it on to the pineal gland, a pea-sized structure at the base of the brain—sometimes called "the third eye"—which secretes the hormone melatonin. The amount of melatonin fluctuates cyclically, peaking at night (when it helps us to sleep) and ebbing during the day.[21]

Norman Rosenthal, the South African psychiatrist who gave a name to the condition, has himself been troubled by deep depressions since 1976, and now directs a company that makes ten-thousand-lux lamps (retailing at $250 each) to help the light-deprived; he and his students have produced many books and papers on SAD, reinforcing the impression of its ubiquity. Rosenthal has a website and runs clinics for sufferers—in the United States alone estimated at 14 million "severely affected" and a further 33 million with winter blues.[22] Forty-seven million Americans seasonally depressed? Common sense and experience, of course, tell us darkness *is* depressing—even in some of the sunniest parts of the world.[23] For most people, dark days and long, cold nights do lower the spirits, but as one of the people I interviewed told me, "Most people just get on with things." Which raises the question, Is SAD really a disease? There are studies that suggest that this is so, others that deny it, and yet others that argue that it exists but that the correlation between its occurrence and the absence of sunlight, or between it and the degree of latitude, is less direct than Rosenthal and others believe to be the case.

I became curious about what life was like in a place where the condition should be endemic—in the chilly polar wastes—and to find out, I traveled to the island of Tromsø, 250 miles north of the Arctic Circle, some two hours' flight from Oslo. The capital of the Norwegian province of Troms, it is the largest permanent community so close to the pole, with about sixty thousand inhabitants. Even though it receives the same amount of sunshine in a year as the tropics, it does so in a single concentrated stretch during the summer. For ten weeks each year, from mid-November until the end of the third week in January, virtually no sunlight falls: a season that people call *mørketiden*, or "the dark time," during which (at least, according to many press reports) depression soars, as do the figures for mental and physical illness, divorce, arrests for brawling, and suicides.

I arrived expecting to be burdened by the oppressive gloom of a place in which, as *The New York Times* put it, "The darkness and cold are enough to stiffen a soul nearly dead."[24] To my surprise, I found almost the opposite. For a start, there is no perpetual night. From about ten A.M. until noon most winter days the sky is silvery gray, quite enough to see by, and the abundant snow reflects what light there is. I arrived late in the evening, and it was odd the next morning reading a local newspaper in which all the exterior photographs seemed to have been taken at night.

I had arranged to see several experts at Tromsø University: the noted astronomer Truls Lynne Hansen; a senior librarian; and various members of the psychology department. Four were Norwegian, the others being a Mexican and a researcher from Ohio (who said she had left her home state because she found it so depressing). I also read three doctoral dissertations on SAD. And found, to my further surprise, that conclusive evidence is definitely not yet in.

A Norwegian research paper of 1991, involving 128 participants, neatly introduces its conclusions with a quotation from Hippocrates: "Whoever wishes to pursue the science of medicine in a direct manner must first investigate the seasons of the year and what occurs in them." It mentions that from the late eighteenth century, scientists regularly described patients with seasonal disorders, but ends by admitting that although SAD manifests at times when people are exposed to less daylight, "no definite proof of a causal relationship" can be found. Again, "It is not yet clear whether what is identified as . . . SAD represents a distinctive affective syndrome, a subtype of recurrent affective disorder, or the most severe form of a widely distributed population trait."[25] Another paper, completed in 1997, records, of the tests carried out in the Oslo area, that the examining psychiatrist was struck by the "non-depressive" look of her patients, who "generally scored low on the item 'apparent sadness.'"[26]

Judith Perry, one of the Tromsø psychologists, has studied whether sensitivity to seasonal changes has led to eating problems, expecting to encounter SAD more prevalently the farther north she went. But the results suggested the opposite: 20.7 percent of those tested in Nashua, New Hampshire (42° N), for instance, exhibited symptoms of SAD, against a figure of 11.3 percent in Iceland (62–67° N). "The interpretation of this discrepancy remains obscure," she admitted. "Nevertheless, there is reason to question whether latitude per se is critical to the development of SAD."[27] Another research initiative concluded: "This is the second study to find the prevalence of SAD and sub-SAD to be lower among Icelanders or their descendants than among populations along the east coast of the U.S."[28]

Four other psychiatrists reported on a study of one hundred participants from Tromsø, whose abilities over a range of cognitive tasks (they were all prospective airplane pilots) were tested in both summer and winter: "The conclusion is negative. Of five tasks with seasonal effects, four had disadvantages in summer. . . . Although the ideas in the SAD literature and anecdotal evidence and the extreme latitude were all reasons to expect a winter deficit in cognition in this study there was little evidence for one: there was more support for a *summer* deficit."[29] None of these doctors was denying the existence of SAD, just questioning whether it could be purely correlated with coldness and lack of light.

It was one of the Tromsø psychologists who put SAD into perspective. Her view was that it affected only those already suffering from clinical depression: it further depressed the already depressed. The researchers I talked with in Tromsø did not belittle the agonies depression can bring; but they believe that SAD has laid its grip on far fewer than is claimed by those who evangelize on the subject. On the other hand, there are those figures for crime and divorce, which may or may not be linked to depression. . . . It seems to me that we are still in the early days of understanding how we react to the loss of light.

I left Tromsø recalling a research trip to Heidelberg just a month before, when I had escaped from a depressingly rainy day into a local tavern. I was served by a Spanish girl, who had symbols of earth, fire, wind, and water, all framed within an image of the Sun, tattooed on the small of her back. I asked why she had chosen that design. "It is because I love the Sun," she replied. "And while it may not be here in Heidelberg, it is always *here*"—and she vigorously patted her backside.

CHAPTER 17

THE BREATH OF LIFE

Dear Proffesser, We are in the sixth grade. In our class we are having an argument. The class took sides. We six are on one side and 21 on the other. . . . The argument is whether there would be living things on earth if the sun burnt out. . . . We believe there would be. . . . Will you tell us what you think? Love and lollipops, Six Little Scientists

Dear Children: The minority is sometimes right—but not in your case. Without sunlight there is: no wheat, no bread, no grass, no cattle, no meat, no milk, and everything would be frozen. No LIFE.

—Correspondence between schoolchildren
and ALBERT EINSTEIN, 1951[1]

The Sun, with all those planets revolving around it and dependent on it, can still ripen a bunch of grapes as if it had nothing else to do.

—GALILEO GALILEI[2]

SEVERAL TIMES A MONTH, BILL ALBERS, HEAD DOORMAN OF MY NEW YORK apartment building, pushes under my door envelopes marked "From the Man at Re-write," each containing some item from a magazine or newspaper, which range from "Ripley's Believe It or Not: Pistol Shrimp have claws that fire shock waves to stun their prey, creating a wake of bubbles as hot as the surface of the Sun!" to recondite articles on solar energy or the latest theories about sunspots. One evening I told him I was writing about the Sun's effects on animals and plants, and while I would be discussing animal behavior, I had to begin with explaining photosynthesis. He asked how this was going. "Slowly," I told him. "The whole process is so complex." "Complex!" he echoed with a laugh. "It's simple—just repeat what you learned in primary school."

Maybe he is right. I well remember being taught that our fuels and foods derive from plants, and in turn their energy comes from sunlight, with the main difference between us and plants being that we (and all other animals) absorb energy from the Sun indirectly, in the form of food, while plants receive it directly. The process by which plants do this is called photosynthesis ("putting together with light"), which occurs both in bacteria and in "photo-autotrophs"—that is, any plant organisms that can synthesize their own food by using sunlight, converting energy from a physical activity into a chemical process. It is under that latter heading that most plants are classed.

By means of a compound called chlorophyll ("leaf green"), plants use the hydrogen of water to transform carbon dioxide into more complex chemicals based on carbon, including sugar molecules such as glucose; the water's oxygen is then liberated as a gas—a plant waste that, while precious to us, certainly has its drawbacks. As Bill Bryson observes in *A Short History of Nearly Everything,* oxygen, although critical to animal life on Earth, is more often than not toxic: "It is what turns butter rancid and makes iron rust. Even we can tolerate it only up to a point. The oxygen level in our cells is only about a tenth the level found in the atmosphere."[3]

Carbon is passed into leaves through the energetic effect of sunlight, and those leaves pass on the materials to form the wood of a tree or the petals of a flower, the chlorophyll within them capturing energy from the violet and red parts of the solar spectrum and transforming it by driving electrons around certain molecular circuits in a series of chemical reactions. The Sun engages over 25 million square miles of leaf surface every day. But just 1 to 3 percent of the light falling on a green plant becomes biological energy; the rest is lost by transmission, reflection, or ineffective absorption.[4]

Photosynthesis takes place in structures within cells called chloroplasts, typically only a few thousandths of a millimeter in width. They contain the chlorophyll and other chemicals, especially enzymes ("levelers": proteins that control specific responses). Scientists don't yet entirely understand the complex biochemistry of photosynthesis, although it is the most important metabolic innovation in our planet's evolutionary history. Each day in midsummer, an average acre of corn creates enough oxygen to meet the respiratory needs of about 132 people. The rates at which living beings die and consume one another are so high that we would all disappear within a human lifetime were it not for this process.

As far back as the 1640s, researchers suspected that plants required air and water to grow, and by the 1700s had begun to identify the individual gases involved in combustion, respiration, and photosynthesis. Then a major step forward took place—thanks to a city brewery. In 1772, plans were well advanced for Cook's second voyage in search of the hypothetical Great Southern Land. Several scientists were given permission to join the expedition, and the Royal Society at first approved the inclusion of an astronomer-botanist named Joseph Priestley (1733–1804); but his notoriety as a religious and political freethinker led to the withdrawal of his invitation. Priestley instead took up the post of salaried literary companion to Lord Shelburne, a Whig grandee. While in Shelburne's service, he settled down to his experiments.

Priestley's previous job had been in Leeds, where his house abutted a brewery, and he had been drawn to experiment with the fumes given off by the fermenting beer, which he noticed remained in the vats to a depth of a foot or so and did not mix with the air above it. This gas was carbon dioxide (or "fixed air," as he called it), and his tests showed that lighted candles would go out when put into it. He knew that the air we breathe differs from carbon dioxide, but wasn't sure how many other forms there might be. Soon after taking up his post with Shelburne, he demonstrated that air was a mixture of gases, not an element (as the ancient Greeks had propounded). He argued for the existence of a near-weightless substance called "phlogiston," which he posited had to be present to create fire. On surer ground, he also identified nine separate gases, all of which he believed to be polluted versions of "normal" air. These gases would later be named nitrous oxide ("laughing gas"), ammonia, sulfur dioxide, hydrogen sulfide, carbon monoxide, chlorine, silicon tetrafluoride, hydrogen chloride, and something he called "de-phlogisticated air"—an awkward term for what came to be called "oxygen" (from the Greek, "the acid maker").

Priestley showed that a candle would burn more brightly when this gas was pure, and that even a mouse could survive in it (previous experiments had sent several mice, happily drunk on beer fumes, to an early grave).* The gas was "between four and five times as good [i.e., for respirability] as common air," he said; given that we now know the air we breathe is 21 percent oxygen, the rest being mainly nitrogen, this estimate was remarkably accurate. He showed that a candle in a closed container eventually spluttered out, but after a sprig of mint was added, the flame would be invigorated: he had shown that plants release oxygen.

Priestley went on to demonstrate that oxygen is taken up by blood in the lungs, and that water is a compound of hydrogen ("water maker") and oxygen, in the proportions of two to one, as measured by gas volume. But that was the

---

* In 1773, a poem by one Anna Barbauld described the experiment from the mouse's point of view, "The Mouse's Petition to Dr. Priestley, Found in the Trap where he had been Confined all Night"—perhaps the first animal-rights manifesto ever written. Commentators on Priestley have suggested that his interest in asphyxiation was spurred by reports of the Black Hole of Calcutta (1756), the horror in which forty-three of sixty-nine people in hot, airless incarceration had died. Around the world, caves and similar confined spaces collect carbonic acid naturally, to deadly effect; twelve miles outside Naples, on a mountainside near Lake Agnano, is the Grotta del Cane (Grotto of the Dog). Carbonic acid gas, being heavier than normal air, gathers at the bottom of the grotto, so while taller animals—such as humans—can breathe easily, smaller animals perish, or are thrown into convulsions. In the nineteenth century it became a loathsome local industry to push dogs into the grotto to regale tourists with the gas's effect.[5]

end of his original work in this area. While he had demonstrated that plants revived the air after animals vitiated it, he failed to realize that they needed sunlight to do so—which seems odd, as he was well acquainted with the work of Stephen Hales (1677–1761), who had speculated that leaves were the "lungs" of a plant, and asked, "May not light also, by freely entering the expanded surfaces of leaves and flowers, contribute much to ennobling the principles of vegetables; for as Newton puts it in a characteristically probing query, 'Are not gross bodies and light convertible to one another?'"[6]

Unfortunately, Priestley soon became less interested in science than in preaching against the divinity of Christ and campaigning against slavery. In 1791, a Church-and-King mob razed his meetinghouses to the ground and moved on to his home, reducing it to ashes, laboratory and all. Eight rioters and one special constable were killed. The great scientist took himself off to America, and the baton passed to the Dutch botanist Jan Ingenhousz (1730–1799), whose research revealed both that stale air could be revitalized only by the green parts of plants, and that this required the aid of sunlight.*

Ingenhousz also discovered it was the Sun's energy in the form of light, not warmth, that was essential for plant respiration. He demonstrated that under the action of light, plants absorb carbon dioxide through minute pores in the surface of their green parts, giving off tiny bubbles of oxygen; in the dark, the bubbles eventually stop. "It seems to be more than probable," he wrote, "that leaves are destined to more than one purpose":

It is also probable, that the tree receives some advantage from the leaves absorbing moisture from the air, from rain, and from dew; for it has been found a considerable advantage to the growth of a tree, to water the stem and the leaves now and then. . . . It will, perhaps, appear probable, that one of the great laboratories of nature for cleansing and purifying the air of our atmosphere is placed in the substance of the leaves, and put in action by the influence of the light.[8]

As the revolutions of science gathered speed, so, too, did the unraveling of the mystery of photosynthesis. In 1845, Julius Robert Mayer (1814–1878) ex-

---

* A debate has raged through the centuries over what degree of credit Priestley deserves. Thomas Kuhn, in his study *The Structure of Scientific Revolutions*, favors either Antoine Lavoisier (1743–1794) or Carl Wilhelm Scheele (1742–1786) as the discoverer of oxygen and Jan Ingenhousz of photosynthesis; but his main point is that such breakthroughs are never as simple as legend or history suggest.[7] Priestley himself, writing to a friend about Ingenhousz's work, mentions the considerable difference between the day and night activities of plants as something "he hit upon and I missed." It was Lavoisier who measured another form of air called simply *gaz* (a word not yet assimilated into English and meaning "ghost, spirit") and renamed it "hydrogen."

plained that plants converted light into chemical energy; just how this came about, however, was to preoccupy scientists for over a century. In the 1920s, photosynthesis was confirmed to be a number of successive steps that combined two separate and opposite uses of sunlight: the photooxidization of water and the photoreduction (i.e., deoxidation) of carbon dioxide.

During the 1950s and 1960s, the American chemist Melvin Calvin demonstrated that light reactions generating chlorophyll were not a protracted process but transformed the Sun's energy instantly.* Working with green algal cells, he was able to identify at least ten intermediate products formed within a few seconds. Over the past decade, researchers have pieced together the story of how the photosynthesis necessary for a flourishing biosphere first evolved on Earth.[11] Other researchers are trying to work out how to duplicate the chemical processes involved, in order to find new sources of energy. A pan-European collaboration called Solar-H has been set up to find ways of extracting hydrogen by means of sunlight, to help store energy as a fuel. A consortium in Sweden is studying artificial analogues to photosynthesis. Other groups are developing algae with small antennae to generate local, low-tech biomass energy. And so on.

Photosynthesis may at one level be easy to comprehend, but (as Bill Albers and I came to agree) it is not simple. Indeed, studies of it are a growth industry, and one's head spins at some of the questions that have occupied recent years: How do plants respond to too *much* light? What are photosynthetic bacteria doing inside grains of desert sand? If one reverses the processes of photosynthesis, can water molecules be put back together? How do the north and south faces of a vineyard differ in their sugar production? Some of these questions, at least, can now be answered.

THE RESPONSES OF the natural world to the Sun are wondrous. Several species of mollusk that register daylight lay down a certain thickness of cells,

---

* The biochemist George Wald has argued that chlorophyll is so well suited to photosynthesis that it is the only pigment that can function in the role, and that life on other planets would have to evolve chlorophyll in order to photosynthesize. However, in April 2008, *Scientific American* conjectured what plant life might look like elsewhere and reckoned that photosynthesis not necessarily involving chlorophyll was very likely on other planets, quoting H. G. Wells in *The War of the Worlds*: "The vegetable kingdom in Mars, instead of having green for a dominant color, is of a vivid blood-red tint." Mars, the magazine argued, has no surface vegetation; but light of any color from deep violet through the near infrared could set off photosynthesis, which adapts to the spectrum of light that reaches organisms, which in turn is the result of the parent star's radiant spectrum.[9] When photosynthetic organisms first appeared on Earth, the atmosphere lacked oxygen, so they could not have used chlorophyll. Not even Earth was always green.[10]

which thickness—a sandwich of daily imprints—directly correlates to the number of hours of daylight to which the mollusk was exposed, and one can gauge their age by the number of layers. Coral fossils found in Devon, in the south of England, reveal an extraordinary periodicity in growth rings—about four hundred in each annual set—by which evidence we can calculate that some 370 million years ago, there were about four hundred days in the year, each lasting about twenty-two hours.[12]

Many myths tell of the power of the Sun to affect nature. "The Sun breeds maggots in a dead dog, being a god kissing carrion," says Hamlet, reflecting one superstition. *Antony and Cleopatra* repeats another: that the Sun's rays engender serpents by spontaneous generation, an old mummy's tale disproved only in the seventeenth century. A hundred years on, the French inventor Joseph Nicéphore Niépce (1765–1833) coined the term "actinism" to describe the Sun's ability to produce chemical effects in nonliving things; some minerals, such as white marble, can become phosphorescent—light-emitting—by long exposure. Niépce noted that "granite rocks, and stone structures, and statues of metal, 'are all alike destructively acted upon during the hours of sunshine.'"[13] More, the Sun starts fires, sinks ships by warping their planks, and in the world's hottest zones can "roast [rocks] into disintegration."[14] In 1814, the British scientist Humphry Davy subjected a diamond to intense heat using a huge solar lens: the gem eventually burst into flame, leaving a fine crust of charred carbon, proving that it was little more than a lump of coal.

Julius von Sachs (1832–1897), one of the greatest nineteenth-century German scientists, systematized what he called "phototropism," from the Greek for "light turning"—the tracking of the Sun by organisms. (I saw this in action when in July 2006 I visited the most advanced tomato farm in southern Spain, where seedlings are turned on their sides twice daily, first one way, then the other, so that their stems will always bend toward the Sun, making them stronger.)* Plants position themselves with extraordinary ac-

---

* Later in the day, the seedlings are taken out again and stroked with small hairbrushes, building up the stems' resistance in order to strengthen them still further. A side effect of absorbing a great deal of sunlight is absorbing a great deal of heat: most plants' leaves are prone to heat stress and, if overexposed to ultraviolet rays, shrivel and die. So the roofs of the greenhouses are now made of light-dispersing plastic and no longer of glass, which can lead to sunburned tomatoes. This is ironic, because recent research has suggested that eating tomatoes can protect against sunburn and premature aging of the skin. Experts at Manchester and Newcastle universities in England have found that the fruit improves the skin's ability to protect itself against ultraviolet light: the effect is thought to be caused by the pigment that makes tomatoes red.

curacy to catch as much sunlight as possible: gaze up at a forest canopy, and the leaves overhead will usually form a near-continuous ceiling, like the pieces of a jigsaw. They are not acting as some neighborly cooperative, but are instead fierce competitors for light. The more of it they can catch, the better their chance for survival, and some have evolved particularly ingenious mechanisms for doing so. Thus the giant edible aroid that flourishes in marshy parts of the Borneo rainforest not only has leaves ten feet across and a surface area of more than thirty square feet; it coats the underside of its leaves with a purple pigment that catches the light after it has passed through the thickness of the leaf, giving the chlorophyll a second helping, as it were. Begonias found on the same forest floors have transparent cells in the upper surface of their leaves that act as tiny lenses, gathering and focusing light onto the chlorophyll within.

Generally, competition drives plants to grow taller, but to do that they need structures to prevent them from toppling, so roots become thicker, either spreading laterally through the surrounding soil or plunging downward. Trees offer a particularly efficient solution to the problems of photosynthesis, but we should be careful not to be earthbound in our thinking. As Oliver Morton writes:

> Think of a birch tree in winter, its leaves lost, its architecture revealed in dark lines against cold gray cloud. . . . Take away the tree's established "common sense" context by turning round, bending over and looking at it upside-down through your legs. Its growth looks less like something pushed from the earth than it does something drawn from the sky. . . . The bulk of the tree is not made from the soil beneath it—indeed, the soil is in large part made by the tree. . . . Trees are built from sun and wind and rain. The land is just a place to stand.[15]

The Sun plays an even greater part in the life of those organisms that are constructed around a central core. This includes the largest family of flowering plants, the composites, especially sunflowers and other flowers that are shaped like daisies (the word itself meaning "day's eyes"). The sunflower, whose turning to follow the Sun was first scientifically described by Leonardo da Vinci in his botanical studies, was imported into Europe about 1510 from the Americas by the Spanish. The flower was sacred to the Aztecs, and had been used as an emblem of the sun god by the Inca; within a few decades of its arrival in Europe it had become an accepted symbol of loyalty because of the way it followed the Sun so closely.*

David Attenborough begins *The Private Life of Plants* by declaring, "A shoot kept in the dark will creep towards a single chink of light. The plant can see."[16] Pardonable hyperbole. The pursuit of sunlight obtains even under extreme tem-

In 1745 the Swedish botanist Carolus Linnaeus proposed a flower clock to complement sundials: it would indicate the time within a half-hour of accuracy by means of flowers known to open and close at certain times of day. This dial was painted in 1948.

peratures: certain polar animals regularly mark out their territory by locating the Sun against particular landmarks, while *Lecidea cancriformis,* an Antarctic lichen, can carry out photosynthesis at −4°F (−20°C). The Arctic poppy will face east from midmorning, bending westward from late afternoon (the motion is carried out by motor cells in a flexible segment just below the flower, the pulvinus). The high-altitude snow buttercup aligns itself in a similar way, sunlight helping the flower to maintain ideal levels of temperature and humidity and enabling it to attract insects more effectively. Nature forever adapts.

---

* Several flowers do not welcome sunlight. There are at least three thousand species of nonphotosynthetic plants, many of them parasitic on other plants, often feeding off fungi, which in turn draw their energy from trees. Such plants are an odd tribe. Both peanuts and truffles ripen underground (as succulent roots, such as potatoes and carrots, obviously do). In 1729 the French astronomer J. J. de Marian noticed that certain plants in darkened rooms opened up by day and closed at night independent of any sunlight. The ghost orchid, which entirely lacks chlorophyll, spends so much time underground, flowers so irregularly, and grows in so few places that in some countries it has been declared extinct, only to reappear. A Western Australian orchid, *Rhizanthella gardneri,* blooms underground and never emerges above the soil at all. At the start of the autumn rains, it extrudes a tulip-shaped structure that grows upward, lifting the soil above it so that cracks in the earth form, from which drifts a faint insect-attracting perfume.

In 1920, researchers working for the U.S. Department of Agriculture discovered that the flowering of many plants was controlled by the amount of daily light they receive, and introduced the term "photoperiodism" to describe this response. Plants were subsequently categorized as being short-day, which will not flower when daylight exceeds a certain critical length; long-day, which will not *unless* the days exceed a critical length; and intermediate, which bloom whatever the duration. It was next learned that for plants to flower, the duration of daily *darkness* is also crucial. Thus short-day plants bloom when nights are long, and long-day plants when nights are short or nonexistent. One wonders if talking to any informed countryperson would not have led to the same conclusions, for the names of many plants have long reflected this sensitivity to light and dark. At least fifty varieties of flower observe regular hours of opening or closing, some of which are named accordingly: hence calendulas ("little weather glasses") became first "gold-flowers" (as a quite different flower, chrysanthemums, are now known), then, through an association with the Virgin Mary, "Mary's gold," and finally "marigolds," celebrated for opening their flowers only during the brightest hours of sunlight. It is, as Perdita acknowledges in *The Winter's Tale,*

> *The marygold that goes to bed with the Sun*
> *And with it rises weeping.*

The "shepherd's sundial" (also called "the poor man's weatherglass") opens in summer a little past 7 A.M. and closes just after 2 P.M.; when rain is imminent, it doesn't open at all. But many plants without such names also open and close on a schedule. The garden lettuce spreads itself at 7 A.M. and draws itself in at 10 A.M.—and so on. This orientation to the Sun of what are known as "horary" ("hour-keeping") plants was put to practical use in 1751, when Carolus Linnaeus (1707–1778) suggested that a clock be made of flowers, so the time could be reckoned by seeing (or smelling) which variety had just opened up. The rotation was complex, but at least one roster of openings went:

5–6 A.M.: morning glories and wild roses

7–8 A.M.: dandelions

8–9 A.M.: African daisies

9–10 A.M.: gentians

10–11 A.M.: California poppies

noon: morning glories close, goat's-beard opens

4 P.M.: four o'clocks open

4–5 P.M.: California poppies close

6 P.M.: evening primroses and moonflowers open

8–9 P.M.: daylilies and dandelions close

9–10 P.M.: flowering tobacco opens

10 P.M.–2 A.M.: night-blooming cereus opens

When Linnaeus planted a version of the clock in the garden of his summer home 6° north of Uppsala, he took account of the difference in latitude—for instance, he reckoned that goat's-beard would open there at 3 A.M., to meet the first rays of the midnight sun—and shuffled his flowers accordingly.[17] In recent years, various forms of flower clock have been planted in places as disparate as Tehran, Iran, and Christchurch, New Zealand; but their accuracy is doubtful, as flowering time is so much at the mercy of weather.*

Research initiated in 1960 discovered that each variety of plant reflects light slightly differently, so a satellite in space should be able to identify the plant life of any given area on Earth. In 1972 the United States, then under Nixon's presidency and preoccupied with Soviet plans for world dominance, created a task force to reevaluate Soviet crop expectations. As Dan Morgan tells us, "Information about Soviet crops was seen as vital economic intelligence with a bearing on the economic security of the United States."[19] Several months later, LACIE (Large Area Crop Inventory Experiment) was set up, and by 1977 a U.S. satellite accurately predicted the Evil Empire's wheat crop six weeks prior to harvest. LACIE seems to have been curtailed shortly afterward; but it is possible that some new form of agrospacial espionage has been initiated since. We do know that in 1995 the U.S. Navy researched whether blooms of bioluminescent algae could be used to track submarines. (They couldn't.) But between 1992 and 2001 a scientific group known as MEDEA (Measurements of Earth Data for Environmental Analysis) advised the federal government on environmental surveillance, and after energetic lobbying by Al Gore for its revival, it was reported in January 2009 that "the nation's top scientists and spies are collaborating on an effort to use the federal government's intelligence assets—including spy satellites and other classified sensors—to assess the hidden complexities of environmental change." So the use of satellites

---

* During the 1920s a young scientist named John Nash Ott, experimenting with time-lapse photography, discovered that different wavelengths had differentiable effects on photosynthesis. He extrapolated that varied light frequencies might also affect human well-being.[18] Thereafter he worked on the links between light and cancer, and by the late 1960s the U.S. Congress had passed a Radiation Control Act, its coauthor crediting Ott with "getting us all started on the road toward control of radiation from electronic products."

Ott's work was sufficiently well known that decades later he was asked by Paramount to make a time-lapse photo sequence of flowers for Barbra Streisand's film *On a Clear Day You Can See Forever*, in which her singing is meant to cause the flowers in her house to bloom within seconds.

seems never to have stopped, only now it is espionage helping the environment, not the other way around.[20]

AS WITH PLANTS, so, often, with the animal kingdom. Sunfish, strikingly ugly, seemingly tailless creatures that can grow to six feet long—they are the heaviest bony fish in the world—live deep down in the oceans during stormy seasons (they are known as the "couch potatoes of the sea"), then rise to sunbathe on the surface in bright weather. In the Sahara, foraging ants orient themselves by registering the polarization of light as well as the Earth's magnetic field and then employing their spatial memory; animals such as albatrosses and turtles that live almost all their lives in or over the sea use the Sun as a navigation beacon in much the same way. The minute beach crab *Talitrus*, with barely a millimeter of nerve substance, is able to calculate the time of day, even the course of an hour, from the angle formed by its body and the Sun's position. In response to changes in sunlight, scores of animals change color with the seasons, altering pigmentation and camouflage along with their environment.

The Sun also plays its part in reproductive activity. As sunlight fades, herring shoals will swim ever closer together and move into shallow water, where they will develop their eggs, protected by their strength in numbers. As the Sun rises, so the shoals will move apart.[21] Many of the more brilliant tropical birds live in the upper canopy of rainforest, where they are bathed in abundant light and can show themselves off to prospective mates to maximum effect. Other lushly plumaged creatures take advantage of the stippled sunshine that breaks through to lower parts of the forest to perform acrobatic mating displays, in which their colors glimmer in the fractured light, "like dancers beneath a spinning disco ball."[22]

Heliconius butterflies use polarized radiation to choose mates, their employment of visual cues in sexual selection being an example of light exploitation that may also have adaptive value in dense forest, where illumination varies greatly in color and intensity. This raises the question of what Sun-inspired activities exist beyond our powers to observe. "Some birds can simply see what humans can't," says Dr. Miyoko Chu of the Cornell Laboratory of Ornithology: for instance, blue tits differentiate among one another in ways inaccessible to humans. It has been known for some time that birds (along with some lizards, fish, and insects) can see into the ultraviolet, but in 1998 researchers discovered that certain plumages reflect frequencies invisible to human eyes: while we have three cone types in our eyes, birds have four, admitting them to zones of the electromagnetic spectrum outside our ken and greatly extending their range of color combinations.

In the summer of 1944, Karl von Frisch (1886–1982), who in 1973 would share the Nobel Prize in Physiology with Konrad Lorenz, realized that when bees wagged their behinds in the hive they were signaling where to fly. Bumblebees do two dances, one circular, the other in a figure eight, which Frisch interpreted to mean—to give one of his own examples—"nectar 1.5 kilometers away and at thirty degrees from the Sun"; he also identified the bees' modes of communication, showing their sensitivity to both ultraviolet and polarized light. They could take off on a bearing coordinated with the Sun's passage, even setting up flight paths to take in sunny resting places. This ability to keep at a constant angle to the Sun regardless of the passage of time Germans charmingly call *Winkeltreue*.[23]

When a bee arrives back in the hive after visiting a newly opened flower, she dances on the landing platform in front of the colony entrance (a colony will number twenty thousand in winter, sixty thousand in summer), first making a circle, then bisecting it, waggling her abdomen and buzzing energetically. Then she enters the hive and sets to again: sufficiently stimulated, she may carry on dancing for close to four hours. The farther into the hive she goes before dancing, the farther away the pollen source. As the combs of the nest are vertical, the waggle-steps cannot point directly to the flower, and instead refer to the Sun. If the bee crosses the circle vertically, then the food source is in line with the Sun. If the target is, say, 15 degrees to the right, then the dance will be 15 degrees to the right of vertical. The workers surrounding the dancer assimilate the message and fly away to find the flower. When they return with the pollen, they, too, will dance, so that most of the workforce will soon be actively gathering.[24] Flushed with his discovery, Frisch set himself to work out how bees could possibly convey the Sun's position. Hard as it was to believe, he concluded that they could even predict where the Sun would be all the way around the clock, so that if they kept dancing, changing their pattern regularly, they would reproduce its movements.[25]

Among the insects that use the pattern of polarized light as an optical compass are ants and spiders, the latter being equipped with a specialized pair of secondary eyes just for this purpose. These do not "see" in the normal sense but have a built-in filter that determines the direction of polarization. Spiders are mainly active after sunset, and use this aid to find their way back to their nest after foraging trips.[26]

Of the thousands of species of ant, several are similar to honeybees in the way they use the Sun to orient themselves, and have a similar spectral sensitivity: what the great biologist E. O. Wilson calls "the almost fantastic ability" to "memorize the path and angular velocity of the sun."[27] What goes on in the brains of both insects during foraging trips is extraordinary: an outward-bound worker ant typically winds and loops in "tortuous searching patterns" until it finds food; during each of the twists and turns on its journey it registers the constant light of the Sun and is aware of the angles it takes relative to it. On its return home, it simply reverses its mean angle by 180 degrees—hardly a simple feat: a human traveler would require compass, stopwatch, and integral vector calculus.

Other creatures orient themselves by the Sun when they migrate, a feature of life for many animals, from caribou, which travel some two thousand miles, the longest such movement overland; to the baby loggerhead turtle, embarking on its eight-thousand-mile trek around the Atlantic; to the Zambian mole rat, whales, salmon (famously), eels, homing pigeons, the common toad, and birds that fly from Alaska to New Zealand nonstop.[28]

Every summer, throughout North America, 650 different species of birds feast and nest; come the autumn, 520 of those species migrate south, to return the following spring. The direction in which they fly depends largely on the length of daylight, but also on the bird's temperature sensitivity.[29] They journey partly in pursuit of food; but as much an impetus is the decreasing (or increasing) daylight.

Gliding consumes only one-twentieth of the energy required for sustained wing beating, so soaring species like broad-winged hawks ride warm air columns—"thermals"—generated by the solar heating of the Earth, sometimes a mile aboveground, covering as much distance as possible in long shallow sweeps.[30] They can gauge where thermals are powerful and reliable, and wait for the best conditions before setting off. Because thermals are fueled by sunlight, they are most common on long summer days, although a blazing Sun can seriously overheat fliers—a danger that geese avoid by traveling at night.

For centuries, scientists (like most people) thought birds unintelligent—hence the dismissive "birdbrained"—but within the last five years we have come to understand the processes by which these creatures orient themselves, and learned a new respect for them.[31] Day fliers, for example, which use the Sun to orient themselves while airborne, must possess some form of internal clock to keep track of time, because the position of a point on the Earth relative to the Sun changes by 15 degrees every hour—so in order to orient itself consistently, a bird must determine the Sun's apparent trajectory relative to its own direction at various times throughout the day; in other words, its solar compass has to be time-compensated.[32] To give a bird's-eye view: if the Sun at its present location is higher than it would be at the bird's goal at that time of day, the bird would have to fly away from the Sun; if lower, fly toward it.

JUST AS SOME animals seek out the Sun, others avoid it. The tiny Selevin's mouse, or desert dormouse, a Russian rodent discovered only in 1939, is unable to withstand more than about eight minutes' continuous exposure without falling ill. Not possessing sweat glands, reptiles are especially sensitive and will pack themselves tightly into any space where there is shade.[33] They are called "cold-blooded," but this is a misnomer: they actually regulate their temperature quite precisely by moving into and out of the sunlight. An albino king snake will often not survive in the wild because it can't get warm enough. Australian "magnetic" termites build large, thin mounds with the flat sides facing south and north so that they can go to the north side to make use of the Sun's

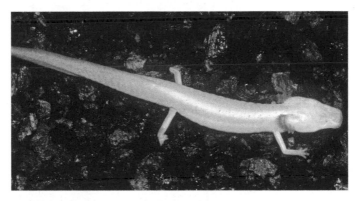

The Texas blind salamander lives in underground streams in caves and will never see the Sun. Fewer than a hundred are known to exist.

warmth, and then, when the heat gets to be too much, cool off by moving south.[34]

Certain animals seem to do without the Sun completely. One such, the olm, or *Proteus anguinus,* is found in the subterranean waters of southern Europe, notably those of the Soča River basin near Trieste. It has lungs, four feet, tiny teeth forming a sieve to keep larger particles inside its mouth (and so is presumably a predator), an eel-like head and snakelike lower body, and no fins or eyes; it eats, sleeps, and breeds underwater. A transparent fleshy white, if exposed to light it changes to an olive tint.

Several species have lost their useless light-adapted organs; others (such as bears, bats, and owls) much prefer darkness—one recalls Professor Higgins complaining of Eliza, "She's an owl sickened by a few days of my sunshine"— but if completely deprived of sunlight, these animals would soon die.

———

IN THE EARLY 1960s, a British scientist, Dr. James Lovelock (b. 1919), who had been working for NASA on detecting life on Mars, advanced the Gaia Theory, named after the Greek goddess of Earth. "The biosphere is a self-regulating entity with the capacity to keep our planet healthy by controlling the chemical and physical environment," he argued in a bestselling book, *Gaia: A New Look at Life*.[35] The living parts of the Earth, runs the theory, regulate the inanimate parts (atmosphere, oceans) to their own ends, stabilizing the environment and keeping it benign, a complex system that can be thought of as a single organism, and thus our world regulates itself to make itself hospitable to the amazing number of interacting species that constitute its "life." The globe, in other words, works to keep everything in ideal balance.

Among a battery of criticisms from other scientists was that such a theory assumed that evolution had foresight and planning, whereas the evidence suggests that life is mainly a matter of chance. Lovelock refined his ideas in a second book (he has since written four more), developing a mathematical model he calls Daisyworld: on a hypothetical otherwise unpopulated planet, two different types of plants, black and white daisies, are so organized as to maintain a balance of temperature and atmosphere ideal for overall daisy growth: the black absorbs sunlight and warms the Earth, the white reflects light and cools it, with each species growing or declining as necessary. In other words, living systems stabilize their global environment. It is not yet clear how well the Daisyworld formula can catch the full biospheric and climatic complexity of the Earth, and scientists, particularly evolutionary biologists, still regard Lovelock with suspicion. However, some of his arguments have been accepted. As Oliver Morton says, "The idea that life is content passively adapting to environments over which it has no sway—which really was the dominant paradigm just forty years ago—has gone for good."[36] We are now more ready to accept the idea of the Earth as the Sun's collaborator.

CHAPTER 18

THE DARK BIOSPHERE

The ocean looks like a fairly homogeneous place; it is all water, after all. . . . Yet that sea water, like the seafloor, only seems to be homogeneous. . . .

—ROBERT KUNZIG, *Mapping the Deep*[1]

Why did Dr. Ballard, on his journeys to photograph the Titanic, *habitually play classical music during the descent and rock music on the way back up? It is unthinkable that it could have been the other way around.*

—JAMES HAMILTON-PATERSON,
Three Miles Down[2]

ACCORDING TO LEGEND, ALEXANDER THE GREAT HAD HIMSELF LOW-ered into the Mediterranean in a glass cage whose door was fastened with chains, and during his vigil observed a fish so large that it that took three days and three nights to swim past. He is said to have recorded: "None of the men who have been here before me, and none of those who shall come after me shall see the mountains and the seas, and the darkness, and the light that I have seen."[3]

That darkness covers many unknown worlds. Even now, 95 percent of our oceans remain unexplored,[4] and oceanographers are fond of pointing out that we know more about the Moon's surface than the seafloor. We do know that throughout the great seabeds wind a series of volcanic gashes—about forty thousand miles of carbon-dioxide-spewing mountain ranges pierced by immense scalding springs. The oceanographer Bruce Heezen has named it "the wound that never heals."[5] Through these fissures surge great quantities of water at temperatures up to 750°F, almost four times the boiling point at surface pressure, but here prevented from passing into steam by the down-weighing miles of ocean above. Laden with inorganic chemicals of all kinds—sulfates, nitrates, and phosphates, along with compounds of hydrogen oxide, carbon dioxide, and methane gas—the water smothers the seabed with untold amounts of

A digital re-creation of a hydrothermal vent. While nearly all life on Earth depends on solar energy, the many thousands of creatures that live around these vents survive on organic material produced through chemosynthesis. However, even at these depths some still depend upon oxygen liberated by photosynthetic organisms.

insoluble salts that can never rise to the surface. The one exception comes through vents known as "black smokers"—furious chemical reactors so hot that they leach copper, iron, and zinc out of the Earth's crust and erupt above the surface to form columns of rock belching what looks like black smoke. "Godzilla," off the coast of Washington State, is sixteen stories tall.*

Some bacteria manage to exist in these immensely hot places, using iron as a means of respiration—just as we employ oxygen—metabolizing it into the black magnetic mineral magnetite. More than a hundred different species dependent on these bacteria have been discovered around the vents, not only microbes but holothurians—"sea cucumbers"—eight-foot-long tubeworms that grow in thick clumps, with bloodred heads like rosebuds; shrimp and clams, each of which can be as much as a foot across; reefs of six-inch mussels; and scrabbling porcelain-white crabs—all using the energy derived not just from

---

* John Maddox, *What Remains to Be Discovered* (New York: Free Press, 1998), p. 150. On January 16, 2006, the *Chikyu* ("Japan's Apollo mission") set off to drill 4.3 miles below the seabed, more than three times deeper than ever before, 125 miles off the coast of Nagoya, southwest of Tokyo. What it dredges up may show whether the energy that originated the first semblance of life on Earth was geothermal rather than solar. It may also help us understand why the Earth's magnetic poles repeatedly switch.

the oxidization of ferrous iron salts but also from sulfur, hydrogen sulfide, and molecular hydrogen. The iron-devouring bacteria feed the clams and shrimps, which in turn are eaten by the crabs and by a sinister-looking five-foot-wide whitish-gray hooded octopus. Nearly all these beasts were new to science and cannot live anywhere else. This part of our world has been named "the dark biosphere."

What has all this got to do with the Sun? These creatures and habitats are unique, constituting the only ecosystem on Earth that derives its energy from chemosynthesis, and exists not on trickle-down sunlight but by blast-up geo-chemistry.[6] Yet even here photosynthesis plays a role, as the vast bulk of the oxidants used by these creatures comes from the light-driven ecosystems at the ocean surface. Even the sulfur-eating bacteria can live only in places where they also have access to oxygen, stealing its electrons and latent energy, be-cause they need it to oxidize hydrogen sulfide. If the Sun were to disappear, most of these deep-sea communities would not long outlast those higher up (although a small number, bizarrely, would continue to exist, sustained by the light of the hot lava of the vents themselves).[7] So the Sun indirectly provides life even here.

The creatures photosynthesizing at the surface, which can extract hydro-gen from water, are blue-green algae—cyanophytes, or "blue growers"—found wherever there is constant moisture. "The arrival of the blue-greens marked a point of no return in the history of life," writes David Attenborough. "The oxygen they produced accumulated over the millennia to form the kind of oxygen-rich atmosphere that we know today."[8] These photosynthesizing or-ganisms, made up of single cells called phytoplankton (Greek for "wandering plants"), which measure at most a few thousandths of an inch across, are con-fined to a thin layer near the ocean surface; for light is quickly absorbed, and below about seven hundred feet normal photosynthesis is impossible. (The greatest depth at which conventional plant life has been found is 710 feet, off the Bahamas, where a clump of maroon algae prospers in exceptionally trans-parent water). But that thin layer is rich in photosynthesizing agents; and all animal life in the sea, from jellyfish to whales and even the denizens of seafloor hot springs, depends on these cells. With the end of winter, chlorophyll, the molecule that plants use to absorb sunlight, turns the entire North Atlantic green. No one can number the many species of phytoplankton, but these seaborne plants (along with land-dwelling forests) draw out of the atmosphere half the carbon monoxide that we put in—and make life possible not just on land but at the greatest depths.

The oceans can be divided into two major realms, the shallower seas that fringe the continents and the deeper oceanic waters. The former, the continental shelf areas, are like submerged shoulders of the landmasses: it is here that the vast majority of marine fish live. At depths of 300 to 600 feet, the shelf suddenly gives way to great slopes that descend to what is called the abyssal floor, cut by winding V-shaped formations that, if exposed to terrestrial eyes, would rival the Grand Canyon. Huge trenches have been formed, in some cases plunging nearly seven miles down and throwing up the Mid-Oceanic Ridge, foremost of all the mountain ranges on Earth, which stretches in an unbroken chain from the Arctic through the Atlantic to the Antarctic, Indian, and Pacific oceans, a distance of some forty thousand miles. The greatest depth to be sounded so far was measured in 1962 in the Mindanao Trench, off the Philippines, where the seafloor lies 37,780 feet down, a gash that could drown Mount Everest with almost a mile and a half to spare. About 86 percent of the world's oceanic water lies below three thousand feet.[9] One recalls that the Latin word *altus* translates as both "high" and "deep," a meaning that lingers in the English phrase "the high seas."

In 1951 the naturalist Rachel Carson (eleven years later to publish her celebrated study of environmental dangers, *Silent Spring*) wrote *The Sea Around Us.* "Down beyond the reach of the Sun's rays," she begins poetically, if not with total accuracy,

> there is no alternation of light and darkness. There is rather an endless night, as old as the sea itself. . . .
>
> If we subtract the shallow areas of the continental shelves and the scattered banks and shoals, where at least the pale ghost of sunlight moves over the underlying bottom, there still remains about half the Earth that is covered by miles-deep, lightless water, that has been dark since the world began.[10]

Light is filtered out layer by layer, wavelength by wavelength, as it penetrates downward, beginning with the ultraviolet and infrared, which are absorbed in the uppermost three feet.[11] In the night sky the lights of an aircraft are visible many miles away; even when overhead, the same lights under water fail to reach beyond six hundred feet.[12] Below the first three hundred, the bottom of the "euphotic," or "well-lit," zone, red wavelengths are completely absorbed, and with them all the orange and yellow warmth of the Sun. By five hundred feet, only around 1 percent of sunlight remains. Then the green goes too, although in the clearest waters blue-green illumination is still detectable at nearly three thousand feet. By a thousand feet down there is only a very dark blue. Hamilton-Paterson, in a diving bell that eventually descended to 4,978 feet, recorded at seven hundred: "Surprised there's any light outside, but there is. . . . The sea's intense violet color is strangely piercing. . . . It's a quality of light I've never quite seen before. It doesn't exist on the Earth's surface and perhaps can't even be produced artificially."[13]

As sunlight falls away, myriad new creatures fill the darkness—inhabiting zones known respectively as twilight, sunless, and bottom-living. In 1818, Sir John Ross, exploring the Arctic seas, brought up from six thousand feet mud embedded with worms, "thus proving there was animal life in the bed of the ocean." In 1860 the survey ship *Bulldog* found evidence of life far beyond the Sun's assumed reach: shrimp, lanternfish, squid, and arrowworms. Carson recounts the remarkable voyage of HMS *Challenger*, the first vessel equipped for oceanographic exploration, which set out from England in 1872—although she is not specific about details. In 2006, the *New Yorker* writer David Grann, on a hunt for the giant squid, retraced the *Challenger*'s voyage. The ship had roamed the globe for three and a half years, dredging the ocean floor in a zigzag stroll covering 68,930 miles at the rate of two knots—the equivalent of a slow walk. The work was repetitive and brutal—two men went insane, another killed himself—but by journey's end the explorers had landed 13,000 different kinds of animals and plants. It took the next nineteen years to process this booty, which included 4,700 new animals, 2,000 of them living below 820 feet—about one-tenth of the (even now) known species of fish. It was evident that between the waves and the floors of the deep ocean basins there thronged the largest and perhaps most remarkable biological communities anywhere, a diversity rivaling that of the tropical rainforests.*

The oceans total 320 million cubic miles of water and 140 million square miles of seafloor stretching over seven-tenths of the planet. Their depths have been inhabited for a comparatively short time, for they present awesome challenges to life: the extreme cold, as well as the crushing weight, forbid easy adaptation.[15] For the unprotected human body, six hundred feet is the absolute limit; one-atmosphere-pressurized diving suits allow man to go as deep as 2,500 feet. But for the often transparent creatures of the deep, the pressure inside their tissues is the same as that without, and so is not a problem. However, how sperm whales endure the pressure—close to 1.6 tons per square inch in dives of more than seven thousand feet—remains a mystery. The concentration of oxygen in the lower depths is only one-thirtieth of that close to the sur-

---

* In November 2009 scientists reported that the number of species known to be living beneath the ocean waves was now 17,650. Some 5,722 species live deeper than the "black abyss" of 3,280 feet (1,000 meters), including the whalebone eater, the see-through cucumber, the "wildcat" tubeworm, the jumbo dumbo (a specimen of octopod), and the yeti crab, which lives on hypothermal vents at roughly 7,200 feet down.[14]

face, but within the zone of lowest oxygen a unique group of animals has adapted to meet such challenges.

Darkness might seem yet another obstacle, but the creatures of the deep have answers to that, too. While at fifteen hundred feet the unaided human eye can see only a coarse pattern of silhouettes and shadows, fish can discern the subtlest differences. Such sensitivity may regulate their depth, with changes in light levels controlling the morning and evening migrations of various fish and krill. Even eyeless creatures may register the Sun's presence: molecules of water polarize the Sun's light, which helps many animals in their hunting, as the tissues of their prey absorb or rotate the light passing through them. Squid,

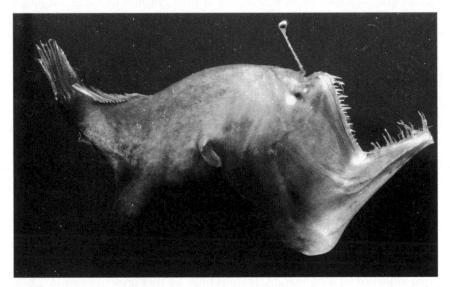

The deep-sea angler fish makes its own sunlight. Searching for food at between 3,300 and 5,000 feet down, it will lure its prey with a bioluminescent glow at the end of a long rod. The impressive teeth will then impale the curious.

employing special cells embedded within their skin, also use polarized light (which would appear indistinguishably black to human eyes) to regulate their vertical migrations and to signal one another.[16]

Where there is not enough light, many create their own. Between six and seven hundred species have this facility, but they employ it conservatively, as every time they create light they expose themselves. However, some creatures can vary the wavelengths they emit, thereby using light itself as camouflage. Several have modified cells that act like flashlights, to be turned on or off at will (presumably to find or pursue prey), while others have rows of lights over their bodies or at the end of their legs or tentacles, another effective misdirection. Some fish have developed light-producing organs on their undersides, whose weak downward glow erases the shadow cast when they are viewed from below against the relatively better lit waters above. Others use biolumines-cence as a booby trap, coating an advancing predator with sticky, glowing tis-sue that makes the attacker vulnerable to other visually cued hunters, "like bank robbers marked by exploding dye packets hidden in stolen currency," as one marine biologist colorfully puts it.[17]

Some eyeless creatures light up brightly when touched. Others concentrate red body pigment, which absorbs whatever blue-green light falls on them and reflects nothing, an effective "visual stealth" strategy. The predatory *Malacos-teus niger*, known as the lightless loosejaw, generates long-wavelength infrared (in addition to blue bioluminescence), invisible to other deep-sea animals, giv-ing it vision in the same light zone as a military sniperscope. The fish's optic retinas contain derivatives of chlorophyll (harvested from its diet of tiny crus-taceans), which in modified form allow it to generate infrared. During the Cold War, the U.S. military considered using these creatures as sentries to signal the passage of Soviet submarines. In the end, however, Operation Lightless Loose-jaw never materialized.[18]

Transparent animals, the better to conceal themselves from predators, have developed the smallest guts possible, as the intestines are the only visible part of their anatomy.[19] In some, the stomach is needle-shaped and always points down, whichever way the animal is oriented, minimizing its presence against the surface light. Others cloak the stomach in reflective tissue, which in the open ocean reflects light that is indistinguishable from behind it.

All these highly varied creatures, be they fish, crustaceans, algae, or other organisms, inhabit the deep oceans well away from the Sun's glare. Yet even the minutest microorganisms living on or near the seafloor owe something of their existence to the Sun. Swimming through H_2O along the darkest frontier of the solar realm, they rely indirectly on the oxygen released by the photo-synthesis activated at the surface.*

THE OCEAN, WRITES Robert Kunzig in his magical book *Mapping the Deep*, "is Earth's primary distributor of heat. Sunlight causes water molecules to strain

---

* The Sun, however, primarily energizes organisms living in the shallows, not the depths. Among the functions that certain coralline algae perform is to manufacture a chemical that acts like a high-factor sunscreen shielding both themselves and their polyp hosts from ultraviolet rays. Alginates from brown seaweed offer considerable protection from radiation, since they se-crete iodine compounds that help treat cancer in humans. Japan has one of the world's lowest rates of breast, endometrial, and ovarian cancers, attributed to the kelp konbu, which contains U-fucoidan, which interferes with tumor cell colonization, causing these cells to self-destruct. In protecting itself from the Sun, seaweed has provided us a way to protect ourselves.

at their hydrogen bonds, pushing and pulling their neighbors, and the heat that is stored in those vibrations can be transported great distances—by ocean currents that are driven by the heat, by variations in the amount of salt dissolved in the water (which affects its density), and by winds."[20] The ocean's liquid topography is constantly shifting—which leads us on to those mighty agents of the Sun, currents and tides.

Currents, which operate at all sea levels, are of two types, surface and deepwater. Sometimes they flow simultaneously in different directions at different depths. There are nine such major ocean streams:

1. The Gulf Stream, which originates south of Florida and flows north along the eastern coast of the United States, crossing the North Atlantic to the Norwegian Sea. It has a huge impact on the long-term weather of any landmass that it touches—the overall climate of Norway and the British Isles, for example, is about 18°F (10°C) warmer in winter than continental interiors at the same latitude.

2. The Labrador Current, which cuts south from the Arctic Circle and travels along the Canadian coast, cooling the Canadian Atlantic provinces and, by journey's end, coastal New England. Some of it runs off into the Gulf of St. Lawrence, but most of its chilling mass continues southwest.

3, 4, 5, and 6. The North Equatorial Current and South Equatorial Counter Current (3 and 4), which travel east-west near the equator. Both are about six hundred miles wide and reach roughly 4° to 10° on either side of the equator, but never touch it, because they're deflected by their counterparts, the South Equatorial Counter Current and the Equatorial Counter Current (5 and 6).

7. The Kuroshio Current, which separates from the North Equatorial Current and sweeps the eastern coasts of Taiwan and Japan. It then splits to form an eastern branch flowing nearly to the Hawaiian Islands and a northern branch that skirts the coast of Asia to merge with the cold Oyashio Current to form the North Pacific Current.

8. The California Current, which runs from the Gulf of Alaska along the western coast of the United States and is partially responsible for the relative coldness of the water there.

9. The Doldrums, which are centered slightly north of the equator, where the intensity of solar radiation makes the air especially humid, lowers air pressure (by the heated atmosphere expanding outward), and creates clouds, light variable winds, and thunderstorms, squalls, and other severe weather patterns. The Doldrums are also characterized by long periods in which the winds disappear, becalming sailing vessels for days or weeks—thus giving the language one of the few nautical phrases that has a nonnautical meaning.

Two main factors determine how currents form and re-form: the Sun's great heat and the Earth's rotation (a third and lesser factor being the gravitational pull of the Sun and the Moon). By warming the air above the oceans, the Sun creates winds that drive the water along through friction (surface currents take up the upper quarter mile of any ocean, about a tenth of our total hydrosphere). If water cools or is made saltier through evaporation, it becomes denser and sinks, setting off fresh currents, which in turn transport heat from one location to another, both upward and outward, further altering temperature patterns.[21] Aeschylus wrote of the "multitudinous laughter of the waves," but the Sun's power to drive winds across open water can build ocean surges up to a hundred feet, trough to crest, and at any given moment there are ten such monsters churning through the oceans: not quite the light-hearted image conjured up by Aeschylus. Lying in front of such waves can be a deep depression, known as a "hole in the sea," which can fling any ship that it traps on a fatal roller-coaster ride. These rifts form regularly in regions swept by powerful currents: the Agulhas, off South Africa; the Kuroshio, off Japan; and the Gulf Stream, off the eastern United States—which last also flows through the Bermuda Triangle, that mythical devourer of ships and aircraft.[22]

Some ocean layers can be no more than a few inches thick, but each is distinguishable from its neighbors by sharp changes in temperature and saltiness, which together govern its movements. Water nearer the equator—and thus more intensely heated by the Sun—can be 40°F warmer on the surface than farther down. The warm currents at the equator, in part because they are lighter than currents to the north and south, sweep toward both poles.

The Earth's rotation acts on ocean currents through what is called the Coriolis force, named for the nineteenth-century French engineer and mathematician Gustave-Gaspard Coriolis (1792–1843), which deflects currents northeastward in the Northern Hemisphere and southeastward in the Southern. Such movements are affected by friction with the Earth only at the seafloor. The Earth's eastward rotational velocity decreases from its maximum of one thousand miles per hour at the equator to zero at the poles. Equatorial water is deflected at the speed of the planet's fastest rotation; as it migrates across progressively slower latitudes, it is still traveling more quickly than the waters through which it passes, and given its greater momentum, it has to move diagonally north- or southeast, in the direction of the planet's turning.

Beyond currents, there are monster waves, and these too have a solar cause, if indirectly. In 1997–98, people around the world watched satellite images showing winds change direction and a huge pool of unusually warm water roll across the tropical Pacific to mass into an area about one and a half times the size of the continental United States. This was the famed El Niño, now blamed for rearranging the weather patterns of countries as far apart as Chile and Australia and bringing storms, flooding, drought, and brush fires. "El Niño" means "the little boy" in Spanish and refers to the baby Jesus (the effect usually reaches its peak around Christmas). It has a twin sister, a rebound of unusually cool water called "La Niña," which shows up on satellite images in brilliant blues and purples stretching across the mid-Pacific. Either sibling can bring extremely hot weather or extremely cold. The ferocious winter chill of 1941–42 that inflicted such a setback on the German invasion of the Soviet Union (when temperatures dropped to –40°F, machinery froze, and a quarter of a million troops died of cold and disease) was set off by La Niña.[23]

FROM THESE COLOSSI it may seem literally a comedown to talk about tides— although those of the Minas Basin in Nova Scotia rise and fall by more than fifty feet twice a day. And the total power of waves breaking around the world's coastlines exceeds 2 trillion watts, enough to supply the daily electricity needs for 200 to 300 billion homes—more than a hundred times as many as there are in the world.

Many people see tides as the Moon's province alone, but as is so often the case, the Sun cannot be ignored. The word "tide" (from the Germanic root for "time" or "season," not for anything watery) has two meanings: first, the variation of the sea level at any point along a coast, which depends on shoreline topography and on currents near the shore; second, the deformation of the Earth's land and water caused by the pull of Moon *and* Sun. Almost four hundred times farther from the Earth but of 30 million times greater mass than the Moon, the Sun exercises by far the greater pull—by a factor of 178. However, because tides relate not to the strength of a gravitational field but to its gradient, it is the Moon that is responsible for 56 percent of the Earth's tidal energy, the Sun for 44 percent—still very significant.

When the Earth gets as close to the Sun and the Moon gets as close to the Earth as each can (putting them "at perigee"), they exert their maximum pull, causing the highest possible tides; when the Earth is at its greatest distance (apogee), tidal ranges are smaller; and at a new moon, when Sun, Moon, and Earth are all in alignment, the Sun and Moon pull on the Earth in the same direction (a condition bearing that Scrabble player's dream word, "syzygy"), the solar tide augmenting the lunar, heightening high tides and lowering low ones.

The tidal cycle is 24 hours, 50 minutes long, during which time most shores on Earth will experience two high tides and two low. Depending on local conditions, however, some places may have no observable tides, or a high tide that does not peak when the Moon is overhead but lags behind by several hours. One can appreciate the dilemma of the Breton lady in *The Knight's Tale*, when, as Chaucer has it, having promised to give herself to her ardent suitor only if the rocks in the sea disappear, she looks over the cliff to find they have done so. Fortunately for her (since she is happily married), her suitor proves chivalrous and does not hold her to her promise.

Tides have been a source of speculation for millennia (although the oldest surviving tide table, for London Bridge, dates only from the twelfth century). The Romans, generally having known only the equably flowing Mediterranean, had little experience of them, and Caesar describes how his expeditionary forces twice paid a heavy price for their ignorance when invading England along the Dover coast. The Greeks wrote about "motions of the sea," but not until Kepler, in *Astronomia Nova* (1609), and Galileo, in *Dialogue* (1632), was there any concerted attempt to explain them. The first well-documented version was given in 1687 by Newton, although Laplace's *Mécanique céleste* (1799) is the major text before the nineteenth century; thereafter the clearest descriptions of tides are by two remarkable British scientists, Sir William Thomson (1824–1907) and Sir George Darwin (1845–1912), fifth child of Charles.*

Thomson was one of the most brilliant of all the great nineteenth-century men of science (he lies next to Newton in Westminster Abbey), and in 1882 he addressed the British Association on the working of tides. "If I were asked to tell you what I mean by the tides," he began, "I should feel it exceedingly difficult to answer the question. . . . The truth is, the word 'tide' as used by sailors at sea means horizontal motion of the water; but when used by landsmen or sailors in port, it means vertical motion." He judged the mariners' notion correct, "because before there can be a rise and fall of the water anywhere it must come from some other place, and the water cannot pass from place to place without moving horizontally, or nearly horizontally."[25]

Often what we see are "more properly wind waves than true tides," but there is another rise and fall, caused directly by the Sun's heat, that affects at-

---

* A third great writer on tides was William Whewell (1794–1866), whose "tidology," as he called it, charted the movements of the world's oceans, with the object of showing all those points where, throughout the globe, high water occurred at the same time. Whewell, who went on to become a celebrated master of Trinity College, Cambridge, introduced the term "scientist" into English in 1834. Spurning the alternative "savant" as too unassuming and too French, he offered the word to denote the common enterprise of all those who studied the natural world, even as their disciplines were becoming more specialized and less unified.[24]

mospheric pressure: when the barometric reading is high, the water is pushed down by the greater pressure of the air; it swells upward when this drops. This Sun-mediated pressure affects land, too; but because continents are much more rigid than oceans, the effect is much smaller. Even so, Thomson explained, whole tracts of a continent may rise and fall as much as sixteen inches at high or low tide (we do not realize this, as everything around us rises or falls together). Local lakes, swimming pools, baths, cups of coffee, or indeed the human stomach will not show any signs of rising or falling, but still tides do funny things: for example, tidal friction is lengthening Earth days by 1.6 milliseconds per century, while tides themselves stretch each one of our bodies, making us taller.*

Sir George Darwin's early education had been at his father's side, and although he was called to the Bar, it was always likely he would turn to science. In 1898 he published *The Tides and Kindred Phenomena in the Solar System*, a scientific bestseller, and later contributed the long entry on tides for the eleventh edition of the *Encyclopaedia Britannica*. Thomson (or Lord Kelvin of Largs, as he became—the first scientist elevated to the peerage) had treated the Earth as an incompressible sphere, one that would not be drawn out of shape by solar and lunar pullings, whereas Darwin saw our world as having a liquid core responsive to both the Sun's and the Moon's powers. (A pound coin in Edinburgh, he liked to point out, is slightly heavier than the same coin in London, because slightly closer to the Earth's center.) He calculated that in the remote past the Earth must have been rotating much faster and the Moon much closer.[27]

Darwin set out to apply Newtonian principles to calculating the Sun's effect on other planets. "The efficiency of solar tidal friction is very much greater in its action on the nearer planets than on the farther ones," he concluded. Speaking of his own planet, he showed that the Earth's high tides occur at local noon and midnight, the low tides at sunrise and sunset, although in calculating tidal behavior "the irregular distribution of land and water and the various depths of the ocean in various places produce irregularities in the oscillations of the sea of such complexity that the rigorous solution of the prob-

---

* See Mikolaj Sawicki, "Myths About Gravity and Tides," *Physics Teacher* 37, October 1999, pp. 438–41. One should not get too excited by this enhancement, though. High tides change our height by a factor of 10^{-16}, a distance a thousand times smaller than the diameter of an atom. In comparison, while in space, astronauts grow taller by as much as three inches: as the skeleton is unrestrained by the pressure of gravity, the vertebrae relax, elongating the spinal column—which soon returns to normal on Earth. Similarly, earthquakes make us lose weight: when the world is dilated by a giant shock, gravity at the Earth's surface decreases by about 0.0000015 percent, meaning that a 150-pound person loses about 4×10^{-6} of an ounce.[26]

lem is altogether beyond the powers of analysis"—a most un-nineteenth-century admission.[28]

Beyond the powers of analysis it may be, but sailors have not waited for the findings of science and have been navigating the seas for thousands of years as best they can; and for thousands of years that has meant learning how best to harness the Sun.

H*ARNESSING* the SU*N*

When Elizabeth I of England received the papal bull of 1582, astronomers urged her to make the change to the new "Gregorian" calendar, while Protestant bishops opposed it. This engraving from c. 1641, Father Time Carrying the Pope Back to Rome, *catches the popular mood. As the poem beneath the image puts it,* "This load of vanity, this pedler's packe / This trunk of trash. . . ."

CHAPTER 19

THE HEAVENLY GUIDE

*Ahab soon calculated what his latitude must be at that precise instant.
Then falling into a moment's revery, he again looked up towards the Sun
and murmured to himself: "Thou sea-mark! Thou high and mighty Pilot!
Thou tellest me truly where I* am*—but canst thou cast the least hint
where I* shall *be?"*

—HERMAN MELVILLE, *Moby-Dick,* 1851[1]

*The sailor cannot see the North,
but knows the needle can.*

—EMILY DICKINSON, 1862[2]

CAPTAIN AHAB CAN BE FORGIVEN HIS MOMENT OF FRUSTRATION.
"The emptiness and homogeneity of the sea," writes Daniel Boorstin, "the vast
sameness of the oceans on the surface, naturally drove sailors to seek their
bearings in the heavens,"[3] and the Sun's role in guiding such travelers has
a long history. The oldest proven sea travel anywhere is the migration of
anatomically modern *Homo sapiens* to Australia, beginning about sixty thousand years ago; but the most impressive ocean voyagers of old were the Polynesians, who by around 2000 B.C. had already made themselves formidable
astronomers. The Sun was their principal marker, but usually in relation to
other stars, the positions of which they memorized in the hundreds. The Chinese were not far behind, calling their blue-water junks "starry rafts." The ancient Greeks, too, became seafarers (although Homer engagingly introduces "a
talkative, bald-headed seaman who set the crew laughing and forgot his
course," and the Greeks had no native word for "sea" when their invasions
broke into the Mediterranean). But sailors who left sight of land were constantly groping their way forward. Even on short voyages men were often "at
sea," as we still say. Seafarers gloomily swapped stories of those who had fallen
afoul of Mediterranean weather—from the Spartan king Menelaus, whose

homebound fleet was blown from Troy to Egypt, to Saint Paul's shipwreck on Malta. The lack of a means to determine longitude (east-west positioning) presented an often insurmountable challenge.

The directions by which we position ourselves—north, south, east, and west—are set from two coordinates, horizon and zenith. The horizon (Greek, "defining" or "separating") is the line where Earth and sky appear to meet. The zenith (Arabic, "over one's head") is the point in the sky immediately above the observer. These two—lateral plane and vertical point—provide our initial frame of reference. Man learned to assign the values east and west by the risings and settings of objects in the sky, calling the directions of the line at right angles to their course "north" and "south." He eventually invented "meridians"—imaginary lines that curve through any given point on the Earth's surface to mark the shortest distance between the poles.

One's own meridian is nothing more than the north-south line that runs through the particular spot where one happens to be. At a given moment, one of these meridians is exactly underneath the Sun, making it noon for everyone along it. For those to its east it is afternoon or later, for those to the west morning or before, and halfway around the world in either direction it is midnight. Points on the same meridian have the same longitude, longitude being measured by its distance east or west from a standard meridian (which, since 1884, has been accepted as passing through Greenwich, in east London).

Early navigators could roughly determine east and west by the Sun's rising and setting. Later, seafarers developed other aids. Knowing the latitude (i.e., north-south alignment) of a ship's position was easy, for the Sun's altitude was an almost perfect guide. (One degree of latitude is about 65 nautical miles, 1 nautical mile equaling 1.15 "statute miles" or 1.85 kilometers.) At the equinoxes, the noonday Sun is directly above the equator, or at an altitude to it of 90 degrees, while at the poles it is invisible in a hemisphere's winter, but clearly seen in its summer. In between equator and pole, the Sun's noon altitude can be established and compared with astronomical tables to determine one's position. The only device needed is a sighting instrument to measure altitude in degrees. The Greeks did not even use that, but simply measured the height of the circumpolar stars above the horizon at night.

The art of navigation evolved over the centuries, with certain notable breakthroughs, not all of which related to the orientation of the sun. In A.D. 40 the Greek merchant Hippalus sailed from Berenice, on the Red Sea coast of Egypt, to Madras and back in a single year, a journey that had previously taken two, thus helping to transform oceanic commerce: he had discovered that monsoon winds (from *mawsim*, Arabic for "season") reversed twice a year, and that by running directly before them, vessels traveling the China Sea and the Indian Ocean could cover vast distances at much greater speeds. Thereafter, the Greeks became so accustomed to using the names of the winds to indicate the directions from which they came that "wind" became a synonym for direction, with the puffed cheeks and exhaled strong breath of the cherubs on early maps not mere decorations but the main direction markers (hence "Australia"—"land of the south wind"). Columbus's Spanish crews thought of direction not in terms of degrees on a compass but as *los vientos*, the winds, while the Portuguese called their compass card a *rosa dos ventos*, a wind rose—duly marked off with the most significant aspects of the sky.*

As the Middle Ages opened (say, by Charlemagne's time, around the year 800), Arab navigators gauged latitude by either a *kamal*, a wooden rectangle on a string, or a crossbow on which two rods were hinged at one end to measure the angle of declination when the observer leveled the bottom rod on the horizon and the upper on the Sun (or a star). Then in 1086 the magnetic compass was pioneered by a Chinese waterworks director, Shen Kua; at least, he offers the first written source. For the first time, Chinese sailors were able to find an absolute direction anywhere on the globe without complicated calculations, and were provided with—in Boorstin's words—"a worldwide absolute for space comparable to that which the mechanical clock and the uniform hour provided for time."[4]

Shen Kua further advised that if one rubbed a thin piece of steel on a lodestone ("the stone that leads") and hung it by a thread, the magnetized needle would point almost south. Though a lodestone is simply a piece of ore possessing magnetic properties, its remarkable power was for a long time associated with the dark arts (in China, such stones were used in fortune-telling), and common seamen were wary of it. Saint Augustine recounts his amazement on seeing that magnetite could not only attract iron but give it the power to attract yet more, creating a chain held together by an unseen force: hence its association with magic.

---

* Critics have pointed out several errors in Dan Brown's *The Da Vinci Code*, but his note about the Rose Line is to the point. "For centuries," he writes, "the symbol of the Rose has been associated with maps and guiding souls in the proper direction. The Compass Rose—drawn on almost every map—indicated North, East, South, and West. Originally known as the Wind Rose, it denoted the directions of the thirty-two winds, and sixteen quarter-winds. When diagrammed inside a circle, these thirty-two points of the compass perfectly resembled a traditional thirty-two-petal rose bloom. To this day, the fundamental navigational tool was still known as a Compass Rose, its northernmost direction still marked by an arrowhead . . . or, more commonly, the symbol of the fleur-de-lis." *The Da Vinci Code* (New York: Doubleday, 2003), p. 106. Winds, of course, are primarily solar.

The origins of this wondrous new instrument are still hotly debated, historians disagreeing on whether it was invented more or less simultaneously in several places or whether the Europeans and Arabs learned of it from the Chinese. But wondrous though it was, its authority could be undone by extreme weather conditions, a truth that rang through the centuries.

From at least the eleventh century, most ships' pilots had manuals to help them, the best "pilot book" to survive from such times being the *Konungs Skuggsja* (*The King's Mirror*), written as a dialogue between a Viking father and son. The youth is told he must "observe the course of the heavenly bodies, recognize the quarters of the horizon, mark the movements of the ocean and understand the rise and fall of the tides."[5] However, the ability to establish the Sun's position is not in itself sufficient for good navigation, but must be supplemented. When crossing the open sea, the Vikings would gauge latitude by means of a "Sun-shadow board"—a wooden disk with a pin or gnomon at the center that could be adjusted upward or downward according to the season. The disk was floated in a bowl of water and the Sun's noon shadow noted. "If the ship was on course," explains Robert Ferguson in his panoramic history of the Vikings, "then the shadow would reach a circle marked on the board. If it passed beyond, the ship was north of this latitude. Should it fail to reach the circle she was south of it and the skipper could make the necessary adjustments."[6] Nevertheless, without a sure reckoning of latitude, the navigation of the Vikings' great voyages principally depended on deep acquaintance with the seas they were crossing. For longer voyages, a pilot would put himself on the latitude of his destination, then try to hold course by keeping the observed angles of land (where available) and celestial bodies uniform. Not surprisingly, he often missed his destination, which is how Vikings ended up on the "Long and Wonderful Beaches" of Greenland and Labrador.[7]

Another skill the Vikings may have employed was using a "sunstone" (a clump of natural crystal) to polarize light and thereby locate the Sun even when it was hidden from view. The Vikings' sunstone is believed by many to have been a form of cordierite, a mineral found almost exclusively on the pebbly beaches of Norway, which can split a light beam into two shafts of color; when held up against a clear patch of sky and rotated, it changes from yellow to blue at the moment it is pointed toward the hidden Sun. A short story, "Raudulfs Thattr," preserved in a manuscript from the early fourteenth century, tells of a visit by Olav Haraldson, king of Norway, to a rich farmer. Olav asks his host's son, Sigurd, if he has any special talent, and the boy replies that he can tell the time even when there is no celestial body visible. Intrigued, the king the next morning challenges Sigurd to demonstrate his skill against the overcast sky. Once the youth has done so, Olav orders him to hold up a sunstone in the direction that he believes the Sun to be. Sunlight streams through the prism, verifying Sigurd's gift.*

More conventional aids were improving all the time. A compass with a graduated scale that worked over 360 degrees is mentioned by Peter of Maricourt in 1269. Judging from British ship account books, most vessels had at least two compasses; Magellan carried thirty-five replacement needles aboard his flagship. The fact that the north arrow of a compass does not, in most places, point to the true pole, but either east or west thereof by a spread that sailors term "the variation," was early detected;† but such inexactitude was not properly quantified until about 1490, and precautionary bearings were still taken of the Sun at its rise and setting well into the seventeenth century.

From about 1480 on, a mariner's astrolabe was used to calculate the Sun's noon altitude, so determining the observer's distance north or south of the equator. Such a device—with a ring marked in degrees for measuring celestial altitudes—was followed by the cross-staff (a nearly three-foot-long piece of wood marked with graduated measurements, with a perpendicular vane that slid back and forth upon it) and the backstaff, or back-quadrant, and together were the three main navigational instruments until the invention of the octant in 1731. Timothy Ferris brings such sightings to life:

---

* During the Second World War, the Sahara's colossal deposits of iron ore seriously misdirected Allied compasses, and instead soldiers had to go back to working out their position by the Sun. However, there are several ways to determine direction using the Sun (or the stars). One is the shadow-tip method. Push a straight stick about three feet long into the ground so it stands upright. Mark the tip of the shadow so cast, then wait until the shadow moves one and a half to two inches—between ten and fifteen minutes—then mark the tip of the second shadow. Draw a line from the first mark and about a foot beyond the second, then stand with your left foot on the first and your right on the end of the line drawn: in the northern temperate zone, you will be facing in a northerly direction; if in the southern, the opposite.

† The Arctic has *five* north poles: (1) the North Geographical Pole, at 90°, the fixed axis-head of the Earth; (2) the North Magnetic Pole: compasses point to it, but it wanders about; (3) the North Celestial Pole, in the sky directly above (1); (4) the North Geomagnetic Pole, which centers the Earth's magnetic field, and which also shifts; and (5) the Northern Pole of Inaccessibility—the point in the Arctic Ocean farthest from land, approximately 680 miles (1,100 kilometers) north of Alaska, at 84° 05′ N, 174° 85′ W. Why the magnetic poles should be located where they are remains a mystery, while the Earth's magnetic field has reversed its polarity tens of thousands of times over billions of years. The field inverts about once every 300,000 years, the North Pole becoming the South and vice versa, although the last switch occurred 780,000 years ago. Over the last 2,500 years the field has declined 40 percent in strength, so another reversal soon is a real possibility. The field is generated by currents of molten iron flowing within the Earth's outer core, in response to the gravitational pull of the Sun. Traces of these ancient journeys are to be found in rocks worldwide.[8]

At local noon on any clear day aboard a ship of the line, three officers could be seen helping to shoot the Sun—one holding the astrolabe steady, another sighting it, and the third reading the elevation—while deckhands stood by to catch the navigator when he fell or to retrieve the astrolabe if it were dropped and went scuttling across the rolling deck.[9]

In *Don Quixote* (1605), Cervantes makes his feelings plain about people's general ability to use all this newfangled gadgetry. Quixote (whose surname, in Catalan, translates as "horse's ass") is busy expounding to his long-suffering groom how Ptolemy remains the great authority on navigation, and how well he, the Don, might pinpoint their position had he the right equipment at hand. "If only I had an astrolabe here," he proclaims, "with which I could take the height of the Pole, I could tell you how far we have gone." Sancho Panza gloomily retorts: "My God, but your worship has got a pretty fellow for a witness of what you say, this same Tolmy or whatever you call him, with his amputation."[10]

For once, contemporary readers may well have laughed in sympathy with Quixote, not at him. The new instruments—possibly because, like compasses, they were associated with the black arts—were slow in coming into general use: it has been said that navigators were the only surviving upholders of the Ptolemaic theory, not because they believed the Sun went around the Earth but because it was simpler to consider that it did. And Jonathan Spence makes the point that by the beginning of the seventeenth century, "no benefits from the exploration of the heavens opened up by Copernicus had yet been applied to the art of navigation."[11]

Astrolabes passed out of fashion because they were never very precise, to be replaced by orreries, clockwork models of the planetary system. But be it compass, astrolabe, quadrant, sextant, telescope, or orrery, none was perfect, and sailors still needed to know how to use their own firsthand observation of the Sun to calculate their position. Even the advent of the telescope and its sister instruments did not put an immediate end to naked-eye astronomy, since it was not easy to make accurate positional observations until crosshairs were incorporated into telescopes in the 1660s.

Thus a huge weight of responsibility lay on the lonely pilots. Spence goes on: "Armed with whatever experience they had of winds and currents, fish movements and birds' flight, carrying simple maps and the narratives of previous voyagers when these existed, with compass, astrolabe, and quadrant, the pilots took responsibility for vessels of a thousand tons or more, with more than a thousand passengers and crew jammed aboard."

As Fernand Braudel describes sixteenth-century voyages, "Navigation in those days was a matter of following the shoreline, just as in the earliest days of water transport, moving crabwise from rock to rock, from promontories to islands and from islands to promontories . . . as a Portuguese chronicler puts it, 'of traveling from one seaside inn to another, dining in one and supping in the next.'"[12] In exceptional circumstances a ship might lose sight of the coast, were she blown off course or set on one of the three or four direct routes long known and used; but rarely did ships take voluntarily to the open sea. The threats to avoid under even greater penalty were storms and piracy (regarded for centuries as a quite honorable profession).[13] The Mediterranean, adds Braudel, "has never been inhabited by the profusion of sea-going peoples found in the northern seas and the Atlantic."[14]

Braudel's view has come under challenge, historians pointing out that as early as the days of Alexander the Great (356–323 B.C.), a new era in navigation was beginning with the construction of lighthouses, while during the same period rigging and steering gear were also improved, so that sea captains began to sail direct instead of hugging the coast.* But whatever the route, sea travel remained unpredictable: a 1551 record has a "certain ship which, without ever striking sail, arrived at Naples from Drepana, in Sicily, in 37 hours" (a distance of two hundred sea miles), the writer ascribing "such swift motion" to the power of "violent floods and outrageous winds."[15] Without such unusual conditions, Sicily to Rome (eight hundred sea miles) still took twenty to

---

* Braudel's classic first appeared in 1949. In 1955 a commodore at the British Admiralty, in a foreword to a history of sea navigation (not mentioning Braudel by name) delivered this broadside against the "rock-scratching" theory: "In these pages you will find disproved once and for all that persistent myth that the first sailors navigated by 'hugging the shore.' Those words could never have been written by a sailor. Nothing is more fraught with peril, and therefore the more assiduously avoided on a little-known coast, than hugging the shore. The myth is based on the assumption that the mariner has neither the means nor the ability to find his way out of sight of land. Now this assumption is shown to be groundless. Also it takes no account of the extra sense and undoubted ability of the small-ship sailor to know where he is with remarkable accuracy without instruments or observations. It is a fact that many deep-sea fishermen have this ability in some measure today, and it may well have been more commonplace in days gone by. Nothing is more sure, by whatever means they achieved it, than that the sailors of all ages have navigated in deep waters." Nonetheless, the grand French historian made no alteration in later editions. Another attack is made in David and Joan Hay, *No Star at the Pole: A History of Navigation from the Stone Age to the Twentieth Century—for All Who Enjoy the Sea* (London: Charles Knight, 1972), p. 9: "I disagree most strongly with the impression I get from existing works on this early classical period that they were by nature 'coastal creepers'; nothing can be further from the truth once reliable sea keeping boats could be constructed by the average shipyard and once . . . man learned to use the feel of the wind and the sun and stars to help him lay and keep a course of sorts. The odds are heavily that they were better, not worse, seamen than we are today." See also E.G.R. Taylor, *The Haven-Finding Art* (London: Hollis and Carter, 1971), pp. x–xi, and also Hay, *No Star at the Pole*, p. 125.

twenty-seven days, and outside the Mediterranean traffic was even slower: the 2,760 miles from Berenice, Egypt, to peninsular India, for instance, consumed up to six months—much less than a mile an hour.[16] During such voyages, the Sun remained one's reliable helpmate; and the geographer Richard Hakluyt could still write in 1598 that "no kind of men in any profession on the commonwealth pass their years in so great and continual hazard of life. . . . Of so many, so few grow to gray hairs."[17]

INSTRUMENTS THAT MADE use of the Sun had another ally in safe navigation: reliable—or semireliable—maps. Such aids had been employed in one form or another for centuries; yet the first mention of a chart being used aboard ship comes only in 1270, when the sainted Louis IX, king of France, attempting to cross directly from the south of France to Tunis on a Crusade, found himself driven by storm to take refuge in Cagliari Bay, on the Sardinian coast. To allay his fears, his crew produced a map, showing him exactly where they were.

As the world shrank and trade grew, accurate versions of the Earth's surface were as prized as they were rare, and most mapmakers mixed fiction, fact, and supposition. Some six hundred charts survive from the Middle Ages— Ecumenical Maps, as they are known, for they aimed to show the "Ecumene," the whole inhabited world, all circular, with Jerusalem at the center. Ptolemy's twenty-seven charts, drawn up for his *Geography* in A.D. 150, had been lost, but were rediscovered in the fifteenth century and widely copied; yet they treated the Earth as if it were a flat surface; these "plane charts" (from which we erroneously get "plain sailing") regularly led sailors astray, a failing for which not all the beautiful embellishments of artists could compensate. Gerard de Kremer (1512–1594), who became Gerardus Mercator Rupelmundanus ("the merchant from Rupelmonde") when he matriculated at the Flemish university of Leuven, resolved the question of how most nearly to reproduce a spherical world on a flat piece of paper: his famous projection extended the curves of latitude and longitude onto plane surfaces. There were other projections before Mercator's, but his allowed lines of fixed bearings to be rendered as straight lines, which made course plotting easier. His maps, like those of his contemporaries, were often highly classified, considered to be instruments of imperial power; to copy or share them with foreigners was treated as a capital offense.

By the year 1600 the revival of science and the development of printing had led to the widespread publication of astronomical maps, globes, and books. In 1665 the Jesuit scholar Athanasius Kircher made the first world map—*Mundus subterraneus*—to show currents, volcanic phenomena, and valleys: thematic cartography was born. But even as late as 1740, fewer than 120 places in the world had had their coordinates accurately established; cartographers simply marked off whole areas with the single word "Unexplored."*

Perhaps in part for that reason, throughout Western Europe, astronomy was viewed not so much as a science that describes the universe but as an aid to navigation (a word that, taken from the Latin *navis*, "ship," and *agere*, "to drive," for centuries meant the demanding art of conducting a ship across the sea). Astronomy was the first state science, and the story is told that the English king Charles II learned from the pillow talk of a mistress from seagoing Brittany that a Frenchman had devised a method of using the positions of the Moon to determine longitude at sea. No matter that there was scant evidence for this: Charles decided that in order to seize the maritime initiative, he should invest in the latest astronomical technology, and in 1675 established the Royal Observatory at Greenwich. Navigation had already become so much part of the English consciousness that the poet laureate John Dryden could look back on its progress in "Annus Mirabilis: The Year of Wonders 1666":

> *Rude as their ships was navigation then;*
> *No useful compass or meridian known;*
> *Coasting, they kept the land within their ken,*
> *And knew no North but when the Polestar shone.*

> *Of all who since have used the open sea,*
> *Than the bold English none more fame have won:*
> *Beyond the year, and out of heaven's high way,*
> *They make discoveries where they see no sun.*

Nonetheless, during the Third Anglo-Dutch War (1672–74), the English navy's run of poor performances was ascribed to the low level of astronomical data available to enable sea captains to navigate and maneuver efficiently.[18] In 1731 John Hadley, a fellow of the Royal Society, and Thomas Godfrey, a glazier from Philadelphia, independently invented the reflecting quadrant, or octant, in which a 45-degree arc is divided into ninety parts, the ends of the arc being joined by two arms. In 1757 John Campbell made the instrument less cumbersome by enlarging it to one-sixth (the *sextant*) of a circle but in an instrument that was itself a circle, and more accurate by adding filters and a telescope.[19]

---

* David Grann, *The Lost City of Z* (New York: Doubleday, 2009), p. 51. The power of print can be seen in the fact that the naming of "America" (after the Florentine banker Amerigo Vespucci rather than Columbus) seems to have originated in a woodcut globe of 1507. See also Rodney W. Shirley, *The Mapping of the World* (London: New Holland, 1993), p. xii.

Well into the nineteenth century, despite the profusion of instruments, observatories, and maps, there continued to be times when the Sun was all that sailors had to go by. At one point in *Moby-Dick,* as Captain Ahab gazes through the ship's quadrant, waiting for the Sun to reach the meridian, he starts to rant at the strange equipment. His outburst reflects a ship's company's mixed awe and anger at the Sun's power, their doubts about how much science could ever help them, and their dread belief that they might lose their way forever:

Foolish toy! Babies' plaything of haughty admirals, and commodores, and captains; the world brags of thee, of thy cunning and might; but what after all canst thou do, but tell the poor, pitiful point, where thou thyself happenest to be on this wide planet, and the hand that holds thee: no! Not one jot more! Thou canst not tell where one drop of water or one grain of sand will be to-morrow noon; and yet with thy impotence thou insultest the sun! science! Curse thee, thou vain toy.[20]

Ahab well understood how fallible such "toys" could be. Later in the narrative, our obsessed captain asks his helmsman how the ship is heading, and is told east-southeast. "Thou liest!" shouts Ahab, smiting his man with his fist; but when he consults his two compasses they both point east, even though the *Pequod* is going west:

The old man with a rigid laugh exclaimed, "I have it! It has happened before. Mr. Starbuck, last night's thunder turned our compasses—that's all."[21]

Ahab is up to the task; he seizes a steel rod from the ship's rigging and hands it to the mate with word to hold it upright, without its touching the deck; then remagnetizes the needle to the Earth's field with a few well-directed hammer blows before taking out two needles from the binnacle and suspending one over his compass card. When finally the needle settles, he points at it and exclaims, "Look ye, for yourselves, if Ahab be not lord of the level loadstone! The Sun is East, and that compass swears it!"

Because even the best compasses could be temperamental, the most modern instruments faulty, science was always looking for better ways to measure. In the eighteenth century, the search was focused on the need for more accurate determinations of longitude. While measurement of latitude (north-south positioning) was relatively simple, longitude (east-west positioning, which was most easily measured as a function of time) was a far greater challenge. A typical ship's clock in the early seventeenth century was accurate to

In 1740–44, during one of England's wars with Spain, six British ships under Commodore George Anson set out to circumnavigate the globe. Anson's map of his journey movingly shows what befell his force as a result of their lack of knowledge of longitude (their east-west position). The straggly thin line to the west of the continent tracks the squadron's progress, and shows how often Anson and his crews lost their way. Only one ship—500 men out of an original 1,900— completed the voyage. The charts they carried typically placed Juan Fernandez 135 miles west of Valparaiso on the South American coast. In fact, it is 360 miles west. Unsure of his maps, Anson set off in the wrong direction; by the time he reversed his course, it took nine days to get back to his starting point, during which period 70 men died.

no better than a few minutes a day—which after several days at sea could mean a miscalculation of several miles. In her book *Longitude*, Dava Sobel reminds us of the blood price of poor navigation. Sir Clowdesley Shovell, returning in 1707 from having captured Gibraltar three years before, piled four of his ships directly onto the Scilly Isles, at the far southwest tip of England, when he thought he was safely rounding the coast of Brittany about 120 miles to the west. Sixteen hundred of his men perished with him. The shock of so great a disaster so close to home aroused Parliament to establish, in 1714, a special Board of Longitude, and also a £20,000 prize for the accurate determination of longitude.

As early as 1610, Galileo had proposed optimistically that absolute time could

Jonas Moore
(1627–1679), a leading
mathematician of
Charles II's time, helped
persuade Parliament to
offer a large cash prize
to the first person to
solve the longitude
problem. An illustration
from Moore's
A New System of
Mathematicks (1681)
shows navigators and
astronomers at their
instruments.

be measured at any point on the Earth by bearings from the moons of Jupiter, and had even devised a helmet with a telescope attached that the observer could don while seated in a chair set in what were called "gimbals"—similar to those used to keep a ship's compass horizontal. This proved effective for surveying distances on land, but never worked at sea. Around 1710, the Yorkshireman Jeremy Thacker introduced the word "chronometer" into the English language. In *Gulliver's Travels,* published in 1726, Gulliver imagines himself living so long that "I should then see the discovery of the longitude, the perpetual motion, the universal medicine, and many other great inventions brought to the utmost perfection." To Swift, an accurate measure of longitude obviously seemed about as impossible as the other marvels on the list.

Not to those who entered the competition, however. In theory, the problem sounded simple: one had to compare the local time with the time at a given place, such as Paris or Greenwich, relying on the regular, "clockwork" nature of the motions of heavenly bodies, then make an easy mathematical calculation. But to do that one had in the first place to produce a clock that could maintain accurate time on a long, rough sea voyage with widely varying conditions of temperature, pressure, and humidity.[22] In 1736, suitably motivated by the handsome prize—£20,000 then would be worth $4.5 million, or nearly £3 million, today—John Harrison (1693–1776), a lowly clock maker, unveiled his "marine timekeeper" or "sea clock," accurate to one-tenth of a second per day. The board, set up specifically to assess claims for the prize, granted Harrison £500 for further development, and in 1759 he came up with an even better version—a succession of sturdy portable clocks he called "watches," a term taken from the practice of dividing a ship's day into six watches of four hours each. Unfortunately for him, the board was dominated by well-established professional astronomers, who refused to believe that a man from so modest a background could have accomplished such a feat, handing over but half the prize, and that much only in 1765—more than three years after tests had shown his invention lost just 5.1 seconds over eighty days at sea. When in 1773 Cook took HMS *Resolution* on one of the first crossings of the Antarctic Circle, among the voyage's signal accomplishments was the vindication of a chronometer based on one of Harrison's models. By the time the magnitude of his achievement was finally recognized, Harrison had but three years to live.

In 1884, after almost a century of nearly every great nation's running the meridian from which time was calculated through its own capital, a conference in Washington awarded the prime meridian to England, the consideration that tipped the scales being that Greenwich had developed the necessary instrumental facilities and compiled the observational data over two hundred years. By then, as the jingoistic English poet William Watson (1858–1935) put it in his ode on the coronation of King Edward VII,

> Time and the ocean and some fostering star
> In high cabal have made us what we are.[23]

But the development of the prime meridian is more than the story of sailors' use of the Sun. It is part of the whole history of timekeeping, and how the Sun was put to measure its own passing.

CHAPTER 20

OF CALENDARS AND DIALS

Time must never be thought of as pre-existing in any sense; it is a
manufactured *quantity.*

—HERMANN BONDI[1]

A calendar is a tool which cannot be justified by either logic or astronomy.

—E. J. BICKERMAN[2]

E VEN THE MOST SOPHISTICATED PEOPLE LOSE THEIR HEADS OVER
the mystery of counting time," wrote Umberto Eco in the run-up to the current millennium. He was stating a general truth, although he was specifically
talking about the confusion over whether the millennium would arrive on December 31, 1999, or the same day in the year 2000—which went to the question of whether year one began at zero or a year from zero—the zero referred
to, of course, being the moment of Christ's birth, that itself being debatable.[3]
This system of counting time from the birth of Christ was initiated in the sixth
century A.D. by Dionysius Exiguus; before him, dates were calculated from the
reign of Diocletian (A.D. 284–305) onward, or from the beginning of the world
(measured with an unreal precision), but Dionysius got the year of the Nativity wrong, so calculations were off from the word go.

In 1956, many school examinations in Britain, as well as plays and other
celebrations marking the two thousandth anniversary of Caesar's assassina-
tion, caused substantial embarrassment when people realized that without the
year zero the deed was only 1,999 years ago. By the same reckoning, Jesus died
at thirty-two, not thirty-three. There are of course other systems for creating
calendars and measuring years. On September 12, 2007, Ethiopians ushered
in their own millennium, based on a calendar more than seven years behind
the Gregorian. Then there is the town of Hamelin, which long dated its municipal records not by the year of grace but from July 26, 1284, when, say its
archives, 130 children were taken from the city—apparently by the Pied
Piper—never to be seen again.[4]

There are yet more possible sources of confusion. Calendars still divide about equally on when the seasons begin: is the vernal equinox March 20 or 21? It is still a matter of contention among historians when Britain entered the First World War, since her ultimatum said hostilities would begin at midnight, but neglected to say whether by London or Berlin time. More recently, the raid on Pearl Harbor in 1941, which is commemorated as having taken place on December 7 in Hawaii, occurred on December 8 in Japan, which lies on the other side of the International Date Line.

Ethiopians celebrating the new millennium—our September 12, 2007

We can put to one side the business of correctly reckoning time zones, eras, and millennia; but what of years and months? All calendric calculations involve some form of astronomy, whether observations of the Sun or the Moon or both, and every major civilization has developed at least one form of calendar acknowledging its own commemorations and festivals. Not to mark time is to be lost out of time: in the early 1940s, an anthropologist studying the Sirionó tribe in Bolivia would tellingly note, "No records of time are kept, and no type of calendar exists." The tribe was, he concluded, "man in the raw state of nature."[5]

Almost all early civilizations started with lunar calendars—the Babylonians, Greeks, Jews, and Egyptians in the Middle East; the Aztecs and Inca in the Americas; and the Chinese and Hindus in East Asia—then transformed them into lunar-solar hybrids. The Moon, after all, before the advent of sophisticated mapping systems, was more reliable than any star (*moon*, meaning "the measurer"; thus *month*, "the time measured"); but it often differed from the natural calendar of farmer and shepherd. As a result, any given society might employ three calendars, one for government use, one for religious observances, and yet a third for day-to-day purposes.

Of all the myriad forms known to history, only four calendars have been based on the Sun alone: the Egyptian (eventually), the Achaemenian Later Avestan (as used in Persia from 559 to 331 B.C.), that developed by the Mayas and adopted by the Aztecs, and our own Julian/Gregorian. Even in these, the lunar element was never suppressed completely.[6] Whatever the community, its main religious dates shifted yearly against the seasons, be it Easter, Eid, Deepavali, the Chinese New Year, or Yom Kippur.

The difficulties confronting the early calendar makers, whether using Sun or Moon, were stupendous. The only way to tell the time was by the Sun during daylight hours and by the stars at night, and in the course of history each people, sometimes each generation, attempted its own solution. An added pressure came shortly before 3200 B.C. with the invention of writing: as literacy spread, people sought to assign dates to records, letters, and inventories that would be commonly recognized and understood.

Whenever a calendar had to be created, such was its importance that it acquired a dimension of real power. In ancient and medieval China, since its emperors were considered the embodiment of the heavens' will, each change of reign—even more important, each change of dynasty—required that a fresh calendar be created, fixing different festival days, and yet new dates to plant or harvest, in order to show that a new disposition of celestial influences had asserted itself. (One might wonder just how many such changes could be made.) This tradition was well established by Han times (206 B.C. to A.D. 220) and the era between the early Han Dynasty and the Ming Dynasty of A.D. 1368 gave rise to some forty new calendars. "For an agricultural economy," observes Joseph Needham of all this, "astronomical knowledge as a regulator of the calendar was of prime importance. He who could give a calendar to the people would become their leader. . . . The promulgation of the calendar by the emperor was a right corresponding to the issue of minted coins, with image and superscription, by Western rulers. The use of it signified recognition of imperial authority."*

---

* Joseph Needham, *Science and Civilization in China* (Cambridge: Cambridge University Press, 1959), p. 189, and Christopher Cullen, *Astronomy and Mathematics in Ancient China: The Zhou Bi Suan Jing* (New York: Cambridge University Press, 1996), p. 6. Cullen, who currently oversees the Needham Research Institute in Cambridge, reckons that Needham badly underestimated the importance of calendrical astronomy in China; his own writing makes only a partial case.

From about 2000 B.C., the Babylonians set their calendar from direct astronomical observation. The official day began at sunset, except for the first of each new month, as reckoned from the sighting of its young crescent moon. If this couldn't be seen—owing to poor visibility or because it was too close to the Sun—the beginning of the month was simply deferred, although common sense required that no month last more than thirty days. Strictly, a lunar month is the time it takes the Moon to pass through each of its phases—new moon, half, full moon—and return to its original position: 29 days, 12 hours, 44 minutes, and 3 seconds. At some point in the fourth century B.C. yet another system was devised, the Babylonians conceiving the notion of a "mean" sun—a fictitious sun moving at a uniform rate—which, without essential change, is what guides our reckonings today.

The Chinese, like the Babylonians, relied wholly on observation, and seem to have drawn few deductions from the errors that resulted: their almanacs were continually falling off course. Over two thousand years they made more than fifty revisions, not all of which marked changes of reign. Their first calendar went by the Moon, with alternating months of twenty-nine and thirty days, shortchanging the year by eleven days. By about the sixth century, their astronomers, again echoing the Babylonians, recognized a nineteen-year cycle after which the phases of the Moon recurred on the same day of the solar year, so they took to adding extra days as necessary, to "save the phenomena"—that is, to keep calculation in step with nature. By the first century B.C. they had adopted a system of twenty-four fortnightly periods, each such period corresponding to a 15-degree motion of the Sun along the ecliptic (its apparent path in the sky). Their year began on February 5, and the days of their months had names that, in imagery at least, put other cultures to shame—for instance, "The Awakening of Insects" (March 7); "Grain in Ear" (June 7); and "Descent of Hoarfrost" (October 24). By the sixth Christian century, the irregular apparent motion of the Sun had been taken into account, and eventually the motions of Sun and Moon were incorporated.*

The Islamic calendar, at first lunar, then changed to a lunar-solar mix (the rightward-facing crescent, the sign of a new moon, appears on the flags of Muslim nations), also achieving a degree of accuracy by adding days where necessary. However, in A.D. 632 Muhammad, angered that certain communities were altering the sacred months in which fighting was prohibited, forbade

---

* Chinese rulers had it both ways. Foreign calendars were regularly introduced, an Indian one arriving during the Tang Dynasty (A.D. 618–906), a Muslim one under the Yuan (A.D. 1279–1368), and the Gregorian in the seventeenth century. But the Chinese systems were not officially supplanted until 1912, and even today most Chinese calendars give both the Gregorian year and the year in the old, sixty-year cycle.

all such intercalation, saying that the addition of days to make the calendar conform to the solar year violated the commands of God. He then introduced a purely lunar calendar, but this makes the Islamic year only 354 or 355 days long, so that over the course of thirty-four or so of our years, Muslim holidays will slip through all the seasons.

Strict adherence to the lunar calendar also means that the declaration of a new month depends on the actual *sighting* of its new moon, so the calendar cannot be determined ahead of time and the months begin unpredictably. New moons, says the Qur'an, "are fixed times for the people and for the pilgrimage [to Mecca]." Each month, the official declaration of the Moon's sighting is eagerly awaited—although as Islam has dispersed geographically, it has become harder to adhere to an exclusively lunar calendar. In 1971, the shah of Iran switched his country from the Islamic (lunar) calendar to the Persian (solar), the better to celebrate the 2,500th anniversary of the pre-Islamic Peacock Throne, prompting an intensive program worldwide to create an international Islamic calendar, a mixture of calculation and sighting; but no version has yet been universally accepted.

In Jewish observance, the day divides into six watches—as reflected in the Psalms—from midnight to midnight. To keep pace with the solar year, Jewish leap years add an extra month in the third, sixth, eighth, eleventh, fourteenth, seventeenth, and nineteenth years of each nineteen-year period. Every twenty-eighth year (known as a Nisan, when the position of the Sun is calculated to have been the same as it was on the day of Creation), Orthodox Jews engage in a blessing of the Sun (the last one was in 2009). As with the Muslim calendar, the months of the Jewish register are reckoned by the Moon, beginning when its crescent first appears in the western sky.

The Egyptians began with a lunar model, but it, like others of its time, proved so inaccurate that (just as later) a year of twelve lunar months was a full eleven days apart from a solar year, and festival dates again drifted badly. They elected to divide their calendar into three seasons—the flooding of the Nile, its subsidence, and the harvest—each lasting four lunar months. To make a full year, they added another month whenever Sirius rose late in the twelfth month. Eventually, once they recognized that the length of the year was close to 365 days, they simply tacked five extra days onto the final month.

After moving to a solar year, they devised a more sophisticated calendar using a roster of thirty-six stars located around Sirius, the fresh sighting of each star marking a new day. Each of the thirty-six "decans" (now so called because they came into view ten days apart) would be invisible for seventy days before rising. At any one time, eighteen decans covered the period from sunset to sunrise, three each were assigned to the in-between periods of dusk and twilight, and the remaining twelve marked total darkness (a word rendered in Egyptian by the powerful phrase "what grips the bowels"). From this emerged the twenty-four-hour day, with hours that varied in length through the year—daylight hours being longer in summer, for example—until a new dynasty, beginning the Egyptian New Kingdom (1539 B.C. on), introduced the sixty-minute hour. They added an extra day every four years.

The Mayan method of measuring time was impressive but mystifying. They had a calendar that, by A.D. 800, used both agricultural and solar cycles, one of which was made up of eighteen months, each of twenty days—followed by an intercalary "month" of five days, called the *uayeb*—with names such as Pop, Zip, Zec, Mol, Yax, and Zac, which have a fine ring; but these were later subsumed into a calendar of thirteen twenty-eight-day months based on the Moon. Their initial, solar version remains accurate to within about three seconds per year—closer to the true year than the Gregorian.[7] Its "long count" ends on December 21, 2012, which coincides with various unusual alignments in our solar system sufficiently well to have inspired myriad theories (under the general rubric of "the Mayan Prophecy") about a coming apocalypse.

The Aztec calendar consisted of a 365-day cycle called *xiuhpohuali* (year count) and a 260-day ritual cycle called *tonalpohuali* (day count). The former made up the agricultural calendar, since it had to take account of the Sun; the latter was the sacred calendar. The Inca likewise had two versions. When in 2004 I visited the mountain city of Cuzco, I was told that its massive sun pillars (sadly destroyed by the Spaniards) had been used to fix the planting of crops. More recently, a range of thirteen stone towers within a 2,300-year-old ruin 250 miles north of Lima has been revealed as a solar index (something like Stonehenge), employed by the Inca to help run their empire; but there also co-existed a star-determined agricultural calendar consulted by the lower classes.

In astronomy, the Inca, like the Aztecs, were not as developed as the Mayans; yet they had an elaborate calendar of twelve lunar months, which they corrected from time to time according to sightings from Cuzco. The four months bracketing the solstices honored the Sun; those around the equinoxes were governed by the cults of water and the Moon Goddess; and the four left over were dedicated to agriculture, death, the Thunder God (deity of war and of all climatic events), and the goddess of the planet Venus.[8] Otherwise, the Inca ways of measuring time were not very different from those of many of their counterparts—imaginative, imprecise, regularly tinkered with, and driven by a mix of solar and lunar observations.

The innumerable Greek states had a whole mix of indices. One, used in Athens in the fifth century B.C., organized itself from the summer solstice, beginning the New Year with the following new moon. In Athens, as in most

Greek jurisdictions, the day began at sunset, a natural arrangement with a system that reckoned time by the Moon; but magistrates were free to repeat dates—several times if they so wished, such as having a succession of December twenty-fifths. There were limits to this power, for the year had to end on the right date; if one day was repeated, another had to be omitted, and by the last month there was little freedom left for manipulation. The calendar was finally reformed around 432 B.C. by Meton of Athens, the lunar month being aligned with the solar year by intercalating seven months into every nineteen-year cycle, and by varying months between twenty-nine and thirty days, thus making the lunar month overlong by only two minutes.*

· The first significant change in European calendars following the Athenian innovations of 432 B.C. came about as a mixture of practical necessity and political factors (and perhaps a little lust). Under the Roman Republic, the endlessly recurring need for a correcting month was assessed annually, often in the light of blatantly political considerations. Julius Caesar was both pontifex maximus (a senior priest of the state religion) and a proconsul (provincial governor), and in his former role oversaw the calendar. However, he was at grave contention— that ultimately passed into civil war—with the faction led by his son-in-law and former ally, Pompey; and during this time the calendar year had contracted anarchically from a 365-day span, with January falling in the autumn.

Once he had triumphed over his enemies, Caesar commissioned a Greek astronomer, Sosigenes of Alexandria, to devise a new calendar, and he suggested increasing the lengths of some months by one day each and February by one day every three years only (a "leap year"), creating a new calendar, the Julian, of about 365.25 days per annum, with the year starting on January 1. One difficulty with this was that too many leap days were added with respect to the astronomical seasons. On average, solstices and equinoxes advance by eleven minutes per year against the Julian calendar, causing the Julian to drift backward one day about every 128 years. Caesar happily dubbed 46 B.C. "the last year of confusion," but with 45 B.C. lasting a full 445 days, the citizens of Rome wittily reversed this to "the year of confusion." They were nevertheless jubilant because they believed that Caesar had extended their lives by an extra three months.

---

* If one takes the year 432 B.C. as reckoned in the Gregorian calendar, it is daunting to realize that this same year appears in other calendars as follows: *Ab urbe condita* (from a massive history of Rome, c. 753 B.C.), 322; Bahai, −2275/4; Berber, 519; Buddhist, 113; Burmese, −1069; Byzantine, 5077–78; Chinese (sexagenary cycle), 2205–66; Coptic, −715–14; Ethiopian, −439–38; Hebrew, 3329–30; Hindu, Vikram Samvat, −376–75; Hindu, Kali Yuga, 2670–71; Holocene, 9569; Iranian, 1053–52 BP; Islamic, −1085–84 BH; Korean, 1902; and Thai, 112. This list omits the Armenian, Japanese, Mayan, Aztec, and Inca calendars, which do not cover such early dates.

Several of our months reflect the Julian counting method, the seventh month being September (*septem*), the eighth October (*octo*), and so on. The new version mirrored the Greek-Egyptian calendar of 238 B.C., which Caesar was happy to introduce, as at the time he was passionately involved with Cleopatra, and also because before the civil war the Senate had refused to intercalate the extra months that the old calendar needed, as that would have prolonged Caesar's term of office.

The new calendar was still seriously out of kilter, for it misreckoned every third rather than fourth year a leap year, so that by 11 B.C., a mere thirty-three years after Caesar's death, the year was starting three days late. Caesar's great-nephew, adopted son, and more tactful successor as dictator, Augustus, corrected this by skipping three leap years, not having another one until A.D. 8. During his reign, the fifth and sixth months, Quintilis and Sextilis, were renamed July and August in honor, respectfully, of Caesar and Augustus.*

The advent of Christianity made new demands on the reckoning of time. The Julian calendar was used to "fix" certain events such as Christmas, the Epiphany, and the Annunciation, and also to determine the movable-feast sequence of Easter, Pentecost, and Lent. Easter is assigned as the Sunday after the first full moon to fall on or after March 21; at least a dozen other festivals follow from that dating. As the Gospels state unequivocally that Jesus was crucified at Passover, Easter depended on the complicated lunar calculations by which Jewish authority set that festival. Many early Christians held that Jesus had died on a Friday and risen two days later; but if one followed the Jewish calendar, there was no assurance that Easter would fall on a Sunday. This led to a major schism between the Eastern Orthodox Church and Rome, the former observing Easter on the fourteenth day of the lunar month regardless of the day of the week. Different faiths still celebrate Easter on different Sundays.

---

* Over the centuries other attempts to rename the months have been made, but all have failed. In 1793, for example, the French Republic declared a new Revolutionary calendar, with twelve thirty-month days beginning with Year One of the Revolution, September 22, 1792. The new months were Vendémiaire, Brumaire, Frimaire, Nivôse, Pluviôse, Ventôse, Germinal, Floréal, Prairial, Messidor, Thermidor, and Fructidor. Every tenth day was a holiday, and there were five *sans-culottides* (intercalary) days (named after the Revolutionary lower classes, who did not wear fashionable culottes). It lasted just over a decade before it was abolished, since a ten-day week gave workers less rest; also, every new year started on a different date (because fixed to the equinox), a major source of confusion for almost everybody, not least because it was incompatible with the secular rhythms of trade fairs and agricultural markets. The Revolution also attempted to introduce a decimal clock, with each *heure* twice as long as the traditional pre-Revolutionary one. On Fructidor 22, Year XI (September 9, 1803), this innovation too was repealed. (More successfully, the English, from 1880 on, would yoke together the last day of the Christian week—Saturday—and the first—Sunday—into "the weekend," which, after the Great War, more and more countries adopted both as a term and as a practice.

How best to determine an ideal calendar was exhaustively examined in *On the Theory of Time-Reckoning*, written in 725 by the Venerable Bede, who calculated that the 365.25-day Julian year was longer than the solar year by eleven minutes and four seconds, and that a more accurate measurement was closer to 365.24 days. Still nothing was done, and over time Caesar's calendar got more and more out of step with the seasons. When at last reform came it took a pope to do it. The story that has been passed down is that when it became apparent that the Easter of 1576 would be particularly ill timed, Gregory XIII went to the Vatican's Tower of the Four Winds, where a papal astronomer showed him the solar image on the meridian line across the floor of the Calendar Room, and the pontiff could see that the Sun was ten days away from where it had to be to reach equinox on March 20–21. At that moment, it is said, he decided that the calendar must be aligned with the unyielding heavens. Maybe the story is true, but the need for reform had long been recognized: aside from Bede, as early as the thirteenth century Roger Bacon had sent Clement IV a treatise on the calendar's shortcomings.

In a matter of weeks, a new plan, devised by a well-known Calabrian physician and amateur astronomer named Aloysius Lilius (c. 1510–1576), was presented to the pope by the astronomer's brother (Lilius himself having recently died). Gregory asked the Jesuit mathematician Christopher Clavius, a Bavarian living in Rome, where he was acclaimed as "the Euclid of the sixteenth century," to review it. Clavius endorsed the innovations and added some of his own. Over the next few years all Catholic countries were instructed to omit the extra ten days. The pope agreed to 1582 as the changeover year, and October was chosen as the changeover month, as containing the fewest Church feasts, so imposing the minimum disruption.

There was also a good political reason for the choice, for Easter 1583 would thus fall on the same day in both the Julian calendar (March 31) and the Gregorian (April 10), a happy conjunction that would not repeat itself for many years. One might ask why Gregory didn't delete fifteen days rather than ten, shifting the spring equinox back to its traditional date of March 25. Had he done so, however, the winter solstice would have moved to December 25, by this time a major Christian feast day. As Duncan Steel notes, "By allowing Christmas and the solstice to coincide once more, the Church would have walked into trouble. Christianity had successfully pinched the solstice festival of the pagan religions over twelve hundred years before . . . and was not about to hand it back."[9]

In Spain, Portugal, and parts of Italy the new calendar was taken up immediately; France and the Low Countries followed by the end of that year. In what is now Belgium, the calendar went from December 21, 1582, straight to January 1, 1583, depriving everyone of Christmas. Catholics in Germany followed suit in 1584, Denmark-Norway by 1586—although Sweden held out until 1753. Most other non-Catholic Christians scorned the new calendar; not until 1700 did the German Protestant states adopt it. In Grisons, the most easterly of the Swiss cantons, Catholics and Protestants, even those living on the same street, kept different calendars, a situation that continued down to 1798, when the French invaded and imposed the Gregorian on everyone.

Protestant England was likewise suspicious of any pronouncement from Rome, and although Elizabeth I was not unsympathetic to the proposed change, the Spanish Armada of 1588 effectively scuttled any chance of its being accepted. Voltaire would scoff: "The English mob preferred their calendar to disagree with the Sun than to agree with the Pope."[10] More likely, well aware of the Gregorian calendar's flaws, English scientists thought that if they developed a superior calendar it would help effect a rapprochement with those European nations fence-sitting in the quarrel with Rome. One interpretation of *The Taming of the Shrew* is that Katharine represents the Protestant religion, Petruchio pre-Reformation Catholicism; thus his order that she should call the

The 1751 bill bringing in the Gregorian calendar was so unpopular that William Hogarth introduced a stolen Tony placard, "Give Us Our Eleven Days!" (it lies on the floor, under the foot of the man with the knob kerry) into The Election Dinner *(1755), his painting of a tavern meeting organized by Whig candidates with Tories protesting outside.*

Sun the Moon no longer appears senseless when one considers that at the time the play was written (1592), England was already ten days out of sync with the Roman calendar.*

Britain and its colonies delayed two centuries before making the adjustment, by which time their calendar had to be moved forward another day, Wednesday, September 2, 1752, being followed by Thursday the fourteenth. When the British Parliament debated the change, rumors spread throughout the country that salaried employees were losing eleven days' pay and that everyone was surrendering eleven days of their life. The protests were not unreasonable: people who had paid rent for a full month found they received no rebate, while bankers refused to pay their taxes on March 25 and put off payment by eleven days—hence the British tax year still begins on April 5. As recently as 1995 an Ulster politician attacked the Vatican for meddling with the calendar. Yet ultimately it is not the objections that are remarkable, but the fact that there was so much agreement.

Japan waited until 1872 to adopt the Western calendar, when the reform led to peasant riots. Turkey was last, accepting the inevitable in 1927. Russia may boast the most confusing history of all. Until the end of the fifteenth century, new years began every March 1, then on September 1 until 1700, when Peter the Great changed the date to January 1. In 1709 the Julian calendar, as favored by the Orthodox Church, came in, more than 127 years after the Gregorian had been introduced into Western Europe. For most of the nineteenth century, the Department of Foreign Affairs used the Gregorian, as did the expansionist Russian navy; finally, in 1918 Lenin decreed the adoption of the Gregorian throughout the country—but this held only until 1923, when both Gregorian and Julian were discarded in an attempt to eliminate the Christian reckoning. In their place came the "Eternal Calendar" (with special cards distributed to workers, to increase production). By 1929 a week of only five days and therefore months of six weeks were in place, the former remaining so until 1934, when the Gregorian calendar was reimposed, although the seven-day week did not make a return until 1940. The potential for confusion was amply realized.†

---

* According to the Gregorian version, Shakespeare died on Friday, May 3, 1616, his fellow writer Cervantes on Tuesday, April 23 (said to be Shakespeare's birthday). But by the older Julian calendar that England used, Shakespeare had died on its April 23, Saint George's Day. Although the Spaniard's death was thus ten days earlier than Shakespeare's, the two men are often said to have perished together. In honor of this conjunction, UNESCO established April 23 as International Day of the Book. Anyway, Cervantes probably died on April 22 (Gregorian), but was buried on the twenty-third.

† The great variety of calendars over history was to provide Tolkien with rich material. *The Lord of the Rings* devotes a seven-page appendix to the "Shire Year," with its twelve months Afteryule,

The new Gregorian calendar had both virtues and drawbacks. It did not perfectly track the equinoxes and solstices, but it did more accurately indicate the time of year in respect to the seasons, being based on the tropical/solar year (i.e., the time it takes the Earth to orbit the Sun, measured between two spring equinoxes) of 365 days, 5 hours, 49 minutes (approximately). As each year is reckoned as 365 days exactly, it means that the following new year begins 5 hours and 49 minutes ahead of the Earth's completing its circuit, so that after any four years the calendar is four times 5 hours 49 minutes ahead of the Sun. To get back in sync with the Sun, every fourth year, the so-called leap year, an additional twenty-four hours (the intercalary day) is added, yielding a year of 366 days. (Islam makes no such adjustments, and so Ramadan, for example, floats through spring, summer, fall, and winter: particularly confusing since it is from the Arabic word for August—*rams*, meaning "parched thirst"— that Ramadan takes its name).

In fact, the calendar has really been held back eleven minutes too much each year—forty-four minutes every four years. In the course of a hundred years, this adds up to nearly a full day; if allowed to accumulate, four hundred years would embrace 146,100 days instead of the actual 146,097. Consequently, the three centurial years out of every four whose first two digits are not exactly divisible by four—so far, 1700, 1800, and 1900—are not leap years; by this arrangement the calendar again inches back toward the path of the Sun, and all is well with the world.

Gregory's calendar was meant to get Easter back to its old sequence—an adjustment that was made for devotional rather than scientific purposes. Yet working out exactly when that festival (and others) will fall continues to engage our attention. In October 2003 I traveled to Heidelberg to see Dr. Reinhold Bien at the Astronomisches Rechen-Institut. Comfortably dressed in brown-checked shirt, pleated pants, and scuffed shoes, he was small and rounded, a little like a friendly hedgehog. At the institute, Dr. Bien measures the heavens, charts the motions of the stars, and each day posts the times for the morrow's sunrise and sunset. But he also advises the German government as to when certain key celebrations should take place each year. "One can't frame a calendar for all time," he patiently explained. "The Earth's orbit is slowing down, so what is accurate today will not be in the future." He was soon telling me with

---

Astron, Afterlithe, Winterfilth (winter–full moon), Solmath, Thrimidge, Wedmath, Blotmath, Rethe, Forelithe, Halimath, and Foreyule. Each year begins on a Saturday, and Mid-Year's Day has no weekday name. "In Middle-Earth the Eldar also observed a short period or solar year, called a *coranar* or 'sun-round' when considered more or less astronomically," and so on; J.R.R. enjoying himself.[11]

gusto stories about Christopher Clavius, and the railleries of John Donne, a great poet but also an Anglican convert, who dismissed Clavius as a glutton and a drunkard. As Dr. Bien defended Clavius, I found my gaze wandering to a small notice pinned just above his cluttered desk. It read, in English: THIS YEAR CHRISTMAS WILL BE ON 25 DECEMBER. A joke; but only just.

THE CALCULATION OF the year's passage is one kind of timekeeping. Measurement of the hours, and of lesser units of time, probably began much later, though nobody knows quite when. So long as people lived by raising crops and herding animals, there was little need for measuring smaller units of time. From at least the ninth century on, most cultures measured in weeks, marking out the year with dates drawn from folklore, their own needs and observations, and (in the more sophisticated cities) the liturgical calendar. "The understanding of a savage," wrote William Hazlitt in 1827, "is a kind of natural almanac, and more true in its prognostication of the future."[12] Leaving aside what is implied by that loaded word "savage," we know that the Konso people of central Africa still reckon their day by function rather than by calculation: 5 P.M. to 6 P.M., for instance, is *kakalseema* ("when the cattle return home"), with periods of the day named after the activity performed then.*

The simple concept of an "hour" as having a uniform duration took more than two of the five millennia of known history to develop. For the Egyptians, an hour in January and an hour in August, or an hour in northerly Alexandria and an hour in southerly Memphis, had markedly different lengths. The most natural division of time is into two parts—"day" and "night." The Romans, until the end of the fourth century B.C., divided the former into before midday (*ante meridiem*) and after midday (*post meridiem*). Since all court work took place before noon, they posted a civil officer to detect the moment of the Sun's crossing the zenith and proclaim it in the Forum.† They also distinguished be-

---

* Many sense small passages of time naturalistically. I have corresponded with several blind people about their experiences of the Sun, and certain responses seem widespread: the aviator Miles Hilton-Barber, for instance, uses the feeling of the Sun on his face to help when flying (but not at the controls) to get an idea of time. The writer Ved Mehta told me that when he left his native India in 1949, aged fifteen, to attend a college for the blind in Arkansas, many at the school didn't wear a Braille watch, as they preferred to reckon the time from the atmosphere, "and the Sun of course is part of the atmosphere."

† In court proceedings, *klepsydrae*, or clepsydras (meaning "water thieves," in which a pierced, measured vessel sank in a tub of water), were used like egg timers, and their span measured the time that advocates could speak: the phrase *aquam dare*, "to grant water," meant to allot time to a lawyer, while *aquam perdere*, "to lose water," meant to waste time. If a speaker in the Senate spoke out of turn or for too long, his colleagues would shout that his water should be removed.

tween *dies naturalis*, the natural day that runs from sunrise to sunset, and *dies civilis*, the civil day that runs from the completion of one Earthly rotation to another—for them, from midnight to midnight. The word "day" was ever ambiguous.

Finer subdivisions were made: the night into four "watches," each named after its last "hour" and proclaimed by constables. Were greater accuracy needed, a range of time-based words came into play: *occasus solis* (sunset), *crepusculum* (twilight, hence "crepuscular"), *vesperum* (appearance of the evening star), *conticinium* (the falling of silence), *concubium* (bedtime), *nox intempestum* (timeless night, when nothing is done), and *gallicinium* (cockcrow), among others.

Before the Industrial Revolution, and the advent of better lamps and lanterns, work hours for most European societies were set by sunrise and sunset. From about the twelfth century on, church bells were sounded indicating when to begin and end toil, announcing curfew (from the French *couvre-feu*—"cover fire," i.e., lights out), and so on. For centuries in Europe, the period between midnight and cockcrow was seen as "dead time," thus inspiring the phrase "the dead of night."

Regardless of the era, we have remained transfixed by the instant of noon. Noonday demons were the merciless torturers of the hermit Desert Fathers of the early Christian Church. Both France and Italy have whole regions—the Midi and the Mezzogiorno—named for the noonday Sun. While ancient Rome had timekeepers to shout out its arrival, one Parisian inventor fitted a lens into his sundial to act as a burning glass that precisely at noon set off a small cannon, and a "noon gun" is still fired daily in places as far apart as Cape Town and Santiago, Chile. For years, the lighthouse keeper at Brockton Point, in Vancouver, marked midday by detonating a stick of dynamite. During the nineteenth century, to give navigators a precise visual signal by which to reset their chronometers, certain major harbors would drop a huge time ball at 1 P.M. (not at noon, as that is when observatories took their readings).*

To tell the time—more or less—throughout daylight hours, not just at noon, or sunrise, or sunset, was a challenge, but once again the Sun provided the necessary means. Gnomons (from the Greek word for "indicators") were first used to measure height, but later employed as proto-sundials, with the length of the shadow cast by the gnomon determining the hour of the day. A

---

* The duel fought out in the American West was traditionally held at "high" noon so that the Sun's glare would not be in either man's squinting eyes: part of the myth of fairness. Fred Zinnermann's classic 1952 film *High Noon* eschews images of dwindling shadows while Sun and tension mount, and instead we see inexorable clock hands as time drains away—a sign of how these machines had usurped the Sun's ancient role.

Borneo tribesmen measure the length of the Sun's shadow at summer solstice with a gnomon. Joseph Needham included this photo in his Science and Civilization in China *(1953).*

gnomon could be anything vertical, including the human figure. As Chaucer writes:

> *It was four o'clock, according to my guess,*
> *Since eleven feet, a little more or less,*
> *My shadow at the time did fall,*
> *Considering that I myself am six feet tall.*[13]

To measure the hours after dark man turned to water clocks, used at night from at least 1450 B.C. on in Egypt, a millennium before they came to Rome. However, they were not particularly accurate until the third century A.D., when the Alexandrian Tesibius (c. 285–222 B.C.) invented a device to ensure a uniform flow. China had water clocks from at least 30 B.C., and ended by adopting a whole series of small vessels on a rotating wheel. "And so," writes Joseph Needham, "the great breakthrough in accurate time measurement came about."[14]

Besides clepsydras, notched-candle clocks (invented by Alfred the Great, says legend), sandglasses (the necks of which became worn from repeated use, passing the grains too quickly, thus shortening their hours), fire clocks, the Egyptian *merkhet* (made from plumb line and palm leaf), incense clocks banded in different aromas (by which one could "smell" the hour), and a score of other methods all served to reckon the time. But the most popular was the sundial.

Made of stone or wood, it fixed a piece of metal parallel to what we now know to be the Earth's axis and at right angles to a circle divided by as many lines as were deemed necessary, so that when the Sun shone, the shadow of the metal blade moved around the circle (the "heliotrope"—sun-turning) according to its motion through the sky.

The earliest known example is Egyptian, dated to about 1500 B.C. By the sixth century B.C., sundials were being used in Greece, and it was Anaximander who invented the discipline of "gnomonics" as a science. For at least the next ten centuries the sundial would be the world's most accurate timekeeping device,[15] but it took time for it to spread, partly because it was not always used properly. No one seems to have understood, for example, why a dial looted in the sack of Syracuse (latitude about 37° N) in 212 B.C. failed to show the right time when shipped to Rome (about 42° N).[16] Dials, it began to be realized, had to be specially made for different latitudes, because the Sun's altitude decreases as one moves poleward, producing longer shadows. "In order to make the fallen shadow even approximate the correct time," explains Dava Sobel, "the dial must be laid out with regard to latitude north or south of the Equator where it is to be used, respecting the changing high point of the sun in the sky from day to day over the course of the year and the variable speed of the Earth's movement around its orbit. There is nothing obvious about the construction of a proper sundial."[17] Of course, for centuries it was not realized that these were the transactions involved.*

Not until about A.D. 1371 would there be a polar-oriented sundial—at the Great Mosque in Damascus—tilting its gnomon at an angle corresponding to its latitude and thus, as we now know, the curvature of the Earth. Thus time was measured not by the gnomon shadow's length, but by its angle. This was a major breakthrough, but one not actually needed until mechanical clocks were invented. Until then, almost everyone used unequal hours.

The approximations used to design ancient portable dials could introduce errors of up to a quarter of an hour. The inconsistency of measurement between two or more sundials may be guessed from the words Seneca (c. 4 B.C.– A.D. 65) puts into the mouth of one of his characters, talking of the death of the emperor Claudius: "I cannot tell you the hour exactly: it is easier to get agreement among philosophers than among clocks."[18] Inaccurate as they were, sundials struck some people as only too modern, changing their relationship to time in ways that made them nostalgic for the "happy life" that ex-

---

* Even so, Boy Scouts learn to tell direction by their watches, holding them face upward and pointing the hour hand toward the Sun. In the Northern Hemisphere, the south alignment will then lie halfway between the hour hand and twelve o'clock: a rough but ready reckoner.

isted before time could be measured. A play attributed to Plautus (c. 254–c. 184 B.C.) has one character exclaim:

> The Gods confound the man who first found out
> How to distinguish hours! Confound him, too,
> Who in this place set up a sundial,
> To cut and hack my days so wretchedly
> Into small pieces! When I was a boy,
> My belly was my sundial—one surer,
> Truer, and more exact than any of them.

Evidently, sundials were still something of a novelty. But it was not long before people learned to love them, and appreciate what they brought.

Around 1400 the chiming clock was invented, and by the 1600s it had become so pervasive that theatergoers might not even have registered the anachronism in *Julius Caesar*, when, in response to Brutus's "I cannot, by the progress of the stars / Give guess how near to day," and asking the time, Shakespeare has his fellow conspirator Cassius reply, "The clock hath stricken three." A similar mistake appears in *Cymbeline*, when again a clock strikes three; but in *Richard II* it is a dial, not a clock, that Shakespeare has his king invoke, to make the passing of time a visible movement:

> I wasted time, and now doth time waste me;
>   For now hath time made me his numb'ring clock:
> My thoughts are minutes; and, with sighs, they jar
> Their watches on unto mine eyes, the outward watch,
> Whereto my finger, like a dial's point,
> Is pointing still, in cleansing them from tears.[19]

The increasing number of weight-driven clocks intensified the preoccupation with time, paradoxically sparking a boom in sundials, which became so profitable a business that design methods were closely guarded. The "art of dialing" formed an important branch of mathematics and was the subject of many textbooks.[20] The making of sundials remained the province of astronomers rather than clock makers, as one had to take into account the Earth's rotation and elliptical motion as well as the angle of its axis.

Even the arrival of precision timekeeping in the form of pendulum clocks and the balance spring did not dent the sundial's popularity. As Dava Sobel says, "A clock or watch may keep time, but only a sundial can *find* time [by interrogating the external world]—a distinctly different function."[21] Charles I

(1600–1649) carried a silver sundial, which he entrusted to an attendant on the eve of his execution as a last gift to his son, the duke of York (after whom New York is named). Thomas Jefferson in old age would distract himself from his chronic rheumatism by calculating the hourlines for a dial. Instead of a watch, George Washington used a silver pocket dial given him by Lafayette.

Over the years sundials have been T-shaped, portable, perpendicular, sunken, cubical, and flat (the common or garden variety). Vitruvius, architectural theorist of ancient Rome and a contemporary of Julius Caesar's, counted at least thirteen styles already in use in Greece by 30 B.C., and says he could not invent any new types, since the field was exhausted. But that was by no means the case. Universal sundials, adjustable for use in any latitude, came in during the eighteenth century. Design became ever more demanding as standards of clock making rose; many became prized works of art.

The Samrat Yantra, the giant sundial at the Jaipur Observatory, one of a family of massive instruments built under Maharajah Jai Singh II (1686–1743). These devices had no telescopes and relied upon naked eyesight and extremely precise construction.

Where once there had been nostalgia for the era before sundials, the persistence of the device may have had something to do with the pastoral sense of rustic peace that seemed to attach to them in the era of the clock. Shakespeare's stricken Henry VI exclaims: "O God! Methinks it were a happy life / To carve out dials quaintly, point by point, / Thereby to see the minutes how they run."[22] "Of the several modes of counting time," wrote Hazlitt, "that by the sundial is perhaps the most apposite and striking, if not the most convenient or comprehensive. It does not obtrude its observations, though it 'morals on the time,' and, by its stationary character, forms a contrast to the most fleeting of all essences."[23]

"Morals on the time" is a reference to the custom of sundials being inscribed with mottoes. There are myriad such sayings. An eighteenth-century English verse runs: "Read the riddle that I've found, / Come, answer it to me, / What is it travels o'er new ground, / And old continually?" Answer: a shadow. Two others much quoted are "I tell only sunny hours" and "The clock the time may wrongly tell, / I never if the sun shines well," although the latter, while celebrating its dial's accuracy, underlines its one drawback: it works only when the Sun is out. Nonetheless, when NASA landed a spacecraft on Mars in January 2004, it deposited a suitable timepiece: two aluminum sundials, no larger than a human palm, built into two Mars Exploration Rovers and carrying the motto "Two Worlds, One Sun."[24]

Back in the 1930s, the flamboyant movie mogul Sam Goldwyn visited his New York bankers and spied a sundial. Swinging around on his companions, he exclaimed: "What *will* they think of next?" *Not* a joke; but only just.

CHAPTER 21

HOW TIME GOES BY

Somewhere in the east: early morning: set off at dawn. Travel round in front of the sun, steal a day's march on him. Keep it up for ever, never grow a day older technically.

—LEONARD BLOOM in *Ulysses*[1]

PETRUCHIO: It shall be what o'clock I say it is.
HORTENSIO: Why, so this gallant will command the sun.

—SHAKESPEARE,
The Taming of the Shrew[2]

IN THE EARLY 1960S, MY FATHER RETIRED FROM HIS FAMILY FIRM to run a pub in Cornwall, in the far southwest of England, and during the school holidays I would work alongside him. Twice each day, at 2:30 in the afternoon and again at 11:00 in the evening—in keeping with the licensing laws whereby pubs had to close first for three hours, then again for nearly twelve hours until reopening at 10:30 the following morning—he would intone in his basso profundo: "Time, gentlemen, please!" It was a peculiar social ritual, a polite command to stop drinking that had the feel of a teleological pronouncement. Patrons would know they had a minute or two to drink up. In contrast to that sixty seconds of silence imposed by my English teacher at the end of his classes, which we'd so longed to be over, the hardened locals stretched out every last moment of tippling as long as they could, for the announcement was never welcome. As Oliver St. John Gogarty (1878–1957), the Irish poet who inspired the character of Buck Mulligan in *Ulysses*, wrote:

> *No wonder stars are winking*
> *No wonder heaven mocks*
> *At men who cease from drinking*
> *Good booze because of clocks!*

> *It makes me wonder whether*
> *In this grim pantomime*
> *Did fiend or man first blether*
> *"Time, Gentlemen, Time!"**

"Time" has continued to vex mankind since the beginning of . . . well, the word is difficult to avoid. Its complexities feed on mankind's subjective experiences of time versus our objective measurement, and the impossibility of ever reconciling the two. Anthony Burgess hints at this in his essay "Thoughts on Time":

> In the first great flush of the establishment of universal public time [Greenwich Mean reckoning], there were certain works of the imagination which . . . said something about the double reality of time. Oscar Wilde wrote *The Picture of Dorian Gray*, in which the hero transfers public time, as well as public morality, to his portrait while himself residing in a private time which is motionless. . . . The experience of time during both the First and Second World Wars was of a new kind to the average participant. . . . The start of battles had to rely on public time, but soldiers lived on inner time—the eternities of apprehension that were really only a minute long, the deserts of boredom, the terror that was outside time.[3]

Time as experienced subjectively may indeed be mysterious: in the *Iliad*, its quality has one value for the victors, another for the vanquished. Saint Augustine said sourly that he knew what time was until he was asked to explain it. But no matter how it is parsed, it is the Sun that determines time, and our employment of that guiding star to monitor time's passing has been the most pervasive of all the ways in which civilizations harness the Sun.

It has ever been critical for two groups—astronomers and navigators—to measure time with precision; but throughout the Christian world it has been the Church that has provided the main impetus toward timekeeping. Much the same is true for Muslims and Jews; Islam requires its faithful to pray five times a day, Judaism three. Christians would pray according to the movements of the heavens, and it was Saint Benedict who, in his Rule of A.D. 530, laid down exact times for devotion: Matins, Lauds, Prime (first), Tierce or Terce (third), Sext

---

* Quoted in Kevin Jackson, *The Book of Hours* (London: Duckworth, 2007), pp. 164–65. The modern history of pub opening times in Britain begins with "DORA," the Defence of the Realm Act, passed during the First World War to reduce hangovers among munitions factory workers.

(sixth), None (the ninth hour after sunrise, which survives, though three hours backtimed, as noon), Vespers, and Compline. Lauds and Vespers, the services of sunrise and sunset, are specifically Sun-related, while the others are linked to set hours. This timetable was widely adopted, so much so that Pope Sabinianus (605–6) proclaimed that church bells should be rung to mark the passing of the hours. In the years that followed, a large number of forms of civil life came to be regulated by time. "Punctuality," writes Kevin Jackson, "became a new obsession, and constant research into timekeeping mechanisms eventually resulted in the making of the clock."[4]

In the later Middle Ages, from roughly 1270 to 1520, by far the bestselling

An illuminated manuscript page from Les Très Riches Heures du Duc de Berry *(1412–16) by the three brothers Limbourg, showing June, a curious time for haymaking, with the Hôtel de Nesle, the duke's Parisian residence, in the background. The book is a collection of sacred texts for each liturgical hour of the day.*

volume throughout Europe was not the Bible but *The Book of Hours*, a guide to Benedict's Rule. During these years, the practice was maintained of defining one hour as the twelfth part of the day or the night, so that in the summer, daytime hours were longer than those of darkness, the opposite being true in the winter—a custom ending only when the newly invented mechanical clock, with its uniform motion duplicating that of the heavens, gradually made people familiar with the "mean solar" way of measuring used by astronomers. Mechanical clocks driven by weights and gears would seem to have been the invention of an eleventh-century Arab engineer and were introduced into England around 1270 in experimental form; the first clocks in regular use in Europe and for which we have sure evidence were those made by Roger Stoke for Norwich Cathedral (1321–25), and by Giovanni de' Dondi of Padua, whose one-yard-high construction of 1364, with astrolabe and calendar dials and indicators for the Sun, Moon, and planets, provided a continuous display of the major elements of the (Earth-centered) solar system and of the legal, religious, and civil calendars. These clocks were made not to *show* the time but rather to *sound* it. The very word "clock" comes from the Latin word for "bell," *clocca*, and this hour-checking machine was for years also known as a horologue ("hour teller")—although medieval striking clocks were thoughtfully designed to remain silent at night. This device, Daniel Boorstin notes, was

> a new kind of public utility, offering a service each citizen could not afford to provide himself. People unwittingly recognized the new era when, noting the time of day or night, they said it was nine "o'clock"— a time "of the clock." When Shakespeare's characters mentioned the time, "of the clock," they recalled the hour they had heard last struck.[5]

In 1504, after a brawl in which a man was killed, the Nuremberg locksmith Peter Henlein (1479–1542) sought sanctuary in a monastery, staying there for several years, during which time he invented a portable clock—the first watch— devised, as the Nuremberg Chronicles of 1511 record, "with very many wheels, and these horologia, in any position and without any weight, both indicate and strike for forty hours, even when carried on the breast and in the purse."[6]

Yet until well into that century, people had to set their timepieces according to the shifting sunrise each day, and stop to adjust them every time they were caught running fast or slow. The expected accuracy was no better than to the nearest fifteen minutes—Tycho Brahe's clock, typically, had just an hour hand. Cardinal Richelieu (1585–1642) was showing off his collection of clocks when a visitor knocked two onto the floor. Quite undismayed, the cardinal observed, "That's the first time they have both gone off together."

Toward the end of the sixteenth century, the Swiss clock maker Jost Bürgi constructed a piece that could mark seconds as well as minutes; "but that," observes Lisa Jardine, "was a one-off, difficult to reproduce, and reliable seconds' measurement had to wait another hundred years."[7] That is probably an exaggeration: by 1670, minute hands were common, and the average error of the best timepieces had been reduced to about ten seconds a day. (The noun "minute," derived from *pars minuta prima*, "the first tiny part," had entered the English language in the 1660s; "second" comes from *pars minuta secunda*.) By 1680 it was common for both minute and second hands to appear.

Punctuality and timekeeping soon became the fashion, even fetish: Louis XIV had not one clock maker but four, who, along with their armory of devices, accompanied him on his "progresses." Courtiers at Versailles were expected to order their days by the fixed times of the Sun King's rising, and of his prayers, council meetings, meals, walks, hunts, concerts. One of the six classes of French nobility was the *noblesse de cloche*—"nobility of the bell"—drawn largely from the mayors of large towns, the bell being the embodiment of municipal authority. Timepieces were exact enough that philosophers from Descartes to Paley took to using clocks as a metaphor for the perfection of the divine creation. Lilliput's commissioners report on Lemuel Gulliver's watch: "We think it is his God, because he consults it all the time." With both Frederick the Great (1712–1786) and that hero of Trafalgar and Navarino, Admiral Codrington (1770–1851), having their pocket watches smashed by enemy fire, it seemed the mark of eminent commanders "up at the sharp end" to have their accessories thus shattered.

In order to meet the new demand for individual timepieces (not only watches, but clocks small enough to fit into a modest house or a craftsman's workshop), clock makers were forced to become pioneers in the creation of scientific instruments: their products, for instance, required precision screws, which in turn necessitated the improvement of the metal lathe. The mechanical revolution of the nineteenth century was the result in large part of ordinary people's wish to tell the time for themselves. And yet there continued to be holdouts even into the twentieth: Virginia Woolf, in *Mrs. Dalloway*, lets out almost a cry of pain at the ubiquity of the clock:

> Shredding and slicing, dividing and subdividing, the clocks of Harley Street nibbled at the June day, counseled submission, upheld authority, and pointed out in chorus the supreme advantages of a sense of proportion, until the mounds of time were so far diminished that a commercial clock, suspended above a shop in Oxford Street, announced, genially and fraternally, as if it were a pleasure to Messrs. Rigby and Lowndes to give the information gratis, that it was half-past one.[8]

When in 1834 a clock was commissioned as part of the rebuilding of the fire-gutted Palace of Westminster, the government called for "a noble clock, indeed a king of clocks, the biggest the world has ever seen, within sight and sound of the throbbing heart of London." The astronomer royal further insisted that it should be accurate to within a second. The outcome was Big Ben (strictly speaking, a bell), finally completed in 1859.[9]

But once people could tell the "exact" time, what time were they telling?

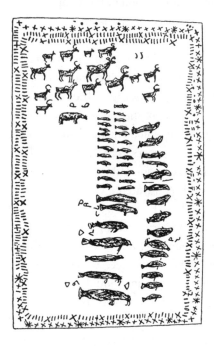

The calendar used by an Inuk hunter in the 1920s, following the introduction of Christianity to the Canadian eastern Arctic: days of the week are represented by straight strokes, Sundays by Xs. The calendar was also used as a hunting tally, including counts of caribou, fish, seals, walrus, and polar bears.

There were so many to choose from. In 1848, the United Kingdom became the first country in the world to standardize time across its entire territory—to the signal of the Greenwich Observatory (with Dublin Mean Time set just over twenty-five minutes behind). That year saw the publication of *Dombey and Son*, in which a distraught Mr. Dombey complains, "There was even a railway time observed in clocks, as if the sun itself had given in." In 1890, Dr. Watson records traveling by train with Sherlock Holmes to a case in the West Country and being startled by the master's gauging minute fluctuations in the train's velocity by timing the telegraph poles, set at standard sixty-yard intervals, as they flashed past their window, with the train acting as the Sun, the poles as longitudes—a powerful image of the way in which the concept of the standardization of time had insinuated itself into people's consciousness.

Not everyone's, of course. Oscar Wilde (1854–1900) once arrived exceptionally late for a dinner, his hostess pointing angrily at the clock on the wall and exclaiming, "Mr. Wilde, are you aware what the time is?" "My dear lady,"

Wilde replied, "pray tell me, how can that nasty little machine possibly know what the great golden sun is up to?" But it did.[10]

Einstein, of course, tells us there is no such thing as absolute time: salutary to recall that the office where he worked for so long was a clearinghouse for patents on the synchronization of clocks. And he himself recalled: "At the time when I was establishing how clocks functioned, I was finding the utmost difficulty in keeping a clock in my room."*

While the Sun appeared to have been replaced, in fact it continued to hold sway throughout the civilized world. Even by the mid-nineteenth century, the towns and cities of most countries still had individual Sun-based systems of timekeeping. Every French city, for instance, held to its own local time, taken from reading the Sun at its equally local zenith. Time bowed to space; there was nothing God-given in the starting point of a second, or a minute, or an hour. The result was that the prospect of a globe united by the telegraph, fast trains, and steamships had so far been stalled, because the spinning, angled Earth and the apparently moving Sun made nonsense of the idea of universal time. The clock on the wall told the hour for the family, the town hall clock that for the citizenry; but on the other side of the hill the clocks were strangers, and the imposition of universal time standards might even be dangerous: Field Marshal Helmuth von Moltke (1800–1891), head of the Prussian, then of the German, army for nearly thirty years, lobbied for a unitary time system for Germany as an aid to making the trains run on time so that its troops might be mobilized more effectively, but this was opposed by those who feared that an integrated rail system would provoke Russia to invade. Nevertheless, it was becoming obvious that uniformity could not be put off indefinitely. "Societies were moving faster than their ability to measure," as the historian Clark Blaise puts it.[13] Writing of the First World War, when wristwatches came into common use (they were much favored by sentries), Burgess says:

> It was a war of railway timetables. For the first engagements in August 1914, about two million Frenchmen were deployed in 4,278 trains, and only nineteen ran late. The wristwatch, which before the war had been considered effeminate, became the badge of masculine leadership. "Synchronize your watches." And then over the top.[14]

---

* Similarly, in *The Sound and the Fury*, Faulkner has the Harvard student Quentin, desperate to escape civil time, shatter his pocket watch, "because Father said clocks slay time. He said time is dead as long as it is clicked off by little wheels; only when the clock stops does time come to life."[11] Others have found the very ticking of a clock a thing of solace: T. E. Lawrence, tied down and viciously lashed by his Turkish captors during the Great War, yet notes, "Somewhere in the place a cheap clock ticked loudly, and it distressed me that their beating was not in its time."[12]

—

IN THE UNITED STATES, the same problem that had bedeviled the rest of the world was troubling individual states. Following the Civil War, railroad expansion had been prodigious. In the forty years after 1860 (by which year the United States already had the world's largest railroad system), the amount of train track that had been laid down multiplied sixfold. By the turn of the century, nearly every town of any size had its own railway station, if not several. However, as in most of Europe, the reckoning of time was a local matter, set by local noon, which, at the latitude of New York, is about one minute later for each eleven miles one travels westward. Noon in New York was 11:55 A.M. in Philadelphia, 11:47 in Washington, 11:35 in Pittsburgh. Illinois used twenty-seven different time regions, Wisconsin thirty-eight. There were 144 official times in North America, and a traveler in the 1870s going from the District of Columbia to San Francisco, had he set his watch in every town through which he passed, would have had to adjust it more than two hundred times. If a passenger wondered when he might arrive at his final destination, he had to know the time standard of the railroad that was taking him there and make the proper conversion to the local times at his boarding and at his eventual descent. Two cities, set a hundred miles apart, maintained a ten-minute temporal separation, even though a train could cover that distance in less than two hours—so which town's time was "official"? The train itself might have set out from a city five hundred miles away, so who "owned" the time—the towns along the route, the passengers, or the railroad company? No wonder that Wilde noted that the chief occupation of a typical American was "catching trains." He visited the States in 1882; one wonders how often he missed his connection.

When people had traveled no faster than the speed of a horse, none of these considerations came into play, but in a railway-framed economy, scheduling was a nightmare. As Blaise writes, "It was the slow increase in speed and power—the fusion of rails and steam—that undermined the standards of horse- and sail-power and, eventually, the sun itself in measuring time." That quotation comes from Blaise's biography of the enterprising nineteenth-century Canadian Sandford Fleming, who in June 1876 happened to miss his train in Bandoran (set on the main Irish rail line between Londonderry and Sligo) because the timetable he had consulted had the misprint "5:35 P.M." for "5:35 A.M.," condemning him to wait for sixteen hours.[15] Fleming was, among other things, the chief engineer of the Canadian Pacific Railway, and his "monumental vexation" at this delay inflamed a desire to number the hours from one to twenty-four. "Why should modern societies adhere to ante meri-diem and post meridiem, why double-count the hours, one to twelve, twice a day? [Hours] ought not to be considered hours in the ordinary sense, but sim-ply twenty-fourth parts of the mean time occupied in the diurnal revolution of the Earth." He would make it his mission to bring about a twenty-four-hour clock by which 5:35 P.M. would become 17:35 H. On further reflection, he set himself the much greater task of relating all time zones worldwide to their re-lationship to longitude, and of introducing "terrestrial, non-local" time.*

Standard time was the best gauge in the world: it would convert celestial motion to civic time. By 1880, England had been on standard time for more than thirty years, reform having started with the railways; why not America? Because the U.S. Congress, fearing a municipal uproar if it initiated change, had procrastinated, and the railroad industry, though well aware of the negative im-pact of a lack of standardization on its profits, dithered, too. From 1869 on, this question was debated, until at last popular discontent forced the railroad barons' hand: as of Sunday, November 18, 1883, they bypassed Congress and adopted the Greenwich system, dividing the country into four zones—Eastern, Central, Mountain, and Pacific—with noon falling an hour later in each. That day came to be known as "the Sunday of Two Noons," since towns along the eastern edge of each new time belt had to turn their clocks back half an hour, creating a second midday, in order to mesh with those closer to the western edge of the same belt. Within a few years, this system became standard, but not without considerable bitterness, and some towns, such as Bangor, Maine, and Savannah, Georgia, still would not cooperate, from either religious principle or stubbornness. Detroit, perched on the crack between the Eastern and Central time zones, could not make up its mind, so that for many years its inhabitants had to ascertain, in making appointments, "Is that solar, train, or city time?" Congress itself did not ratify standard time until forced to do so by war in 1918.

As the historian Mark Smith records, "the telegraph, not the Sun, now com-municated time to a temporarily unified nation and, in the process, helped pave the way for the globalization of abstracted, decontextualized world time."[16] Sim-ilarly, people were coming to accept that if a calendar were to serve a practical, global use, all dates had to be interpreted through a solar dateline. The question

---

* Written before the twenty-four-hour clock had entered popular consciousness, the opening line of Orwell's *1984*—"It was a bright cold day in April, and the clocks were striking thirteen"— was intended to make a British reader, unused to continental measurements, feel uneasy. The standard Italian translation (clocks that strike all twenty-four hours having been commonplace in Italy from as early as the fourteenth century) runs: "*Era una bella e fredda mattina d'aprile e gli orologi batterono l'una*"—in English, ". . . were striking one."

was, Where would that dateline be?* In 1884 twenty-five countries sent represen-
tatives to a conference in Washington, D.C. Eleven national meridians (through
St. Petersburg, Berlin, Rome, Paris, Stockholm, Copenhagen, Greenwich, Cádiz,
Lisbon, Rio, and Tokyo), as well as additional contenders fixed on Jerusalem, the
pyramids of Giza, Pisa (to honor Galileo), the Naval Observatory in Washington,
and the Azores—the original point of definition for the age of exploration—
competed for primacy. The French, represented by their chief of mission and by
the great astronomer Jules-César Janssen, were intransigent, insisting (without
much evidence) that their *ligne sacrée* was a better scientific choice.

Fleming argued for a notional clock set 180 degrees from Greenwich,
which would fall in the middle of the Pacific, thus bypassing national sensibil-
ities but still making use of Greenwich, though without involving England; but
every longitudinal meridian touches land at some point in its arc, and were
Fleming's solution to be adopted, England would be split every noon between
two days. His proposal soon foundered, and the debate ended only when Sir
George Airy, Britain's irascible astronomer royal, wrote that the prime meridian "must be that of Greenwich, for the navigation of almost the whole world
[even then, 90 percent] depends on calculations founded on that of Greenwich." France abstained from voting, and true to its delegates' defiance,
"Greenwich" has never appeared on any of its charts. (By pure coincidence,
the anarchist who tried to blow up the observatory in 1894 was French.) Once
the vote came through, the meridian was moved from the famous obelisk on
Pole Hill to a point nineteen feet farther east. Sundial time was thereby banished, and a sophisticated abstraction rose to take its place.

And so in 1884 the Earth was sliced into twenty-four time zones, each an
hour apart, with eastbound travelers adding successive hours, westbound
ones losing them. Of course, there has to come a moment—which happens to
confront us in the western Pacific—when the logic of the system subtracts a
day from eastbound travelers and gives one to those heading west (Verne's
Phileas Fogg learns this just in time to win his bet in *Around the World in Eighty
Days*). One by one, countries adopted Greenwich Mean Time (GMT)—even
France, in 1911. However, in 1972, the French, still unhappy at what they perceived as Britain's unjustified supremacy, carried a resolution at the United Nations to put GMT alongside Coordinated Universal Time, or UTC—to be
regulated by a signal from Paris (naturally). While GMT is reckoned by the
Earth's rotation and celestial measurements, UTC is set by cesium-beam

---

* When the United States bought Alaska in 1867, before standard time was implemented, the
Russian Orthodox inhabitants suddenly found themselves having to observe the Sabbath on the
American Sunday, which was Monday by Moscow reckoning. They were forced to petition for
guidance as to when to celebrate Mass—on the Russian Monday or the American Sunday.

atomic clocks, which are less accurate but easier to consult.* In point of fact, the two systems are rarely more than a second apart, as UTC is augmented with leap seconds to compensate for the Earth's decelerating rotation.

Physicists Jack Perry and Louis Essen adjusting a cesium-beam atomic clock, which they developed in 1955. One second is about 9,193 million oscillations. This clock led to the replacement of the astronomical second with the atomic second as the standard unit of time.

THOSE WHO SOUGHT to meddle with the Sun for their own purposes and pleasures were not finished: saving daylight was next. This was a project that dated back to Benjamin Franklin (1706–1790), who, on April 26, 1784, while serving as U.S. minister to France, proposed (at the age of seventy-eight and in a moment of whimsy) that Parisians conserve energy (in the form of candlewax and tallow) by rising with the Sun rather than sleeping in with their shutters closed against the day.

* Cesium clocks measure time by counting the ticktock cycles of atoms of the metallic element cesium, each cycle being the exceptionally fast vibrations of the atoms when exposed to microwaves in a vacuum. The fifty-five electrons of cesium-133 are ideally distributed for this purpose, only the outermost being confined to orbits in stable shells. The reactions to microwaves by the outermost electron (which is hardly disturbed by the others) can be accurately determined. As the microwaves hit, the electron jumps from a lower orbit to a higher one and back again, absorbing and releasing measurable packets of light energy. This corresponds to a time measurement inaccuracy of two nanoseconds per day, or one second in 1.4 million years. The experts make the necessary corrections.[17]

The idea never caught on, and more than a century would pass before it would find a receptive audience. Then, in July 1907, the successful London builder William Willett (1857–1915), a keen golfer and horseman, self-published *The Waste of Daylight*, in which he argued that more people should enjoy the early morning sunshine as he did, and complained about how annoying it was to have to abandon a game of golf because of fading light. Might clocks be shifted forward or backward by twenty minutes on four successive weekends, to make the change easier? It was not just a question for sports enthusiasts:

> Everyone appreciates the long light evenings. Everyone laments their shrinkage as the days grow shorter, and nearly everyone has given utterance to a regret that the clear light of early mornings, during Spring and Summer months, is so seldom seen or used.[18]

Scientists and astronomers were divided on the question, although the press chirped, "Will the chickens know what time to go to bed?" and the editors of *Nature* ridiculed the idea by equating the time change with the artificial elevation of thermometer readings:

> It would be more reasonable to change the readings of a thermometer at a particular season than to alter the time shown on the clock . . . to increase the readings of thermometers by 10° during the winter months, so that 32°F shall be 42°F. One temperature can be called another just as easily as 2 A.M. can be expressed as 3 A.M.; but the change of name in neither case causes a change of condition.[19]

But Willett was not easily put off, and within two years a Daylight Saving Bill had been drafted, finally achieving passage as a wartime economy measure in 1916. Earlier that year the Germans had already passed such a law, hoping it would help them conserve fuel and allow factory workers on the evening shifts to work without artificial light. Willett himself had died the previous year; but his neighbors erected a handsome memorial to him—a sundial, positioned to read an hour in advance of the time on any ordinary face.

The 1916 law was made permanent throughout Britain in 1925. America had passed a similar measure in 1916, too, but it proved so unpopular that Congress had to repeal it three years later. (Farmers, whom the measure was designed to help, hated DST, since they had to wake with the Sun no matter what time their clock said and were thus inconvenienced by having to change their schedule to sell their crops to people who observed the new system.) Then in

1922, President Harding issued an executive order mandating that all federal employees start work at 8 A.M. rather than at 9. Private employers could do as they pleased. The result was chaos, as some trains, buses, theaters, and retailers shifted their hours and others did not. Washingtonians rebelled, deriding Harding's policy as "rag time." After one summer of anarchy, he backed down. It wasn't until World War II that DST was adopted again, with only the governor of Oklahoma holding out. However, it was repealed with victory, and in the decades following remained in use in the United States only by local option, with predictably deranged results: one year, Iowa alone had twenty-three different DST systems. Jump to 1965, and seventy-one of the largest American cities had adopted DST, while fifty-nine had not. Confusion reigned, particularly over transport routes, radio programs, and business hours. The U.S. Naval Observatory dubbed its own country "the world's worst timekeeper."

The problem was finally resolved by the Uniform Time Act of 1966, although Indiana, (most of) Arizona, and Hawaii still don't observe DST. In 1996 the European Union, faced with a puzzle map of time zones, standardized DST, while in the United States, as part of the Energy Policy Act of 2005, DST was extended to begin on the second Sunday of March instead of in April, and to end on the first Sunday of November. As of today, DST has been adopted by more than one billion people in about seventy countries—slightly less than one-sixth of the world's population.*

Like Robinson Crusoe notching a stick, or inhabitants of the Gulag scratching a line for every day of their imprisonment, we remain enmeshed in time, unable to leave it alone. As recently as August 2007, Venezuela's president, Hugo Chávez, announced that, to improve the "metabolism" of his citizens, he had ordered the country's clocks to be moved forward half an hour "where the human brain is conditioned by sunlight"—reversing an opposite decision made in 1965, but one that aligns Venezuela with Afghanistan, India, Iran, and Myanmar, all of which offset time in half-hour increments from Greenwich reckoning. Gail Collins, writing in *The New York Times* about Chávez's diktat, likened it to the scene in Woody Allen's film *Bananas*, in which a revolutionary hero becomes president of a South American country and announces that from that day on, "underwear will be worn on the outside."[21] But Newfoundland is also on the half-hour system, defying the rest of Canada, while Nepal is fifteen minutes ahead of India, five hours and forty-five minutes ahead of Greenwich. Saudi Arabia supposedly has its clocks put forward to

---

* When clocks move forward or backward an hour, the body's clock—its circadian rhythm, which is governed by daylight—takes time to adjust. In a study of fifty-five thousand people, scientists found that on days off from work, subjects tended to sleep on standard, not daylight, time.[20]

midnight every day at sunset. As one commentator has quipped, "Keeping one's watch properly attuned aboard the Riyadh-Rangoon express must be an exhausting experience."[22]

Nor has fine-tuning stopped there. A second was at one time defined as 1/31556925.9747 of the solar year. But it is nearly sixty years ago now since the National Physical Laboratory at Teddington in England invented the atomic clock, happily to discover that it was more accurate to base timekeeping on vibrating atoms than on the orbiting of the Earth. "It was slightly embarrassing," recalls David Rooney, curator of timekeeping at the Royal Observatory. "When clocks diverge, it isn't good. By the seventies, we needed another fudge factor. So the leap second was introduced, to push together Earth rotation time and atomic vibration time." Such seconds are not inserted every year: the decision to add or subtract one (so far, it has always been added) is made by the International Earth Rotation Service in Paris. The last insertion was on January 1, 2006; it added an extra pip to the BBC time signal.

Now that measuring precise time is the responsibility of agencies such as the U.S. Naval Observatory in Washington, D.C., the International Earth Rotation Service at the Paris Observatory, and the Bureau International des Poids et Mesures in Sèvres, France, all of which define a second as 9,192,631,770 vibrations of the radiation (at a specified wavelength) emitted by a cesium-133 atom, the Sun has been officially deprived of its long-term role as our timekeeper. This definition of a second—the first to be based not on the Earth's motion around the Sun, but on the behavior of atoms—was formally endorsed in 1967. But those "leap seconds" periodically have to be added to keep our clocks in sync with the planet's turnings, because the Earth goes its own way in space heedless of atomic time, ensure that we can never entirely turn our backs on the Sun's guardianship.

One can go on fiddling almost indefinitely (the old joke runs that even a stopped clock is right twice a day). Back in 1907, Einstein came up with the equivalence principle, which states that gravity is locally indistinguishable from acceleration and diminishes as distance increases from the center of mass, so that time goes faster in, say, Santa Fe, high up in New Mexico, than in Poughkeepsie, down low in New York, by about a millisecond per century. A recent experiment on a westward around-the-world jet flight showed that clocks gained 273 nanoseconds, of which about two-thirds was gravitational.[23] Meanwhile, a clock eight feet in diameter has been constructed atop Mount Washington in Nevada, made to last ten thousand years (the period of time in which cesium-ion atomic clocks are said to "lose" a second); and a French clock, made by the engineer and astronomer Passement, exhibits a perpetual calendar showing the date through the year 9999.[24] An advertisement lauds "the Ultimate Time-keeper," a watch based on "complex astronomical algorithms" that provides "local times for sunrise and sunset, moonrise and moonset, lunar phase as well as digital, analog and military time for wherever you are," in A.M./P.M. or 2400 format. It is programmed for 583 cities worldwide and automatically adjusts for DST. Constructed of titanium or steel with a sapphire crystal, it offers "the broadest interpretation of time that money can buy," and can be purchased for $895.[25]

The Swiss watchmaker Swatch has proposed a planetwide Internet Time that would enable users throughout the world to bypass individual zones and rendezvous in the same "real" time. Meanwhile, the scientists who run the atomic clock at Teddington, as well as competitors in the United States and Japan, are at work on an even more accurate machine, known as the "ion-trap" and due to be realized by 2020. Experts say that if it were activated now and were still running at the time predicted for the end of the universe, it might be wrong by half a second, if that—twenty times the accuracy of the current most advanced model.[26] In 2006, the United States suggested that world time should be switched entirely to the atomic clock, which would involve the dropping of leap seconds—a notion that had the British Royal Astronomical Society up in arms. Had the proposal been accepted, says David Rooney, it would have been the first time in history that time was not dependent on the rising and setting of the Sun.[27]

Then there is that famous exchange from *Waiting for Godot:*

VLADIMIR: That passed the time.
ESTRAGON: It would have passed anyway.

CHAPTER 22

THE SUN IN OUR POCKET

*The first man I saw . . . had been eight years upon a project for extracting
sunbeams out of cucumbers, which were to be put into vials hermetically
sealed, and let to warm the air in raw inclement summers. He told me,
he did not doubt in eight years more, that he should be able to supply the
Governor's gardens with sunshine at a reasonable rate.*

—JONATHAN SWIFT, *Gulliver's Travels*[1]

He will literally have the sun in his pocket.

—The villain in *The Man with the Golden
Gun*, talking about buyers of his solar
energy converter

*T*HE MAN WITH THE GOLDEN GUN WAS IAN FLEMING'S LAST NOVEL,
published posthumously and incomplete. Nevertheless it became the ninth
James Bond film, released in December 1974, at the height of the 1970s energy
crisis, as worldwide enthusiasm for alternative forms of energy was coming to
a boil. Bond has to recover a solar agitator, vital to a special converter, "95 percent efficient, a device that will harness the Sun's radiation and give awesome
power to whoever possesses it." His principal adversary is a professional assassin played by Christopher Lee (Fleming's cousin, and the author's original
choice to play Bond); the story climaxes in the destruction of the solar installation on an island off the China coast.

In the decades since the film was released, solar power has become an ever
more popular subject. The Sun's role as supreme timekeeper may have been
overtaken by the atom, but it is still an open question whether the atom will
also preempt the Sun in mankind's search for a usable source of energy. The
Sun is the great self-renewing resource, the creator of coal, peat, oil, hydro-
electricity, and natural gas (methane). It raises moisture into the atmosphere,
to return as the downpours that drive turbines; it powers the winds and the
waves, and all their effects; it shows no signs of dying out (except to those

thinking in numbers of astronomical magnitude); and it lavishes itself over the entire planet, delivering to the Earth's surface more energy in just forty-four minutes than we use in a year. About 35 percent of what actually reaches our world is reflected back into space by clouds, and 19 percent is absorbed by the atmosphere; but this still leaves twelve thousand times as much energy as that used in all man-made devices. Only two forms of renewable energy are not the result of solar radiation: geothermal and tidal (the Sun raises tides by its great mass, not its radiance). Yet only in the last thirty years has this plentiful source been taken seriously by those in power. Ian Fleming was ahead of his time.

The notion of harnessing the Sun occurred to human beings as soon as

According to legend, around 212 B.C. the Greek astronomer and mathematician Archimedes used mirrors to focus the Sun's rays onto a Roman fleet in an attempt to set its ships on fire.

they began to experiment with their surroundings. By the third century B.C. both Greeks and Romans employed "burning mirrors"—handheld concave reflectors—to focus sunlight on enemy warships. Archimedes (287–212 B.C.) reportedly devised just such a battery in 212 B.C. to defend Syracuse from a blockading Roman fleet by burning the enemy's sails "at the distance of a bow-shot" (about fifty yards). While the story is likely a myth,* it does show that the

---

* In 1992, researchers at Leicester University noted that Roman galleys traditionally furled their sails in battle, so setting them alight was not an option. The combined efforts of 440 men and

Greeks knew early on that sunlight was of its nature pure energy—and dangerous.[2]

Around A.D. 100, Pliny the Younger (61–113) built a home using glass for the first time to retain heat; and over the next several hundred years, Roman public baths were designed with large south-facing windows. The Romans were also the first people to build greenhouses. In the sixth century the emperor Justinian even enacted a law protecting public and domestic sunrooms from the erection of buildings that obstructed sunlight.

The great tenth-century Persian scientist Ibn al-Haytham (c. 965–c. 1031) produced an important study, *On Spherical Mirrors,* which retold the legend of Archimedes and the reflectors at Syracuse as fact; by 1270 it had been translated into Latin and so came to the notice of Roger Bacon, who warned his pope that curved reflectors might be used by the Saracens against Crusaders in the Holy Land. "The idea of transforming the beneficent rays of the Sun into a fierce military weapon to incinerate human beings," writes Frank Kryza of the Vatican response, in his history of solar power, "was seen as a deviant and evil idea, the work of witchcraft and the devil."[3]

Early in the sixteenth century Leonardo da Vinci proposed a monstrous mirror four miles wide, to be employed commercially as a source of heat, not as a weapon. For whatever reason—lack of funds, of capacity (he was proposing to use more glass than then existed!), or of time (as ever, he was engaged on many other projects)—nothing came of this idea, but his concept marked a shift in focus from destructive to industrial uses, and also boosted solar research; interest in mirrors and lenses flourished.

During the reign of Louis XIV—perhaps inspired by the Sun King himself—numerous solar experiments were made. In 1747, under Louis XIV's successor, Georges Buffon (1707–1788) used 140 flat mirrors to ignite wood placed two hundred feet away—so demonstrating that Archimedes' feat was at least a possibility. Then came the Industrial Revolution and the new thinking it engendered. "Mastery of the Sun," Kryza observes, "seemed tantalizingly close in the age of steam. . . . The engineers of the nineteenth century worked with forces large enough to give them the sense, for the first time in history, that they were the masters of nature, in possession of the instruments they

---

shields could ignite a damp patch of wood at fifty yards, the project concluded, but not do serious damage, and would have been a poor use of manpower. A "mirror platoon" of fifty men could have inflicted deep burns on targeted steersmen, or even the Roman high command (whose purple battle cloaks would have made them stand out); but such a tactic, if successful, would have been used again, which no historian has reported.

needed to change the conditions of life profoundly. Why not tame the Sun's energy?"[4] Solar pumps, heat engines, and stills were just some of the secondary products.

By the 1830s, during his stay in South Africa, Sir John Herschel invented the "actinometer," essentially a bulb of water that, when exposed to sunlight, could be used to calculate the energy received from the Sun. Further, as Stuart Clark relates,

> he also performed some rather more eccentric experiments such as the day he placed a fresh egg in a tin cup and then laid a pane of glass across the top. Returning with his wife and six children some time later, he retrieved the cooked egg, burning his fingers in the process. Ceremoniously, he cut the egg into pieces and doled it out, so that all could say they had eaten an egg boiled hard by the South African sun. Suitably impressed by his newly found culinary skills, a week later he cooked a mutton chop and potatoes the same way. "It was thoroughly done, and very good," he recorded in his diary.[5]

Herschel went on to build a solar cooker out of blocks of black-painted mahogany, yielding a maximum temperature of about 240°F, 11 percent above sea-level boiling point.

One long-running ambition had been to produce a viable solar-powered engine. Since the early seventeenth century, when Salomon de Caux, employing lenses, a frame, and a metal vessel containing water and air, had constructed an early prototype, there had been attempts to make such a machine; but people considered them more games than practical projects. In 1861, however, a French mathematics teacher, Augustin Mouchot, poured water into an iron bucket, which he surrounded with reflectors. As the water evaporated, it produced enough steam to propel a small motor. Within four years, he had succeeded in creating a conventional steam engine. A few months later, he displayed this device to Napoleon III, who, impressed, offered financial assistance. Mouchot enlarged his invention's capacity, refining the reflector into a truncated cone, like a dish with inward-slanting sides. He further constructed a device that enabled the entire machine to face the Sun continuously. After six years' research, he amazed spectators with his brainchild, one reporter describing it as an inverted "mammoth lampshade . . . coated on the inside with very thin silver leaf" with the boiler sitting in the middle as an "enormous thimble" made of blackened copper and covered with a glass bell. At the 1878 Universal Exposition in Paris, he displayed a solar-powered printing press using a parabolic mirror, steam engine, and piston; it was to take another 122 years for solar energy to make its comeback at a world's fair—at Expo 2000, presented by the city of Freiburg.

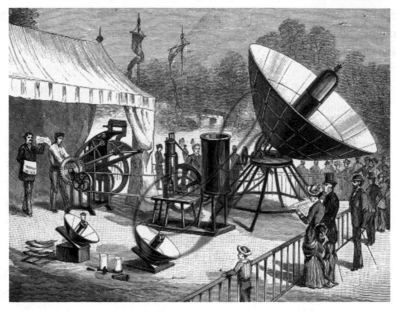

Augustin Mouchot's solar-powered printing press at
the Paris Universal Exposition of 1878

Eager to exploit these inventions, the French government decided that the most suitable venue would be Algeria, a colony bathed in almost constant sunshine yet entirely dependent on coal, which was prohibitively expensive there; the willing Mouchot was immediately dispatched. "Eventually industry will no longer find in Europe the resources to satisfy its prodigious expansion; coal will undoubtedly be used up. What then?" He soon invented a portable solar oven for French troops and a solar engine that could operate a printing press; but the high cost of these contraptions, coupled with the increasing cheapness of English coal, condemned them to idleness, and the Industrial Revolution roared on unimpressed; in those days no one anticipated global warming.

In 1891, the Baltimore inventor Clarence Kemp, "the real father of solar energy in the U.S.," patented the Climax, the first commercial sun-powered water heater, which combined the old practice of exposing metal tanks to sunlight with the scientific principle of the hotbox, so increasing their capacity to absorb heat.[6] By 1897, his biographers proudly state, "thirty percent of heaters used in Pasadena, California, were of Kemp's design"; but this merely underlines the truth that such inventions were never successful outside the state, and when in 1902, using on a reflector made up of 1,788 individual mirrors, he created enough energy for a fifteen-horsepower* solar pump to irrigate a Pasadena ostrich farm, it was reported as no more than an offbeat experiment.

For centuries already, the rich of Europe had taken to setting trees against "fruit walls," which built up solar heat, then slowly released it when the Sun went down, and England and Holland pioneered the development of green-houses with south-sloping faces. The first commercial use of solar energy was probably using evaporation to remove salt from brine, and the first large-scale application was probably the distillation of drinkable water into appropriate containers by evaporation to remove from brackish wells or isolated tracts of seawater. A still erected in Chile in 1872 produced six thousand gallons of water per day from fifty-one thousand square feet of collecting surface for more than forty years.

The story of harnessing the Sun continued to advance by fits and starts. In the late 1870s, William Grylls Adams, deputy registrar for the Crown in Bombay, wrote an award-winning book, *Solar Heat: A Substitute for Fuel in Tropical Countries,* and tried to introduce the technology into British India—without success. The torch then passed to France and the engineer Charles Tellier, the "father of refrigeration," who in 1885 installed a collector on his roof similar to the solar panels used today. Instead of using water to produce steam, he chose liquid ammonia, which passes into vapor at a lower temperature. Exposed to the Sun, the containers emitted enough gaseous ammonia to drive a water pump that could lift three hundred gallons per daylight hour. But Tellier decided to pursue his refrigeration interests instead (there was more money in preserving food), and France bade adieu to the last major development of solar mechanical power on her soil until well into the twentieth century.[7]

A few years later, in 1900, an enterprising Bostonian, Aubrey Eneas, formed the first solar power company and began building sun-powered machines to irrigate the Arizona desert. In 1903 he moved to Los Angeles, closer to his most promising customer base, and the following year sold his first complete system for $2,160. After less than a week, a windstorm tumbled the boiler

---

* The word "horsepower" dates back to the Scots inventor James Watt (1736–1819), who created the modern steam engine. Finding that potential buyers had difficulty understanding what the devices could achieve, he restated their relative strengths in terms that could be more easily understood. His customers were mine owners who used horses to haul coal and pump groundwater from the mines. By testing a number of horses at mine sites, Watt calculated that the average English horse could haul coal at the rate of twenty-two thousand foot-pounds per minute for about ten hours a day (a foot-pound being that amount of work needed to lift one pound a distance of a foot). Watt arbitrarily increased the figure by 50 percent, and so was born the "horsepower" unit—a rate of work equal to thirty-three thousand foot-pounds of labor. It is still used to compare the relative power of anything from lawn mowers to spacecraft.

rigging into the reflector. Accustomed to setbacks, Eneas constructed another pump. In the fall of 1904, a rancher in Wilcox, Arizona, bought the improved model, but it, too, was destroyed, this time by a hailstorm. It became clear that the massive parabolic reflector was too vulnerable, and the company folded. Other entrepreneurs were to follow (notably Henry E. Willsie, operating out of St. Louis and Needles, California, who created a system that allowed his machines to function at night using the heat gathered during the day), but their companies also failed to turn a profit.

Despite this dismal history, proponents continued to believe that if they could find the right combination of technologies, they could produce unlimited energy. One who shared that dream was a Brooklyn-born engineer, Frank Shuman (1862–1918), whose first solar motor, built in 1897, performed poorly, because even at respectable pressures the steam exerted insufficient force. Rather than trying to generate more heat, he replaced the boiler pipes with a flat metal container similar to Tellier's original design and devised a low-cost reflector: two rows of mirrors strung together to double the amount of sunshine intercepted. He then constructed the largest conversion system ever built, capable of delivering fifty-five horsepower and driving a water pump that gushed three thousand gallons per minute—at a cost of $150 per horsepower, compared with the $80 of a conventionally operated coal system: a respectable outlay, he reckoned, considering that the investment would be quickly recouped because the fuel was free. Another reason he was not concerned about the fact that the energy delivered by his machine cost more than that from coal- or oil-fired engines was that, like the early French entrepreneurs, he planned to ship his invention to vast, sunburned North Africa.

In 1912, he started work on the world's first solar power station—fittingly in Egypt, once the center of sun worship. The site was at Meadi, then fifteen miles south of Cairo, and it boasted seven curved reflecting troughs, each 205 feet in length, with a thousand-horsepower steam engine. But the beginning was also the end. Two months after the final trials, Archduke Franz Ferdinand was assassinated, igniting the Great War. The engineers operating Shuman's plant returned to their respective countries to perform war-related tasks, and before the armistice was signed, Shuman was dead. After the war, with the fall in the price of oil, interest in solar experiments once again evaporated.

By this time, oil and coal companies had established massive infrastructures, stable markets, and ample supplies of carbon fuels. Solar pioneers, by contrast, were still trying to perfect their technology, and they had the additional difficult task of convincing skeptics that energy from the Sun was something more than a curiosity. In North America, the discovery of huge quantities of natural gas in the Los Angeles basin during the 1920s and 1930s killed off the local solar water-heater industry there. George Gamow, writing in 1940, could comment dismissively: "The direct utilization of solar heat . . . is employed only in a few tricky devices—to run the refrigerators of cold-drink stands in the Arizona desert, or to heat water for the public baths of the oriental city of Tashkent."[8] Solar water heaters did find some renewed enthusiasm in Florida, and by 1941 some sixty thousand were in use in the Sunshine State. *Your Solar House* (1947), a bestselling book by forty-nine solar architects, reflected a real demand. However, after the Second World War, Florida Power and Light campaigned aggressively to increase electricity consumption by offering water heaters at bargain prices, and again solar power retreated.

Elsewhere the situation was similar. In Japan, where rice farmers urgently needed cheap hot water, a company began marketing a simple heater made up of a basin covered by glass, and by the 1960s more than a hundred thousand were in use; but the industry collapsed under a glut of cheap oil, just as it had in California and Florida. Even in Australia, with its abundant light, solar heating devices numbered just a few thousand. In Israel, in the early days after its founding, electricity was rationed, so people sought to meet their needs in other ways, and by the mid-1960s one household in twenty had solar heating; but then cheap oil from the fields captured in the Six-Day War once again put solar power on the back burner.

A revolving solarium at Aix-les-Bains, France, September 1930

In the United States, the first solar-heated office building had been erected in the early 1950s, the first solar-heated and radiation-cooled house had started up (at a retrofitting cost of nearly $4,000—about $30,000 in today's money), and some companies had returned to producing solar cells and water heaters. Then, in 1953–54, researchers at Bell Labs (now subsumed into AT&T) made a startling discovery based on an earlier technology. In 1839 the French physicist Alexandre-Edmond Becquerel had established that if two electrodes were immersed in acid and one of them was activated, a current flowed between the two. In 1873 the British engineer Willoughby Smith observed that the element selenium changed electrical resistance when subjected to sunlight, but achieved only moderate efficiency. Bell Labs, experimenting with various materials, discovered that silicon possesses five times that efficiency, so the most effective way to convert the Sun's rays into electricity would be through silicon photovoltaic cells.

Bell was soon producing thin wafers of ultrapure silicon, to which were added small amounts of arsenic and boron to improve conductivity. When struck by the Sun's rays, electrons within the silicon were knocked loose by the penetrating heat and moved toward the wafer's upper surface, creating an imbalance between the front and back of the cell. When the top and bottom surfaces were joined by a connector—usually a wire—a current passed between them. "A solar cell is altogether a much simpler structure than a green leaf," wrote the solar chemist Mary Archer (wife of the popular novelist Jeffrey), "but it is like a green leaf in that one side is designed for looking at the Sun."[9]

The New York Times lauded Bell Labs' discovery as "the beginning of a new era, leading eventually to . . . harnessing the almost limitless energy of the Sun for the uses of civilization." And indeed it marked a significant improvement: in bright sunlight a cell had an energy conversion rate as high as 22 percent. Even so, photovoltaic cells were still not economical, costing $300 ($2,200 in 2010 dollars) per kilowatt. But this was the time of the space race, and the U.S. government's budget for research on solar cells skyrocketed once it was realized that satellites could generate electricity from solar paneling, which needed no restocking. In 1958, *Vanguard 1*, the first satellite to be equipped with cells, was lofted into orbit. In the decades since, costs have dropped dramatically—on average, 4 percent per annum over the past fifteen years.

Photovoltaic cells protect pipelines from freezing; they power lights, radios, roadside emergency telephones, refrigerators, air conditioners, water pumps, and village electrification. They are to be found in even the smallest devices, such as pocket calculators and watches, iPod chargers, cameras, and car mirrors. In 2003, around 50 percent of all such cells were manufactured in

Japan,[10]* while the United States accounted for about 12 percent. In 1985, annual installation demand worldwide stood at 21 megawatts (that is, 21 million watts); in 2005, it was 1,501 MW, an increase of more than 7,000 percent.

AFTER COLLECTING ABOUT a dozen books and more than 140 articles in just five years, I realized that solar initiatives are being made everywhere, in ever greater numbers, from China to Tanzania, from South Africa, where solar-powered stoplights free traffic from the whims of an enfeebled power grid, to Abu Dhabi, capital of the United Arab Emirates, which, despite its reputation as an oil metropolis and the highest per capita emitter of CO_2, is planning a research facility and a five-hundred-megawatt solar power plant. So how far has the revolution gone?

In 2004 and 2006, I made two research trips to find out. My first stop was Freiburg, a city of roughly 215,000 people in Baden-Württemberg, between the Black Forest and the Rhine Valley. It was badly damaged during the Second World War, in 1940 German planes mistakenly dropping sixty bombs near the train station, then in November 1944 an Allied air raid destroying 80

---

* Japan's history is illuminating. During the 264 years after 1603, known as the Edo Period, a policy of near isolation from the rest of the world was enforced. Edo at the time was the largest city on Earth, with between a million and a million and a quarter people (compared to London's 860,000 in 1801). Virtually nothing could be imported (with the notable exception of through the Dutch trading post in Nagasaki Harbor), so the country had to be self-sufficient in all resources. Japan has few fossil fuels, coal, for example, being mainly used for salt extraction. Because of this and other scarcities, the Japanese were forced to reemploy and recycle all they could. Therefore everything was treated as a potential resource, even ash, candle drippings, and human waste, ever a valuable fertilizer.

Tinkers repaired pots and pans, kettles and buckets; ceramic specialists glued broken china and glass; cobblers stitched and soldered: hardly anything was thrown away, but carefully put together again or given a new use. There were used-paper buyers, used-clothes dealers (all clothes at the time were handwoven, so especially precious: there were about four thousand old-clothes dealers in Edo alone), used-umbrella-rib buyers, used-barrel brokers, and tinkers who worked the city singing "Let's exchange, let's exchange," offering toys and candies in exchange for old nails and other scraps of metal. Rubbish dumps didn't exist. Above all, Edo had one source of energy: the Sun. Almost everything was made directly or indirectly from it, with the exception of stone, metal, ceramics, and other mineral-based materials.

This was even true of lighting. Commercial power started up in November 1887, twenty years after the fall of the shogunate, with the first fossil-fuel-driven generator, but until that moment, all artificial light came from paper lanterns and wax candles. Oil was extracted mainly from sesame seeds, but also from those of camellia, rape, and cotton. In coastal areas, whale and sardine oil were used. Firewood was widely employed, but people consumed less than the annual increment of growing trees, so they were never eating into their energy supply. Was Edo an irrelevant, impoverished Shangri-la, or a model for today?

percent of the inner city. But this meant that after 1945 an extensive program of reconstruction was put in place, and in recent years, Freiburg has attracted a wealth of solar industries and research; no other German city has as many environment-related entities. Given that, and the fact that it is the country's sunniest city, it is not surprising that Freiburg proudly calls itself "the environmental capital of Germany."

At SolarRegionFreiburg,* the center for solar energy in the area, I talked to three of the city's experts: Franziska Breyer, a trim blond woman in her late thirties, a forester by training; Tom Dresel, a sociologist turned publicist who oversees solar projects; and Otto Wöhrbach, the director of the city's planetarium.

"It really all began by accident," Frau Breyer explained. "There's a village about fifteen kilometers from here called Wyhl, and in the early 1970s it was planned to build a nuclear power plant there. Students, farmers, and vintners (we're a wine-growing area) staged a sit-in, and the moat of the building site became a discussion center. Eventually the plans were withdrawn, but not before somebody said to the protesters, 'You don't want nuclear energy, fine; but what are you going to put in its place? Being destructive is easy; try to be constructive.' That set people thinking."

The first hesitant steps toward solar power began in 1976; then in 1981 the Fraunhofer Institute for Solar Energy Systems was founded in Freiburg. Ridiculed at the time by the science community, who saw it as little more than an offshoot of hippie enthusiasms, the center is now the largest of its kind in Europe, with a staff of more than 350. Its success was fueled in part by changing political realities. In 1983, for the first time in thirty years, a new political party won the 5 percent of the vote necessary under West Germany's proportional representation system to obtain seats in the federal parliament, and the Greens, Die Grünen, entered the Bundestag. Pictures of bearded, long-haired deputies without ties sitting next to the very proper Chancellor Kohl in parliament swept around the world. The Greens' success galvanized Freiburg's citizenry, and an entire solar economy developed. In 1992, the city council resolved that only the construction of low-energy buildings would be allowed

---

* During my visit I encountered a research group from Madison, Wisconsin, Freiburg's U.S. sister city, there to learn from Freiburg's initiatives so as to implement them back home. Another of Freiburg's nine twins is Besançon, a town of 130,000 inhabitants, one of the few in France to manage its own energy projects. In 1991, the city installed photovoltaic panels on cars in its highway department fleet. Beforehand, the warning signs indicating roadworks were connected to car batteries that required regular recharging—usually the motor was kept running during the day to prevent the battery from going flat. Solar panels have entirely solved this problem and save the money previously spent on gas, battery recharging, repairs, and maintenance.

on municipal land. In addition to solar panels and collectors on roofs, many passive features, such as superinsulation, south-facing windows with low-emissivity glass, and foam insulation panels became popular. "The future starts every day," said Tom Dresel. "You see the movement of the city step by step."

Back in 1945, the town had been occupied by the French. In the late 1990s, on the site of a former French army base, construction began on a new neighborhood of six thousand people, Vauban (so named by the French occupiers after an eighteenth-century marshal), with the aim of making it a "sustainable model district." Solar energy was used to heat the water of many households, and the neighborhood was developed according to a principle of sustainability. "The road to sustainability is paved with innovation," said Frau Breyer, smiling. Plainly, Vauban works.

The four of us talked for nearly two hours, then I set off to walk around the town. It seemed at first glance to be like any other university city in the area—prosperous, clean, filled with students. Then I began noticing all the innovations. On the outskirts, atop a long-disused silver mine, there now stands a solar observatory. Five wind turbines operating within the city boundaries have doubled the percentage of renewable electricity. By the end of my walk I had counted thirty installations: a technology park, several solar power stations, a "zero-emissions" hotel, the solar-towered railway station, and a number of houses with solar panels on their roofs. Thirty-year-old buildings are undergoing solar-friendly redesign with backing from the savings and loan societies, and private companies and public facilities are making their roofs available for solar modules. Locals buy shares in the panels and are reimbursed when the power is sold to the city electricity grid. There are eighty-four thousand square yards of cells on Freiburg's roofs, while large solar scanners are programmed to realign with the Sun every twelve to fifteen minutes in order to to maximize intake. The city has also constructed the world's first autonomous solar building, not connected to any main grid. The schools have a solar training center and "sun-power stations"—this in a country that gets an average of only 1,528 hours of sunshine a year.

"Freiburg has been a pacemaker for other German cities," Franziska Breyer told me. Gelsenkirchen, for example, on the northern Rhine, was the most important coal-mining and steel town in Europe early in the twentieth century—the "city of a thousand furnaces." It is now reinventing itself as the "city of a thousand suns," borrowing many of its innovations from Freiburg. "Our changes are ones that China and the two Koreas could adopt too." On that note, Beijing and Freiburg already have one feature in common: twice as many bicycles as cars. Only the reasons differ. In Beijing, the bicycle is the primary means of transport; in Freiburg, the use of bicycles is a symbol of an environmentally conscious community. It's only a matter of time, I was told, before students are roaring off to their classes on solar-powered motorbikes.

MY NEXT SUCH trip, to Almería, in southern Spain, took place nearly two years later, in July 2006. Less than an hour's drive from the city lies Tabernas, the only surviving sand desert in Europe. Not only does the terrain experience 355 sunny days a year, it is ideally suited to action movies; among those shot here have been *Patton*, *The Magnificent Seven*, *The Wind and the Lion*, *Indiana Jones and the Last Crusade*, and, memorably, *Lawrence of Arabia*. Fellini's masterpiece *8 1/2* was partly filmed here, as was Sergio Leone's trilogy of Westerns, *A Fistful of Dollars*, *For a Few Dollars More*, and *The Good, the Bad and the Ugly*.

Little of this success benefited Almería. By the early 1970s the area was one of the poorest in Spain. Then underground water tables were suddenly discovered, and an agricultural revolution began. Hundreds of greenhouses were put up to take advantage of the combined possibilities of the water and the ever-present sunlight ("Almería is where the Sun spends the winter," went the saying: the average temperature is 62.6°F [17°C]); soon locals were claiming that it was not the Great Wall of China that could be seen from outer space but their greenhouses, one vast sea of plastic. By the end of the century the city had been transformed into the richest in southern Spain, and settlers were flooding in.*

The period directly overlapped the oil crises of the 1970s, and by the early 1980s, the International Energy Agency, with the participation of nine countries, set up a small power station there to test two separate solar energy installations: one, a field of ninety computer-controlled mirrors (heliostats) that followed the Sun's rays and concentrated them on a single central tower, where they were converted into thermal energy; the other, three fields of curved troughs, which tracked the Sun and reflected its energy onto metal tubes filled with oil. These tubes would be slowly heated to 554°F (290°C), and the oil fed into a steam generator. A third project was added, entirely a Spanish affair, in which a central tower was supplied by three hundred heliostats whose mirrors reflected the concentrated heat onto black absorption panels atop the tower, with a water/steam receiver and a thermal storage system of molten-salt tanks. The hope was that at least one project could be taken up commer-

---

* Since my visit in 2006, Spain's real estate bubble has burst and its construction industry has crashed. Almería now has an unemployment rate of almost 25 percent, one of the country's worst. High-rise apartment buildings lie unfinished on the city's outskirts in an area known as "Pueblo de Luz"—Town of Light. Yet half the solar power installed globally in 2008 was in Spain.

cially, but by the late 1980s, Spain's partners had become frustrated with the lack of progress, and left; only Germany stayed the course. Since 1999, just one project has kept going, under separate management, but that has thrived: the Plataforma Solar de Almería (PSA) has since gone on to become the largest R&D center for solar energy in Europe (worldwide, its only rivals are the Weizmann Institute in Israel, the Sandia Laboratories in Albuquerque, New Mexico, and the nearly two thousand giant mirrors in the Mojave Desert outside Barstow, California).

One of the first lessons I learned on my trip to Almería was that concentrated solar power and photovoltaic panels represent very different technologies. PV panels use solar photons to excite electrons into generating a current, while solar thermal energy, the kind in which Almería has been a pioneer, uses photons to heat fluid molecules. Such conversion requires long metallic mirrors that focus sunlight onto a pipe, whose water runs through a heat exchanger, generating enough steam to turn a turbine. Since the whole process requires a lot of land and copious sunlight, its ideal location is in sunbaked deserts.

I was taken around the site by José Martínez Soler, a cheerful member of the staff in his early thirties who was completing a doctorate on the marketing of solar energy. He proudly announced that PSA would soon be celebrating its twenty-fifth anniversary and the fact that two Spanish companies had at last made commercial ventures out of the station's research. A town like Freiburg can process only relatively low heat from its sunlight, but the state-of-the-art silicon panels and silvered mirrors used in Almería can intensify sunlight to temperatures up to a thousand times its original level. (As José was explaining all this, we passed an outdoor sign reading, in English, DANGER—CONCENTRATED SUNLIGHT.)

I left impressed by all I had seen, but unconvinced that solar energy can be more than a minor contributor to the world's needs. The next day I met Alfonso Sevilla Portillo, a local expert on energy matters who had been the first director of PSA, having earlier worked for several years in California. Fashionably turned out and in his early sixties, he disagreed profoundly with his old company's strategy. PSA was developing a technology it could sell, when "what we have to do is *shape consumption.* If we continue to live as we do now, we will never be able to supply the energy for all our needs." Solar energy R&D should be integral to an overall philosophy of how to live, he said. Dr. Sevilla had left PSA to work on a project in Kronsberg, on the outskirts of Hannover, where some six thousand units would house fifteen thousand people, making up five compact neighborhoods. The town will be using 40 percent less energy for the same quality of life.

Is this the way of the future? Whereas in the past, inventors were more interested in the scientific and philosophical ramifications of "capturing the Sun," much of the present enthusiasm arises from fears about global warming and the exhaustion of natural fuels—exhaustion, or the fact that most oil is found in politically unstable regions such as the Persian Gulf, Nigeria, and Venezuela. The sources of fossil energy may have increased—gasoline, kerosene, propane—but techniques of converting other sources have multiplied, from solar power to nuclear fission, fusion, wind, waves, and biomass processing (translating material from dead plants and animals into ethanol, biogas, and biodiesel fuel). Nearly every day, newspapers carry a fresh story about such alternatives, not all of which are necessarily feasible. As one skeptic noted about wind, "Not since Don Quixote have so many windmills presented such an orgy of illusion."[11]

But even the most fantastic-seeming ideas have a way of becoming reality. In the 1980s, the great science fiction writer (and master skin diver) Arthur C. Clarke argued that we could obtain unlimited electricity from the sea "without huge masses of spinning hardware," making a case in "The Shining Ones" for using heat engines "to tap the thermal gradient between the warm surface layers and the near-freezing waters of the abyss."[12] Sure enough, the Pelamis, a snakelike machine—at five hundred feet, almost as long as a passenger train—now generates energy by absorbing wave motion, while horizontal turbines are being mounted on the seabed, similar to underwater windmills. Britain alone could generate up to 20 percent of the electricity it needs from waves and tides.[13]

The Helios prototype flying wing soaring on solar power over Hawaii, July 2001. This first test flight lasted eighteen hours.

In 1981 a solar-powered plane flew the English Channel, and now Omega, the watch company, is developing a sun-powered aircraft (which accumulates energy in lithium batteries on its wings) to circle the globe. In March 2007 the Swiss boat *sun21* crossed the Atlantic in sixty-three days; and sailing of a more ambitious kind is now part of space exploration, which uses sunlight beating on a giant reflective area to power the craft. Meanwhile, scientists at the celebrated furnace of Odeillo Font-Romeu in the Pyrenees, the largest in the world, have found that reflectors taken from antiaircraft searchlights can concentrate temperatures to 6,332°F (3,500°C). In Udaipur, Rajasthan, the local maharaja has introduced sun-powered rickshaws, while for the last two decades, engineers have been working on a solar-powered car, and although such vehicles are not yet commercially viable, a biennial 1,877-mile race for these cars takes place across central Australia, from Darwin to Adelaide.[14] Hybrids such as Volkswagen's Eos (named after the Greek goddess of dawn) and the French carmaker Venturi's Eclectic have retractable hardtops with energy-translating sunroofs. However, sunlight can provide only so much—enough power for about fourteen miles a day; even then, the vehicles have to be extremely light and aerodynamic, and normally carry just one person.

Ingenuity is certainly not wanting wherever one looks. In May 2009, the fashion magazine *Visionaire* published a "solar" issue, with a black-and-white cover whose photochromic inks broke into full color when exposed to the Sun. "Solio," a portable charger about the size of a mobile phone, fans out a photovoltaic trefoil to capture sun power, which it can then transfer to cell phones, handheld computers, game consoles, or music players. Such personal devices can also be wired into sports bags or into a "solar jacket" (made from a fabric called Microtene), which has two-by-three-inch panels embedded in its removable Nehru-style collar.[15] Trash cans being introduced on American beaches have solar-powered sensors that send an email to the public works department when a container is three-quarters full. While chemists are currently working to produce a paint that can transform sunlight directly into electricity, in 2008 scientists at the Idaho National Laboratories came up with a plastic that did just that: Solar Skin uses a thin film of copper-indium-gallium selenide that can be applied directly to glass or metal. And a self-healing paint on cars and furniture is being developed that would restore a damaged coating in a matter of minutes by exposure to the Sun.

On a larger scale, engineers in New Jersey have patented a device for switching the state's electricity system from conventional power to backup solar power within seconds of a failure. Solar ovens are now able to cook six hundred meals twice a day;[16] methane that builds up in garbage landfills is being sold as a power source; and biological systems are being developed to use sunlit algae to convert exhaled carbon dioxide and water into oxygen and protein-rich carbohydrates—and into fuel.

It was recently argued that the Moon's slow axial rotation, unclouded skies, and abundant local materials would enable installations to be built there to harvest the Sun's energy. Solar power collected through hundreds of lunar panels would provide a clean and reliable energy for space-based applications and ultimately for earthlings as well.[17] Another initiative, space-based solar power, or SSP, would entail launching satellites equipped with massive photovoltaic surface areas that would unfold (or inflate; the technology is still experimental) once in orbit. Sunlight is roughly eight times as intense in our neighborhood of space than it is on Earth; but full-scale tests of SSP have yet to be conducted, and an attempt in 2005 to launch the first solar sailing craft, *Cosmos 1,* failed.[18]

All these initiatives make it difficult to know how to gauge the progress of solar energy—after all, by the early years of the Reagan administration, government enthusiasm for solar power had largely disappeared, along with the panels Jimmy Carter had installed on the White House roof. Reagan also slashed the budget of the Solar Energy Research Institute and allowed tax incentives for renewables to collapse. It was not just the Gipper: between 1980 and 2005, the fraction of all U.S. research and development spending devoted to energy declined from 10 to 2 percent, and the 2007 budget committed just $159 million to solar R&D, barely half of that on nuclear energy ($303 million) and one-third of that on coal ($427 million). At the end of 2009 legislation was introduced into Congress to scuttle thirteen solar plants and wind farms planned for the Mojave Desert in California, "arguably the best solar land in the world," according to one environmentalist. In a world expected to add 2.5 billion people by midcentury, government and industry investment in energy technologies has been falling, not rising.*

There is also the question of how long subsidies will continue: solar projects of all kinds have been hugely helped by tax credits, grants, and programs

---

* See Andrew C. Revkin, "Budgets Falling in Race to Fight Global Warming," *New York Times,* October 30, 2006, A1 and A14, and Todd Woody, "Desert Vistas vs. Solar Power," *New York Times,* December 22, 2009, B1 and B5. In 2009, half the electricity in the United States was generated from coal, which yields energy more cheaply than does oil or natural gas and is competitive with the uranium used in nuclear power plants. America, often called the "Saudi Arabia of coal," has enough, at present rates of consumption, to last 250 years: every day, on average, each person in the United States consumes twenty pounds of coal to keep electricity flowing. Yet China consumes more "black gold" (coal, not oil!) than the United States, the European Union, and Japan combined. India, too, is stepping up its construction of coal-fired power plants. Ironically, the consequent sulfur pollution is so pervasive as to exercise a temporarily beneficial side effect, its tiny airborne particles deflecting the Sun's radiation back into space and so slowing global warming. See Jeff Goodell, "Black Gold or Black Death?" *New York Times,* January 4, 2006, A15.

in which power companies must compensate subscribers who channel energy back into the national grids from solar appliances. But in Denmark, where 17 percent of electricity comes from wind turbines, new projects have virtually ground to a halt since subsidies fell victim to fluctuating political priorities. In Spain, whose government funding is more generous than anywhere else in Europe, a program of cutbacks was set in place in 2008. Other nations may follow—indeed, in October 2009, companies importing solar panels into the United States were hit by up to $70 million in unexpected tariffs.

Germany, too, is politically divided by squabbling on this matter (conservatives argue that subsidized solar power is growing so fast that it will raise electricity bills). Despite the row over subsidies, however, the country continues to be a leader in solar power: fifteen of the world's twenty largest plants are based in Germany,[19] generating 750 megawatts of sunpower, more than five times the 2006 U.S. figure. The Japanese are not far behind—a million and a half buildings in Tokyo have solar water heaters, more than in the entire United States. In January 2010 it was reported that China has leapfrogged Japan and the West to emerge as the world's largest manufacturer of solar panels, as well as the leading maker of wind turbines.[20]

In 2005 a Spanish law decreed that all new dwellings should provide for solar appliances. Israel uses solar water-heating systems in about 30 percent of its buildings, and these are mandatory in all new homes. China imposes similar provisions, and Sweden intends to give up fossil fuels entirely.[21] Solar power plants are under construction in Mexico, South Africa, Egypt, Algeria, and Morocco. Several U.S. states have legislated to protect the "right to light" for town gardens—a late bow to Justinian*—while California, home of the "Million

---

* Justinian's law protecting sunrooms from constructions that block their light is echoed in the "ancient lights" provision of English law, which gives home owners the right to a certain level of natural illumination. In effect, the owner of a building with windows that have received daylight for twenty years or more is entitled to forbid any obstruction that would deprive him of that illumination. (In the center of London, near Chinatown and Covent Garden, particularly in back alleys, signs saying "Ancient Lights" mark individual windows.) The question "Who owns sunlight?" has become pertinent worldwide. In 1959, a Florida appellate court stated that the "ancient lights" doctrine has been unanimously repudiated in the United States; but in 1975, a committee of the California legislature decided that if persons had rights to sunlight and were able to sell them, institutions would be more willing to finance solar projects. Although several witnesses warned that the creation of such rights would create problems, such as disputes over how much these rights were worth, in 1978 California enacted a law protecting home owners' investments in rooftop panels. Trees that impeded the panels' access to the Sun brought upon their owners fines of up to $1,000 a day. California now has a history of litigation on the issue. Meanwhile, in 2005, three heliostats were imported into New York City, at a cost of $355,000 each, to redirect sunbeams into a vacant lot in Lower Manhattan. Teardrop Park lies in the shadow of three skyscrapers, but the heliostats, installed on the roof of a twenty-three-story building nearby, follow the movement of the Sun, intercept its light, and redirect it into the park.

Solar Roofs" initiative, alone produces 54 percent of the world's wind electricity. A decade ago, only five hundred California rooftops bore solar panels; today, there are nearly fifty thousand such installations, providing the equivalent energy of a major power plant.

Worldwide, removable roof panel insulation is a relatively new phenomenon, having become part of the original architecture of houses only in the 1980s, on the initiative of the remarkable American engineer Harold R. Hay, who got the idea on a mission for the U.S. government to India in the 1950s, when he noticed that many people were living in rusty sheet-metal shacks that were hot by day and cold at night. He designed roof panels that could be taken off during the day and replaced at night in the winter months, and the reverse could be done during the summer. This one idea made a huge difference to hundreds of thousands of Indian homes. Back in 1976, Hay offered some prescient advice:

> We're a Mediterranean—not an omni-climatic—animal. We belong in the Earth's temperate zones . . . but our technology has made it possible for us to heat the planet's arctic regions and cool its tropics enough to make ourselves comfortable there. That, in one sense, is what the energy crisis is all about. We've learned to use enough energy to make ourselves comfortable in areas that we're not really physically adapted to live in. . . . We can both use and misuse even solar energy.[22]

Whatever the philosophical implications, solar energy evangelists point out that at least costs are coming down. "Thirty years ago solar power was cost-effective on satellites," says Daniel Shugar, president of SunPower Systems, a California company. "Today it can be cost-effective for powering houses and businesses."[23] What in 2007 was an $11 billion market is growing at more than 25 percent a year. Even so, solar panels are still comparatively expensive.

Early in 2008, *The New York Times* asked four companies to provide quotes on installing such panels on a Manhattan apartment house roof.[24] The lowest price, after negotiation, was $370,000 for a fifty-kilowatt system. State grants and tax credits brought in $265,000, and the balance was financed with a low-interest ten-year loan, while the 266 panels were set to run for a good twenty-five years. Then there is what is known as "net metering": home owners with solar electric systems can see their meters spin backward when the Sun is shining, as they earn credit that erases part—sometimes all—of their power costs. So solar power, while initially expensive, may arguably make sense in the long run.

It may be a long run indeed. In 2009, solar energy generated 0.5 percent of Germany's electricity, renewable resources as a whole 14.2 percent; in the

United States in 2001, renewables accounted for 6 percent of energy consumed, a figure that is expected to remain about the same for the next quarter century. Solar input currently meets just 9 percent of humanity's primary energy demands, and is projected to decline to 8 percent (these figures vary according to source: *National Geographic* puts the figure at less than 1 percent).

Some long-hoped-for options are still at the planning stage. Fusion—producing energy by combining hydrogen atoms into helium, the process that fires the Sun—has been talked about for decades as a potentially limitless energy source, but has still to be harnessed.[25] While the harvests from renewable sources of energy may be growing rapidly, they are starting from such a small base that their overall place in the energy supply chain remains minimal.

"The big problem is big numbers," says Neil deGrasse Tyson of the Hayden Planetarium. "The world uses some 320 billion kilowatt-hours of energy a day."[26] Within the next century we will be using three times that much. Americans are particularly heavy consumers; they constitute less than 5 percent of the people in the world, yet consume more than one-fifth of its energy resources. A true switch to sun power in the United States would require $420 billion in subsidies—a huge figure, but only about eleven days' worth of the U.S. gross domestic product. In his inaugural address, President Obama declared, "We will harness the Sun and the winds and the soil to fuel our cars and run our factories," and the following month he signed a $787 billion stimulus package intended to double the amount of renewable energy produced over the next three years. Enthusiasts speak of at least a quarter-million square miles of land in the southwestern United States alone that would be suitable for solar power plants, or point out that if only 0.35 percent of the Earth's land surface (an area the size of France) were covered with solar cells, this would be enough to cater to all our energy needs. If only . . .

Dr. Tyson puts the whole question into startling perspective with one simple story. In 1964 the Russian astronomer Nikolai Kardashev proposed three different levels of civilization in respect of energy use. A Type I civilization harnesses the energy on or within its home planet. It controls all the sunlight that falls on the surface, and can, if it so desires, also reach into a volcano or a hurricane to tap their respective energies. A more advanced Type II civilization can utilize the entire energy production of its host star, making it ten billion times more powerful than a Type I; and Type III uses all the energy of all the stars in its host galaxy, increasing its power ten billion times over Type II.

"So where do earthlings fit in?" concludes Tyson. "Sorry to break the news, but any civilization so fragile that its members must stockpile fossil fuels, run away from erupting volcanoes, evacuate cities in advance of hurricanes, and rush to high ground during tsunamis is not in charge of its own planet, and can be none other than Type Zero." Perhaps there is some hope. As the biologist Oliver Morton suggests,

> the challenge that faces us is to find new technologies that sit in the space between the photovoltaic cell and the leaf—new hybrids of industry and nature. To make leaf-like things that generate alternative fuels, or even, conceivably, electricity. . . . We need to work on a whole gamut of solar conversion technologies.[27]

Finding such solutions is not impossible. In 1931, not long before he died, Thomas Edison, the godfather of electricity, told Henry Ford, "I'd put my money on the sun and solar energy. What a source of power! I hope we don't have to wait until oil and coal run out before we tackle that."[28]

PART FIVE

INSPIRED by a STAR

Swastikas date back to the Neolithic period (c. 9500 B.C.) and were used all over the world as sun symbols long before the Nazis appropriated them. This "black sun" or "sun wheel" is a symbol in Nazi mysticism.

CHAPTER 23

THE VITAL SYMBOL

I see the sun, and if I don't see the sun I know it's there. And there's a whole life in that—in knowing the sun is there.

—MITYA, in *The Brothers Karamazov*[1]

"What," it will be Questioned, "When the Sun rises, do you not see a round disk of fire something like a Guinea?"

"Oh, no, no, I see an Innumerable company of the Heavenly host crying, 'Holy, Holy, Holy is the Lord God Almighty.'"

—WILLIAM BLAKE,
A Vision of the Last Judgment[2]

ANNE FRANK WOULD REGULARLY CONFIDE TO HER DIARY HOW IN-spired she felt when she looked at the old white horse chestnut outside the small house where her family was in hiding from the Nazis, especially when the tree was lit up by the morning Sun: "As long as this exists, and I may live to see it, this sunshine, the cloudless skies, while this lasts I cannot be unhappy."[3] The Sun, either in itself or as an embodiment of what is good in nature, in-spires in all kinds of ways. It is the most ubiquitous of symbols, appearing fre-quently in heraldry (when represented as giving light, it is a "sun radiant"; with a human face, a "sun in splendor"), and as one of the most popular insti-tutional emblems, for everything from oil conglomerates to the Smithsonian. In Britain alone, 497 trademarks employ it, against 366 that use the crown and 239 the Union Jack.

One wonders which world culture has the most elaborate solar imagery. Twenty nations display the Sun on their flags: Antigua, Argentina, Bangladesh, Ivory Coast, Japan, Macedonia, Malawi, Namibia, Nepal, Nicaragua, Niger, the Philippines, Rwanda, Taiwan, Tunisia, and Uruguay (a full-faced sun in splen-dor), as well as little-known countries such as Bergonia, Kazakhstan, Kiribati, and Kyrgyzstan. The newly designed "Ausflag," the national ensign of Aus-tralia, features a black kangaroo leaping across the Sun, while several countries have solar stripes. India's flag has a golden wheel representing the twenty-four hours of the day, while the Oval Office boasts a huge yellow sunburst on its carpet.

Japan takes its flag from a rising-sun banner first displayed at court in 701. In 1870 the scarlet-on-white design, known as *Ninomaru*—the sun flag—was adopted. The Imperial Japanese Navy (the army too, until 1945) was given its own emblem, showing the Sun and sixteen rays: *Kyokujitsu-ki.* Yet among the several hundred family crests in Japan, while fifty-one depict the stars and seventeen the Moon, only seven bear the Sun—as if even its greatest noblemen felt they could never be on such intimate terms with so imperial an entity.

Royalty feels no such qualms. Early on February 3, 1461, during the fifteenth-century struggle between the rival houses of Lancaster and York known as the Wars of the Roses, Edward IV (as he was about to become; 1442–1483) and his army reached Mortimer's Cross, a hamlet in South Shropshire, ready to do battle against the Lancastrian foe. Edward, only eighteen years old, had just heard that his father and youngest brother had been captured, mocked, and slaughtered. His troops were battle-weary and in low heart. As dawn broke, three suns appeared in the sky and suddenly joined together in one.

Hilary Mantel, in her novel *Wolf Hall*, brings what happened to dramatic

Fireworks on the Grand Canal at Versailles during the 1674 festival that Louis XIV organized to celebrate one of his military victories

life: "[Edward could see] three blurred discs of silver, sparkling and hazy through particles of frost. Their garland of light spread over the sorry fields, over the sodden forests of the Welsh borderlands, over his demoralized and unpaid troops." What took place was most likely a parhelion, the dispersion of light through ice crystals, although it could also have been the morning sun multiply reflected off the wetlands to the east. Whatever the cause, his men cowered back from so awesome a sight, but Edward was inspired to cry, "This is a good sign, for those three suns betoken the Father, the Son and the Holy Ghost!" At this, his whole army, reportedly between five thousand and ten thousand strong, sank to their knees in prayer—then rose to storm to victory. "His whole life took wing and soared. In that wash of brilliant light he saw his future."[4] Edward went on to have the Sun in splendor woven into his standards.

The royal personage most closely identified with the Sun is certainly Louis XIV, *le roi soleil* (1638–1715). In 1653, at just fifteen years of age, he appeared before his court wearing a golden wig, an embroidered tunic, and a pink-and-white-plumed headdress crusted with rubies and ablaze with solar rays, while little suns blazed forth from his garters and the buckles of his high-heeled shoes.[5] Nine years later, he staged an equestrian ballet between the Louvre and the Tuileries before five thousand guests. The opening extravaganza featured a troupe of noblemen turned out in costumes representing the great civilizations, Louis himself dressed as a Roman emperor, with a golden cloak and a shield carrying a solar emblem. After him came the Persians, led by his brother, whose escutcheon bore the emblem of the Moon and the motto *Uno soli minor*—"Lesser to the Sun alone." Eight years on, Molière depicted Louis as Apollo in *Les Amants magnifiques*, in which the king, once more caparisoned as the Sun, appeared onstage in an opera that climaxed in volcanoes and fireworks.

On another occasion, in the hours before dawn of August 18, 1674, Louis led his court down through the palace gardens, where they suddenly came upon a rocky island some seventy feet across, crowned by an eighty-foot obelisk topped by a sphere of fiery light. A bas-relief depicted the king crossing a river at the head of his army. Below, to one side, sprawled a lion crushed by defeat; to the other, an eagle in submission.[6] The whole display burned silently, until at a signal fifteen hundred charges of powder were detonated as firework fleurs-de-lys blazed along the canal. The triumph of both Sun and the Sun of Justice sovereign was complete.

Identifying himself with the Sun was a politically astute move, for it shed a quasi-divine magnificence over Louis—after all, the Sun was the heavenly body associated with Apollo, god of peace and the arts, as well as a symbol of military glory, which suited the warrior-monarch's ambitions. Louis's decision was received with joy by contemporary chroniclers, the historian Father Claude-François Ménestrier, perhaps the most famous Jesuit in France, rejoicing:

> What grander and more heroic effort could be wished for than the action of the Sun lighting up the whole world and working incessantly to maintain all that lies upon it? What could be chosen more worthy of a king who promised such great things as soon as he started governing his states as his own?[7]

Then there was Versailles, the spectacular palace that Louis made his supreme residence. The Sun was not the only symbol on display, but it cast all others into shadow. Throughout, decoration combined images and attributes of Apollo (laurel, lyre, tripod) with the king's portraits and emblems. The Grand Apartment glorified this Most Christian King as the Sun, and within it the Apollo Salon was designated the main room.

Yet another point of comparison between monarch and Sun is that both the star and the institution of royalty have been devalued in the intervening years. From at least the nineteenth century on, the diminishing influence of the diurnal cycle on the developed world's daily life, and the evolution of scientific knowledge, have made the Sun less and less a mythological object. No longer regarded as a god with divine powers (the easiest way to channel one's respect for it), it became associated with a new, more impersonal pantheon: it was the clock of our existence, maker of our weather (a late discovery), regulator of the seasons, and a force with the power over life and death—but in an impersonal sense, not as an anthropomorphized deity that humans could appeal to or placate. And as we have come to understand it differently, its symbolic value has changed, too.

THE SUN HAS always played a role in religion. For Jews, the seven-branched menorah originally represented the Burning Bush seen by Moses on Mount Sinai, and thus the light of God, but in later readings its arms signified the five then known planets plus the Moon and, as the central candle, the Master Star. In Japan, the devout are advised that the best time for meditation is in the moments before sunrise or after sunset. To describe the Transfiguration of Jesus, evangelists could find no better simile than the Sun's brilliance, the whiteness of its light an apt image of the transition of the Son of God from the world of the flesh to that of the spirit.

The Sun has also appeared in visions. Between May 13 and October 13, 1917, "a beautiful lady who said she came from heaven" was said to have appeared six times to three young shepherds near the town of Fatima, on the outskirts of Lisbon, the children reporting a "dancing Sun" cavorting across the

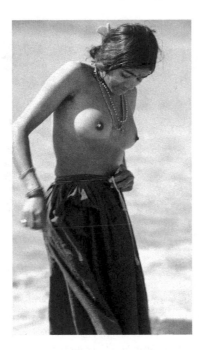

For thousands of years, Indian women who long to get pregnant have stood naked in water, facing the Sun.

sky, a sure instrument of God. As Fatima grew into one of the staging posts of Christian pilgrimage, on the following October 13, crowds of more than twenty-five thousand are reported to have witnessed how the atmosphere became "orange" as the Sun dimmed, then the clouds suddenly parted and a bright light like a halo emanated from the gloom.[8] The Sun began to spin and change color, then hurtled earthward.

From the 1960s on, Sun-inspired visions of another kind arose from that part of the West's counterculture involved in taking psychedelic drugs. The "father" of LSD, Albert Hofmann (1906–2008), for instance, became fascinated by the mechanisms through which plants turn sunlight into substances that transform mental states. "Everything comes from the Sun," he said, proclaiming LSD the gateway into new forms of consciousness.[9] Hofmann took to studying hallucinogenic substances found in Mexican mushrooms. The songwriter-singer Donovan wrote one of the first psychedelic songs, "Sunshine Superman," while the hippie musical *Hair* ended with "Let the Sunshine In." And people did.

Of course, drug-enhanced celebrations of the Sun are not unique to any one group or country. Mircea Eliade describes the mythology of the Desanas, a small tribe living in the equatorial forests of the Vaupes River in Colombian Amazonia that persists in its hunting-based culture, and whose religious mythology revolves around the creative power of the Sun-Father. "To the Desanas," says Eliade, "the soul . . . possesses its own luminosity, be-

Freewheelin' Frank Reynolds, secretary of the San Francisco branch of the Hells Angels. At the Human Be-In at the Polo Fields at Golden Gate Park in 1967, which drew a crowd of twenty thousand and was a major counterculture event, Frank spent as long as he could gazing heavenward. Larry Keenan, the photographer who recorded his self-imposed ordeal, wrote: "When I saw Freewheelin' a few weeks later, he told me his eyes had gotten burnt staring into the sun on acid."

stowed by the Sun at the birth of every human being." This association is reinforced by the psychedelic drug *yagé*, the drinking of which brings the initiate to a subterranean world inhabited by luminous beings. "Taking *yagé* is expressed by a verb meaning 'drink and see,' and it is interpreted as a regressus [*sic*] . . . to the primeval moment when [the] Sun Father began the creation." Light, says Eliade, is also associated with sexuality—another aspect of the creation myth involving souls originating from semen that falls with the Sun's rays:

> If everything which exists, lives, and procreates is an emanation of the Sun, and if "spirituality" (intelligence, wisdom, clairvoyance, etc.) partakes of the nature of solar light, it follows that every religious act has, at the same time, a "seminal" and a "visionary" meaning. The sexual connotations of light-experiences and hallucinatory visions appear to be the logical consequence of a coherent solar theology.[10]

Perhaps. Solar myths confirm, reassure, inspire. It is up to individual taste whether they do more than that. There have ever been plenty of innocent and practical ways of taking the Sun as inspiration. The wheel, first seen on a pictogram in Uruk, in Mesopotamia, around 3500 B.C. is one example. Yoga practitioners perform the Sun salutation. Ted Hughes demanded that his publisher bring out his poems only on days when the Earth's magnetic field was propitious. Mata Hari (1876–1917), the Dutch erotic artist shot by the French as a German spy, took her stage name from the Malay for Sun, literally "eye of day." The game of polo descended from a competition among Mongol tribesmen, who would whack a human head up and down a field, a pastime said to have a solar origin. A similar version of the game appeared in Persia in the fifth century B.C.—horsemen driving not a skull but a "lighted fireball" across a "sky field" in "the sun game."

I realize that the list above will appear arbitrary, but that is how it is with the Sun. It finds its way into odd and uncategorizable areas. The rest of this chapter details four of them: how the Sun has affected language, and its common symbols: blond hair, gold, and mirrors.

WORDS BASED ON the Sun, or influenced by it, reflect cultural attitudes. The names of both people and places frequently have something of the solar. In Persia, such first names as Afrouz, Afshid, Dalileh, Dori, Farimehr, Jala, Jahantab, Khorshed, Kurshid, Mehrasa, Mehrshid, Shams, Shidoush, and Talayeh, all meaning "Sun," are all still used, and there are as many again meaning "light" or "dazzling." In Sanskrit, *Asia* means "sunrise"; *Jayaditya* translates as "victorious Sun," and *Khorvash* as "Sun-lovely." Both *Ravi* and *Ravindra* mean "Sun," the syllable *ra* in many languages being used both for people's names and to cover important Sun-related concepts. For the Japanese, *Nippon* translates as "the origin of Sun."

In the Northern Hemisphere, the farther north or west one goes in the Old World, the less likely one is to find such equivalences. Sun-related positional vocabulary is still used in celestial navigation; but geographical terms— "Orient," "Occident," "Levant"—have died away, though so much of our culture is Mediterranean-connected that there is still a repository of such metaphor, in English especially, not because of the climate but because its writers enjoyed thinking up new solar phrases. Shakespeare coined several, "to burn daylight," an expression for wasting time, occurring in both *The Merry Wives of Windsor* and *Romeo and Juliet*.

Street slang makes full use of the Sun. The chilling phrase "a Harlem sunset" describes a wound given in a razor fight—introduced by Raymond Chandler in a 1940 novel and used since in a poem by Seamus Heaney. A "sundowner" is both a tramp in the Australian bush—someone who comes to a sheep station at sunset for food and shelter—and a strict naval captain, originally one who compelled midshipmen to return from shore leave by sundown. It is said to be time for drinking when "the Sun is over the foreyard" or "over the yardarm"; since in home waters and northern latitudes the Sun would be over the yardarm—one of the horizontal spars mounted on the masts of sailing ships—toward noon, the drinking hour evidently started earlier.[11]

Many proverbs revolve around the Sun. Some of the more imaginative include "The same Sun that will melt butter will harden clay" and "The Sun doesn't shine on both sides of the hedge at once." "Flying into the Sun" means being so passionately committed as to destroy oneself, while a "sunflower" is a pretty girl, i.e., someone unfolding for the sons/suns (c. 1959); and "sunbeam" means tableware that hasn't been used and so doesn't have to be washed (c. 1950). "Sunfisher" is slang for a bucking horse; but to "get the Sun into a horse's coat" is to allow a horse a rest from racing.

Most derivations are easy enough to tease out—"solar plexus" is so named because the ganglia radiate from the stomach area throughout the body, and when Archie Goodwin in the Nero Wolfe books describes things as "sunfast," he means that they are certain, because sunfast dyes do not bleach out—but others do mystify: to be intensely nervous is said to be "walking on the Sun." The legal world has its own way with words: a "sunshine law" is one that forbids government to pass edicts without there being a certain amount of time for them to be considered by the public; "daylight" is a legal term stipulating that the period of time after sunrise and before sunset is to be considered part of the day and not the night, often invoked when defining burglary. One of the oddest names for the Sun is "Spanish faggot," which appears in *Grose's Dictionary* of 1811: this may have come about because from the fifteenth century and continuing through the reign of Carlos I (1500–1558), coins from Spain or its dominions were stamped with the image of faggots—sheaths of arrows bound loosely together—which were seen as similar to the Sun's rays.

If there is one color that symbolizes the Sun it is yellow. "Just look at that yellow sun!" exclaims Nellie Forbush, the heroine of *South Pacific*. The Sun can be orange, of course, or red, or white, but yellow is its primary color. Given that fact, however, the various shades that the Sun can take on have challenged human ingenuity. The varieties of solar yellows stocked by paint stores range from the time and weather tints (high noon, low sunrise, sun shower, sun porch, and on) and the golds (from fool's through to golden pond) to such possibly solar colors as yellow brick road, mellow yellow, and ho-hum yellow (I list, I do not explain). It seems hardly surprising, in the light of such an inventory,

Louis XIV, age fourteen, as Le Roi Soleil in Ballet de la Nuit, *performed in Paris on February 23, 1653—the first time Louis was called the Sun King*

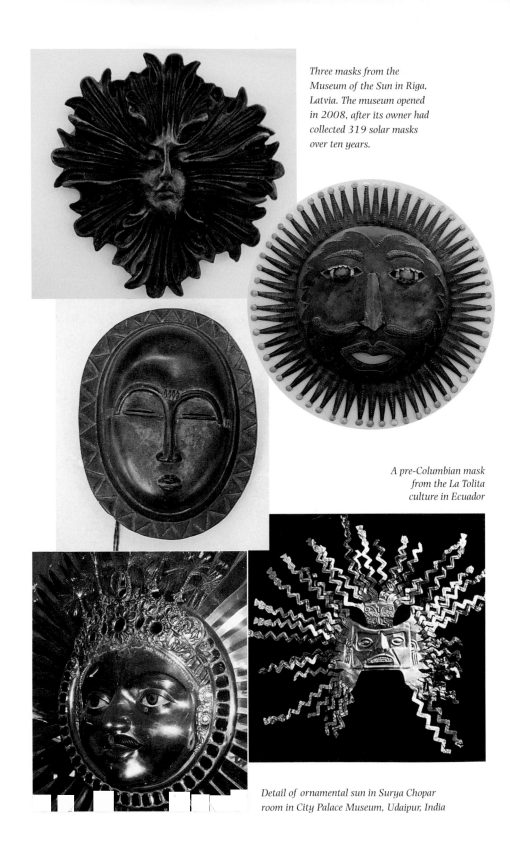

Three masks from the Museum of the Sun in Riga, Latvia. The museum opened in 2008, after its owner had collected 319 solar masks over ten years.

A pre-Columbian mask from the La Tolita culture in Ecuador

Detail of ornamental sun in Surya Chopar room in City Palace Museum, Udaipur, India

The Sun *(2002), by Yue Minjun, is unusual in taking a humorous approach—all this award-winning Chinese artist's paintings show laughing characters.*

Girl with a
Sunflower *(1962),
a painting from the
Croatian naïve school
by Ivan Rabuzin*

Inti sacrificing a goat to the Sun, while a band of conquistadors looks on from a distance.

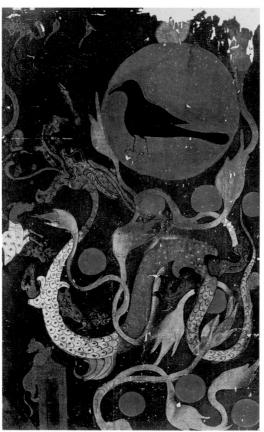

The Chinese myth (maybe the first fable about global warming) tells of ten suns, nine of which were shot down to save life on Earth. Here, only nine suns are shown.

New Year's celebration at the Futamigaura seashore, Japan. In the background can be seen the Wedded Rocks shown in the first color insert. As part of the ceremony, bronze frogs are specially stationed to combine Sun with water to symbolize fertility.

Sunset at the Crimean Shores *(1856), by Ivan Aivazovsky. He was the
most prolific Armenian painter of his time, with more than six thousand works.*

J. M. W. Turner's Heidelberg Castle *(1840–45). These ruins are considered among the most important Renaissance structures north of the Alps. Turner completed this painting in the decade before his death, when he saw the Sun not so much as a resplendent yellow but as an inspiring white.*

Van Gogh painted a number of variations on the theme of a man sowing before a setting sun. With The Sower *(1888), he finally captured "the most vital yellow" of sunlight.*

Roy Lichtenstein's Sinking Sun, his rendering of a California sunset, was acquired by the actor/director Dennis Hopper and his then wife Brooke Hayward in 1964; in 2006 it was sold at Sotheby's for $15.7 million.

Edward Hopper's Rooms by the Sea *(1951) uses sunlight to emphasize loneliness. Hopper produced many works in which a figure is near or gazing out of a window, all of them preoccupied with sunlight, while according to one of his patrons, "He always said that his favorite thing was painting sunlight on the side of a house."*

David Hockney also has favorite subjects, one being California swimming pools. A Bigger Splash was painted in the summer of 1967. One critic noted its "salient eruption of white chaos," but set next to the paintings of Turner's last years, it suggests the two artists had remarkably similar visions.

that the British artist Sir Terry Frost (1915–2003), captured by the Germans when they overran Crete in 1941, spent his captivity experimenting with yellow and came up with 381 shades. All are part of the Sun's wardrobe.

Human beings come nearest to the Sun in the color of their hair. The women of ancient Rome would bleach themselves to blondness with quicklime and wood ash, jealous of the German slave women brought back by soldier-husbands. A blond head typically has many more strands than a dark or red one, and blondes have long been thought more fertile. In her cultural history of the species, Joanna Pitman points out that the majority of heroines in fairy stories are blond, from Beauty in "Beauty and the Beast" to Cinderella, Goldilocks, and Rapunzel. "The men of [ancient] Greece were bewitched by blond hair," she adds, for it represented fantasy and wealth: "'Golden' was the master epithet for Aphrodite in all of Homer's work."[12]

Blondness, says Pitman, became a prejudice in the Dark Ages, an obsession in the Renaissance, a mystique in Elizabethan England (Elizabeth herself was auburn-haired), a subject of myths in the nineteenth century, an ideology in the 1930s—both for Hitler's Germany (where in fact at most only 8 percent of women were natural blondes) and Stalin's Russia (where blondes were almost as scarce)—and a sexual invitation in the 1950s, with Hollywood's glorification of glamorous but unintelligent blondes. As Raymond Chandler points out in *Farewell, My Lovely,* "There are blondes and blondes and it is almost a joke word nowadays" (and he wrote that in 1940).[13] In the early 1960s, California set the trend of tanned blond beauties, and the Beach Boys rhapsodized over such golden girls. We seem to be back in the realm of never-ending paint colors, but the message is the same: yellow, of whatever hue, is the Sun's color, evoking awe, even worship.

Just as yellow is the color, gold is the prime metal of the Sun. Pitman makes this connection: "The color gold had long been established in the classical canons of beauty and power. Almost two thousand years before Homer, in the time of the Proto-Indo-Europeans, the color was connected to the worship of Sun and fire, and to the adoration of a yellow dawn goddess."[14] The association has endured through the centuries. Thus James Joyce fuses solar rays, blondness, and gold into one perfect tableau when Leo Bloom, suddenly bathed in morning sunshine, reflects: "Quick warm sunlight came running from Berkeley road, swiftly, in slim sandals, along the brightening footpath. Runs, she runs to meet me, a girl with gold hair on the wind."[15] He was writing more truly than he knew. Gold can indeed be found in human hair (almost the only part of the body to contain it). On average, it is concentrated at the rate of around 8/1,000,000,000 of a gram—not enough to make a barber smile—and more in adults than in children, and more in men than women.[16]

The Sun's association with gold is easy to understand, and not just as a matter of color. Long before the days of alchemy, gold was regarded as the most perfect thing on Earth, and thus indissolubly linked with the great star.[17] The alchemist in *The Hunchback of Notre Dame* exclaims: "Gold is the Sun; to make gold is to become God!"[18] The only metal that never tarnishes, gold is a suitable symbol of purity; it can be beaten out so thinly that it has hardly any solidity left, approaching a state of pure evanescence.[19] It has often been associated not just with divinity but, because of its rarity and worth, with royalty.

Every culture seems to have valued gold. In 624 B.C., Periander, tyrant of Corinth, invited the city-state's nobility to a dinner and had his soldiers strip the women of their gold ornaments and gowns adorned with gold thread, which booty financed his rule for decades. By 560 B.C., the first gold coins were minted in Lydia, a kingdom of Asia Minor, and in 1284 Venice introduced the gold ducat, which became the most popular coin in the world. Columbus, sending gold back to his queen, exulted: "O most excellent gold! Who has gold has a treasure with which he gets what he wants, imposes his will in the world, and even helps souls to paradise." The Inca considered gold "tears wept by the Sun" (or, more coarsely, its sweat), employing it only for ceremonial purposes, although Atahualpa sued for his freedom from the Spanish by having his cell filled with gold higher than a man could reach, and Montezuma II sent Cortez a sunlike gold disk the size of a cartwheel. The fabled city of El Dorado was named for its king, "the gilded man," whom the sixteenth-century chronicler Gonzalo Fernández de Oviedo (1478–1557) describes as going about "continually covered in gold dust fine as ground salt."[20]

Durable, flexible, and beautiful, gold has conjured cities out of the ground almost overnight, from San Francisco to Johannesburg. Above all else, it is extremely scarce: an ounce of the stuff is today (May 2010) worth more than nine hundred dollars. During humankind's time on Earth, only an estimated three hundred thousand tons have been won from the earth: only enough to form a single thirty-three-yard cube. Yet it is estimated that 8 trillion tons are dissolved in the oceans alone, still more in the Earth's crust. As for the Sun, about six ten-billionths of its mass consists of gold, which totals 1,320,000,000,000,000 tons—enough to cover Scotland half a mile deep.

In an essay he wrote in 1930 on the return to the gold standard in international markets, John Maynard Keynes considered what associations the metal evoked. In the days before the evolution of representative money, he argued, it was natural to choose a metal as the most suitable commodity for representing value. Quoting Freud, he said there are reasons deep in our subconscious why gold should "satisfy strong instincts" and serve as a symbol. "The magical properties, with which Egyptian priestcraft anciently imbued the yellow metal,

Five young Gypsies perform their salute to the Sun in the mid-1950s.

it has never altogether lost."* It does duty, so to speak, for the Sun on Earth. The Latin word for gold, *aurum* (thus the chemical symbol Au), is cognate with the Greek, "golden goddess of dawn," Aurora.

It is no surprise that the coin considered the most beautiful ever struck in the United States (known as a double eagle, as it was twice the value of the ten-dollar coin known as the eagle), contained an ounce of 22-karat gold. This coin was in circulation until 1933, when, in response to the Depression, President Franklin D. Roosevelt banned private ownership of gold. It was revived in 2009, slightly smaller than the original but containing exactly an ounce of 24-karat gold, with Liberty striding toward the viewer on one side, the Sun gilding the Capitol dome behind her, and on the other an eagle in flight over a blazing Sun. Flip the coin, and the Sun still lands heads up.[†]

* John Maynard Keynes, "The Return to the Gold Standard," in *Essays in Persuasion* (New York: Harcourt, 1932), p. 182. Money has a long association with the Sun. The first coin issued by the U.S. government, on July 6, 1787, was called the Fugio ("I flee"), with the Sun at its zenith. Over the centuries, Muslim sovereigns have held solar ceremonies in which they were weighed against a quantity of gold (representing the Sun), and the poor received money accordingly.

[†] See Matthew Healey, "Century Later, Gold Coin Reflects Sculptor's Vision," *New York Times*, November 25, 2008. A tradition stretching back three hundred years dictates that British coins (and later stamps) show an incoming monarch's profile facing in the opposite direction to his or her predecessor's. On the death of George V in 1936, the appropriate mandarins set to work. George V had faced left, so his successor Edward VIII would be looking the other way—but the

———

MIRRORS GIVE US back the Sun and are another of its most potent symbols. The Egyptians invariably styled their mirrors after the solar disk, a slightly flattened circle representing the Sun as it often appears on the horizon. This divine connection made the mirror a religious symbol that was used during festivals and ceremonials and even placed in tombs, often in front of the face or on the breast of the corpse, to ensure the presence of Ra.[21] The Chinese hung mirrors from temple ceilings, reflecting the wisdom of heaven and drawing fire from the Sun. Other cultures linked mirrors to the Sun through language—the Aztecs, for example, named one of their sun gods Tezcatlipoca, meaning "Smoking Mirror."

The Japanese were unique in the prominence and ubiquity of their Sun-linked mirror imagery. According to legend, when their sun goddess, Amaterasu, hid herself in a dark cave, a mirror hung on a tree outside gave her a glimpse of her own light, encouraging her to come out. Thereafter she ordered her grandson to Earth to create the "Cradle of the Sun"—Nippon—giving him a mirror to hand down to his descendants (and sending with him a whole company of mirror makers). In this myth, the mirror's identification with the reflection of the disembodied spirit of divine light made it the terrestrial representation of the Sun, and thus the most important treasure in the imperial regalia.

Burnished gold shines back well enough, and some of the earliest Roman mirrors were lined with it, but generally, more reflective surfaces have been used: bronze, tin (mixed with mercury), and eventually silver. By the sixteenth century the main Venetian mirror makers would back a plate of glass with a thin sheet of reflecting metal, which remained the dominant method for three hundred years. The modern process—backing a glass surface with metallic

---

new king refused, believing that his left side showed him to better advantage. As a compromise, it was suggested that his left profile be deployed but photographically reversed, so that the royal head would still be gazing right. Again the new king exercised his veto: the manner in which he parted his hair on the left side was one of his better points. All this created turmoil at the Royal Mint and at the Post Office, but the first issue of stamps emerged with his left profile facing left, as decreed. Unfortunately, the background prepared for the stamps had already been engraved to show the monarch looking right; when the authorities learned of Edward's wishes, they changed the direction of his gaze, but not the background. Hence the king is shown peering into the shade rather than the Sun—the opposite of an inspiring omen. After Edward abdicated that December, the London *Times* commented, "Even the strong minds can yield to this weakness, the superstitious anxiety of those who shook apprehensive heads at the new stamps, because the head of King Edward VIII was turned away from the light, and looked forward into the gloom—apt symbol of a reign that began with everything in its favor and moved onward into calamity." See Ted Schwarz, *T. H. Paget of the Royal Mint* (New York: Arco, 1976), pp. 190–91.

silver—had to wait until the 1830s, when the discovery of immense deposits of silver in Australia, Central America, and Europe (which drove down its value against gold by a factor of five) and the inventive powers of a great German chemist, Justus von Liebig (1803–1873) combined to make it practical. Until then, looking glasses, particularly large ones, had been extremely expensive— one reason why, at a cost of 654,000 livres, Louis XIV created the most famous of all halls of mirrors, the Galerie des Glaces at Versailles. With seventeen windows overlooking the gardens matched by seventeen vast mirrors along the wall, all crafted in a Paris workshop in an attempt to outdo anything Venice could produce, it further affirmed the greatness of France—and its king.

When in 1682 the hall was displayed to the awed court, it was a commanding success, "a dazzling mass of riches and lights, duplicated a thousand times over in just as many mirrors, creating views more brilliant than fire and where a thousand things even more sparkling came into play."[22] There were 306 panels in all, blending together so that each seemed part of a much larger sheet. The gallery was not only the symbolic focus of the kingdom during the ancien régime, but also continued to play a role after the Revolution—in 1919, the Treaty of Versailles was signed there. In her history of the mirror, the historian Sabine Melchior-Bonnet links the hall to the Sun. "Everything at Versailles," she writes, was

> specular magic, not only the castle reflected in the water of the canal, not the symmetry of an unfolding architecture, nor the repetition of gesture in the mirror, but first and foremost all the rules of etiquette by which the audience of the courtiers bowed in unison. . . . The court thought itself a spectacle. Each person wanted to see, to see himself and be seen, to be narcissistically bedazzled as all gazes converged upon the eye of the Sun King, from whom all light radiated.[23]

To a contemporary observer, standing inside the gallery on a bright summer day, it could seem that the room had been merged into the Sun, light bouncing endlessly from mirror to mirror.

CHAPTER 24

DRAWING ON THE SUN

Light is not only glorious and sacred, it is voracious, carnivorous, unsparing.
It devours the whole world impartially, without distinction.

—J. M. W. TURNER, note appended to
The Angel Standing in the Sun, 1846

The essential thing about Matisse's painting is not to judge it except by eye.
You have to look at it as you would look at sunshine through the window.

—JULES FLANDRIN, 1871–1947

IN THE ART OF THE WESTERN WORLD, IT WAS A LONG TIME BEFORE painters depicted the Sun with any real interest. For centuries, they illuminated their canvases with indirect light, all objects near and far equally detailed and vivid, so that without internal clues it is hard to determine the time of day. Sunshine, in any real sense, is out of the picture. Painters explored candle glow long before they turned to daylight.

In medieval times, artists rendered landscapes as idealized backgrounds that represented the panoply of God's abundance, a filling-in behind the main subject. They generally ignored the world of clouds, storms, rain, and Sun except for freaks of nature—an earthquake, a volcanic eruption, or a bolt of lightning, all of which they could use to evoke divine displeasure or to illustrate some event from the Bible. The Sun often appears in the sky, but it is left hanging there as talisman or accessory. In the Sistine Chapel, Michelangelo's Sun is a blotchy yellow ball being willed skyward by the pointing finger of omnipotence—it certainly has no majesty of its own.

Between 1300 and 1650, however, as Renaissance man began to analyze the world around him, the Sun gradually lost its power as a religious and astrological entity. What light consisted of, how it operated, what its effects were, all came under investigation. A similar change occurred in the world of art, as painters of the era began to take an interest in nature for its own sake. By the mid-1600s, landscape (the very word being a coinage of that era) had become a fit subject for art.* While Rembrandt (1606–1669) often used the Sun's rays to intensify his portraits, by the 1650s, artists such as Jacob van Ruysdael (c. 1628–1682) and Claude Lorrain (1600–1682) had become interested in depicting sunlight itself. Ruysdael's attention to meteorology was extraordinary; his *Winter Landscape with Two Windmills*, for instance, shows a shaft of light radiating from a sun pillar created in convincing detail by ice crystals in the freezing air.[1]

Claude (one of the last artists to be known by his first name alone) was a friend of Nicolas Poussin's, and they would travel the Roman *campagna* together, sketching; but whereas Poussin subordinates the landscape to biblical and allegorical figures, Claude does the reverse: his subjects are the land, the sea, the air. Light, the key feature of his seaport pictures, derives from a Sun just above the horizon, a feature he introduced into *Harbor Scene* (1634)—the first time in the history of art that the Sun clearly illuminates an entire picture.[2] He had so little interest in anything besides the natural world that he engaged others to paint in the small figures that appear on his canvases, and he once remarked to a potential buyer that what he was selling was the landscape; the figures came as an extra.

Elsewhere, artists were beginning to depict natural phenomena with an almost scientific concern for accuracy. In the early eighteenth century, for instance, the German portraitist Aegid Quirin Asam, having been commissioned by the Benedictine abbey of Weltenburg to paint their order's founder, Saint Benedict (said to have been inspired to initiate his Rule during a solar eclipse), pursued eclipses all around Europe. His portrait shows the saint on a high tower, a thin ray emerging from an otherwise occluded Sun. No other painter has gone to such lengths to get eclipse details right.

---

* Leonardo da Vinci (1452–1519) distinguished among three kinds of perspective. First is *linear*, the apparent convergence of parallel lines as they recede into the distance—the kind that depends upon drawing skills rather than color mixing. Second is *aerial*, where, as da Vinci wrote, "The impact of the appearance and of the substance of things diminishes with every successive degree of remoteness; that is, the farther the object is from the eye, the less will its appearance be able to penetrate the air." (Since scientists tell us that a single cubic mile of air over the Earth's more fertile temperate areas may contain as many as 25 million organisms, little wonder that light must fight to reach us.) Finally he considers *color perspective*: "The greater the depth of the transparent layer that lies between the eye and the object, the more that object's color will be modified by that of the intervening transparent layer." In other words, the eye is affected by the light scattered in the air between it and the object viewed.

CLAUDE AND RUYSDAEL had in common that they were much admired by an artist who, perhaps more than anyone else, was to make the Sun his main subject: Joseph Mallord William Turner (1775–1851). John Ruskin recorded that weeks before dying, the old artist announced, "The Sun is God," a remark that sums up his painting so appositely that, although perhaps apocryphal, it has been little questioned.

Our star was Turner's primary inspiration from a very early age. At the evening classes on the nature of perspective that he attended from 1783 to 1789, he insisted on putting reflections in the windows he drew onto an architect's perspective sketch of a mansion. When the architect told him to paint the panes a plain dark gray and the bars white, as this was the established practice, Turner remonstrated, "But it will spoil my drawing." The *Annual Register* would later note that the young artist was "always on the alert for any remarkable phenomenon of nature. He could not walk London streets without seeing effects of light and shade and composition, whether in the smoke issuing from a chimney pot, or in the shadows upon a brick wall, and storing them

An altar painting in the Benedictine abbey of Weltenburg, in southern Germany, showing their founder, Saint Benedict, at the moment he was inspired to initiate his Rule, during a solar eclipse.

in his memory for future use." By his early twenties, rather than producing imitations of his elders, as other young artists did, he was turning his canvases into visions of air and light.

But he was still not the equal of the master. In 1800, invited by the picture's owner to view Claude's *Seaport with the Embarkation of the Queen of Sheba,* he was left on his own to appreciate it. When the owner returned, breaking the spell, Turner burst into tears: "I shall never be able to paint anything like that." By then he had completed three hundred watercolors and was already an extraordinarily tactile artist: the fingertips were for him as capable of creating expression as any brush. When years later a young man claiming to be a painter called upon him, Turner commanded, "Show me your hands." They were clean. "Turn the fellow out," he snapped to his manservant. "He's no artist."

In 1802, he made his first trip to the Continent. In the brief peace between Napoleonic wars, Paris was packed, "fluttering all the gay colors of the rainbow, and harmonized and softened by the glowing tints of the setting Sun." He came home to produce a series of wonderful watercolors and oils. At the Royal Academy exhibition of 1807, he showed *Sun Rising Through Vapor: Fishermen Cleaning and Selling Fish.* As one art historian has remarked, the first part of the title "could easily be the name of half his work."[3]

How to put onto canvas the true nature of light and shadow? A sketchbook of 1806–8 illustrates Turner's musings: "Reflection in water[:] tho' the real shadow is nearly the same from the plane of the Horizon in near Objects, yet when the whole of the light lays behind it frequently streaks a shade 3 times its hight." In 1808, seeing an object floating on the river Dee, he observes: "Yet the reflection of the white body had not any light or white reflection but on the contrary had its reflection dark." There are many times, he realizes sadly, "When panting art toils after truth in vain."[4]

In 1810 he bought a house in Twickenham, southwest of London, the Thames a short stroll away. Impelled by his desire to depict such elemental forces as water and air, he devoted himself to analyzing sunlight and its portrayal on canvas, reading Goethe on the subject, who argued that color was a product of both light and darkness, and that yellow and red carried active and affirmative symbolic associations. He himself would write several essays on the nature of light, drawing on recent research to show that "light is color." Around 1817, he adopted a lighter palette, helped by a range of new pigments, particularly shades of yellow.

He stayed put in Britain until 1819, then began a series of forays to the Continent, particularly to Italy. The quality of light he encountered there changed his art. He began to paint "golden visions, glorious and beautiful," as Constable put it.[5] Around this time, his friend the amateur painter James Skene contributed an empathetic article to Brewster's *Edinburgh Encyclopedia* in which he wrote:

> Painting can but approximate to all the niceties, combinations, and intricacies, of direct and reflected light, . . . modified by the almost imperceptible gradation of intensity as it recedes from the eye. . . . [Turner's] scrutinizing genius seems to tremble on the verge of some new discovery in color.[6]

Nevertheless, not all was praise. In 1826, the *British Press* caviled at *Forum Romanum* and *Cologne: The Arrival of a Packet Boat*, "All is yellow, yellow, nothing but yellow, violently contrasted with blue." Another reviewer said he was "desperately afflicted with what we may call a 'yellow fever.'" Turner cheerfully dismissed such criticisms, tripping off to a friend's wedding in yellow stockings and quoting Malvolio (depicted in *Twelfth Night* as "cross-gartered" and "yellow-stockinged")—perhaps that unfortunate's speech "I'll be revenged upon the whole pack of you." By now yellow was famously his favorite color, and he rejoiced (or so he told friends) in the nickname "the Yellow Dwarf."

By autumn 1828 he was back in Rome, the inspiration for his great painting *Regulus*. As in many of his later works, the Sun here consumes nearly all other elements, including humans. The painting refers to the grim story of the Carthaginians destroying the intransigent Roman Marcus Atilius Regulus by keeping him for days in a darkened room, then cutting off his eyelids and turning his face to the Sun. As one critic commented, the star had become "a lump of white standing out like the boss of a shield," while another noted that the painting could not be looked at from up close; one had to step well back, and even then one was greeted by "a burst of sunlight." Viewers found themselves compelled to use words such as "blazing," "fiery," and "resplendent."

Canaletto's cityscapes had captured the special light of Venice; Turner, in the paintings of his final years, caught another quality of that luminous city: the way it casts its light over water. Some who saw these works were bothered and bewildered by these explosions of color: "Here is a picture that represents nothing in nature beyond eggs and spinach." Mark Twain was to describe one canvas as "like a ginger cat having a fit in a bowl of tomatoes." But Turner's most discerning admirer, Ruskin, argued that his works were distinguished by the special intensity

> of the light which he sheds through every hue, and which, far more than their brilliant color, is the real source of their overpowering effect

The Fighting Téméraire—*the painting Turner called "My Darling"*

upon the eye . . . as if the Sun which they represent were a quiet, and subdued, and gentle, and manageable luminary and never dazzled anybody, under any circumstances whatsoever.[7]

A particularly glorious representation of sunlight, even by Turner's standards, is his 1839 painting *The Fighting Téméraire, Tugged to Her Last Berth to Be Broken Up, 1838*—one of the best-loved works of all time. Behind the old warship a low Sun fires the sky and, in reflection, the river. As Anthony Bailey has observed, to paint that sun, Turner packed his palette with "his most ferocious pigments: lemon yellow, chrome yellow, orange, scarlet, vermilion and red lead, hot paints that he laid over an already warm ground of earth colors."[8] The *Téméraire*, which had served heroically at Trafalgar, offered not only an occasion for elegy but also a political point. As Turner's other main biographer, James Hamilton, describes the work:

It brings and balances facts together, sail and steam, air and water, past and present, setting Sun and New Moon; it balances qualities: [the] old age and the new, dignity and presumption, silence and noise, steadiness and urgency, the temporal and the eternal; and it balances geometric forms: the horizontal, the vertical and the diagonal. Where these lines rush towards the setting Sun, the black tug and its ghostly white charge move inexorably out into our space.*

Turner referred to *The Fighting Téméraire* as "My Darling," and refused to sell it. As he grew older, he became ever more obsessed with the Sun, getting up early to watch it rise downriver from his rooftop balcony: the darker the day, the more he craved its brightness. A friend who happened to be with him on one of those mornings saw him gazing unflinchingly at the fully risen star. When he expressed wonder, Turner replied, "It hurts my eyes no more than it would hurt yours to look at a candle." Now his yellows were becoming white: he wanted to capture pure light and seemed to understand intuitively what astronauts and space cameras have only recently confirmed—that even though the midday Sun looks yellow, it is actually white. Once he tried, without success, to sketch an eclipse.

Twelve years after painting the *Téméraire* on her way to the breakers, Turner, too, was reaching his final berth. In December 1851, he is recorded as saying, "I should like to see the Sun again," and it was this request that Ruskin (not present at the time) turned into the declaratory "the Sun is God."†

TURNER WAS FAR from being the only important painter to take an interest in the Sun. Artists like Jean-Baptiste-Camille Corot (1796–1875) were particularly fond of sunrises and sunsets, moments rich with color harmonies and symbolic potential. Sunrise, for example, tended to speak of hope, while the intensities of sunset could imply such emotions as anguish or desire. Sunset also

---

* Hamilton, *Turner: A Life*, p. 283. A letter to the London *Times* of August 24, 2005, pointed out that the *Téméraire* was being towed upstream, so Turner, in sketching the scene, must have been facing east as the ship was tugged west. Thus the painting may not portray the nostalgic sunset of the age of sail at all, but symbolize instead the dawn of the exciting age of steam. The first *Téméraire* was a French seventy-four-gun ship of the line captured in 1759. It was considered deadly bad luck to rename a taken vessel (hence the presence of a *Swiftsure* in the French fleet, from which it was recaptured in 1782 at the Battle of the Saints). Names were then passed on, and a second *Téméraire*, launched in 1798, was twice painted by Turner, the first time shown closing with the enemy at the height of the Battle of Trafalgar. It is interesting to compare this powerful narrative with Dickens's painterly vision of the morning Sun rising over the London docks in *Our Mutual Friend:* "The white face of the winter day came sluggishly on, veiled in a frosty mist; and the shadowy ships in the river slowly changed to black substances; and the Sun, blood-red on the eastern marshes behind dark masts and yards, seemed filled with the ruins of a forest it had set on fire." Charles Dickens, *Our Mutual Friend* (London, Penguin, 1997), pp. 80–81.

† In 1993, I asked the art historian James Hamilton if he would write Turner's life. When four years later his book came out, one of the points he made was that, given the painter's religious preoccupations, he might have said, "the *Son* is God": we have no way of knowing.

suggested the fleetingness of this world, or perhaps the life eternal beyond death, and portraitists sometimes used it to suggest something elevated about their subject: thus Gainsborough's *Mrs. Sheridan* shows her at sunset to emphasize her contemplative side, and Henry Raeburn's *Sir John and Lady Clerk* (1792) sets them against the evening light to give them an otherworldly glow.

This culminated in the romantic nature paintings of Europe and America in the nineteenth century. In contrast to Turner's vertiginous sunlight, the German Romantics preferred silent vistas in which luminosity pervades vast space. Influenced by the pantheistic doctrines then popular in Germany, Caspar David Friedrich (1774–1840) and his peers sought to express a transcendental vision through the effects of sunlight as well as moonlight. The landscapes of the American Luminists of the mid-nineteenth century are also filled with atmospheric effects capturing the wonders of the natural world, though in a very different mood. Although they did not articulate any particular philosophy, their paintings of vast mountains suffused with light expressed the optimism of their new country, and the feeling that America was the "New Eden," a mighty nation specially blessed by God.[9]

Similar schools were to be found in Britain. Samuel Palmer (1805–1881) was the period's most successful realizer of solar light, an artist who, while not a successor to Turner in style or early subjects, was inspired by him. Living in Shoreham, on the West Sussex coast, in his twenties he was one of "the Ancients"—a group that included Blake and Edward Calvert (1799–1883)—whose nighttime wanderings led locals to call them "extollagers," a caustic version of "astrologers." Palmer achieved his most famous paintings before he was thirty, most of them showing their subjects under moonlight, but he was fascinated by sunlight, too, especially that of the fading day: *Late Twilight* (1825), in which the Sun has already set, but its gloaming suffuses the sky, *Yellow Twilight* (c. 1830), *The Golden Valley* (c. 1833), and *Landscape Twilight* (c. 1824).

Departing from such works, the Impressionists of the 1860s through 1890s had something quite different in mind: the depiction of light as a natural phenomenon. They had learned from scientific studies to consider white sunlight as the composite of the colors of the spectrum, and, realizing that sunlight and its refractions reached even into shadows, they banished black from their palettes. Nor did the Sun itself appear often on their canvases: rather, the Impressionists devoted themselves to capturing the way light changed at different times of day and different seasons. They made a point of going outdoors—*en plein air*—to paint, until then a surprisingly rare practice: artists of earlier times might sketch out their landscapes in the open, but they then carried the canvases back to the studio to be completed. When in 1876 Pierre-Auguste

Impression: Sunrise. Claude Monet's 1872 painting hung in the first exhibition of the Impressionists (as the participating artists came to be known) in 1874. From its title a French critic coined the disparaging term—which its followers immediately took up with enthusiasm.

Renoir (1841–1919) exhibited his masterpiece *Le Moulin de la Galette, Mont-martre*, critics recognized that "the effect of strong sunlight falling through foliage on the figures" was the artist's primary concern.

Of all the Impressionists, Claude Monet (1840–1926) was the most obsessively concerned with the varying qualities of sunlight: it was his *Impression: Sunrise*, exhibited in 1874, that earned him and his circle their nickname. Little is to be seen of the Sun's disk in Monet's picture of the sea reaching up to Le Havre, but a good deal of its light spills across the canvas. Monet discovered that he required a whole sequence of paintings to record the nuances of light, working on each for just half an hour a day, even then finding it difficult to capture its fleeting changes, and writing in despair, "The Sun goes down so quickly that I can't follow it." He came to devote himself to two series of paintings, and in both the Haystack and Rouen Cathedral series (the latter twenty-seven paintings, the former thirty, all made between fall 1890 and summer 1891) scrutinized the nearly imperceptible shifts in the quality of light, hour by hour, from dawn to dusk, throughout the seasons. When in 1879 his first wife, Camille, was dying, he spent hours sitting by her bedside, until one afternoon he realized with horror that he was preoccupied not with her health, but with the changing light on her bedspread—and fled in anguish from her room. But his self-reproach was unjust: he was simply following his lifetime's passion. As his obituary in the London *Times* declared: "A pioneer in the discovery of 'colour in shadow,' he painted colour for the sake of light rather than light for the sake of colour."[10]

Light was the main subject for Paul Cézanne (1839–1906), too—not the Sun itself, but its effects. "I have been satisfied by my work since the day I realized that the Sun cannot be portrayed," he wrote, "but must be represented by another medium, by color."[11] When in 1876 he moved to the Riviera town of L'Estaque, he urged his friend Camille Pissarro (1830–1903) to join him.* "The Sun is so startling, it makes it look as if objects could be lifted off in their outlines, which are cast not just in black and white, but in blue, red, brown, purple." Pissarro didn't come, but Cézanne's own love of the region—its red roofs and blue sea, the greens of pine and olive and the glare of the Mediterranean sun—prolonged his experiments. Exploring sunlight in its many iterations, he depicted it as filtered through leaves, on water, in shadows cast by the noonday Sun, and as the background to laborers returning home at sunset or fishermen chatting in the shade.

In an essay on painters and their relation to the Sun, the art historian Valerie Fletcher argues that as the Impressionists exhausted the possibilities of sunlight, a general reaction set in. The Neo-Impressionists (a title coined in 1886 by the critic Félix Fénéon) were led by Georges Seurat (1859–1891), whose fascination with the principles of optics and longing to depict sunlight more accurately led him to experiment with a complex mix of colors. Seurat had at first worked only in drawings, without using color at all, deliberately leaving patches of paper unmarked. The values would alternate between dark and light, resulting in a glow or halo around a figure, an effect he called "irradiation." But his fullest explorations of light came through his radical use of color. Believing that his predecessors, while looking at a coastal sunrise or a mountain sunset, had lost sight of the fact that color is merely the brain's interpretation of variable wavelengths, he determined to make painting more objective by taking it back indoors. Thus, in his great work *Sunday Afternoon on the Island of La Grande Jatte*, which was painted entirely in his studio, he simulated the vivid midafternoon sunlight by meticulously scattering tiny dots of a high-keyed solar orange among the other colors to induce a shimmering sensation, with more orange in

---

* Pissarro painted with a dry, hard brush, developing a granulated, thickly encrusted surface that captured the refractory character of dappled sunlight as it filtered through dense foliage— an approach that couldn't have been more different from that of the Luminists, who believed in concealing their brushwork. The ways in which sunlight was rendered really did vary hugely during these years.

the sunlit areas and less in the shade. (Constable had done something similar with small points of red, but as subliminal foci, not to emphasize sunlight.) His purpose was not immediately understood, and when his *Models* was exhibited in 1868, one critic wrote of Seurat's figures "daubed in all the colors of the rainbow . . . suffering, it would seem, from some horrible skin disease."

Dedicating himself to the translation of light into scientifically determined combinations of pure color, Seurat (still only thirty-one when he died) would wander the poorer districts on the outskirts of Paris at dawn and at dusk. He called the paintings that this produced "Chromo-Luminism," to emphasize his commitment to catching the Sun's colors. By treating his subject analytically, however, he removed himself from the immediacy of experience.

A close contemporary of his, Vincent van Gogh (1853–1890), would take a very different approach, his paintings of the Sun—his central subject—exploding with the intensity of his inner life. Although early on he was, like Seurat, enthusiastic about gray, "or rather the absence of color," in February 1888 van Gogh left his native Holland at age thirty-four for southern France in search of a "stronger sun." Thereafter it is a constant, in both his pictures and his letters. "The fact is," he wrote, "the Sun has never penetrated us people of the North. . . . Nature and fine weather are the advantages of the South. . . . The difference in the stronger light and in the blue sky teaches you to see, and especially, or even only, when you see it for a long time."

While living in his little sunlit house at Arles (whose outside walls he painted sunflower yellow), he painted a series of golden sunflowers, as if through them he could gather the Sun's life-giving powers into his home. The flowers were a vehicle for the Sun's energy—so too were cornfields—as comes through in his description of these paintings: "Under the blue sky the orange, yellow, red splashes of the flowers take on an amazing brilliance, and in the limpid air there is something or other happier, more lovely than in the North. It vibrates."[12] He adds, "The peony is Jeannin's, the hollyhock belongs to Quost [two painter contemporaries], but the sunflower is mine."

One of van Gogh's tenets was that an artist should "exaggerate the essential." Accordingly, his drawings suggest the Sun's heat and light by concentric circles of dashed lines. Similar visual shorthand brings to life his painted suns, in such works as *Wheatfield with Reaper* (1889), in which a disk of intense golden yellow dominates the horizon. Ascribing a strongly optimistic significance to the golden color of the Sun, he experimented endlessly with color combinations. Writing to Émile Bernard, he details his paint choices: "Field of ripe wheat in a yellow ochre tone with a little crimson. The chrome yellow 1 sky almost as bright as the Sun itself, which is chrome yellow 1 & 2 mixed, very yellow, then." In his quest for "the most vital yellow" of sunlight, he began to take stimulating drugs—mainly digitalis, derived from foxglove leaves and at that time a treatment for his epilepsy (he also suffered from, besides much else, acute intermittent porphyria), but also camphor and terpenes, a dangerous chemical used in paints. These he would augment with absinthe, the liquor that contains thujone, known even then to poison the nervous system. Some doctors believe these various concoctions brought on xanthopsia (yellow vision), whose sufferers see everything as if through a yellow filter. But it was passion, not sickness, that produced his great late paintings.

Like most of the Romantics, he had moved away from traditional Christianity, and though he still aspired to believe in an all-powerful deity, his canvases were not religious statements. He painted the Sun stripped of allegory or personification, to capture, he said, the *sensation* of its intensity. As Robert Mighall observes, "Even Turner, self-pronounced sun-worshipper, swathed his god in mists or trailing clouds of sublimity; approached it with a due reverence of poetic allusion, yet never quite rent the veil of its temple aside."[13] But if the Sun was for van Gogh the source of life, he was no mere idolator: he caught the threat of this great burning mass, the pitilessness of the universe, the evanescence of our existence.

In a letter to his brother Theo, van Gogh explains what he is trying to accomplish (his easel anchored with iron pegs against the mistral winds):

> I am struggling with a canvas begun some days before my indisposition, a "Reaper"; the study is all yellow, terribly thickly painted, but the subject was fine and simple. For I see in this reaper—a vague figure fighting like a devil in the midst of the heat to get to the end of his task—I see in him the image of death, in the sense that humanity might be the wheat he is reaping. So it is—if you like—the opposite of that sower I tried to do before. But there's nothing sad in this death, it goes its way in broad daylight with a sun flooding everything with a light of pure gold.[14]

Then: "There! The 'Reaper' is finished, I think it will be one of those you keep at home—it is an image of death as the great book of nature speaks of it—but what I have sought is the 'almost smiling.' . . . I find it queer that I saw it like this from between the iron bars of a cell."*

---

* There is a surprisingly appropriate passage in Aristotle's *Poetics* linking sowing with the Sun: "To cast forth seed corn is called 'sowing'; but to cast forth its flame, as said of the Sun, has no special name. This nameless act, however, stands in just the same relation to its object, sunlight, as sowing to the seed-corn. Hence the expression in the poet 'sowing around a god-created flame.'" Van Gogh was an omnivorous reader of the classics, but seems not to have come across this observation.

Because we know of his bouts of derangement and his confinements to a sanatorium (where the scientific treatment for his illness was to leave him soaking in a tub for hours at a time), it is easy to forget that in his letters (he was a compulsive letter writer, his *Collected Letters* taking up more than fifteen hundred pages) van Gogh comes across as the best and most lucid of correspondents. Humorous, endlessly curious, sympathetic, wise about literature as well as painting ("There is no writer, in my opinion, who is *so* much a painter and a black-and-white artist as Dickens: his figures are resurrections"), indeed about all the arts, he was wonderfully sane on the page. His letters are almost unbearably moving, the more so because of the serenity of the writing. His last years saw him paint some of his greatest works, in such an outburst of creativity that he seems to know his time is short. Then one high summer day, July 29, 1890, the Sun's blaze on a cornfield near Auvers finally consumed his will to live. After he shot himself, a note was found stuffed inside his shirt: "The truth is, we can only make our pictures speak. . . . Well, my own work, I am risking my life for it, and my reason has half-foundered because of it—that's all right."

MANY WORKS BY twentieth-century artists center on the Sun—Georges Rouault's *Crucifixion,* in which the Sun stands for the blood of Calvary, Paul Klee's *Leaf from the Book of Towns* and *Ad Marginem,* Max Ernst's *Dada Suns,* Paul Nash's enraged painting of the western front in 1918, *We Are Making a New World,*[15] several paintings by Wassily Kandinsky, Graham Sutherland's *Sunrise Between the Hedges,* and Joan Miró's *People and Dog in the Sunlight,* to cite just a few. But one artist actually advanced our understanding of what painting could achieve with the help of the Sun: Henri Matisse (1869–1954). His childhood was spent under the cool light of Normandy, and only when he moved to Nice in 1917 did color truly erupt into his work; yet those first years shaped him, and the whole story of his artistic development resides in that overcast beginning, in how the luminous subtlety of his originally limited brown-and-gray northern palette was transformed by his years in the Sun into a radical new way of depicting light.

Matisse and van Gogh were not the first painters from the north to experience an epiphany once translated to the vivid south: Delacroix had recharged his vision eighty years before beneath the intense heavens of Morocco. But Matisse was arguably the most transformed by the experience. Writing about his move to the south of France, he mused: "What made me stay was the great colored reflections of January, the luminosity of the days." When Matisse moved south he was uniting his northern culture to one almost completely foreign to him, and his response brought color to painting in a way never achieved before.

Even into late middle age, he was constantly discovering new modes. A letter from his daughter Marguerite, written in 1925 (when he was already fifty-five), describes her response to his latest paintings: "The subtlety in tone of the new canvases in harmonies of mauve and pink is extraordinary, it's the light sliding over and caressing the objects . . . that constantly astonishes you." At the beginning of 1930, Matisse traveled to Tahiti and other Pacific islands, where he would "find dawn and darkness unlike anything he had seen before." Visiting the atoll of Fakarava, he went snorkeling and discovered "that underwater light which is like a second sky." In this new world seen through goggles, he experimented with focus, depth, and angles, staring down into the green floor of the lagoon, looking up at a watery ceiling opaque and wavy as medieval glass, and plunging repeatedly between the two, schooling his retina to compare the different luminosities.

Back in France, his painting became increasingly abstract, his canvases multiplying and dividing light, making it glimmer and glow in compositions that reduced their minimal elements to strips, swaths, and swatches, broad bands and tilted panes of pure color. Pierre Bonnard, to whom he lent some of his paintings, staring in Matisse's studio at the expanses of flat uninflected color, complained wonderingly, "How can you just put them down like that, and make them stick?" But so he did.

Sometimes even Matisse himself needed time to absorb the implications of what he was doing. *Interior with Bars of Sunlight* (1942) expunges his model almost completely, leaving her as a young-woman-shaped space of bare canvas in an otherwise geometrical arrangement of colored bars and rectangles. He kept this apparently unfinished work by him, staring at it from time to time, "as if pondering a problem," finally signing off on it in 1945. "The painting is a time capsule," glosses Hilary Spurling, his greatest biographer, "whose contents would make sense only long afterwards in the light of optical experiments conducted by a generation of abstract artists who had not yet emerged."[16] Matisse wrote a note to any future owners of the painting, urging them not to color in the unpainted figure in the armchair at the bottom right of the canvas: "The figure, just as it is, has its own color, the color I wanted, projected by the optical effect caused by the combination of all the other colors." Like Turner before him, who had etched out rainbows in *Kilchurn Castle, Scotland, with a Rainbow* (1802) and *Crichton Castle, with a Rainbow* (1818) by rubbing away pigment and letting bare canvas conjure up their searing luminosity, Matisse knew that white was often the Sun's most vibrant shade.

His oculist in Nice (who had also treated Monet) explained that the eye could not synthesize pigment fast enough to keep up with the speed and intensity of Matisse's response to color. The artist's last pictures, thirty views of his studio at Vence, just outside Nice, project light and color with a vitality so intense as to seem almost independent of its physical origins.

In 1991, when I was in book publishing, I asked Hilary Spurling to write a biography of Matisse. The resulting two transformative volumes would consume her efforts for more than fifteen years. In May 2003, Hilary invited me to the New York opening of the Matisse-Picasso exhibition. She was the ideal guide, pointing out here a Picasso woodland scene, there the same theme by Matisse, the great blocks of black that massed to form the trees somehow drawing one's eye even more surely to the shafts of light that illuminate the forest floor. It was the same with Matisse's interiors: blocks of furniture—table, bed, chair, a slash of black representing a curtain rod—expressing how sunlight lit up the room. Then she told me a story.

At the end of December 1917, Matisse, just a week shy of forty-eight, had moved to Nice, carrying little more than a suitcase, his paints, and his violin, leaving his wife and three children behind in Paris. He knew that Pierre-Auguste Renoir lived a few miles up the coast at Cagnes, and in his first week in town resolved to pay the master a visit. But as he approached Renoir's house, a roll of canvases under his arm, he found himself walking up and down the street, trying to summon his courage. In the end he tossed a coin, which came down in favor of knocking on the great man's door.

Once inside, he found Renoir hardly pleased to be interrupted. His host was in his late seventies, widowed and so wasted that he had to be carried from his bed to his studio in a special chair. He lived to paint, but here he was—frail, gaunt, and so crippled by arthritis that he could not hold a paintbrush, and every day a padded handle had to be wedged between his right thumb and forefinger. The dying artist was not interested in playing host to a vigorous rival at the height of his powers. Nevertheless, the two talked, and after a while Matisse spread out his paintings for inspection. The old man stomped from canvas to canvas, muttering to himself, then looked up and said, "I knew from the way you were talking that you were no good, had no talent." Matisse stood aghast. Then Renoir motioned to a couple of canvases lying side by side, the very same forest scene and apartment interior that would be displayed so many years later at the New York exhibition. "Until I saw these," he added. "You use black to capture the quality of light. I could never manage that. You're the real thing."[17]

Olafur Eliasson's faux-Sun installation The Weather Project *in the Turbine Hall at the Tate Modern in 2003. On show was an eighteen-thousand-watt bank of streetlight bulbs behind a translucent semicircular screen that produced an unearthly light, while generators wafted plumes of smoke. A huge ceiling mirror doubled the exhibit's apparent size, casting its reflection upon the viewers below, but also suggesting a huge circular sun. Visitors would lie beneath it as if sunbathing. Eliasson has an ingenious take on the white walls of art galleries: "Chalk is white and chalk was used as a disinfectant and so early modernists decided on white walls as symbols of purification, clean space. But if chalk had been yellow maybe all our galleries would be yellow today, and we would interpret yellow as a neutral color."[18]*

EXACTLY TWENTY YEARS after Matisse's cardinal move south in 1917, a major British artist was born in the northern England industrial city of Bradford, and made a similar journey. Had he painted a century earlier, David Hockney might also have moved to southern France, but instead he migrated to Los Angeles in

1964, transforming his art and his reputation. "I was brought up in Gothic gloom," he reminisced in a 1993 interview conducted at his house in Malibu under drenching California light. "All the buildings are absolutely black, and it rains practically every other day. I'm a bit like Van Gogh . . . he thought there was more joy in the Sun, and I tend to think that as well." *Sun,* part of Hockney's 1973 "Weather Series" and one of his most celebrated paintings, shows a potted plant on a windowsill, bathed in color-awakening light. Dressed in gold-cuffed corduroys, a yellow cardigan and shirt, and an olive knit tie with pink polka dots, he explained all this while walking through his bedroom, which he had painted green and pink after a Matisse drawing. "Van Gogh," he added, "is one of the few artists who used yellow very well."[19]

Before leaving for the West Coast, Hockney had been at the forefront of British Pop Art (the term having been first used by Lawrence Alloway in a 1958 issue of *Architectural Digest*). The movement sprang from a rebellion against Abstract Expressionism, which Pop artists thought pretentious and over-intense. In California, Hockney became known for his depictions of waterhole-like swimming pools, tropically lush Los Angeles canyons, and beautiful young men. He also designed the sets for several operas, from Glyndebourne to the Metropolitan in New York—poignant, as for the last thirty years he has struggled with hereditary deafness. But, like Alexander Scriabin and Sir Arthur Bliss, he is a synesthete, in his case perceiving music visually, then translating it into cascades of color. His deafness, he says, has afforded him compensa-

In 1949, Life *photographer Gjon Mili visited Pablo Picasso and showed him some photographs of ice-skaters with tiny lights affixed to their skates jumping in the dark—and Picasso was soon using a small flashlight in a dark room.*

tions, changing the way he perceives space and giving him a sharper sense of light and shade.

In the late 1990s, in order to look after his ailing mother, Hockney bought a house in the old seaside resort of Bridlington, close to where he had spent his childhood—"East Yorkshire's answer to Malibu," as a journalist who interviewed him for the London *Observer* described the town.[20] Just before that interview, Hockney had been lying in bed trying to capture the Sun as it rose over the bay outside his window by using an iPhone application called "Brushes"—and his thumb. When he finishes such drawings, he told the *Observer*, he emails them to twenty or so friends. On good days each recipient might receive half a dozen original works well before breakfast. Visitors to the house routinely find themselves awakened before dawn to join him in looking at the way first light falls on a particular copse—the obsession of his recent work.

Hockney has been invited to fill an entire gallery at the Royal Academy during the Olympic year 2012, and is planning a "very big" sunrise for the show. "I'm well aware that most pictures of sunrises are clichés, but I'm also aware that a sunrise is never a cliché in nature. So that's the challenge." Now seventy-three, he is still energetically seeking out inspiration, and some months ago he traveled to the far tip of Norway, near Tromsø, where it "never got dark. . . . There's a place where you can watch the Sun at midnight that is like the edge of the world." He also went to see Edvard Munch's *Sunrise* at Oslo University and was fascinated by the techniques Munch employed. "He's got lines in it that cameras could never see, but we could—and of course in Oslo in June, Munch could look at the Sun for a lot longer than van Gogh could at Arles."

Like many other artists before him, Hockney has long been fascinated by the way vision works, which has led him to an interest in photography—so much so that for a while he stopped painting altogether. There is something pleasingly apt in seeing an artist whose paintings are so light-suffused turn to a medium that actually means "writing with light." Hockney's idol (and Matisse's longtime friend and great rival) Picasso did just that one day in 1949, drawing figures in the air with a flashlight to create remarkable but ephemeral works of art that could be captured only by slow-exposure film.[21] It is predominantly during Hockney's lifetime, of course, that the two art forms of photography and film have established themselves, each offering artists new ways to express the Sun's power. Those seeking to investigate the optical effects of the Sun now have another set of tools at their disposal.

CHAPTER 25

NEGATIVE CAPABILITIES

That star . . . that star. It's so far away, by the time the light from it reaches here it might not be up there any more.

—ELI WALLACH, in John Huston's
The Misfits (1961)

Architecture is the skilful, exact, and magnificent play of volumes assembled in light. Our eyes are designed to see forms in light.

—LE CORBUSIER, 1923[1]

LONG BEFORE THE FIRST PHOTOGRAPH, THE GREAT PERSIAN SCIEN-
tist Ibn al-Haytham (c. 965–c. 1031) invented the camera obscura, a "darkened chamber" into which light enters through a tiny aperture. If a reflecting surface is placed so that it intercepts the entering rays, the viewer will see a full-color moving image—upside down—of the scene outside. Kepler added a lens to enhance the image's quality.

The invention of the magic lantern was another stepping-stone on the way to moving images, and although he didn't invent it, a Jesuit alchemist, Father Athanasius Kircher (1601–1680), did publish a study of some of the principles involved in its construction. In 1646 his book on optics, *Ars magna lucis et umbrae* (*The Great Art of Light and Shade*), appeared. Among the novelties he describes are a projecting device with a lens that focused sunlight cast by a mirror, and the use of light from a candle or lamp to project images—the very technique that was later refined into the magic lantern, forerunner of film projectors. By century's end portable instruments of this kind were on sale in the major cities of Europe.

Given that the chemical, as well as the optical, principles of photography were known by the early eighteenth century, it is one of the enigmas of cultural history that photography took so long to be invented. Not until 1826 did Joseph Nicéphore Niépce (1765–1833) produce the first permanent photograph (*View from a Window at Le Gras*) with a process he called "heliography"—

Sun writing. The exposure time was at least eight hours, so that sunlight illuminates the photo from both the right and the left sides.

For a few years thereafter, only objects that remained immobile for several hours could be photographed. Then, in 1829, Niépce teamed up with Louis Daguerre (1789–1851), and although Niépce died four years into their collaboration, Daguerre went on to introduce a process—involving a thin iodine layer applied to a silver substrate in order to create a light-sensitive surface—that required a much shorter exposure time. By 1839 these "daguerreotypes" took only about ten minutes, and by 1842 improvements in the sensitizing step, and more effective lenses, had reduced the time to fifteen seconds. Daguerre also created what he called a "diorama"—a room surrounded by newly devised equipment that produced complex light illusions. One of these machines was the phantasmagoria (literally, "marketplace of ghosts")—one of whose first achievements was to "reproduce" the effects of a sunrise.

The Englishman William Fox Talbot (1800–1877) was Daguerre's main rival. To create *Botanical Specimen* (c. 1835), the result solely of the interaction of sunlight and chemicals, Talbot placed a leaf on a sheet of photosensitized paper and left it out in the Sun: the image of the leaf appeared as a faint, almost evanescent, impression against the Sun-darkened background. Produced without the intervention of a camera, these "photograms," as he called them, seem more the work of the Sun itself than of the photographer. The Sun really was the "heliographer."

Working along the same lines as Daguerre, Talbot had earlier discovered another means of fixing a silver-process image, but kept it secret. After reading about his competitor's invention, he refined his technique to make it fast enough to register people, and developed what he called the calotype process (from the Greek *kalos*, "beautiful"), which creates negative images. The earliest surviving paper negative, dated August 1835, is of the oriel window in the South Gallery at Talbot's country house, Lacock Abbey, in Wiltshire. Talbot describes how he created it:

> Not having with me . . . a *camera obscura* of any considerable size, I constructed one out of a large box, the image being thrown upon one end of it by a good object-glass fixed at the opposite end. The apparatus being armed with a sensitive paper, was taken out in a summer afternoon, and placed about one hundred yards from a building favorably illuminated by the Sun. An hour or so afterwards I opened the box and I found depicted upon the paper a very distinct representation of the building, with the exception of those parts of it which lay in the shade.[2]

The pace of invention now accelerated. The great astronomer John Herschel was another important contributor, making extensive experiments on the light sensitivity of various metal salts and vegetable dyes, developing techniques for color photography, and advancing terms now standard, such as "negative," "positive," "snapshot," and "photographer." The first such images, still suggesting the notion that the Sun was their author, were called "heliographs."*

A rivalry soon sprang up between photography and painting. Talbot, for instance, was fascinated by the camera's ability to catch the qualities of light reflected on glass, a phenomenon that he believed challenged the best painters. Another early photographer, Roger Fenton, pointed his lens straight at the Sun—seemingly in contravention of all known photographic principles—to produce an early Turner effect. This desire to challenge painting, a new technology confronting the techniques of an earlier art, continued throughout the early days of photography, as its practitioners struggled to establish their medium's identity. Fenton's *Salisbury Cathedral, the Nave, from the South Transept* (1858), illustrates how the play of light and dark in an interior space could be captured by the photographer: by leaving the shutter open so long that the edges of the shadows blurred under the Sun's motion, he both made the cathedral more ethereal and registered the nuances of semidarkness peculiar to Gothic space.[3] No wonder Roland Barthes suggested that cameras were originally "clocks for seeing."[4]

Exposure time was beginning to be as much opportunity as nuisance. When in 1867 the great portraitist Julia Margaret Cameron took her celebrated photo of John Herschel, such was the poor quality of the indoor light that he had to sit still for several minutes; but this gave a greater depth to his craggy features. However, the next chapter in the story of photography and the Sun was to be set not in England but in Paris. Over the years, several cities have styled themselves "the City of Light," from Varanasi, in north-central India, to Lyon, in southern France. But by the 1890s Paris had asserted the prime claim. The term *lumière*, which embraces "light," is much stronger than its English equivalent, additionally implying happiness, enthusiasm, and the world of art. Yet there is an irony in the title, for Paris has for most of its history been a city of darkness. In the thirteenth century, it boasted a total of three public lamps; in the fifteenth and sixteenth, the law required each household to place a can-

---

* In the 1850s William Barnes, a Dorset peasant who got a "ten-year" Cambridge degree, tried to purge English of Latinisms, and advocated replacing the word "photograph" with "sun print." He failed, although for many years the *Chambers English Dictionary* included the definition "sun-picture—a picture or print made by the agency of the Sun's rays; a photograph." "Heliograph" held its meaning until the early 1900s; thereafter it signified a military signaling device.

The great Hungarian artist Brassaï (1899–1984) photographed this man on the Riviera in 1936. All the light in the picture—and all the shade—seems not to come directly from the Sun but to radiate from the umbrella.[5]

dle in a street-facing window, but it was never enforced. Paris, the City of Light, was in reality one of shadow.*

Now, however, in the run-up to the 1900 World's Fair, whose literal high-light was the Eiffel Tower, Paris was transformed by twenty thousand gas lamps, making nighttime photography possible, while on the outskirts of Lille, in northern France, a new factory was mass-processing photographs. It was in the newly illuminated capital that the great Eugène Atget (1857–1927) now became one of the first to exploit photography's potential as an art form, doing much to capture the great buildings and gardens of France. Early in his career he would photograph at midday, so that his "documents," as he called them, would catch the finer details. In his later studies, by contrast, he worked in the early morning, when a soft luminosity suffused each subject with a sense of peace. In Roman literature, *lux* meant not only "light" but "life." "Light" is a

---

* James Joyce (1882–1941) spent his final days in Paris, steadily losing his sight, sitting by his window wearing a dentist's jacket to reflect the Sun onto the open page. He would lay his head on his desk the better to see what he was writing—his last manuscript slants heavily to the right, poignant evidence of his attempts to catch every particle of light that he could.

term not of physics but of physiology. Atget's work gave these words new meaning.

The Swiss American Robert Frank (b. 1924) is the nearest that modern photography comes to Atget. His *The Americans* captures a cross section of his countrymen in the mid-1950s, and in the introduction to this classic book, the Beat writer Jack Kerouac acclaims Frank's capacity to convey not only "the strange secrecy of a shadow," but also "that crazy feeling in America when the Sun is hot on the streets . . . and music comes out of the jukebox or from a nearby funeral." About one particular shot (*Restaurant, U.S. 1 Leaving Columbia, South Carolina,* 1955), Kerouac notes how it shows "the Sun coming in the window and setting on the chair in a holy glow I never thought could be caught on film."[6]

HELIOGRAPHY, HELIOGRAMS, SUN pictures, photograms—all such terminology underscores the degree to which the Sun was a cocreator of that art form in its early days. Amplifying its role, in 1932 Edwin Land (1909–1991) perfected the first usable material to polarize light artificially, which in time yielded not only inexpensive filters for that purpose (in 1953 Land's company was producing six million pairs of cardboard-frame 3-D movie glasses each week) but also the Polaroid camera. Yet that is the last of the significant innovations in the technology of photography that concern the Sun,[7] because with the coming of digital photography in the 1970s, a drastic change occurred in the nature of the medium. Digital cameras record and capture images as binary data, which facilitates the storage and editing of images on personal computers, and the deleting of unsuccessful takes. According to one historian, an "existential doubt" has "crept into our understanding of these images," as the digital world "would no longer permit a distinction . . . between the real impression . . . and a 'representation' generated in the darkness of a computer." David Hockney has described Gustave Le Gray's combining of two negatives, in the 1850s, to make a single print as a "willed deception" that amounted to a form of "Stalinist photography."[8] Another commentator has pointed out that "the largely unquestioned assumption that the intrinsic and essential nature of the photographic apparatus is to record factual reality has been completely overturned." In traditional photography,

> it is possible to trace a physical path from the object represented, to the light that reflects off it, to the photographic emulsion . . . that the light hits, to the resulting image. In digital imaging this path is not traceable, for an additional step is added: converting the image into data, and

The Great Wave, Sète *(1857), the groundbreaking photograph by Gustave Le Gray (1820–1884)*
that knits together two exposures, taken at different times of the day. It is just possible to see the join
where sky meets sea.

thereby breaking the link between image and physical referent. Any it-
eration of the image may be altered, and there is no "generational" difference to alert us to the stage at which the change occurred.[9]

Such innovators as Atget and Fenton might have replied that photography
has ever been something of a trick; but others might feel that, while the Sun
and photography have yet to be formally divorced, they are surely going
through a trial separation.

THE DOMINANT MEDIUM of our time has proved to be the moving picture—
the one truly new art form to emerge during the twentieth century. While photography was coming into its own, the technologies that would make possible
the art of the moving image were also being developed. In 1824 the English
polymath Peter Mark Roget had published a paper arguing that the human eye
retains an image for a fraction of a second longer than that image is actually
present; this inspired inventors to demonstrate such a principle in practice, and

it was found that if sixteen pictures were made of a movement that occurred over the span of one second and were shown in comparable time, the eye would register such images as one continuous movement.[10]

In 1832, a Belgian, Joseph Plateau, invented the "Fantascope," a device that simulated motion. A series of pictures depicting a sequence of movements (juggling, dancing) was arranged around the edges of a slotted disk. When this was put before a mirror and rotated, a spectator looking through the slots "perceived" a narrative sequence. Exactly forty years later, the British photographer Eadweard Muybridge (1830–1904) employed twelve cameras to take twenty-four consecutive pictures of a trotting horse—"Occident"—to see whether at any time it had all its feet off the ground simultaneously. By capturing Occident's four hooves in midair, thus verifying the "unsupported transit" theory, the experiment also proved a major breakthrough for photography. In 1879, building on this and other earlier experiments, Muybridge invented the Zoopraxiscope, which projected a sequential series of photographs, taken by multiple cameras in order to capture very small increments of motion, that could be linked together to create moving pictures.

In 1882, the Parisian physiologist Étienne-Jules Marey constructed a "photographic gun" to take multiple images per second (the term "shooting a film" possibly derives from this), and by 1888 was producing the first successful moving picture cameras, capable of recording a flow of very brief shots of a subject's movement on the same plate, rather than the individual images Muybridge had produced. That same year, Thomas Edison (1847–1931) completed the first-ever film, just ahead of the French brothers Louis and Auguste Lumière, who in 1895 combined printer, camera, and projector to cast moving pictures onto a screen for an audience.[11]

In its early years, filmmaking depended almost entirely on the Sun, on natural light falling through open-roofed studios. Edison formed the first movie unit in the United States in 1893 by building the Black Maria, a tar-paper-covered shack at his New Jersey laboratories; its stages were mounted on a platform that revolved to follow the Sun's progress across the sky. In bad weather, he resorted to greenhouselike glass shelters, which retained whatever light was to be had.

In 1897, Biograph, the first U.S. company devoted entirely to the new medium (its star director, D. W. Griffith, made four hundred films under its aegis), opened a studio on the roof of a Manhattan office block, moving a few years later to a converted brownstone at 11 East 14th Street, Biograph's first indoor studio and the first in the world to rely exclusively on artificial light. Early in 1910, Biograph sent Griffith with a troupe including Lillian Gish, Mary Pickford, and Lionel Barrymore to film *Ramona* in Los Angeles, then a town of fewer than a hundred thousand people—the move prompted in part by the search for better light, but more by the need to escape the patent disputes that often disrupted production back east.* Filming completed, Griffith and cast gravitated north to a small community famous, if at all, for its flowers—Hollywood—and were much charmed by it.

By 1911, another fifteen film companies had arrived in Hollywood, traveling "westward, to the land of haunted romance where the Sun shines so fraudulently bright," as the film historian David Thomson puts it—and the area proved so invigorating that by the end of the First World War it had become the industry's capital.[12]

The Sun dictated much of the early grammar of film because in those pioneer days, it was the only adequate source of light. Later, special "sun arcs" were developed to simulate sunshine on demand. One such, featured in *Popular Science* in January 1923, measured a foot in diameter, making it the largest lightbulb in the world. These thirty-thousand-watt monsters consumed one-third as much power as was required to run a trolley car, and each contained enough tungsten filament to operate fifty-five thousand household lamps.

At first the goal was simply to have enough light to film by. However, as moviemakers became more concerned with the aesthetics of what they were shooting, it wasn't just the quantity but the quality of the light that mattered. Cinematographers fondly referred to "the magic hour"—the brief period between twilight and full darkness when the light has a deep, warm tone, sometimes known as "sweet light" because of its wondrous effects. As the cultural historian Michael Sims puts it, "The most overlooked ordinary scenes—a rusty basketball backboard, a sand-buried hurricane fence, a bedraggled starling on a dirty street—suddenly gain new significance, merely because they seem abruptly to be touched with grandeur from the sun's first or last light."[13]

As had happened in painting, the Sun itself started to play a role in the movies. Letting the camera linger on a rising or a setting sun became a beloved motif for generations of directors. As Sims observes, "The symbolic new beginning of dawn is as irresistible to filmmakers as it is to writers. It provides powerful visual metaphors (and blissful cinematography) in such films as F. W. Murnau's *Sunrise* . . . and even in the light of a new black-and-white day washing over Baghdad-on-the-Hudson in Woody Allen's *Manhattan*."[14] Allen has shown himself particularly sensitive to the light that bathes his hometown: New York, as the critic Holland Cotter has noted, is "an island city . . . with an

---

* At the time, Edison owned almost all the patents relevant to motion picture production, and often sued filmmakers acting independently of his Motion Picture Patents Company in the East. In California, filmmakers could work beyond reach of his harassment; when he did send agents there, word usually arrived ahead of them, and the competition could escape to Mexico.

island light, alternately obdurate and romantically moody. It can be too candid . . . but its toughness is democratic: it falls on everybody and everything."[15]

Not only dawn and sunset draw the camera's lens. Many notable films have drawn directly on the Sun: the cinematographer Kazuo Miyagawa (1908–1999) was the first to turn the camera directly into its glare, albeit through forest foliage, in Akira Kurosawa's *Rashômon* (1950); Conrad Hall the first to do so in color, in *Cool Hand Luke* (1967). Other films that feature the Sun for purposes of mood and meaning include Erich von Stroheim's *Greed* (1924), with its final scenes in California's Death Valley superbly conveying the full force of desert heat and the wretchedness and despair of death from thirst; Nicolas Roeg's visually dazzling *Walkabout* (1971), shot in the Australian outback; Michelangelo Antonioni's *The Passenger* (1975), which lands Jack Nicholson in the African desert; even Dennis Hopper's 1969 elegiac road movie *Easy Rider*, to choose but four. My most cherished example of a cinematic depiction of the Sun is the opening scene of "Festival of Beauty," part two of Leni Riefenstahl's *Olympia*, her chronicle of the 1936 Games in Berlin. The shots of athletes preparing for the day against the backdrop of a moody, then resplendent Sun movingly evoke the harmony of sportsmen and sportswomen with nature. The film is in black and white.

ALL THE ARTS connect with nature, but this is particularly true of architecture, for better and sometimes for worse. I have mentioned the ways certain buildings have been constructed to record the Sun's movements, reflect its rays, or catch them as an aid to good health. But architecture can also prove the enemy of the Sun. As early as 1902, for instance, Manhattan boasted 181 edifices ten to nineteen stories high and three over twenty. John Tauranac describes the result in his history of the Empire State Building:

> Streets that had once been bathed in midday sun were becoming dark and narrow canyons, and the Financial District came to be described as the "canyons of Wall Street." . . . The canyons appeared, canyons where sunlight seldom reached the sidewalks below, canyons that rose straight up for two hundred, three hundred feet or more, canyons created not by erosion but by economics.[16]

If the problem was at its worst in America, it was repeated elsewhere, and architects worldwide duly took notice. Le Corbusier (1887–1965), one of the most influential architects of the last century, wrote: "The materials for urban design are: sun/sky/trees/steel/cement—in this order of importance. . . . Space and light and order. Those are the things that men need just as much as they need bread or a place to sleep."[17]

In 1922, he proposed a "Contemporary Town" for three million inhabitants, the centerpiece of which was to be a group of sixty-story skyscrapers. In the years that followed, he reformulated his ideas of urbanism, finally giving

The Roman town of Bram, in France, founded in A.D. 333. Its plan recalls the ancient Zoroastrian townships in central Asia, which were conceived either on a circular design or in the shape of a crossed circle.

them concrete shape in the Unité d'Habitation at Marseille, built between 1946 and 1952: an eighteen-floor apartment block that shared many of the features of a heliopathy clinic. "Doling out cosmic energy, the Sun's effects are both physical and moral, and they have been too much neglected in recent times," he declared—revealing his concern about the dangers of tuberculosis. "The results of that neglect can be seen in cemetery and sanatorium." The Marseille building was open to sunlight at all times of day, with its balconies serving as sunscreens that were designed to allow sunshine into apartments in winter and provide shade in the summer.*

He took those precepts a stage further in the Indian city of Chandigarh, a conurbation that covers forty-four square miles at the base of the Himalayas,

* See Richard Hobday, "Sunlight Theory and Solar Architecture," *Medical History* 42 (1997), 470. The designs of buildings in hot-weather countries have often included sun baffles, or sun breakers, also known as *brise-soleils*, placed on the outside of windows or over the entire surface of a building's façade. Different countries have tried different means to reduce the glare: lattices (*shīsh*, or *mushrabīyah*), pierced screens (*qamariyah*), as used at the Taj Mahal, or blinds of split bamboo as used in Japan (*sudare*). One of Le Corbusier's most lasting innovations was a more substantial *brise-soleil* in 1933, taken up around the world.

much of which he designed, working on his plans from 1951 until he died in 1965. Chandigarh was a new city that had been built from scratch to house the many refugees fleeing Pakistan after the catastrophic partition of British India in 1947. Its designs were meant to embody the ideals of an independent na-

The Walt Disney Concert Hall in downtown Los Angeles opened on October 23, 2003. The reflective powers of its surface were amplified by the concavity of some of its walls.

tion, but it was also important to Le Corbusier to take account of the demanding Punjabi climate, where in May and June temperatures can reach 115°F.

Three elements in his designs, partly sculptural, partly architectural, underscore his preoccupation with the Sun: 24 SOLAR HOURS is inscribed on the fifty-yard-wide inclined face of the city's "Geometric Hill"; the "Tower of Shadows" occupies a square of seventeen yards, its façades designed to give maximum shade; and the "Course of the Sun" consists of two parabolic steel arches standing in a pool of water. Other parts of the city are equally Sun-conscious; the High Court building is designed specifically for the tropically rainy climate—it even has Le Corbusier's signature "parasol" roof, an overhanging roll-up lid of concrete to fend off rain and sun.*

---

* The longing to live in houses facing the Sun has continued over the ages, sometimes with unexpected results. In the eighteenth century, when the city of Bath was almost completely reconstructed, one of its leading architects, John Wood, Jr., created the Royal Crescent, several of whose thirty otherwise uniform Palladian houses could not face south, and for years, such was the social importance of receiving the morning Sun that no one would live in them. (On the other hand, some had rather too much sunlight, the "white glare" that Jane Austen noted in *Northanger Abbey.*)

Of all modern architects, Le Corbusier was the most aware of the Sun's effects on people in their daily life. Current practitioners may wish they were more so. A case in point is Frank Gehry's $274 million Walt Disney Concert Hall in Los Angeles, where the glare from the stainless-steel façade has imperiled drivers and raised the temperature on nearby sidewalks to as high as 138°F—hot enough to melt plastic traffic cones and set off spontaneous blazes in trash bins. Ten minutes' exposure to this state-of-the-art building, under a bright sun such as Los Angeles famously boasts, can inflict a bad sunburn. In 2005, just two years after the concert hall opened, its surfaces had to be sandblasted to dim the brightness. Similar buildings, especially those with concave surfaces, now have to be tested at the planning stage to ensure against overheating. Architects have learned to grade the Sun's effects according to three levels of increasing intensity: veiling reflections, discomfort glare, and disability glare. As of this writing, Gehry is constructing an extension to the Philadelphia Museum of Art—entirely underground.

The 151-foot Statue of Liberty, bearing the crown of Mithras. Since the 1940s, it has been claimed that the seven spikes on Liberty's diadem epitomize not Mithras but the seven seas and seven continents; but equally it could represent the crown worn by Roman emperors as part of the cult of Sol Invictus, or the headpiece of the solar deity Apollo. Nobody knows for sure— so why not accept it as a crown inspired by the Sun?

CHAPTER 26

TALK OF THE DAY

At the rising of the sun, I shall triumph! I shall triumph!
—The Prince in Puccini's *Turandot*

There is a house in New Orleans
They call the Rising Sun.
It's been the ruin of many a poor girl,
And me, O God, for one.
—Attributed to Georgia Turner *and*
Bert Martin, c. 1934

In the fall of 2003 I submitted a letter to *The New York Times Book Review* outlining my search for information about the Sun. A few days after this was published, the telephone rang and I found myself talking to Gabriela Roepke, professor of opera at the New School—the university founded in Greenwich Village in 1919—who with Old World courtesy said she would be happy to advise me about the Sun in opera and classical music. I thanked her, noticing her strong foreign accent, and we agreed to meet. Could I come to where she lived, on New York's Upper West Side?

Her apartment turned out to be three small rooms, with books and memorabilia stacked everywhere. A birdlike Chilean in her mid-eighties, she had been a dramatist: one of her nine plays, "A White Butterfly," appears in the 1960 *Best Short Plays of the Year,* but such writing was behind her now, and opera her love. Not being in the best of health, she found it hard to get around, so over the next few months I would make the trip to the twenty-second floor of her apartment building, waiting while she slowly made tea before we settled down to talk about opera and watch her many video recordings, which she expertly fast-forwarded to the scenes she wanted to discuss.

She explained that opera was a rarefied art form, the preserve of composers from just a handful of countries—Italy, France, Germany, England, and Russia—with the mid- to late nineteenth century its golden age, dominated by

Wagner and Verdi: their works and those of their contemporaries often invoke the Sun, she said, generally in the context of young love. None is more tragic than Nikolai Rimsky-Korsakov's *The Snow Maiden* (1880–81), and no others that she could think of make the Sun an actual character in the action. The doomed heroine, Winter's daughter, falls catastrophically in love with the Sun. In Act IV, as dawn is breaking over the valley of the sun god, the Snow Maiden calls on her mother, who warns her to stay out of his light. But before the Snow Maiden can enter the protecting forest, Mizgir, a local merchant, appears, and soon he is presenting her as his bride to the Tsar; just as she declares her love, a ray of sunlight falls over her, and she melts away.

The Sun and Moon dance in traditional Japan

In his memoirs, Rimsky-Korsakov mentions his interest in folk songs, and how he "was captivated by the poetic side of the cult of sun-worship, and sought its survivals and echoes in both the tunes and the words of the songs. . . . These occupations subsequently had a great influence in the direction of my own activity as a composer. But of that later." Sadly, that is all he writes of his solar enthusiasms until the last of his fifteen operas, first performed in 1909, *The Golden Cockerel* (from a folktale retold by Pushkin), when the Japanese Queen sings a "Hymn to the Sun."[1]

To lead me further into the Sun's association with tragic love, Gabriela put on one of her recordings of *La Bohème* by Giacomo Puccini (1858–1924). Soon I was watching Act I, in which Mimi, the seamstress heroine coughing her life away, has just met Rodolfo, the impoverished poet from the room above. Mimi's candle has blown out; she asks Rodolfo to relight it, then goes back to her room, only to return a few seconds later to say she has lost her key. Both their candles are extinguished by drafts, and the young pair stumble about in the dark, with inevitable consequences. "Yes, they call me Mimi," she begins, "but my name is Lucia [Light]. . . . When the thaw comes, April's first kiss is mine. I am the first to see the Sun." They vow to remain together when the world returns to life, come spring. In Act IV, in her dying aria, "*Sono andati? Fingevo di dormire*" ("Have they gone? I was pretending to sleep"), Mimi relives the moment of their meeting ("The first Sun is mine!"). The scene must have been thrilling performed in La Scala, with its five tiers done up in sunrise red and yellow. Nor is *La Bohème* Puccini's only work to use solar imagery. More than any other composer, he scatters his operas with references to the Sun—such arias appear in *Manon Lescaut* (1893), *Tosca* (1900), *Madama Butterfly* (1904), and (memorably) *Turandot* (1926).

When I next visited Gabriela, she was recovering from a fall, and finding it an effort even to make her way across the few feet from her chair to her video collection. This time I made the tea while she put on Puccini's single-act work *Sister Angelica* (1880). The title character is a girl who has borne a bastard child and been sent to a convent to spend the rest of her life doing penance. The opera opens on a lovely evening in May. Sister Angelica and the other nuns have gathered to celebrate one of the three days of the year when the setting Sun turns the fountain in their courtyard to gold—a representation of the "fair smile of Our Lady." By the end of the opera, Angelica, hearing that her child has died, takes poison—suicide of course being a mortal sin. But just as the opera's opening scene is lit by the Sun, so too is the climax, as Angelica implores the Virgin for forgiveness and a radiance blazes out in a sign of the mercy granted her.

The Sun as a sign of the infinite goodness of God is a theme taken up by the other great composer who made the Sun his leitmotif: Richard Wagner (1813–1883). In *Das Rheingold* (1869), the first of the four parts of *Der Ring des Nibelungen* (*The Ring of the Nibelungs*), the lustful dwarf Alberich sets out to woo the three maidens who guard the gold at the bottom of the Rhine: "Look, sisters, the Sun shines into these depths. . . . Nymphs, what gleams and glitters there?" The maidens know all too well what they are protecting, and, even as they repel the dwarf's advances, intone: "Rhine gold! We delight in your glow!" His desires thwarted, Alberich steals the maidens' precious hoard, from which he casts an enchanted ring. Thus Sun and gold blend into a talisman of infinite power.

When the hero, Siegfried, takes center stage in Part III, he comes upon the sleeping Brünnhilde and wakens her with a kiss just as the Sun rises. Most of *Siegfried* is orchestrated in gloomy minor chords, but at this point the violin section enters on high notes and wind instruments take up the first two chords, leading on to strings in a yet higher register and the notes of a harp, as Brünnhilde, dazzled by the morning light, cries out: "Sun, I greet you, Light, I greet you, Radiant Day." The vivid combination of instruments Wagner employs to celebrate the Sun in all its glory is extraordinarily rare in his music.[2]

And yet in his great early work *Tristan und Isolde* (1857–59), he—alone of all opera composers—has his hero inveigh against the Sun:

> *Accursed Day with your light!*
> *Will you forever be witness to my anguish?*
> *Will it burn forever, this light,*
> *Which even at night kept me from her?*
> *Ah, Isolde, sweet beauty!*
> *When at last, when, oh when will you extinguish the spark,*
> *That I may know my fortune?*
> *The light—when will it be extinguished?*

The Sun may indeed be the symbol of civilization and right living, but Tristan rejects it all, giving himself to *Nachtzicht*, night vision. Here Wagner invokes a familiar conflict, going back at least to Egyptian myth, which opposes the forces of light (Horus) to those of sexuality (Seth).*

Tristan and Isolde's love is doomed by a world that cannot allow it to flourish. Only under cover of darkness can they meet as lovers—amid many exchanges about their hatred for daylight's false values and their devotion to *das Wunderreich der Nacht* (the wonder-realm of night). As the conductor Charles Mackerras has pointed out, the music matches the libretto: "The orchestration is bright and glaring for the vanity and deception of the day, then suddenly it becomes nocturnal and shadowy."[3]

Gabriela explained that the lovers' refrain of cursing daylight so pervades the opera that it acquired its own name: *Tagesgespräch*—"talk of the day." Wagner had been initiated into this light/darkness dialectic by the writings of Arthur Schopenhauer (1788–1860), in particular *The World as Will and Representation* (1819), which Wagner describes in his autobiography as having an

---

* Others follow his lead. In Charles Gounod's *Roméo et Juliette* (1867), the chorus sings of Romeo, "Love thrives in the shadows, let it guide his steps!" Then, to take instances from popular music: "The day is my enemy, the night my friend. When the Sun goes down . . . I'm all alone . . . with you. . . . All through the night"—as sung by Ella Fitzgerald; Sky Masterson in *Guys and Dolls* sings: "My time of day is the dark time"; and that creature of the night, Édith Piaf, catching sight of the morning Sun, *"Voilà, le salop!"* ("There's the son of a bitch!")

impact on him that "was extraordinary and decisive for the rest of my life."[4] *Tristan*, which he composed just two years after he read Schopenhauer, equates light and darkness with two of the book's most important concepts, those of Phenomenon and Noumenon: the former being our (false) represen-

Judith Jamison (b. 1943) as the Sun in the 1968 revival of Lucas Hoving's Icarus. *Although very few ballets feature the star per se, Isadora Duncan (1877–1927) wrote that dance "is the rhythm of all that dies in order to live again; it is the eternal rising of the sun."*

tation of the world, while the real world is Noumenon, where all things are indivisible and one. Most people live in "the world of day"; a small elite, of whom Tristan is one, see beyond that chimera into another world, at once more terrifying, more real, and ultimately offering a deeper satisfaction, for it is there that he can be forever united with his love. Tristan is aware that the realm of Night can be shared only in its fullest sense when he and Isolde die: thus his is a *todgeweihtes Herz* (a death-consecrated heart). He rapturously sings:

> *Now we are become the devotees of Night.*
> *Spiteful day, with envy armed,*
> *Could still deceivingly keep us apart*
> *But never again delude us.*

So one of the greatest pieces of music in all Western culture is a direct attack on the Sun? Possibly, which may be why *Tristan* has been denounced as a "work against civilization." Gabriela suggested that at our next meeting we move backward in time to Mozart (1756–1791), who went so far as to change many elements of the fairy tale that inspired one of his greatest operas, in order to wrest it into being a celebration of the Sun: *The Magic Flute.*

MOZART COMPOSED THIS the year he died. The libretto, by his old friend Johann Josef Schikaneder (1751–1812), is based on a story, set in Egypt, in which the Queen of the Night represents the power of good and the black-skinned and black-hearted Monostatos that of evil. Mozart and Schikaneder reversed the Queen's character, making her a symbol of wrongdoing, and added the figure of Sarastro, the all-wise high priest of the Sun (Isis). Sarastro (his name derives from Zoroaster/Zarathustra) has taken the Princess Pamina to the Temple of the Sun, to release her from the evil influence of her mother, the Queen of Night, who in turn induces the noble young Prince Tamino to search out and liberate her daughter with the promise that he will be given her hand in marriage. Tamino, after manifold trials in which his character is tested by Sarastro, not only learns the truth about the fell Queen, but gains both Pamina's freedom and her love; in the climactic scene, the couple are then welcomed to the Temple of the Sun.

The South African director William Kentridge has declared that there are "references throughout the opera of turning darkness into light, of the light of the Sun banishing the night," and certainly the Sun has more than just a "shine on" role. Act II opens in a forest of palms with golden leaves, amid which rises a great pyramid, surrounded by thrones for the eighteen priests of the Sun; and for the finale, the stage transforms into a sun. The star even has its own key, C major being used throughout to embody solar radiance. Mozart and Schikaneder were Freemasons, and Masonic elements appear throughout—none of them more central to the Masonic system than the Sun. The music historian William Mann argues that Mozart was interested "in forging a link between the Freemasonry of his time and ancient Egypt [from] which Masonry claimed its original derivation."[5] This opera exalts the Sun specifically because of its significance to a long-established secret society. A Masonic encyclopedia, published in 1912, explains why:

> Hardly any of the symbols of Masonry are more important in their signification or more extensive in their application than the Sun. As the source of material light, it reminds the Mason of that intellectual light of which he is in constant search . . . the Sun is then presented to us in Masonry first as a symbol of light, but then more emphatically as a symbol of sovereign authority.[6]

Precisely such thinking informs Mozart's opera—although one may be forgiven for escaping into the music.

ALTHOUGH MANY OTHER works find one way or another to pay homage to the Sun, Gabriela and I kept largely to classical opera (partly because that was her preference, partly because she was recovering slowly from her fall and I was anxious not to tire her): *The Sorrows of Young Werther* by Jules Massenet (1842–1912), from Goethe's novel, in which the main aria in Act II is Sophie's "Du gai soleil, plein de flamme"; Pietro Mascagni's *Iris* (innocence and desire in turn-of-the-century Kyoto, written in 1897, several years before *Madama Butterfly*), with its aria "L'Aurora—Son lo la Vita!"; Umberto Giordano's 1898 opera *André Chenier,* about a brilliant young poet who perished under the Reign of Terror; and *The Cunning Little Vixen,* by Leoš Janáček, with its sunrise sequence. Memorably, in Beethoven's *Fidelio,* the prisoners emerge from their dungeons blinded by the Sun's light: it is seen as life-giving, so that in Act II, Florestan sings of how its absence makes him think he is in hell.

It is of course easier to track connections to the Sun in opera than in musical forms without words—the Sun is hard to represent in notes alone—but many compositions invoke it in their titles. Nature permeates the music of Frederick Delius (1862–1934), whether the ocean in *Sea Drift* or the mountains and meadows in the *Mass of Life.* He was a Sun-worshipping pagan, writing about the passing of nature (*To Daffodils*), severed relationships (*Songs of Sunset*), or simple mortality (*Songs of Farewell, A Late Lark*), and equally obsessed with spring and summer: a sense of the Sun seems present in even his untitled pieces. Delius was composing at a time when it was considered vulgar to produce sensory effects for program music; but that did not daunt his contemporary Jean Sibelius (1865–1957). In the final movement of his Fifth Symphony, there appears to be a powerful representation of the Sun. Alex Ross, in an extended critical appreciation in *The New Yorker,* describes the climax of the "swan hymn" that runs through the symphony:

> The swan hymn, now [in the final movement] carried by the trumpets, undergoes convulsive transformations, and is reborn as a fearsome new being. Its intervals split wide open, shatter, and re-form. The symphony ends with six far-flung chords, through which the main theme shoots like a pulse of energy. The swan becomes the Sun.[7]

The Sun inspired other works by Sibelius, including a song to "Sunrise."

Solar elements appear in many other classical works, including music by

Maurice Ravel (1875–1937), Claude Debussy (1862–1918), and Igor Stravinsky (1882–1971). One afternoon Gabriela and I made up a list: it ranged from *Also Sprach Zarathustra* (the brief opening prologue, "Sunrise") by Richard Strauss, 1896, to *Eternity's Sunrise* by John Tavener (1999), based on the poetry of William Blake, and my own favorites, the opening sunrise movement from Dvořák's Ninth Symphony (1893), Elgar's "Chanson de Matin" (1899), and the "Helios" Overture by Carl Nielsen (1903), all composed within ten years of one another.

The majority of these works are inspired by sunrises rather than sunsets, or by the internal works of the Sun; and just because a work mentions the Sun

The set for the final scene of the second act of The Magic Flute, *c. 1730*

in its title does not necessarily mean that it actually relates to it—Haydn (1732–1809) titled one of his string quartets, Op. 20 (1772), *Sun Quartets*— though it contains nothing particularly Sunlike. Gustav Holst (1874–1934) excised the Sun from his best-known work, the orchestral suite *The Planets* (1914–16). The music was partly inspired by his horoscope chart, which dealt with the "seven influences of destiny and constituents of spirit," but he chose to replace Sun and Moon with Uranus and Neptune. In his earlier *Choral*

Hymns from the Rig Veda (1908–12), however, he did include the "Hymn to Vena (Sun rising through the mist)."

I left it for several weeks before I next went to see Gabriela, but in the meantime I had come across something that I knew would interest her: a curiosity from Sergey Prokofiev (1891–1953), although not a musical work. In 1916, shortly after composing his first two piano concertos, he acquired a sun-colored notebook in which he asked people to write their answer to a single question—"What do you think about the Sun?"—and kept up his inquiries until 1921, by which time *The Gambler* (1916) and *The Love for Three Oranges* had brought him fame. There are forty-eight pages all told (now in a Moscow museum), which include answers not just from his friends and acquaintances but from people he encountered in his travels, and from the likes of Arthur Rubinstein, Feodor Chaliapin, Vladimir Mayakovsky, and Igor Stravinsky. I enjoyed many of the answers, such as Mayakovsky's "From all of you, who soaked in the Sun for plain fun, who spilled tears into centuries while you cried, I'll walk away and place the monocle of the Sun into my gaping, wide-open eye," and also Stravinsky's direct "It is very stupid that in German the Sun is of feminine and not masculine gender."

Looking forward to showing Gabriela some of the material from the notebook, I checked in at the desk on the ground floor of her building and asked if I could go up to her apartment. The receptionist looked at me, perplexed. "Miss Roepke's not here," he said. "She had all her things packed three weeks ago and returned to Chile. She won't be coming back."

I turned and headed for the door. I knew why she had gone home.

THANKS TO THE power of Google, we know that 2,462 copyrighted songs have "Sun" in their title—more by now, as this figure is from May 2009. The roster of composers ranges from Beethoven to Irving Berlin, and includes ninety-nine songs featuring the phrase "rising Sun" alone.

When in the early 1970s the Beach Boys recorded "I'll Follow the Sun" they were reflecting more than a decade of a craze for doing exactly that, principally on the great surfing beaches of Southern California.

To the Beach Boys, the Sun stood for freedom. From the mid-1960s to the mid-1980s, it became the metaphor of choice for expressing the ups and downs of love. Katrina and the Waves' anthem "Walking on Sunshine" (1985) seized upon the charge of sunlight to express the intensities of love; Bill Withers's "Ain't No Sunshine" captured the agonies of love lost; the Beatles' "Good Day Sunshine" joined being in love with a feeling of being merged with the world.

One of the best-known songs of those decades uses a different kind of metaphor. "Here Comes the Sun," composed by George Harrison in 1969 to become the opening track to side two of the Beatles' *Abbey Road* album, had its genesis in a collaboration with his friend Eric Clapton and celebrated the escape he longed for during a time when so many things were sapping his energy—primarily the grueling shifts to finish the album before the Beatles broke up; the phrase "Here comes the sun" reflected Harrison's relief when each session was over. As he put it, the song "was written at the time when Apple was getting like school, where we had to go and be businessmen: 'Sign this' and 'sign that.' Anyway, it seems as if winter in England goes on forever, by the time spring comes you really deserve it. So one day I decided I was going to slag off Apple and I went over to Eric Clapton's house. The relief of not having to go see all those dopey accountants was wonderful, and I walked around the garden with one of Eric's acoustic guitars and wrote 'Here Comes the Sun.'"[8]

A second great Sun classic of the pop era is "The House of the Rising Sun." Originally an English ballad, it took its title from a euphemism for a brothel, so to that extent is not about our Sun at all; but as a cultural reference point it is unique. In 2005, a poll ranked it Britain's fourth favorite song of all time, and worldwide its popularity has not been far off that. What *was* the "Rising Sun"? Possibly a metaphor for the slave pens of the American South, or the plantation house, or even the plantation itself. As for its being a particular place in New Orleans, nothing is certain.[9]

Along with all the music that either invokes the Sun or has been inspired by it, a slew of modern-day compositions attempts to capture how it actually sounds. The Stanford Solar Center in California lists a dozen artists who have attempted representations: Stephen Taylor, for instance, has composed *Shattering Suns*, a symphony "inspired by images of celestial catastrophe" with music "built on the breathing—the sounds of the sun, as recorded by Stanford solar researchers and played by a synthesizer." Musicians have moved from the inner realms of their imagination—Mozart denoting C major to represent the Sun, for example, presumably because he felt that that key resonated with a kind of solar energy and optimism—to representations drawn from science. This doesn't necessarily mean such compositions have any claim to accuracy or that anyone, even now, with all our sophisticated technology, actually knows what the Sun sounds like.

In 1992, the prolific Danish composer Poul Ruders, inspired by Carl Nielsen's "Helios" Overture, created the symphonic drama *GONG*, with a score meant to evoke what scientists have told us. He explained that recent research has shown that the Sun's surface reverberates like a gong, in four different, simultaneous tempi. Formally his composition follows the Sun throughout its life span, from explosive birth through the hyperactive release of energy we know today to final collapse, concluding with a chord taken from the middle of the piece and sustained over several bars, from virtual nothingness to full force. I confess to finding the symphony difficult to listen to; and it turns out that it is not an accurate rendering in any case.

From 2006 on, the new discipline of helioseismology has been measuring the super-low-frequency sound waves that echo inside the Sun, caused by the waves bouncing around inside the giant ball of gas, huge convective bubbles of different notes and vibrations, nearly all of them thousands of times lower in pitch than earthly bells. These oscillations trigger small motions—small for the Sun, that is—on its surface, creating tremendous up- and downdrafts that generate a broad range of low-pitched noises, just as air blown across a bottle's mouth sets off a few notes. Only on the Sun they range into the millions, ringing like bells or vibrating like organ pipes. Researchers have been tracking the waves as they resonate from one side of the star to the other—a journey taking at its fastest about two hours, at four hundred times the speed of sound.[10]

When in 2003 I visited the famous old observatory on the outskirts of St. Petersburg, its director asked, "Would you like to hear the Sun?" A moment later I was in the main observation tower listening to a record that sounded like a nest of alarmed and angry snakes—nothing gonglike at all, or indeed like breathing. Months later I read in Kenneth Tynan's *Diaries*, "The Sun has echoes that are not entirely gentle. It hisses." So maybe in years to come we will be given a musical work for timpani and cobra.[11]

CHAPTER 27

BUSIE OLD FOOLE

Down sank the great red Sun, and in golden, glimmering vapors
Veiled the light of his face, like the Prophet descending from Sinai.

—HENRY WADSWORTH LONGFELLOW,
Evangeline[1]

Apollo still raged away in the brazen sky—as one might say if one had a
taste for that sort of thing.

—DICK FRANCIS, *Smokescreen*[2]

ONE TWENTIETH-CENTURY NOVELIST, POET, AND ESSAYIST HAS MADE extraordinary use of the Sun. He considered himself a scientist—an expert on butterflies: Vladimir Nabokov. I hadn't known his work well, but remembered that in his final novel, *Ada* (1969), the heroine and her brother, Van Veen, wander through the gardens of their ancestral home and begin to play a game. Ada explains the rules:

> The shadows of leaves on the sand were variously interrupted by roundlets of live light. The player chose his roundlet—the best, the brightest he could find—and firmly outlined it with a stick; whereupon the yellow round light would appear to grow convex like the brimming surface of some golden dye. Then the player delicately scooped out the earth with his stick or fingers within the roundlet. The level of that gleaming *infusion de tilleul* would magically sink in its goblet of earth and finally dwindle to one precious drop. That player won who made the most goblets in, say, twenty minutes.[3]

This interested me, because Kurt Vonnegut's 1963 novel *Cat's Cradle* takes its title from a game played by Eskimos, wherein children try to snare the Sun with string: same idea, different method. Then, as I read more of Nabokov, I realized that Sun imagery recurs with persistent urgency throughout his work.

The deceived lover in *Laughter in the Dark* cries out for sunlight at the very moment he discovers his mistress's infidelity; in *Bend Sinister,* the protagonist Krug perceives the word "loyalty" as being like a golden fork lying out in the sun; and *Pale Fire* glows with solar reflections.[4] There are many other instances.

Vladimir Nabokov (1899–1977), Sun enthusiast, ardent lepidopterist, and wordsmith: even the fictional narrator of Pale Fire *(1962) is named John Shade.*

Nabokov was unusually acute about the quality of sunlight, writing in his memoirs, "All colors make me happy, even the gray blood-orange Sun."[5] His consciousness of the Sun was heightened by its ability to lessen the effects of a case of psoriasis on occasion so severe that its "indescribable torments" almost drove him to suicide; sunbathing and radiation treatments helped in later life.[6] He was also synesthetic; his biographer Brian Boyd mentions that as a child Nabokov had a "love of color and light"—both he and his mother saw the letters of the alphabet in vivid tints—and that he linked his first attempts at poetry "with his mother's jewels and with stained glass, prisms, spectra, rainbows":

> It [the theme of materials affected by sunlight] shines in all its glory in the scene of his first poem. After taking shelter from a thunderstorm in a pavilion at Vyra, Nabokov sees the sun return and cast luminous colored rhomboids on the floor from the colored-glass lozenges of the pavilion windows, while outside a rainbow slips into view, and at that moment his first poem begins.[7]

Boyd points out how Nabokov would pay attention "to details of an order not expected: the haze over a frying pan, the colors and shapes of shadows. . . . He had a painter's sense of light." Even so, I was quite unprepared when I took up his masterpiece, *Lolita,* a novel quite brimful with solar references, almost as if its author were having one of his private jokes. But this profligacy is not intended facetiously at all; drawing upon the Sun to an extent unequaled anywhere else in literature, he used it to remarkable effect.

This large claim needs furnishing. The first third of the novel is saturated with images of sunlight, dictating the narrative's mood and directing us to the easy optimism of Humbert Humbert's early hopes. In the opening sentence, Humbert presents Lolita as the "light of my life," and goes on to talk of the "sun of my infancy" having set.[8] When he first spies Lolita (through "sun-speared eyes") it is a "sun-shot moment." She is "a photographic image upon a screen . . . already riding into the low morning sun." While her mother engages in "some heliotropic fussing," Humbert focuses on a sunbathing Lolita "through prismatic layers of light," but he can also take in the "sun-shot" sidewalks and "the reflection of the afternoon sun, a dazzling white diamond with innumerable iridescent spikes [that] quivered in the round back of a parked car." When, a few pages on, Lolita creeps up on him, putting her hand over his eyes as he reclines in a low chair, "her fingers were a luminous crimson as they tried to blot out the sun." Playfully, she tosses an apple "into the sun-dusted air," and Humbert is soon lost "in the pungent but healthy heat which like summer haze hung about little Haze"—Haze being Lolita's surname. "The implied sun pulsated in the supplied poplars"; "the sun was on her lips"—all this, pretty well, in the first sixty pages.

Such references are very different from the metaphor bank of the Romantics, in which the Sun is primarily a standard-bearer of classical mythology. Nabokov is observant to a fault, alive to anything that the Sun lights upon. He may use the Sun for its symbolic power, as when he writes that Humbert dreams of "the red sun of desire and decision (the two things that create a live world)," but—naturalist that he is—he is at his best as a subtle observer of the physical world around him, of the "beauty and animation of the sun and shadows of leaves rippling on the white refrigerator," or "those southern boulevards at midday that have solid shade on one side and smooth sunshine on the other." "I knew the sun shone," says Humbert, "because my ignition key was reflected in the windshield." Nabokov's interest in light melded with his lifelong fascination with butterflies in at least one scene: we note that "a big black glossy Packard"—the one that later kills Lolita's mother—mounts the lawn and stands "shining in the sun, its doors open like wings."

The web of images is effective both in and of itself and for the way it serves as part of a larger purpose: to employ sunlight in all its many forms, and the world of shadow and darkness, to mirror what is happening to the central characters. Astonishingly, references to the Sun appear about every three pages in the first third of the book. "Movie-ladies" have "sun-kissed shoulders"; a moment of sexual epiphany is "the ultimate sunburst"; while Lolita at her summer camp appears "a sun-colored little orphan." But such sunny days are numbered, and as Part I ends, the language reflects the darkening mood. "But somewhere behind the raging bliss, bewildered shadows conferred"— this at the very time when Humbert first exults, "She was mine, she was mine." His "rainbow blood" may be in full flow as Part I ends, but the images of sunshine disappear, and the shadows close in. Significantly, there comes "a tremendous sunset which the tired child ignored," and before the short, brutal near-rape scene Humbert wildly pursues "the shadow of her infidelity." It is a sign that he cannot see his love object clearly: when a sunburst strikes the highway along which he is driving, he has to pull in at a filling station for new sunglasses. But Clare Quilty, Lolita's eventual "savior," is now in hot pursuit— "our shadow." (Quilty's co-playwright is Vivian Darkbloom—an anagram of "Vladimir Nabokov"—while a member of one of Quilty's drama groups makes his debut in a play entitled *Sunburst*.)

Though it is nothing as simple as Lolita representing the Sun, or the Sun her, once she seems to tire of Humbert he confesses, "I hear myself crying from a doorway into the sun." And before long he is admitting, "The sun had gone out of the game." After the love of his life has absconded, Humbert has a moment when he thinks he has glimpsed her—"a trick of harlequin light"—and he writes a poem to remind her of an old perfume called (inevitably) "Soleil Vert." But she has left him forever, and he is forced to accept "that I simply did not know a thing about my darling's mind and that quite possibly . . . there was in her a garden and a twilight." Sunsets now are "blotched," forcing Humbert to drive on and on "through the drizzle of the dying day." When, years later, he finds Lolita again, she is worn out, vast with the pregnancy that will kill her in her teens, her smile "a frozen little shadow of itself." Failing to get her to go back with him, he takes off into the night, intent on tracking down and killing Quilty, and when in ghastly fashion he does so, "the sun was visible again, burning like a man," leaving behind the corpse of her latest abductor while he walks away "through the spotted gaze of the sun."

All this may suggest that the second half of the novel succumbs to an obsessive symbolism, but it doesn't come across as such—there are at least as many other references to the Sun and to shadow as I have quoted that have no obvious symbolic content (I must be the only person to have read *Lolita* for its sun images; curiously, in all the 565 pages of Nabokov's published letters, there is not one reference to the Sun). We can simply delight in Nabokov's fascination with sunlight and his awareness of what it conveys.

IF NABOKOV'S IS a special case, we should recall the judgment of Max Müller that from the time that man first started to tell stories he has used the Sun as an image. By the days of the epic poets, the whole heavens were being drawn upon as metaphors and symbols. In the *Iliad*, for instance, Homer identifies some 650 characters with stars, and finds them in forty-five constellations or star patterns, each identified with a particular warrior, while such items as armor, chariots, and spears are linked to objects in the sky. He even harnesses the precession of the equinoxes: the epic battles and the duels between various heroes have been shown to be allegories of the passage of the equinoctial rising. The *Iliad* may be considered the world's oldest substantial astronomic text, using the zodiac to allegorize the movement of the heavens. Such, anyway, is the remarkable argument of *Homer's Secret Iliad*, written in rough draft over a period of some thirty years by Edna Johnston, a librarian turned amateur literary critic from southeastern Kansas, and published posthumously in 1999.*

Johnston's research convincingly suggests that the Homeric epics might be among the first works of literature to incorporate a knowledgeable study of the Sun's apparent motions, as if Homer were intent on setting down what was known about the heavens for future generations. And it is not only the *Iliad* that displays an extraordinary grasp of astronomical knowledge, she says. In the *Odyssey*, the symbolism of Odysseus' return home after nineteen years may show Homer's grasp of the nineteen-year Metonic cycle, over the course of which the sequence of the Moon's phases returns to the same days of the year. In one of the *Odyssey*'s most memorable metaphors, Homer uses the

---

* Florence and Kenneth Wood, *Homer's Secret Iliad: The Epic of the Night Skies Decoded* (London: John Murray, 1999). Johnston was born in 1916 and studied at Kansas State Teachers College, taking mythology as one of her special subjects and learning ancient Greek. Over the next three decades she read and reread the *Iliad* and the *Odyssey* until she could recite long passages from memory. At some point she made her discovery—she had no wish, she said, to diminish the poems as narratives, dense with well-drawn characters, convincing plots, "pathos, horror, excitement, calm; philosophy, history and so on" (p. 5), but she saw an overarching purpose in the two great works: to preserve astronomical knowledge. Fearing she might be out on a limb, she never made her findings public. After her death her daughter Florence and her son-in-law spent seven years editing and expanding the various notes until they could finally publish the results.

Sun's position to denote the time of Odysseus' departure from home:* "and when the child of morning, rosy-fingered Dawn, appeared they again set sail."[9]

The year after *Homer's Secret Iliad* appeared, the work of another amateur critic on another classic would show that it, too, dealt at least in part with astronomy. Dolores Cullen first encountered Chaucer when she was a middle-aged college student in California, and became caught up by his work. In 1998 she published a book about the religious allegories in *The Canterbury Tales*. But still she was not satisfied. "As I read, I was distracted by questions about the pilgrims. What made this precise group necessary? Chaucer's reputation was too well known, his skills too well recognized for me to think that the group was a haphazard collection. Why was there one pair of brothers—not from a religious order, but two men related by birth? Why not three brothers or no brothers? Why was there a wife—but no husband and wife? Why no children? Why so few women?"[10] In *Chaucer's Pilgrims: The Allegory*, she turns to the poet's interest in astronomy, and argues that each pilgrim corresponds to a body or bodies in the night sky, which reflect, in name and appearance, the mythical characters after whom the constellations are called. The study was well received.

It is generally known that Chaucer was highly versed in astronomy, and in 1391 even penned a treatise on the astrolabe for his younger son, Lewis; Cullen shows that this interest suffuses *The Canterbury Tales*, influencing not only the identities of the pilgrims but also their stories. From "The Franklin's Tale," with its theories of tides, and of the effects of perihelion (the time when the Earth and Sun are closest to each other), through to "The Parson's Tale" (about the uses of trigonometry), "The Canon's Yeoman's Tale" (about a charlatan alchemist), "The Squire's Tale" (which takes in the precise relationship between the Sun and Mars), and onward, Chaucer puts astronomy and its related disciplines to intense use. But unlike Homer, he does not seem to be writing for posterity so much as responding to a subject that interests him, confident that his audience, at a time when familiarity with the movement of the heavens was taken for granted, would share that interest.

---

* More than four hundred years later, at least one major Greek dramatist was to make keen use of the Sun. Euripides has Orestes flee after killing his mother, Clytemnestra, and search out a place on which the Sun did not shine when he committed the deed, as a way of escaping the Furies; then, in *The Bacchae*, the young King Pentheus of Thebes is driven mad by Bacchus and sees two Thebes and two Suns. Fact came to mirror fiction when in 1783 the French Academy of Science sent a balloon up into the sky, its pilot becoming the first person known to history to see the Sun set twice in a single day—and finding it terrifying. See William Longyard, *Who's Who of Aviation History* (Novato, Calif.: Presidio Press, 1994), p. 41.

Today's analyses of astronomical references in classical works have nothing over the ingenuity of those early biblical writers who brought such energy to explaining the natures of celestial bodies. For instance, the two "lights" mentioned in Genesis—Sun and Moon—were interpreted by the Catholic Church as representing the papacy (the greater light) and the empire, i.e., the old Roman Empire or the current Holy Roman Empire (the lesser). Metaphorically, this meant that just as the Moon receives its light from the Sun, so, too, did the empire receive its authority from the Church, and thus was subject to being overridden. Dante Alighieri (1265–1321), in *De Monarchia*, bravely rebuts this interpretation, and in his case gets away with it.* But in his greatest work he shows himself to be steeped in the Christian symbolism of his time— and the contemporary understanding of astronomy that provided much of its imagery. All three books of *The Divine Comedy, Inferno, Purgatorio,* and *Paradiso,* rest upon a structure of medieval astronomical doctrine, so that when in the *Paradiso* the souls are placed in rank, their order reflects the doctrine of the planetary spheres as outlined by Aristotle and Ptolemy, while the organization of the *Inferno* almost exactly mirrors the assumed ninefold layout of the heavens. The Sun itself receives due attention—early in the *Inferno,* for instance, it indicates times of day, through an oblique reference to the dawn light on the hillside ("The hour was morning's prime, and on his way Aloft the Sun ascended"). At another moment, Dante realizes that his journey through the heavens has completed one-quarter of the daily revolution of the stellar sphere around the Earth. He has therefore spent six hours in Gemini, roughly the same amount of time that Adam enjoyed the pleasures of Eden before being cast out. Now directly beneath Cadiz, close to the westernmost shores of the Mediterranean, he sees an expanse of the Atlantic and, to the east, nearly as far as the Phoenician coast. Only this portion of the Earth's surface is visible to him, because the Sun lies two zodiacal signs behind (farther west), beneath Aries.

The *Purgatorio* refers more than once to the Sun's position, since Dante uses it as the central device to establish the timing of events in the poem. In the *Inferno,* by contrast, he takes only the positions of the Moon and stars for this purpose, since the Sun, as an expression of God's power, will not tell the time in hell, the infinity of which is part of the horror. In the *Paradiso,* he attains the

---

* "Lucifer" (Light Bearer, with all its positive connotations) caused a special problem for early Christians. The name appears in the fourteenth chapter of the Old Testament book of Isaiah, verse 12—and nowhere else. It is a Latin name, yet here it is in a Hebrew manuscript, written before there was a Roman language. The explanation is that in the original Hebrew text, it is not an angel who falls, but a Babylonian king, who during his lifetime had persecuted the children of Israel; an early Christian scribe made the change.

heaven of the Sun, whose special virtue is wisdom. Two moments bear mention: Canto 33 of the *Inferno* has the first reference in literature to the Sun's being fundamentally a star ("I said nothing / In answer all that day nor the next night / Until another star rose on the world"); and in the *Purgatorio* (pictured as a huge mountain-island in the uninhabited Southern Hemisphere), the various souls Dante meets are shocked to discover that he is still alive when they notice that, unlike them, he casts a shadow.

OVER THE CENTURIES, two main problems had to be faced in writing about the Sun. First, it is such a major theme that the temptation to grandiosity is hard to resist. Poets employed the Sun at their peril; but some did make use of it with greater finesse than others. Thomas Malory (1405–1471) introduces Gawain (a translation of "bright hair") into *Le Morte d'Arthur* (published posthumously, c. 1485) as a sun spirit, the knight gaining and losing strength with the Moon's waxing and waning, so losing the Sun's light. Edmund Spenser (1552–1599) trod the "big theme" tightrope with great skill in *Epithalamion* ("Of a Wedding," 1595), a poem that can, like Homer's works, be read as an astronomical text. By contrast, Milton's "On the Morning of Christ's Nativity" comes up with an image that is simply absurd:

> *So when the Sun in bed,*
> *Curtain'd with cloudy red,*
> *Pillows his chin upon an Orient wave,*
> *The flocking shadows pale*
> *Troop to th'infernal jail.*[11]

The same simile does service in Marvell's "Upon Appleton House,"[12] but he made up for this with "To His Coy Mistress": "Though we cannot make our Sun / Stand still, yet we will make him run." Robert Herrick (1591–1674) was again decidedly pedestrian in his "The glorious lamp of heaven, the Sun / The higher he's a-getting, / The sooner will his race be run / And nearer he's to setting."[13]

The second part of the challenge in earlier times was that references to astronomy had to conform to Church doctrine on matters pertaining to the heavens. In his account of Creation in *Paradise Lost*, Milton has to resolve the biblical conundrum of where light came from if the Sun were created only on the fourth day, and gets around it by having light shine from its nook under God's throne for those first few days until it can find its fit place in the Sun. He probably decided to use Ptolemaic astronomy because it was more in accord with scripture, as well as being more worthy of epic; but he is obviously torn, and the possibility of a Sun-centered universe is left open.[14] He can be a master of prevarication: in Book 10, Man's Fall thrusts the Sun "from the Equinoctial Road"—which can be read as either deflecting the Sun from its course around the celestial equator (as the Ptolemaic system requires) or tilting the Earth on its axis (as in the steadily gaining Copernican system).

During the fifteen hundred years that Ptolemy's doctrines held sway, writers were generally content to employ the Sun for occasional grandstanding (John Suckling's "The Sun upon a holiday is not so fine a sight"), about a beautiful girl dancing;* as an excuse for retelling famous myths (Sir Philip Sidney's sonnet sequence *Astrophel and Stella*); or as a reference point when examining the heavens.

One exception is the Italian Giulio Camillo (1480–1544), whose influential *L'idea del Theatro* (*The Idea of the Theater*) is nothing less than a history of the universe, in which the Sun holds pride of place. Frances Yates, in her study of that writer, explains how his work "shows within the mind and memory of a man of the Renaissance the Sun looming with a new importance, mystical, emotional, magical, the Sun becoming of central significance. It shows an inner movement of the imagination towards the Sun which must be taken into account as one of the factors in the heliocentric revolution."[15] References to Camillo's work have figured in the texts of writers from Rousseau to Ted Hughes.

The Sun was soon seized upon throughout Europe as a royal emblem, kings being portrayed in allegory as the embodiment of the Sun's glorious attributes— notably in *City of the Sun* by Camillo's countryman the Dominican Tommaso Campanella (1568–1639), a philosophical dialogue fashioned after Plato's *Republic* and written in 1602, just after Campanella had been condemned to life imprisonment for sedition and heresy. The city of the title, set on the island of Taprobane (now Sri Lanka), is benignly presided over by a priest-prince called Sun.[16]

Both Shakespeare (1564–1616) and John Donne (1572–1631) were Campanella's contemporaries. The former's use of the Sun characteristically defies generalization: he is ever a keen observer of the heavens, referring to the Sun more than forty times in the late plays alone. Many of these references occur in a literary form that had its origins in classical poetry but which was brought to perfection by the troubadours of medieval Provence and the Minnesängers of Germany: the aubade, a short lyric of love fulfilled, supposedly uttered at dawn (as opposed to the serenade, sung by a hopeful suitor at evening), and generally lamenting that the lovers must bid farewell to the joys of the night. It has been

---

* I have a soft spot for Suckling, for during his short life (1609–1642) he invented the game of cribbage.

suggested that the form was an ingenious development from the cry of night watchmen announcing the coming day. Shakespeare made glorious use of aubades, typically planting them as hinge points in his plays, classically in *Romeo and Juliet*.[17]

Over the centuries, the aubade was adapted for other purposes, the Victorians, for instance, using it for a graver kind of lament, as Tennyson did in *In Memoriam*. Perhaps the best known of all English aubades is Donne's "The Sunne Rising," which opens seductively:

> *Busy old fool, unruly Sun*
> *Why dost thou thus,*
> *Through windows, and through curtains, call on us?*
> *Must to thy motions lovers' seasons run?*

If this is the first instance of a poet mocking the Sun (I have found none earlier), in Donne's case there was reason. In 1601, he had secretly married the seventeen-year-old Anne, daughter of Sir George More, the diminutive and irascible lieutenant of the Tower of London (a "short man with a short fuse," Donne's latest biographer calls him). He wrote to the livid father: "Sir, I acknowledge my fault to be so great as I dare scarce offer any other prayer to you in mine own behalf than this, to believe that I neither had dishonest end nor means. . . . I humbly beg of you that she may not, to her danger, feel the terror of your sudden anger." Sir George replied by having him thrown into the Fleet Prison for some weeks (along with the priest who performed the marriage), attempting (unsuccessfully) to annul the union, and getting Donne's employer to dismiss him. For the next decade the poet and his family struggled in near poverty; not until 1609 were he and his father-in-law reconciled. But from the perspective of immortality, Donne definitely had the last word. Sir George, borrowing from the Psalms, had published his own vision of sunrise, a very different one from Donne's:*

Who seeth not the glorious arising of the Sun, his coming forth as a bridegroom out of his chamber, and his rejoicing like a mighty man to run his race? . . . The course of the Sun goeth round the Earth, and his

---

* More, a longtime member of Parliament, in 1601 declared, during a debate on monopolies, "And therefore to think we can sufficiently record the same, it were to hold a candle before the Sun to dim the light." In one of the small ironies of history, 250 years on, the French political economist Frédéric Bastiat famously reduced protectionism to an absurdity in "*A Petition of the Candle-makers*," in which his candle makers petitioned to put out the Sun, as its radiance was bad for business.

light will have entrance, wheresoever the body of man can have passage.[18]

With "The Sunne Rising" Donne achieved the ultimate poetic payback.

He was to write a number of other poems deploying the Sun as a metaphor, but he also wrote a satire, *Ignatius His Conclave*, which in the course of attacking the Jesuits (for Donne was a fervent Anglican convert from Rome) ridiculed many of the new scientific theories, and in particular the argument that the Earth circled the Sun. He was to express his views even more forcefully in verse:

> *And new Philosophy calls all in doubt,*
> *The Element of Fire is quite put out;*
> *The Sun is lost, and th'Earth, and no man's wit*
> *Can direct him well where to look for it.*[19]

This appeared in 1611, the same year that Galileo reported the existence of spots on the Sun and valleys on the Moon.

By the 1690s Newton had explained the operation of gravity, the makeup of the Sun's rays, and the origin of rainbows. The Scientific Revolution was in full flood, whether the world of poesy liked it or not, and certainly there were those in that world who didn't. William Blake (1757–1827), for instance, had no use for science, and viewed its advance as one of the evils of the day. One poem mentions the Sun, but without any post-Copernican trappings:

> *Ah, sunflower, weary of time,*
> *Who countest the steps of the Sun;*
> *Seeking after that sweet golden clime*
> *Where the traveler's journey is done.*

This has been variously interpreted, the image usually taken to represent man's yearning to escape the transience of this world for the golden eternity of the next. Blake may also have been alluding to the Greek legend in which the sunflower was created when a woman "pined away with desire" after the sun god and so turned into his flower. But whatever meaning we finally assign, here was a major poet using the Sun for symbolic purposes that were quite separate from any scientific findings.

Samuel Coleridge (1772–1834) was a more typical man of his time. "I shall attack Chemistry, like a Shark," he wrote enthusiastically in 1800, on the occasion of his going into London to attend lectures on science, primarily to feast upon a new stock of metaphors.[20] He, like many of his fellow Romantics, sought to break into Science's magic circle, be it fashioned by mathematicians or physicists, chemists or biologists. Far from ignoring the new truths of cosmology, they wanted to understand them. Of those nineteenth-century poets writing in English, for example, Lord Byron (1788–1824) was vitally interested in scientific matters, especially in astronomy and geology; so, too, was Robert Browning (1812–1889), who called Shelley "the Sun-treader" and commended him for his "Hymn to Apollo," with its radiant sun god. Percy Bysshe Shelley (1792–1822) did work as a chemist, introducing scientific references into such poems as "Queen Mab" and "Prometheus Unbound," while in his "Hymn to Apollo" the god calls himself "the eye with which the universe / Beholds itself and knows itself divine." John Keats (1795–1821), who studied medicine, regularly employs astronomical images, his long poems *Hyperion* and *The Fall of Hyperion* both being named for the classical sun god.

Also Emerson, Holmes, Poe, Tennyson, Arnold, Hardy ("The sun rested his chin upon the meadows"), Coventry Patmore, and Charles Dickens (often in the context of the weather in general, but with particular grimness in *Bleak House* and the opening chapter of *Little Dorrit*) wrote on specifically solar themes. Emily Dickinson even composed a poem that poked fun at the post-Copernican understanding of the cosmos: "The earth upon an axis / Was once supposed to turn," runs her "Sic Transit Gloria Mundi," "By way of a gymnastic / In honor of the sun!"

But the truth is that, for all their professed interest in science, from 1750 on, for almost 150 more years, when poets sat down to write, they continued to prefer the safer pastures of the old myths. The Sun was still decked out as Apollo, Helios, Hyperion, or Phoebus; its identity as a roiling ball of gases that was but one of many stars in the galaxy was not—yet—the stuff of poetry. Yet the new findings of science could not be ignored, and overall exerted what was thought to be a negative influence. Even Macaulay, who was to become the very apostle of progress, accepted the decline of poetry as the unavoidable result of scientific revelations, while Hegel argued that as societies advance in rational achievement, so they tend to lose those great skills that rest on imagination. By the turn of the century the great scholar of the classical world's knowledge of the heavens was A. E. Housman (1859–1936). In old age, he wrote to *The Times Literary Supplement:* "Quintilian says you will never understand the poets unless you learn astronomy."[21] That may have been true of the ancients, Housman's field of study, but not of his fellow poets.

At least one of the old myths, however, has persisted through the centuries. The legend of the phoenix is a metaphor for solar rebirth, the Sun dying its daily death in the west and being reborn in the east, crossing the sky like a heaven-arching bird. Hesiod first assigned the phoenix its great longevity, and he also introduced the idea of cyclical time: as the bird was reborn, so history repeated itself. Until the seventeenth century, many believed that the phoenix really existed. Shakespeare (notably in "The Phoenix and the Turtle"), Apollinaire (by his name making himself a son of the Sun), Byron, Nietzsche (who signed himself "Phoenix" and wrote of himself as a creature reborn), Mallarmé, and Yeats all drew upon the myth. E. Nesbit, in her children's novel *The Phoenix and the Carpet*, makes her bird a creature of great vanity and considerable eccentricity. In more recent times, Harry Potter encounters the phoenix Fawkes, owned by his headmaster Albus Dumbledore, while the mutant super-heroine Jean Grey of the *X-Men* comic book acquires the powerful Phoenix Force. Most notably, Cyrano de Bergerac's unfinished work *Histoire des estats et empires de la lune et du soleil* (c. 1661) recounts a journey whose narrator reaches the Empire of the Sun, where he meets a phoenix. The tale underlines Cyrano's belief that there is a burning soul within human beings that is in touch with the Sun, the great soul of the world.

D. H. Lawrence (at right, 1885–1930) and Aldous Huxley (1894–1963)—
in their different ways both devotees of the Sun

Such a theory would have appealed strongly to D. H. Lawrence (1885–1930), whose poem "Phoenix" explores just that idea. Lawrence brings the myth into his novels, and at one point he even adopted the bird as his personal symbol. He could be ridiculous and even repulsive (Bertrand Russell thought he heralded the Nazis, in his anti-Semitism and phallus-centered philosophy), but he did celebrate the Sun to memorable effect. His preoccupation began early (Nottingham, his birthplace, was a grimy coal-mining city) and endured throughout his life.

The final page of his greatest novel, *Sons and Lovers* (1913), describes the night when Paul Morel comes to terms with having renounced his mistress and buried his mother:

> Everywhere the vastness and terror of the immense night which is roused and stirred for a brief while by the day, but which returns, and will remain at last eternal. . . . Night, in which everything was lost, went reaching out, beyond stars and sun. Stars and sun, a few bright grains, went spinning round for terror and holding each other in embrace, there in a darkness that outpassed them all and left them tiny and daunted. So much, and himself, infinitesimal, at the core of nothingness, and yet not nothing. . . . But no, he would not give in. Turning sharply, he walked towards the city's gold phosphorescence. He would not take that direction, to follow her. He walked towards the faintly humming, glowing town, quickly.[22]

Here, as always for Lawrence, the Sun is regenerative, its gold echoed in the phosphorescence of the city calling Morel back to life. Nowhere do we see this more clearly than in the short story "Sun," a paean to its life-giving powers, which Harry Crosby's Black Sun Press published in 1928. It tells of a refined New York lady dispatched to the Sicilian coast to recuperate from postpartum depression. "You know, Juliet," her mother advises, "the doctor told you to lie in the sun, without your clothes. Why don't you?" So she does, and soon the Sun is "gradually penetrating her to know her, in the cosmic carnal sense of the word,"[23] and is not to be denied. A local peasant, about her age, suddenly appears at the ravine close to her rented garden. His eyes meet hers . . .

> Now the strange challenge of his eyes had held her, blue and overwhelming like the blue sun's heart. And she had seen the fierce stirring of the phallus under his thin trousers: for her. And with his red face, and with his broad body, he was like the sun to her, the sun in its broad heat.

Juliet is loath to make the first approach, and her admirer waits endlessly on her initiative with "the dogged passivity of the earth." Into her sunny retreat arrives, unexpectedly, her husband, the "utterly sunless" Maurice, a man who is "the soul of gentle timidity in his human relations," so that Juliet "being so sunned . . . could not see him, his sunlessness was like nonentity." But she realizes that it is her gray spouse that she will have to settle for: "She had not enough courage, she was not free enough." Even though her lust remains unconsummated, this story must rate as the purest expression of the erotic power of sun exposure in all literature.

But Lawrence was not finished. During Easter 1929, he saw in a Tuscan village shop window a toy white rooster bursting out of an egg, this inspiring "The Escaped Cock," retitled against Lawrence's wishes *The Man Who Died* (1929), the story of a resurrected and sexually potent Son of God. The image of the lotus stands in for the female sex organs: "No other flower . . . offers her soft, gold depths . . . to the penetration of the flooding, violet-dark sun that has died and risen." There is something wonderfully daft about the punning here, in which blasphemy is the least of the provocations:

> He crouched to her, and he felt the blaze of his manhood and his power rise up in his loins, magnificent.
> "I am risen!" Magnificent, blazing, indomitable in the depths of his loins, his own sun dawned.

Shortly after this, Lawrence wrote "The Middle Classes," a poem in which he describes the Maurices of this world as "sunless. / They have only two measures: / mankind and money, / They have utterly no reference to the sun." But there is hardly a work by Lawrence that does not touch on these themes. In his

When Norman Mailer was courting Norris Church, his sixth wife, he wrote to her: "And you, stand up lady, are golden as the sun." She later replied in verse: 'You were there and / I was there / in a pocket of sunshine." The New York Times *commented, "These two lived large, sun-drenched lives in almost every regard."*

last book, *Apocalypse*—written in the winter of 1929–30, when he was dying amid general condemnation of his works—he wrote:

> Don't let us imagine we see the sun as the old civilisations saw it. All we see is a scientific little luminary, dwindled to a ball of blazing gas. In the realities before Ezekiel and John, the sun was still a magnificent reality, men drew forth from him strength and splendor, and gave him back homage and thanks.[24]

In this final work, Lawrence urged mankind to reestablish its primal connection to the cosmos—without the mediation of science, or what he saw as its reductiveness. "Start with the sun and the rest will slowly, steadily happen," he ends the book by saying.

An even more fervent apostle of the Sun (and not coincidentally a fan of Lawrence's and fellow sun worshipper) was a millionaire American, a hedonist poet of the 1920s who died in a suicide pact at thirty-one—"a fugitive from a bad Scott Fitzgerald novel," as one obituarist described him.[25] This is a theatrical way to reintroduce Harry Crosby, but everything about Crosby springs from the pages of melodrama. Both Malcolm Cowley, in *Exile's Return,* and Geoffrey Wolff, in *Black Sun,* wrote at length about him, while to Paul Fussell he was "the ultimately mad American" whose life showed "the power of the sun to take over entirely a malleable mind."[26] Crosby's own writings—avant-garde, experimental, surreal, deeply off-putting—were supported by great wealth, wholly inherited and effectively multiplied by his living in Europe when the franc was very weak. The money also allowed him to refurbish a medieval mill (inevitably, "Le Moulin du Soleil") outside Paris into a luxurious country house, travel to exotic destinations whenever he wished, experiment with photography, push his Bugatti to its limits, and even learn to fly solo at a time when the airplane was "a gadget so new . . . that no one had agreed on its spelling."

Descended from old New England money (J. P. Morgan was an uncle), he enlisted in the Field Service Ambulance Corps early in the First World War, serving at Verdun and on the Somme, receiving the Croix de Guerre, and surviving a wound that should have killed him. After an "accelerated" degree at Harvard and a scandalous marriage to a woman six years his senior whom he induced to leave her husband, he settled in France, buying racehorses, sampling opium, drinking to excess, and traveling. And he started to write. In 1927 he launched the Black Sun Press, had a solar face tattooed on his back as an expression of devotion, and elicited his first short story from Lawrence, whom he paid in "sunny" twenty-dollar gold pieces.

Crosby was developing an obsessive interest in imagery centered on the Sun, which he introduced into his own writing with a vengeance. The "black sun" symbolized his attempt to unify the forces of life and death, but as a visual design it also carried sexual significance. "Every doodle of a 'black sun' that Crosby added to his signature," writes one biographer, "also includes an arrow, jutting upward from the 'y' in Crosby's last name and aiming toward the center of the sun's circle: a phallic thrust received by a welcoming erogenous zone."[27]

Crosby comes across as almost totally unpleasant, yet he is not so far off in his views from Lawrence's cri de coeur in *The Apocalypse:*

> What man most passionately wants is his living wholeness and his living unison. . . . The magnificent here and now of life in the flesh is ours, and ours alone, and ours only for a time. We ought to dance with rapture that we should be alive and in the flesh, and part of the living, incarnate cosmos.

By the mid-1920s Crosby had published an elaborate diary of the years 1922–26, *Shadows of the Sun,* as well as two volumes of poetry, the first entitled *Chariot of the Sun.* Other Black Sun projects included work by Hart Crane, James Joyce, T. S. Eliot, and Man Ray. As a spotter of talent, Crosby was not without gifts.*

By now he was interested in the philosophical applications of suicide, and indeed had been openly obsessing about death for at least five years ("Time is a tyranny to be abolished"). One last outré venture was called for. His second book of poetry, *Transit of Venus,* had been inspired by Josephine Rotch (newly Mrs. Albert Smith Bigelow), the woman with whom he would spend the last week of his life. On December 9, 1929, the day before they died, Josephine wrote Crosby a passionate verse letter that ends by insisting "that the Sun is *our* God / and that death is *our* marriage."

DURING AND AFTER the years that Crosby was gallivanting around his adopted country, the literary world was at the height of its love affair with the Sun. A host of writers now reclaimed it, but without the heavily philosophical trappings. The new message was: the Sun gets into *everything.* W. H. Auden's "Take Icarus, for instance" from "September 1, 1939," about the boy who flew too close to its fire,

---

* The Black Sun has other connotations. Russia's great symbolist poet Osip Mandelstam (1891–1938) records: "I have recalled the image of Pushkin's burial [Pushkin fell in a duel in 1837] to arouse in your memory the image of a night Sun."[28] This language recurs throughout Mandelstam's *Tristia* (1922), with such lines as "for the mother in love / The Black Sun will rise." The Sun's role seems to vary, finally linking with the Christian apocalyptic tradition. The Black Sun image also connects with Mandelstam's ambivalence about the Bolshevik Revolution, which he wanted to endorse but which seemed to be bringing in darkness, not light; it not only represents the death of specific people but becomes a symbol of Russia itself—the Sun as a force of despair.

must be one of the most famous lines about the Sun in literature. Yeats, too, in *The Song of Wandering Aengus*: "The silver apples of the moon / The golden apples of the sun"; *Lines Written in Dejection*; and, unforgettably, *The Second Coming*, "A gaze blank and pitiless as the sun," invoked the star; while in the realm of light verse, pride of place goes to John Betjeman's "A Subaltern's Love Song" (1941), with its unforgettable lines "Miss J. Hunter Dunn, Miss J. Hunter Dunn / Furnish'd and burnish'd by Aldershot sun." Of all Betjeman's poems, none so successfully converted suburban trivia into hilarious, helter-skelter eroticism.[29] Maybe, as Josef Brodsky has said, "all the best poets are solar-powered."[30]

The literature of almost every culture contains its share of solar allusions. Arthur Rimbaud (1854–1891) and Paul Verlaine (1844–1896), for instance, shared an ambivalence about the Sun. As the Verlaine scholar Martin Sorrell comments, "Where Rimbaud's quest for a kind of cosmic revelation made him worship the sun's power to absorb and take him over, Verlaine's poems . . . reveal almost an equal fascination with the sun, but as a star whose force is shrouded, guarded, veiled and intangible. In a fascinating way, the two poets' concern with the sun is the two sides of one coin."[31]

For Rimbaud, who produced nearly all his best work before he was twenty, the Sun was either the unmasker of social evils or the bringer of misery. "The hearth of affection and life, it pours burning love on the delighted earth," he wrote; but his last years were spent in the desert heat of East Africa, which seared him into a decidedly less receptive frame of mind. "I saw the low sun, stained with mystic horrors," he confided wretchedly to a friend. "We're in the steam-baths of springtime. Skins are dripping, stomachs turning sour, brains becoming muddled."[32] But in the end the Sun, "which shines like a scoured cauldron,"[33] won through, and on his deathbed he wept to his then companion, Isabelle, that never again would he feel its light. "I shall go under the earth, and you shall walk in the sun!"

Thinly veiled autobiography continued with the novels of André Gide (1869–1951), whose first book, *The Immoralist* (1902), is narrated by Michel, a scholar who spends his early years in northern France. He marries and honeymoons in North Africa—and finds his world changed. "I came to think it a very astonishing thing to be alive, that every day shone for me." As it did for Lawrence, the Sun becomes a liberating power. Leaving his bride behind, Michael embarks on a Mediterranean adventure. Immediately drawn to the "beautiful, brown, sun-burned skins" of Italian boys, he is "shamed to tears" at his own whiteness and, opening his body to the Sun, sets about his transformation: "I exposed my whole body to its flame. I sat down, lay down, turned myself about. . . . Soon a delicious burning enveloped me: my whole being surged up into my skin." Freed from restraint, he becomes the brown immoralist of the novel's title. Oscar Wilde had once mentioned to Gide that the Sun hates or deters thought, and sure enough Michel is soon concluding that "existing is occupation enough."[34]

In *L'Étranger* (1944), Albert Camus has his protagonist Meursault kill a local North African on a blazingly hot beach, apparently for no reason—that being the point, and Unreason the name, that Camus attached to his sun-blasted Algerian experiences. Declaiming this philosophy in *The Rebel* (1953), he wrote: "Historic absolutism, despite its triumphs, has never ceased to come into collision with an irrepressible demand of human nature of which the Mediterranean, where intelligence is intimately related to the blinding light of the sun, guards the secret." When Sartre was asked if Camus, a close friend, was also an existentialist, he replied, "No, that's a grave misconception. . . . I would call his pessimism 'solar,' if you remember how much black there is in the sun." Camus's own comment was, "To correct a natural indifference I was placed halfway between misery and the sun. Misery kept me from believing that all was well under the sun, and the sun taught me that history wasn't everything."

What becomes clear is that by the mid-twentieth century the Sun could serve almost any literary purpose: symbol, metaphor, inspiration, dramatic force, intimate companion, intransigent adversary, comic butt, tragic endgame, source of redemption or of philosophical belief. *Lolita* is so remarkable because Nabokov makes use of the Sun in nearly all these ways.

And still the literary exploitation of the Sun goes on. In 2008 the American novelist Elizabeth Strout published *Olive Kitteridge*, thirteen interconnected stories about a retired teacher in the fictional coastal town of Crosby, Maine: it won a Pulitzer Prize. An extraordinarily perceptive narrative about love and acceptance, it does not use the Sun metaphorically the way that Nabokov does but as a presence—or absence—that affects everyone in Crosby, particularly the prickly, truculent Mrs. Kitteridge. Strout can be wonderfully observant about, say, "the abandonment of a sunbather," or light emerging through the fog, or making a red glass on a bureau glow.[35] A strip of wooden floor shines "like honey" under "a block of winter sun," while later the Sun lingers "across a snow-covered field and made it violet in color." But it is more than that: in a short book, there are seventy solar references, with the Sun acting as a guide to character and to what characters should do. By the novel's end, Olive, newly widowed, suddenly finds a local widower who seems to love her. Will she seize her chance? She is about to turn away when the Sun gives her "a sudden surging greediness for life. . . . She remembered what hope was, and this was it. . . . She pictured the sunny room, the sun-washed wall, the bayberry outside. It baffled her, the world. She did not want to leave it yet."

CHAPTER 28

THE RISING STAR OF POLITICS

If we allow ourselves a glimpse into that shadowy place we call our soul, isn't that why we're here now? The two of us? Looking for a way back? Into the sun?

>—RICHARD NIXON in Peter Morgan's
>play, *Frost/Nixon*[1]

The light was here before. Why should it be dark again? We had the morning. We must act now as though the light has come again.

>—ALEXANDER DUBČEK, welcoming the
>Velvet Revolution in an address to the
>crowds in Wenceslas Square, Prague,
>in 1989

DURING HIS TIME AT YALE, GEORGE W. BUSH RECEIVED ONE D MINUS, in astronomy.[2] On his being elected president, the following tale went around: one day, keen to improve his standing in the polls, he tells his advisers his new idea: "We'll land on the Sun." It is gently pointed out that the Sun is awfully hot. The president considers this information. "No problem," he says. "We'll land at night." Only this joke did not start with Bush; a decade earlier, it was told about Mikhail Gorbachev, and before that Nikita Khrushchev.

Throughout history, the Sun has been appropriated for political or religiopolitical effect, and hardly a generation has failed to provide examples. (In modern times, Sri Lanka has witnessed the rise of a leader of the "Liberation Tigers" who styles himself a solar deity. Sendero Luminoso (the Shining Path), a terrorist offshoot of the Peruvian Communist Party, was set up in the early 1990s by a scrofulous professor of philosophy who had to keep out of the Sun for health reasons. While Louis XIV was the most extravagant in conscripting the Sun, the star lived on after the French Revolution as a potent symbol of a new social order—as seen in the work of a radical aristocrat, Constantin-François de Chasseboeuf, Comte de Volney (1757–1820), a member of the new National Constituent Assembly. In *Les Ruines, ou méditations sur les révolutions des empires,* part novel and part tract, Volney urged his readers to replace the remnants of a decayed civilization with a world order of reason and equality

The most famous door in the world? No. 10 Downing Street, crowned with a rising sun.

whose dominant symbol would be not Christ but the Sun. The work was banned in France, guaranteeing it cult status and numerous foreign editions. When in 1792 an anonymous translation was published in Britain, thirty-five Anglican prelates condemned it. No less a figure than Thomas Jefferson, who had befriended Volney while in France, translated the first twenty chapters for an American edition, only to lose his nerve and ask that the manuscript be burned.

The book (in several volumes) charges that Christianity had purloined solar beliefs, and urges a return to the Sun as the prime symbol of self-determination—ideas that would find their way into a wide range of works. In the nineteenth century, many artists and thinkers, inspired by the Romantic movement and the rise of nationalism, proclaimed that man would achieve his ultimate fulfillment by breaking free from the confines of religion. In Britain, the "Sun as God" theory became a staple of radical pantheism, and the solar nature of all deities was preached up and down the country. "The history of the Sun is the history of Jesus Christ," declared one typical firebrand, Godfrey Higgins (1772–1833)[3]—by which he did not mean that the Sun "equaled" Christ, but rather that Christianity had appropriated solar symbolism. By rejecting that secondary faith, human beings could rediscover what was most of value in themselves.

In the last quarter of that century, no one was more influential (Darwin excepted) than the great philologist-turned-philosopher Friedrich Nietzsche (1844–1900). In the winter of 1876–77, Nietzsche, then a sickly scholar of thirty-two, traveled to Sorrento, south of Naples, where the sensual culture inspired him to put the human body at the center of his thinking. "I shook off nine years of moss," he enthused, after visiting a cave that housed a relief portraying Mithras, the Persian sun god.* During and after his time in Italy, he would write poems praising the Sun as his lover. In *Ecce Homo,* his last book before insanity overtook him, he wrote: "On this perfect day when everything is ripening and not only the grape turns brown, the eye of the Sun just fell upon life: I looked back, I looked forward, and never saw so many good things at once."[4]

By the time he wrote *The Gay Science* (1883), Nietzsche was arguing that the Sun's rise and culmination at its zenith be employed as symbols for what men must become.[5] (This was a little disingenuous, since his migraines often forced him to keep to the shadows.)[6] In *Thus Spake Zarathustra* (1885), he introduced the *Übermensch,* the superman who has attained self-mastery and self-direction, all of which he linked to the Sun.

These books were part of a general movement of artists, musicians, and writers who lauded the Sun, Nietzsche's contemporary, the painter Edvard Munch (1863–1944), making a vital contribution. A member of the radical Naturalist movement in his native Kristiania (as Oslo was then known), he was invited in 1892 to mount an exhibition in Berlin, and he remained there for three years. His friends included Henrik Ibsen, whose portrait he painted on

---

* Mithraism (from the Persian for "light of day") is a recent term: in antiquity, its Roman adherents referred to their religion as "the secret doctrines of the Persians." The cult reached Rome around the first century A.D., finding its apogee in the fourth, being particularly popular among imperial soldiers. Ceremonies centered on a *mithraeum,* an adapted natural cave or an imitation, arranged as an "image of the universe." The faithful were divided into seven ranks, the sixth of which was that of *heliodromus,* or sun courier, and the cult connected strongly with the Sun, whose passage from solstice to solstice was seen as paralleling the soul's journey from preexistence, briefly into the body, and beyond into an afterlife. However, Mithraism quickly declined, as Christianity began to assimilate the other pagan deities. Apollo the sun god was seen as a precursor of Christ, while Apollo the bringer of sudden illumination was transformed into the Holy Spirit.

several occasions, the first time in 1897 as a poster for a production of *John Gabriel Borkman*. The playwright's powerful head dominates the picture, while on the right a lighthouse casts its beams into space. The symbolism is forthright: Ibsen is the light bearer, irradiating the hidden realms of human experience. Munch also completed a series of sketches for *Ghosts*, whose protagonist, the painter Osvald Alving, yearns for life yet lies under the fatal curse of inherited syphilis. Munch, a fellow sufferer, saw his own feelings reflected in Osvald's yearning for light and redemption, and heard in Osvald's cry "Give me the sun!" the voice of his own torn spirit. He described his painting *Spring* as "the craving of a mortally ill person for light and warmth, for life. The sun . . . in *Spring* was the sunshine in the window. It was Osvald's sun."[7] By this time he had come to believe that his art should aspire to "the heaven of the sun's realm."

While in Berlin he also met the leading German interpreters of Nietzsche, and in 1905–6 painted the great man's demonic sister, Elisabeth. He collaborated with another friend and Nietzsche-intoxicate, August Strindberg, on a series of paintings and poems that took over an entire issue of the magazine *Quickborn*, one devoted to glorifying the Sun. Strindberg even bragged about having exchanged books with the philosopher: "Nietzsche has shot such a monstrous load of semen into my spiritual life that I feel I have the belly of a whore. Nietzsche is my spouse!"[8]

In 1909, Munch was commissioned to paint a cycle of eleven murals for the University of Oslo's Festival Hall. His first idea, "the Mountain of Mankind," an image taken directly from *Thus Spake Zarathustra*, was rejected, so he replaced it with the monumental *The Sun*, also inspired by Nietzsche, in which the main painting is flanked by panels showing how the Sun burns at the core of all forms of creativity. In an undated entry in his notebook, he elaborated: on one side of the hall, the rays become primitive invisible forces, on the other, intellectual impulses:

> The first pair [of panels] are overflowing with light—It travels into the bodies—and in and out of the crystal. . . . There is light that travels like X-rays—the other side painting—Chemistry—represents the hidden energies—the workplace of fire and warmth—On the other side the Sun sends its rays to even greater distances.[9]

It took seven years to finish, the completed panels showing a tousle-headed creator-god sitting before the Sun to which he has given life. Munch's most recent biographer describes how "the enormous burning globe is so sun-like that your eyes instinctively flinch away from the central white circle, just as you avoid directly looking at the original."[10] On seeing the painting, Richard Strauss, author of his own homage to Nietzsche—the tone poem *Also Sprach Zarathustra*—exclaimed that it corresponded exactly with what he was trying to achieve in music.

In 1893, Munch completed his masterpiece, *The Scream,* later recording how inspiration had struck him after a sunset walk high up a hill to the east of Oslo. The city's main slaughterhouse stood there, as did its madhouse, and the shrieks of the animals being butchered combined with the howling of the insane into a cacophony terrible to hear. Munch walked on, across the bridge leading into the city. At that moment:

> The sun was setting. I felt a breath of melancholy. Suddenly the sky turned blood-red. I stopped and leaned against the railing . . . stood there, trembling with fear. And I sensed a great, infinite scream pass through nature.[11]

It is impossible to know how much of an influence Munch's paintings or Nietzsche's books had on the general bubbling up of solar-based philosophies of the time, but it had to have been considerable. Music also had its part to play, not just Wagner's operas or Strauss's tone poem but also the erotic compositions of Alexander Scriabin, whose fourth and most famous symphony, composed in 1905, was entitled "Orgiastic [or Orgasmic] Poem"—yet from the outset he saw it as a work about lust not just for the flesh but for the Sun, telling a friend, "When you listen to it, look straight into the Sun's eye."

Like Munch and Strindberg, Scriabin was drawn to Nietzsche, and he took the hero of his third sonata directly from the Superman. His fifth and last symphony, *Prometheus: The Poem of Fire,* written between 1909 and 1910, also portrays a superman hero. Scriabin instructed that the finale should open in a completely dark auditorium, gradually to be pervaded by light as Prometheus' quest ends with the Sun.

Nietzsche, Munch, Strindberg, and Scriabin were part of a movement of artists and thinkers all proclaiming a similar message: the immanence of the Superman, the emblematic importance of the Sun, and man's will to power. Throw in the mysticism of George Ivanovich Gurdjieff (c. 1872–1949), a distorted interpretation of the already dubious doctrine of the "survival of the fittest," Wagner's notions about the outer reaches of art, and the philosophical tracts of Schopenhauer, and where was it all leading? Inevitably to the Nazis. But one must add some other ingredients to the stew first.

—

THE SECOND HALF of the nineteenth century saw the rise of spiritualism and occultism. Astrologers wrote of the Age of Aquarius, the notion that a new epoch would be inaugurated when the Sun rose in Aquarius on March 21 (the year was left vague) to herald Christ's second coming. Then there was *The Secret Doctrine* (1888) by the larger-than-life Madame Blavatsky, founder of the Theosophical Society, which made a deep impression in Germany and elsewhere in Europe. Her admirers included Mohandas Gandhi, Alfred Kinsey, Rudolf Steiner, Aleister Crowley, James Joyce, and William Butler Yeats (who described her as "the most human person alive").*

"In the shoreless oceans of space," she wrote, "radiates the central, spiritual and *invisible* sun." Our inner beings emanate from the "eternal central sun," to be reabsorbed by it at the end of time.[13] The Sun was the lens through which the spiritual light of the sun unseeable (i.e., God) reaches our senses. Her theories were elaborated by the British writer Gerald Massey (1828–1907), who attacked Christianity for turning the Sun from a symbol of life into one of death. Another devotee, Alan Leo (1860–1917), moved the Sun to the center of astrological interpretation, almost single-handedly reviving that doubtful science after its sharp decline in the seventeenth century.

The final ingredient in the cauldron is the *Volk* movement. Even before the advent of Nietzsche, Wagner, Munch, et al., the northern and central German states, which were finally unified in 1871, were aflame with pride in the ancient German archetype, as described by the Roman historian Tacitus in his great pioneering work of ethnography, *Germania*:

> Personally I associate myself with the opinions of those who hold that
> in the peoples of Germany there has been given to the world a race un-

---

* The Ukrainian aristocrat, adventuress, supposed spy, and undoubted occultist Helena Petrovna Hahn, as she was born (1831–1891), was married off at seventeen to a provincial vice governor of forty. Stealing a horse, she rode over mountain ranges back to her grandfather in Tbilisi. Within weeks she had married the skipper of an English tramp ship taking on cargo in Odessa, and embarked on a decade of travel. She smoked up to two hundred cigarettes a day, ate like a horse, and swore like a trooper. While in Cairo in 1871, she formed a society for the occult. Her next port of call was New York, where she was reputed to have demonstrated physical and mental powers that included levitation, clairvoyance, clairaudience, out-of-body projection, telepathy, and materialization (producing physical objects out of nothing). In 1875 she founded the Theosophical Society, a specimen of New Age thinking that bundled all religions into one, plagiarizing from each. (She had also read Volney.) She moved the society first to India, then to London, where she presided over a range of disciples. As a biographer wrote, the world debated furiously whether she was "a genius, a consummate fraud, or simply a lunatic. . . . An excellent case could have been made for any of the three."[12]

mixed by intermarriage with other races, a peculiar people and pure, like no-one but themselves, whence it comes that their physique, so far as can be said with their vast numbers, is identical: fierce blue eyes, red hair, tall frames, powerful.[14]

The self-styled *Volk* gloried in Tacitus' descriptions of their ancient woodland forebears who dwelt among babbling brooks and picturesque mountains, and from the 1870s on, the notion of the *Volk* gained currency. It was a wide church. As Simon Schama notes, it embraced the Wandervogel youth movement as well as the Ramblers, "who communed, Siegfried-style, around bonfires on forested hills," and also many who would later ardently support the Third Reich.[15] But at the outset this was all still fairly innocent.

Völkisch enthusiasms included nudism (as we have seen, a form of sun worship), as extolled in the aptly named Heinrich Pudor's *Naked Mankind: A Leap into the Future* (1893), which also happened to be brazenly anti-Semitic. Nietzsche was particularly close to the *Volk* tradition, which he treated with surprising respect. It was reciprocated. "Taking their inspirational cue from Nietzsche's proclamation that 'God is dead,'" writes Richard Noll, "[the *Völker*] created their own forms of personal religion."[16] Occultism, the ideas and symbols of ancient theocracies, secret societies, and the mystical apparatus of Rosicrucianism, Cabbalism, and Freemasonry were all woven in.[17] Whatever worked.

The Sun was about to be moved center stage. By the 1890s, many *Völker* believed it was "the sole God of the true Germans"[18] and began resurrecting pagan holidays to replace Christian holy days. Noll notes:

Perhaps the most central neo-pagan element in German völkisch movements was sun worship [which] was extolled as true ancient Teutonic religion, and while it was primarily a literary device and a powerful rhetorical metaphor for the experience of God, actual solar-worship rituals did take place.[19]

After Germany's defeat in World War I, a number of *völkisch* movements were reorganized along propagandizing lines, and members became ever more virulently chauvinistic; *völkisch* thinkers began to contrast Germany, as they saw it a nation of heroes, with Britain, "an island of tradesmen" that epitomized the degeneracy of modern materialism. Hitler, who allied himself with the movement—"the basic ideas of the Nationalist-Socialist movement are populist [*völkisch*]," he wrote in *Mein Kampf*—also embraced solar imagery in the form of the swastika,* which, on August 7, 1920, was duly adopted as the official emblem of the Nazi Party.

Once the Nazis were in power, they banned several traditional Christian holidays and substituted others more appropriate to the "New Germany." The summer solstice was one such. Southwest of Hannover stand the Externsteine rocks, also known as "the Rocks of the Sun." Four weathered limestone pillars dating back 70 million years rise a hundred feet, topped by a Neolithic observatory whose round window lines up precisely with dawn at summer solstice. Hitler, cherishing anything that might glorify the Teutonic race's ancient forebears, was enthusiastic about the place, and the SS promoted it as the old Germanics' most sacred site, holding feasts, weddings, and Hitler Youth ceremonies there at equinoxes and solstices.

All these beliefs required intellectual buttressing, and the National Socialists cast about for suitable candidates. In World War I, 150,000 copies of *Thus Spake Zarathustra* had been printed for the kaiser's army; the Nazis now distributed it to the Hitler Youth, and in 1934 they laid a deluxe copy in the Tannenberg Memorial (which honored German soldiers who had fallen at the great victory of 1914). To their delight, they discovered that the glories of "natural" paganism had also been propounded by Goethe himself, who just eleven days before he died had written:

> If I am asked whether it be in my nature to give [Christ] adoring reverence, then I say, "Completely!" I bow before him as the divine revelation

---

* The swastika (from the Sanskrit *su*, meaning "good," and *asti*, meaning "to be," with the suffix *ka*) has been used in cultures around the world (excepting those south of the Sahara and in Australia). The ancient peoples of the Indus Valley believed the Sun to be square, and conceived of the swastika as its symbol, its bent arms representing the solar wheel on its yearly course through the sky. Madame Blavatsky adopted it as part of the insignia of the Theosophical Society and of her personal crest.[20] In Ireland, farmers affixed swastikas, called "Brigid's crosses," on their doors. During the First World War, swastikas were sewn onto the shoulder patches of the U.S. 45th Division and even after the Second could be found on Finnish air force uniforms. All these societies saw the right-handed (clockwise) swastika as an expression of life and good fortune. The left-handed swastika, on the other hand, represents Kali, the terrifying dark goddess of death, and also has black-magical significance. The Nazis adopted both clockwise and counterclockwise versions, although Hitler is said to have favored the latter. Its popularity in Germany had begun in the 1870s, when the archaeologist Heinrich Schliemann found many swastikas on his digs at ancient Troy and Mycenae and wrote about them in two bestselling accounts of his discoveries. Many anti-Semitic and militaristic groups had started using the swastika by the time that it was adopted by the National Socialists. Usually termed *Hakenkreuz*—"hooked cross"—or *Thorshamar,* it became the official emblem of the German Gymnasts League, while solar disk symbols began to appear on posters, armbands, banners, book and magazine covers, and elsewhere throughout Central Europe. (A German Jew buying toiletries in 1933 noted that even the toothpaste tube was so decorated.[21])

of the highest principle of morality. And if you ask me whether it be in my nature to worship the Sun, then I say again, "Completely!" For it, likewise, is a revelation of the most high, and to be sure the most powerful ever granted us mortals to perceive.*

Not only Goethe and Nietzsche were at hand. From 1901, Carl Jung had expressed enthusiasm for *völkisch* ideas, his interest in them aroused by a patient who believed that the Sun had a gigantic phallus that could control the weather. Jung noted similarities between his patient's delusions and pagan sun worship, and offered them as proofs of the existence of what he called a collective unconscious. During late 1909 and 1910, he became fascinated by the literature on Zoroastrianism and Mithraism and their roots in ancient Iranian solar worship and developed a "Sun as God" theme, adding "libido" and "hero" to the existing list of solar correspondences (light, god, father, fire, heat).[22]

In 1912 Jung published *Wandlungen und Symbole der Libido* (*The Psychology of the Unconscious*), which Noll calls "the Völkisch Liturgy . . . a modern mystical contribution to the solar mythology of Müller."[23] For in it, what should the great Swiss analyst do but resurrect the writings of that solar faithful Max Müller, whose many books and articles could now be seen to fit conveniently with Jung's own preoccupations? "Is everything the Dawn? Is everything the Sun?" Müller had pondered. "This question I had asked myself many times before it was addressed to me by others."[24] And here was Jung writing of exactly these concerns—of the hero as Sun, the Sun as God, the God as self-immolating deity. Müller was resurrected, to become the poster professor of the new Germany.†

---

* Johann Wolfgang von Goethe, *Gespräche mit Eckermann*, 5 vols. (Leipzig, 1909), vol. 4, pp. 441–42. While still a young man, Goethe decided that the greatest symbol of God was the star, and set himself to perform a fire sacrifice, whose flames must be kindled directly from the star. To make the ritual more fundamental, he chose not to use wood or coal, selecting instead incense sticks, because, he said, "this gentle burning and evaporating seemed much better to express what is happening in the mind than an open flame." He fastened some sticks to a small table, kindling them with a magnifying glass as soon as the Sun appeared over the roofs of the neighboring houses. In his absorption he failed to notice that the incense sticks had burned down, ruining the fine flowers painted on the family table. See Dr. Otto F. Schrâder, "The Religion of Goethe," Adyar Pamphlets no. 38, Theosophical Publishing House, Adyar Chennai (Madras), February 1914.

† In contrast, Rudolf Steiner, whose schools were shut down by the Nazis, referred to Jews as *der Sonnenwiese*—the Sun Beings—not a racist slur, but part of a personal philosophy, "anthroposophy," that saw Christ as the ultimate Sun Being. Steiner's system, dating from 1913, was based on the premise that the soul can contact the spiritual world. The concepts of reincarnation and karma were central to it, as was the concept that Christ was a cosmic force, a "Sun Being" who became incarnate at the turning point of man's spiritual evolution. Anthroposophy was condemned by the Catholic Church in 1919.

Müller's writings presented the Nazis with much more than an opportunity to build up solar mythology, because of course there were two prongs to his thought: the centrality of the Sun in language and myth and the supreme place occupied by Aryan civilization. He himself may never have been a racist, but he was the perfect advocate for the new dispensation, and his work on Aryan origins was soon being quoted to bolster Nazi claims for a scientific basis to anti-Semitism.

The appropriation of the Sun manifested itself in several ways, but the most malevolent and far-fetched was the fantasy created by Heinrich Himmler,

Charlie Chaplin in The Great Dictator

Reichsführer SS. It was he who spearheaded the quest for the ancestral heritage and pagan culture that had been suppressed by the Christian Church. It was also he who embraced wholeheartedly the doctrine of the Black Sun*—*Schwarze Sonne*—which reaches back to ancient Sumer and Akkadia, and holds that the Sun manifests in two powers, one "white," at the center of our planetary system, the other "black," hidden yet spiritually illuminating, "the strongest and most visible expression of God." This unseen or burnt-out Sun, a source of mystical energy capable of regenerating the Aryan race, has obvious affinities to the sun in Madam Blavatsky's Theosophy (*The Secret Doctrine* was

* The Black Sun has recently been revived as a rallying point for right-wing extremists. In Austria and Germany, a former SS man, Wilhelm Landig (1909–1997), appropriated the Black Sun symbol, swastikas, and other images from the *völkisch* movement to become by the late 1980s "a major political icon of opposition to democracy and liberalism in the West."[25] His movement continues.

first published in German in 1901), to Jung's "inner sun," and to the contrasting concepts of Noumenon and Phenomenon posited by Schopenhauer. The term *Schwarze Sonne* was also to become a central element in the secret initiation of Hitler's senior SS generals. For the rank and file of the SS (about fifty thousand men), the initials SS stood for *Schutzstaffel*, meaning a special staff or military unit; but for initiates, it signified *die Schwarze Sonne*.

When Charlie Chaplin satirized Adolf Hitler in *The Great Dictator*, playing the demented Adenoid Hynkel of Tomania, he held court behind a vast desk, above which was an even larger image of two black crosses in a circle, representing the party of the "Sons and Daughters of the Double Cross." From this central symbol burst forth the rays of the Sun. Released in 1940, the movie was remarkably prescient.

JAPAN DID NOT need the insights of a Jung or a Goethe to see the Sun's political possibilities. From earliest times, its myths had rested on its founding by the sun goddess Amaterasu and the divinity of its *tenno* (or heavenly emperor), believed to be a direct descendant of the Sun: "Even as the Sun is the center of the universe so is the Imperial Dynasty the center of Japanese race life."[26]* Japan's flag, which has for over a thousand years displayed the rising Sun, affirms the nation's sense of its solar ancestry, as does Shinto, the state religion. With the outbreak of the Second World War, the solar image became the nation's most visible symbol: soldiers departed for battle wearing the national flag tied diagonally across their chests and headbands emblazoned with the rising Sun. Kamikaze pilots were similarly inspired by the god-emperor's bond with the Sun.

The link between sun goddess, emperor, and Japan's predatory ambitions of the 1930s was eloquently described by the famous American reporter John Gunther, who wrote just after the war, "The legend of [the emperor's] lineal descent from the Sun Goddess and consequent 'divinity' . . . made the Emperor

---

* The words and ideographs for the Sun in Japanese represent many attitudes. The Chinese ideograph is *hi*, which in Japanese denotes both "sun" and "day." Etymology for the abstract contemporary ideograph shows it derives from a literal picture of the Sun. Its pronunciation today is the same as for the word for "fire," suggesting that before they used Chinese ideographs the Japanese regarded the Sun and fire as being kindred aspects of nature. The basic ideograph, *hi*, with its close connection to natural forces, appears more frequently than other ideographs for "sun" in Japan's oldest written record, the *Kojiki* (A.D. 712). The word for Japan itself is "*Ni-hon*," a compound of a shortened form of *nichi*, which is an alternative reading of *hi*, and *hon*, meaning "source," so that the given name for Japan can be translated as "the source of the Sun." The Chinese returned Japanese enthusiasm for their culture by mispronouncing the name of their smaller neighbor as "Zipango," which eventually evolved into "Japan."

sacrosanct and united the nation in a blindly intense feeling not merely of loyalty but of actual kinship to the imperial house." Following Japan's surrender
in 1945, the commander of the Allied occupying forces, General Douglas
MacArthur, set on demystifying the dynasty, told the emperor Hirohito that he
must declare to his people that he was not a god. This was no idle request. As
Gunther commented:

> Think back. Before the war a traffic cop gave the wrong signal in an im-
> perial procession; he killed himself in shame. The court tailor could not
> measure the Emperor's clothes except at a distance, because touching
> the imperial person was forbidden. When the Emperor traveled, all
> blinds along the entire route had to be drawn, because of the theory
> that direct view of the Son of Heaven might cause blindness.[27]

Knowing that there were many in the West who wanted to see him tried as
a war criminal, Hirohito apparently submitted to MacArthur's demands. In a
historic "Rescript to Promote the National Destiny," delivered on New Year's
Day 1946, this longest-living monarch in Japan's history (1901–1989) repudiated his divine status.

Or did he? This statement, the first made by the emperor to his subjects
since he commanded them to surrender the previous August, has never been
easily available in Japanese. Originally drafted by the American occupation authorities, it was ingeniously edited by Hirohito's courtiers to appease their conquerors rather than to convey meaning to the emperor's subjects; and while in
the West Hirohito's declaration was hailed as a momentous concession (a *New
York Times* editorial said the announcement dealt the "jungle religion" of
Shinto "a blow from which it can never recover"), even today many Japanese
are unaware that Hirohito ever set aside his divinity. The statement, known as
the emperor's "declaration of humanity" (*Ningen-sengen*), has never been discussed or analyzed in Japan's national media. MacArthur wanted the renunciation to emphasize the break with the prewar system; for Hirohito and his
advisers, the object was to minimize the document's impact. The crucial three
sentences run:

> The ties between Us and Our people have always been formed by mu-
> tual trust and affection. They do not depend on mere legends and
> myths. Nor are they predicated on the false concept that the Emperor is
> divine and that the Japanese people are superior to other races and des-
> tined to rule the world.[28]

These words were printed on one occasion only, in the 1946 New Year's Day newspapers; unlike the emperor's earlier speech (of August 14, 1945) admitting defeat, they were never broadcast. Moreover, the whole statement is deliberately vague and equivocal, the subject of each of the clauses being not Hirohito himself but "the ties between Us and Our people." The premise that the emperor is renouncing his divinity rests on a single noun qualifier: "false." Further to confuse the issue, the language was the archaic formal Japanese employed by the imperial household. The phrase "concept that the Emperor is divine" is embodied in the single obscure word *akitsumikami*, rendered by three characters, or ideographs, that even many educated Japanese would not have recognized: when a near-final version of the script was submitted to the Japanese cabinet on December 30, a phonetic reading was written alongside the word so that ministers could grasp the reference. No wonder Hirohito's advisers likened making sense of the directive's import to "cutting smoke with scissors."[29] In this contest of words it was the imperial myth that had secretly triumphed. Hirohito was allowed to retain the three sacred items of sovereignty: the mirror, the curved jewel, and the sword, which together form the proofs of divine descent. Any effective renunciation would have seen these handed over to the occupying powers, or at least to a recognized museum. The divine wind had blown again. That June, Hirohito was formally exonerated of charges of aggression and war crimes.*

In recent years, traditionalists have set about taking advantage of Hirohito's sleight of hand. His son, the current emperor, Akihito (b. 1933), is reported to have resumed the rituals of homage to the rising Sun in the privacy of his palace grounds. Across the country, conservative politicians have toiled to bring patriotism back into the schools. The nationalist movement has been enthusiastically promoting the imperial mystique while offering a version of its country's past that shows no remorse for World War II. The imperial family, said a former minister at a rally organized by Nippon Kaigi, one of the country's largest nationalist groups, "is the precious treasure of the Japanese race, as well as a world treasure."[30]

---

* Hirohito's struggle was the subject of a compelling 2005 film, *The Sun*, produced by the Russian director Aleksandr Sokurov as the third in a series about the downfall of powerful leaders. In it, the forty-four-year-old emperor wanders around under virtual palace arrest, turning over in his mind how he will respond to MacArthur's ultimatum. "I am no longer a god," he openly tells his manservants at one point. "I have renounced my divine nature"—then later he says, with conscious irony, "A deity in this imperfect world can speak only in Japanese." The sound engineer ordered to work on the radio broadcast commits hara-kiri, and Hirohito, a bathetic Chaplinesque figure in top hat and morning coat, ends by declaring, "The Sun will go before the people completely in shadow." Since there was no actual broadcast, one wonders what other liberties have been taken; not surprisingly, the film, for all the denunciations that greeted its Japanese opening, played to standing-room-only audiences.

Although historians trace the start of Japan's imperial system to the fourth or fifth century, according to myth, the first emperor, Jimmu, the son of Amaterasu, began his reign 2,665 years ago. Politicians are now asserting this myth as fact, all the while promoting the corollary myth that the Japanese are an exceptional race and pressing for a revisionist account of the country's wartime past. "Japanese patriots," wrote one observer in the fall of 2006, after the country had voted in a nationalist prime minister, "are sensing their sun is rising after decades of shame."*

OVER THE LAST CENTURY, the rulers of China have also made full use of the symbolic authority of the Sun. Mao Zedong (1893–1976) came to full power only in 1949, but as early as 1936, following his reign of terror in Yenan (the

Mao's Little Red Book. "The Communist Party is like the Sun," ran the anthem of the Cultural Revolution. "Wherever it shines, it is bright." But Mao was a master of solar imagery long before he came to power.

红太阳照亮了
大寨前进的道路

陈永贵

* Michael Sheridan, "Japan Flexes Its Military Muscles," London *Sunday Times*, September 17, 2006, p. 31. In an often harshly male-dominated society, female voices have occasionally brought the Sun directly into politics. The writer and activist Hiratsuka Raicho (1886–1971)—a founder of the fledgling women's movement in Japan—sought to reclaim the Sun as a feminist symbol. Her declaration that "in the earliest days of our history, the Sun was a woman" became nationally famous, and in her poem "The Hidden Sun" she developed this theme, insisting that "Originally woman was an authentic person; but now woman is the Moon." Japanese women's struggle for liberation was "the struggle to be the sun again."

destination of the eight-thousand-mile Long March), he was planning a personality cult, which included associating himself with solar imagery. In 1948, Liu Shaoqi (1898–1969), his original party vice chairman, referred to Mao's thought being "like the rays of the Eastern Sun."

The chairman encouraged depictions of himself framed within the solar orb, and was consistently represented as "the Sun that never sets." By the 1960s this imagery was showing up nationwide in songs, photographs, woodcuts, and paintings, as well as on such artifacts as teacups, money, and swimming trunks. A typical graphic from the Cultural Revolution displays a red Sun over vast grain fields—Mao as the source of prosperity. His famous Little Red Book (first published in April 1964, with more than 900 million copies printed) shows him in a circle, golden rays spreading outward, both sun figure and god. The de facto anthem of the Cultural Revolution—"The East Is Red"—lends its own special force to the identification of Mao and his party with the Sun:

> The East is red, the Sun is rising.
> China has brought forth a Mao Zedong.
> He amasses fortune for the people,
> Hurrah, he is the people's great savior.
>
> Chairman Mao loves the people,
> He is our guide, to build a new China,
> Hurrah, he leads us forward!
> The Communist Party is like the Sun,
> Wherever it shines, it is bright.

The lyrics were attributed to one Li Youyuan, a farmer from northern Shaanxi, allegedly inspired by the rising Sun. Throughout Mao's decades-long supremacy, they blared over PA systems in every city and village from dawn to dusk. "The East Is Red" became the title of a "song and dance epic" promoting Communism, while a film version was released in 1965. The song's title is also the name of a series of Chinese satellites, the first of which, *Dong Fang Hong 1*, included a transmitter broadcasting the famous lyrics across a somewhat unaware planet.

Because of the anthem's associations with the Cultural Revolution, it was rarely heard after the rise of Deng Xiaoping in the late 1970s, and today it is viewed as an unseemly reminder of the cult of personality. So, too, with visual representations of the Great Leader. In the new, modern China, it may seem bizarre to look back at the promotional films of a "radiating Mao," his image set against a blazing red background with rays shooting outward like so many tracer bullets; but that was the everyday reality for hundreds of millions of Chinese for nearly thirty years.*

Despite its political eclipse, the Sun is in no danger of disappearing from the network of Chinese symbols, for it has long been part of traditional medicine, neatly summed up in the statement "Chinese doctors believed the human body was built up, like all the universe, from *Yin* and *Yang*, etymologically, dark coiling clouds and bright sunshine."[31] While written in the past tense, this describes a tradition that is alive and well in China today.†

NOT ONLY LEADERS of state have sought to align themselves with the Sun. Many a religion has adopted solar imagery, none more effectively through the millennia than the Roman Catholic Church. The year 2000 saw the publication of a book written by Pope John Paul II's most powerful aide, Joseph Cardinal Ratzinger (b. 1927), who in 2005 became Pope Benedict XVI. Ratzinger made his reputation as the foremost theologian at the Second Vatican Council and wrote *The Spirit of the Liturgy* while serving as the prefect of the Congregation for the Doctrine of the Faith, a position second only to the pope's in determining what Catholics must believe.

In the course of his propagation of the liturgy, which he considers to have been handed down by God, Ratzinger discusses the range of solar imagery in Christianity—its source, its meaning, its importance, and, most significantly, its identification with Christ, which in his view is free of any pagan influences. This allows him to argue that since Christ is "represented by the Sun," and "we find Christ in the symbol of the rising Sun," we should change the direction in which we pray, even the orientation of our church buildings, so that "the sym-

---

* Of course, solar imagery had been employed in Chinese politics long before Mao. The activist and artist Lu Hao-dong (1868–1896), one of the leaders of the Kuomintang (the revived China Society, which became the National People's Party), designed two flags, the first for the Kuomintang, showing a twelve-ray white Sun against a blue sky. Later, when the United League was established, a red background was added and the design thereafter adopted as the flag of the Republic of China. Hao-dong became "the first martyr of the people's revolution" when he was arrested by the Manchu government, tortured, and put to death.

† In the summer of 2005, my daughter, Mary, spent several months before college working in two hospitals in Shanghai, one of which had returned to using the old physic. She showed me a book that had been recommended to her, *Basic Theories of Traditional Chinese Medicine*, which early on explains: "The original meaning of the concept of yin and yang is very simple. Initially, the terms referred to whether or not a place faced sunlight. The place that faced sunlight or that was filled with Sun was called yang, while the place that faced away from the Sun was called yin."[32] Methods based on this doctrine are used side by side with modern, Westernized medicine.

bolism of the Cross merges with that of the East." St. Peter's Basilica itself, "because of topographical circumstances," faces west, but theologically this is a misfortune. He understands that current theologians think "that turning to the east, towards the rising Sun, is something that nowadays we just cannot bring into the liturgy. Is that really the case? Are we not interested in the cosmos anymore?" He argues that we in the West must be, and should change our ways to face east to pray.[33]

Joseph Ratzinger, the now beleaguered Pope Benedict XVI

In the following chapter, "Sacred Time," Ratzinger considers how "the first day of the week was regarded as the day of the Sun," and by the same cosmic symbolism "the Sun proclaims Christ." He goes on: "We have seen how deeply Christianity is marked by the symbolism of the Sun," which becomes "the messenger of Christ." With each new year, the feasts of Christmas and Epiphany mark "the dawning of the new light, the true Sun, of history." He adds, with breathtaking disingenuousness, "The complicated and somewhat disputed details of the development of the two feasts need not detain us in this little book. . . . The claim used to be made that December 25 developed in opposition to the Mithras myth, or as a Christian response to the cult of the unconquered Sun promoted by Roman emperors in the third century in their efforts to establish a new imperial religion. However, these old theories can no longer be sustained." And for him that assertion is argument enough.

The same confident rejection of any connection between a pagan winter solstice and the celebration of Christmas also applies to the observation of the

Resurrection at Easter, the time of the spring equinox: "The close interweaving of incarnation and resurrection can be seen precisely in the relation, both proper and common, that each has to the rhythm of the Sun and its symbolism." By chapter's end the Sun has become "the image of Christ." Our old friend M. de Volney would have laughed (or, more likely, been enraged) had he read such sophistry: two hundred years later, Mother Church is still appropriating the ancient pagan ways while insisting on the contrary. But the old religions did not own the Sun any more than those that followed them. It can be harnessed by anyone, and ever will be. Politics, of religion or state, is no more than the art of government: and what better to call on for help, symbolic or otherwise, than that great star that governs us? All leaders, in their own fashion, find ways to land on the Sun.

PART SIX

The SUN and the FUTURE

This whimsical sketch graced the title page of a seventeenth-century manuscript, De thermis (Of Temperatures). The Sun looks suitably reflective.

CHAPTER 29

OVER THE HORIZON

Some think that solar work is pretty well played out. In reality, it is only beginning.

—GEORGE ELLERY HALE, 1893[1]

All that you did on his behalf
Has caused the lovely sun to laugh.

—ALBERT EINSTEIN, 1929[2]

IN THE AUTUMN OF 1930, LORD ROTHSCHILD GAVE WHAT WAS, EVEN for him, an unusually grand dinner at the Savoy Hotel to benefit Eastern European Jewish refugees flocking to Britain to escape their increasingly unsafe homelands. The master of ceremonies, George Bernard Shaw, introduced the guest of honor, winding up: "Ptolemy made a universe which lasted fourteen hundred years. Newton also made a universe, which lasted three hundred years. Einstein has made a universe, which I suppose you want me to say will never stop, but I don't know how long it will last." The guest in question laughed out loud, and when his turn to speak came chided Shaw for invoking his "mythical namesake, who makes life so difficult for me."[3]

The universe Einstein unveiled has endured more than a hundred years so far. It began in 1905, when in less than twelve months he wrote four papers that transformed the scientific landscape. The first, published three days after his twenty-sixth birthday, helped lay the foundations of quantum physics. The second redirected the course of atomic theory and statistical mechanics. The other two introduced what came to be known as the special theory of relativity, to be followed by the general theory of 1915. These papers revolutionized our understanding of the nature of gravity and of the passage of light, and our concepts of space and time. How much longer Einstein's universe will prevail is indeed unclear: *The New York Times* in 2005 quoted a leading physicist as forecasting, "The smart money says that something is going to happen: general relativity won't last another two hundred years."[4] The beginning of the end may have come with a concept that surfaced even before Einstein's papers of 1905 and 1915: quantum theory, which, among so much else, ushered in a totally new approach to understanding solar power.

The new way of thinking had taken form five years before Einstein's annus mirabilis, when the great German physicist Max Planck came up with the idea that all energy is emitted in discrete units, which he called "quanta" (from the Latin, "how much"). It was as if, in the words of George Gamow, "one could drink a pint of beer or no beer at all, but not any amount in between." Other physicists, among them Niels Bohr, Erwin Schrödinger, Wolfgang Pauli, Max Born, and Werner Heisenberg, helped extend this disconcerting insight, which redefined the nature of power.

Just as Einstein's general theory applied to interactions among the largest objects in the universe, quantum physics described what goes on at the atomic and subatomic levels, where events occur that are completely at odds with our everyday experience. For example, a quantum particle (a photon, say, a massless and chargeless discrete particle) is so insubstantial that it will move from one place to another without going through any intervening space: it ceases to exist at one point while simultaneously appearing at another—a "quantum leap." This seems to confound common sense: surely to travel from A to a non-adjacent C involves passing *through* something. As Niels Bohr famously said to one of his Copenhagen students, who complained that quantum mechanics made him giddy, "If anybody says he can think about quantum problems *without* getting giddy, that only shows he has not understood the first thing about them."[5] As Brian Cathcart explains in his book on atomic physics:

> Planck's technique had a significant drawback: it worked only if you set aside an essential component of the classical laws, the principle of continuity. Continuity has an analogy in the kitchen: in baking, milk is "continuous" in the sense that any given amount may be measured out and added to the mixture, while eggs tend to be "discontinuous"—it is a perverse cookbook that asks you to separate one-quarter of an egg. . . . Here was the first inkling of a sensational possibility: that the laws of physics that applied in the observable world might not, after all, be valid at the atomic level.[6]

But—quantum theory worked: it explained and predicted phenomena for which there was no other explanation. Classical physics is deterministic: if A, then B—the bullet fired at the window shatters the glass. On the quantum scale, this is only *usually* true. Quantum physics shows that from time to time quanta, behaving not like particles but like waves, break through the force field "as if a cannonball were to fly untouched through a fortress wall,"[7] a phenomenon that was to become known as "quantum tunneling." This was demonstrated by a key experiment conducted in 1909, in which certain radioactive elements were discharged at a wafer-thin sheet of gold foil and a small number bounced back, a phenomenon that Ernest Rutherford was to describe as "almost as incredible as if you fired a fifteen-inch shell at a piece of tissue paper and it came back to hit you."[8] Over the next forty years scientists went from a minimal knowledge of the protons and neutrons that make up the nuclei of atoms to understanding the primary thermonuclear fusion process that powers the Sun. This is a process that only quantum physics could explain, because it involves the fusing of nuclei, which classical physics says is impossible because all nuclei bear a positive charge, and positively charged particles repel one another. As one leading physicist put it, "According to classical physics, two particles with the same sign of electrical charge will repel each other, as if they were repulsed by a mutual recognition of 'bad breath.'"[9]

Classical theory denies that two protons in a star can go fast enough to break the walls of their electromagnetic force fields and merge into a single nucleus. But quantum tunneling permits protons to traverse the barrier created by electromagnetic repulsion. Under intense heat and density—the result of the gravitational energy generated by the solar mass contracting—protons overcome the usual forces of repulsion and collide, to form stable helium nuclei; this mass is released as energy. So here is the Sun, blazing away, proving that classical physics has it wrong.

After quantum theory was generally accepted, so many different types of particles were discovered that physicists had to consult a booklet, the *Particle Properties Data Handbook*, to keep track. (The outside world eventually reacted by producing T-shirts with the legend *Protons have mass? I didn't even know they were Catholic.*) A table was drawn up: sixteen basic particles, twelve of matter (called fermions) and four quanta (called bosons), which mediated the interactions among particles. Fermions are the basic building blocks of matter, divided into either leptons (from the Greek for "thin") or quarks.*

Quarks are never found in isolation, but only in groups of three (when together they are called baryons) or in doublets (as mesons); bound states of quarks are collectively known as hadrons. More complex things—protons,

---

* A term borrowed from *Finnegans Wake* by their hypothesizer, the Caltech physicist Murray Gell-Mann, from a line as spoken in a dream by a drunken seagull. Instead of asking for "three quarts for Mister Mark," the inebriated bird demands "three quarks for Muster Mark"; since the theory originally posited three quarks, the name made some sense. Thus were quarks immortalized, but their namesakes' lives are incredibly brief, their most enduring estimated to last no more than 1×10^{-24} seconds.

neutrons, atoms, molecules, buildings, people—tend to be made out of fermions. "If I could remember the names of all these particles I would have been a botanist," complained Enrico Fermi, for all that this largest group of particles bears his name. Fermions also comprise neutrinos, which interact so weakly with other particles as to make them exceptionally difficult to detect, and indeed trillions stream through our bodies every second. As John Bahcall, most renowned of neutrino researchers, observed, "A solar neutrino passing through the entire earth has less than one chance in a thousand billion of being stopped by terrestrial matter. . . . About a hundred billion solar neutrinos pass through your thumbnail every second and you don't notice them."[10] No wonder John Updike felt compelled to write

> *Neutrinos, they are very small,*
> *And do not interact at all.*[11]

The last particles identified have been the W and Z bosons in 1983, the top quark in 1995, and the tau neutrino in 2000, so this subatomic world is still very much in the process of being discovered.*

From Galileo's day onward, we have been learning about the Sun's shape, size, rotation, and spots; its mass and density; and how it moves. We went on to assess its age and record its infrared and ultraviolet radiation, dark-line spectrum, radioactive output, activity cycle, prominences and chromosphere, corona, chemical composition, bright-line spectrum, radio transmissions, X-ray emissions, neutrino output, coronal holes, and whole-body oscillations. Despite such a roster, it is humbling that many of the Sun's basic processes are only now beginning to be understood: how its magnetic field is generated, for instance, or its atmosphere heated; why, since the Sun is not itself on fire, flames burst from it. And some are even further beyond our grasp. What creates the corona, and how is it heated to such extremely high temperatures? What switches the solar magnetic poles? Why does the atmosphere's tempera-

---

* Indeed, in this bizarre new world, it has been shown that at the micro-microscopic level of photons (the minute entities that make up a beam of light), our usual expectations of how our globe works once again fall apart. Thus it was no surprise when in the 1970s there was hypothesized a medium called "slow glass," through which light might take months or even years to pass. If light were to take a year in its passage, then everything that had been seen in front of the glass would become visible only a year later on the far side. "Slow glass," which would be created from a complex form of plasma, is not yet a reality, but even though it would delay light by a factor of a quintillion or more, it could be, and soon: in 1999 the Rowland Institute for Science at Harvard reduced light speed to below a meter per second. As the historian of science Brian Clegg says, "If the speed of light can be controlled, then so can reality itself. Extraordinary scientific work is now under way, work that will make technical miracles such as slow glass possible . . . and that will, in a few short years, make light the most exciting area of study in all science."[12]

ture suddenly soar in the thin transition region between the lower corona and the chromosphere? Where is the solar wind produced, and how far out does it spread? What kind of shielding do we get from solar magnetic clusters? Even, why do sunspots exist? We have a long way to go.

Three levels of research seek to answer these questions—literally "levels," as they take place on the ground, in the skies above, and in the earth beneath; together they form the tripartite structure of this chapter.

SCIENTISTS GENERALLY AGREE that the Sun is fired by thermonuclear reactions that fuse light elements into heavier ones, converting mass into energy. To demonstrate this truth, however, has been difficult, since the nuclear furnace lies deep in the Sun's interior, and conventional instruments can record only those particles emitted by the outermost layers.[13] Ninety-seven percent of this energy is made up of photons, but about 3 percent sweeps outward in the form of neutrinos—200 trillion trillion trillion of them every second; and analyzing them provides our best means of learning how this fusion works.

However, it is intensely difficult to register neutrinos, which have minimal mass (they were long thought to have none), travel at the speed of light, and

A neutrino detector built half a mile below ground in a disused Japanese mine. The huge stainless-steel vessel, forty yards in height and width, has been filled with fifty thousand tons of highly purified water, and its walls lined with thirteen thousand light sensors, to pick up the flash generated by electrons recoiling from neutrino collisions in the water—a necessarily complex way of discovering how the Sun works.

change their nature as they go. They pass through ordinary matter (not just human bodies but the great globe itself) as if it were entirely transparent. Not until the late 1960s was the ideal apparatus developed, when Raymond Davis and John Bahcall made a detector out of a disused mine outside the town of Lead, South Dakota: a tank located 4,850 feet underground and filled with one hundred thousand gallons of tetrachloroethylene—a simple cleaning fluid, yet highly sensitive to registering neutrinos, each one of which, should it react with the chlorine, may produce a radioactive isotope of argon. Such exposure is useful only underground, since the detector must be shielded from the endless shower of other subnuclear particles, many from beyond the solar system.

About twenty times a day, a neutrino crashes into a neutron in this dark sphere, creating a faint flash. The network of some 9,600 photomultipliers that coat the vessel detects the flash, which is then analyzed for data about the neutrino that set it off. It was here that Bahcall made his discovery, dubbed "the Neutrino Problem." Simply, not nearly enough neutrinos were reaching Earth—between a third and a half of the predicted number. It was an astounding shortfall: Where were the rest of them? Had the Sun's scientific acolytes miscalculated? Bahcall wrote: "The most imaginative idea [is that] neutrinos have a kind of double identity. . . . It will not be easy to test this unusual hypothesis, but it cannot be lightly dismissed."[14] He went on to say that the most imaginative solution suggested was Stephen Hawking's: that the Sun's core might contain a small black hole.

In fact, the question preoccupied solar physicists, cosmologists, and astrophysicists for the next thirty years. (To complicate the problem, neutrinos had only one disposition: that is, they spun as they traveled, but only backward along the direction of their flight, like left-handed corkscrews. As the longtime editor of *Nature* John Maddox asked, "What is it about our world that caters only to left-handed neutrinos?")[15] A number of answers were put forward, including that solar models were indeed wrong, and that the temperature and pressure in the Sun's interior were substantially different from those posited in current theory; or that the nuclear processes at the Sun's core might temporarily have shut down: since it takes thousands of years for heat energy to move from the core to the outer layers, such a disconnect would not be apparent for thousands of years.*

---

* Modern neutrino detectors grow ever more sophisticated. They register their quarry in real time, and thus, for instance, examine the day-night effect: neutrinos passing through the Earth (detected at night) are slightly different from those that don't (detected during the day). When in 1934 Enrico Fermi submitted to *Nature* a paper on neutrinos, postulating that they escaped the Sun due to certain radioactive decays, it was rejected because "it contained speculations too remote from reality to be of interest to the reader."

An answer was eventually posited in 2001. Neutrinos in the course of their journey to Earth may change from the type expected to occur in the Sun's interior into two other forms undetectable by the old machines. Scientists pondered the three different kinds: electrons—the ones produced in the Sun—muons, and taus. (When the muon was discovered, a physicist famously asked, "Who ordered that?") After extensive statistical analysis, it was found that about 35 percent of the neutrinos reaching Earth were electroneutrinos, the rest being made up of muons and taus. Once all three categories could be accurately tracked, the total number lay well within the range of earlier predictions. Solar physics was vindicated. As one of its practitioners told me, "It used to be believed that we were being dumb because we were doing something wrong. We knew the neutrino particles' reactions, we knew everything, so, if we weren't capable of tracing their journey from the Sun, well, it wasn't a *physics* problem, it was a *solar* physics problem—'These guys can't get these things right!' Well, it turns out it's the other way around: we *were*."[16]

But the solution of 2001 is still a "best theory," and who is to say when or whether it will be confirmed? One further problem is that the materials being examined are *so* minute. As a character in Tom Stoppard's *Hapgood* says, "When things get really small, they get truly crazy, and you don't know how small things can be, you think you know but you don't know. . . . Every atom is a cathedral."[17] Another visitor from the world outside science, Bill Bryson, begins *A Short History of Nearly Everything* by complaining that a proton "is just way too small. . . . A little dib of ink like the dot on this *i* can hold something in the region of 500,000,000,000 of them, rather more than the number of seconds contained in half a million years."* It is daunting to know that the smallest things in nature hold the clue to one of the largest.

ON TO WHAT is done up in the air. The progress of solar astronomy since the Second World War has required overcoming distances so great that the challenge involved is at least as daunting as coming to grips with the subatomic realm. To date, rockets have been our means of getting closest to the Sun. In the 1920s, the prototypes created by Robert Goddard (1882–1945) were derided, *The New York Times* notoriously declaring that Goddard seemed to "lack the knowledge ladled out daily in high schools," and mocking his experiment as "Goddard's Folly." Any such attempt on the heavens was regarded as the

---

* A rare Bryson miscalculation: he should have written 15,000 years, not half a million: $60 \times 60 \times 24 \times 365 \times 15,000$ totals 473,040,000,000. Bill Bryson, *A Short History of Nearly Everything* (New York: Broadway, 2003), p. 9.

stuff of fantasy, best kept for film houses and comic strips: it was impossible to move in a vacuum, as there was nothing to push against. However, on November 15, 1936, a group of students at Caltech, known as "the Suicide Squad," scraped together enough cheap engine parts to make worthwhile a trip to Arroyo Seco, an isolated canyon wash at the foot of the San Gabriel Mountains, where they test-fired a small rocket. As they huddled behind sandbags, the motor burned for about three seconds before an oxygen hose tore loose, bathing the testing area in flame. Over the coming weeks, they tried again, and on their fourth attempt, on January 16, 1937, the motor fired long enough—forty-four seconds—to heat its metal nozzle red. Liftoff.

November 15, 1936: students from Caltech, known as "the Suicide Squad"
from their early mishaps, on the day they first tried to send a rocket into space.
Thanks to their persistence, we are now sending rockets to the Sun.

The adventurers' explosive and noisy experiments in time proved too dangerous for the campus, but by now people who could finance such activities were interested. In 1938 Caltech's aeronautical lab received a surprise visit from the chief of the U.S. Army Air Corps, and within two years a facility was built in the Pasadena foothills. Three test stands and tar-paper shacks marked the first buildings of what would become the Jet Propulsion Laboratory (JPL). The first substantial financing came from the air corps, which wanted small rockets to help lift heavily loaded planes from short runways. After their successful development—and Pearl Harbor—the army called for other models, and JPL began to develop guided missiles. But those were early days, with no vision yet of solar research.

World War II hampered solar physicists in countless ways, but it also prepared for a later flourishing. British and American scientists working on radar discovered that the Sun emitted radio waves (first mistaken for German jamming), while both Axis and Allied solar physicists participated in weather forecasting, some Germans trying to observe the Sun from above the atmosphere with "Retribution Weapon 2"—V-2 rockets. They failed, but their initiative paved the way for much that followed.

In the spring of 1942, Wernher von Braun (1912–1977), technical director for the V-2, seeking to give the program a scientific dimension, asked the physicist Erich Regener to develop a special payload and nose cone.[18] The most sophisticated instrument that Regener considered was an ultraviolet spectrograph, aimed at correlating the amount of atmospheric ozone to altitude. All the while, hundreds of missiles were being readied for their primary, destructive purpose. The first of more than three thousand blasted off in September 1944, and the British War Cabinet discussed, as it never had throughout the Blitz, evacuating London. (The rockets, each carrying about 1,700 pounds of explosive, were particularly feared since, traveling faster than sound, they arrived in advance of audible warning.)

That December, paying no heed to the imminent ruin closing in on the Reich, von Braun scheduled the launch of a research prototype for the following month. By mid-January, the instruments were moved up for launch. It never took place. The last V-2 roared toward England in late March; six weeks later, the Red Army took Berlin. As Germany collapsed, a special U.S. intelligence unit rounded up von Braun's 118 scientists and all available V-2 components—enough for a hundred rockets. The importance of this find was not lost on Western leaders. Leo Goldberg, later NASA's principal investigator, would write to a colleague:

> If anyone asked you what technological development could, at one stroke, make obsolete almost all of the textbooks written in astronomy, I am sure your answer and mine would be the same, namely, the spectroscopy of the sun outside of the earth's atmosphere. . . . [The] V-2 Rocket has attained a height of 60m[iles], and with the host of control mechanisms that have come out of the war it should be possible to point the rocket at the sun.[19]

In 1944, the Jet Propulsion Lab formally became an army facility, operated under contract by Caltech. Between 1945 and 1957, several small devices were indeed sent outside the atmosphere to observe the Sun's extreme-ultraviolet and X-ray spectra (all getting considerably scorched in the process). A V-2 panel was set up, an informal committee that initiated experiments in such subjects as variation in atmospheric pressure, radio propagation, cosmic rays, temperature, and solar radiation. In 1946–47 alone, physicists managed to mount instruments to photograph the Sun's ultraviolet spectrum onto eleven of the twenty-eight V-2s launched; from these beginnings was inaugurated ultraviolet astronomy, and the entire field became known as solar physics.[20]

British and American scientists had meanwhile been drawing on solar data to predict both magnetic storms and the behavior of the ionosphere (that is, that part of the Earth's atmosphere where the Sun affects the transmission of radio waves). The U.S. Navy employed such forecasts to anticipate radio frequencies used in its surveillance of Soviet submarines. Neither activity played a decisive part in the Cold War, but they were significant enough to establish solar physics as a military asset.

By 1953, fourteen of the fifty observatories in the world conducting visual studies of the Sun were equipped with coronagraphs (the special telescopes that, by blocking out certain solar light, enable scientists to see the Sun more clearly); between 1945 and 1951, about 70 percent of all publications in the emerging discipline of radio astronomy were devoted to its solar aspects. In 1955, the U.S. government announced plans to launch a satellite during an "International Geophysical Year," the eighteen months from July 1957 to December 1958, and ninety-five observatories and stations around the world were set to participate. Then a thunderbolt struck.

On October 4, 1957, the Soviet Union launched *Sputnik* (in Russian, "attendant" or "fellow traveler")—a simple sphere weighing 184 pounds and not quite two feet wide that, in the words of *The New York Times*, "changed everything: history, geo-politics, the scientific world."[21] The "beep-beep-beep heard round the world"[22] plunged the United States into a crisis of self-doubt, precipitated a furious missile buildup, and stimulated prodigious amounts of funding for scientific and engineering research, all aimed at maintaining America's security and prestige. After *Sputnik*, politicians and military men set the agenda for man's explorations in space.*

In January 1958, President Eisenhower created the National Aeronautics and Space Administration (NASA), and the JPL was put under its control. That

---

* In 1961, Russia's pioneer cosmonaut, Yuri Alekseyevich Gagarin, the first person ever to see Earth in its entirety, set out to describe the experience. During his 108-minute flight, he had been in orbit for less than an hour and a half, but even that long a view dazzled him: "The Earth has a very characteristic, very beautiful blue halo, which is seen well when you observe the horizon. There is a smooth color transition from tender blue, to blue, to dark blue and purple, and then to the completely black color of the sky." After so much discussion of how the Sun appears to us on Earth, suddenly such visions were reversed.[23]

same month, *Explorer 1* went into orbit, to "the collective sigh of relief from an anxious American public."[24] Circling the Earth every 113 minutes, the craft radioed back data about temperatures, meteorites, and radiation. Even more instruments for studying the Sun were fixed to stratospheric balloons, high-altitude aircraft, and rockets, while solar scientists obtained new telescopes or substantially improved existing ones at ground level.[25] Between 1957 and 1975 the solar physics community roughly doubled; *Solar Physics: A Journal for Solar Research and the Study of Solar Terrestrial Physics* appeared in 1967 and was soon publishing more than two hundred articles a year. From 1959 on, U.S. and Soviet programs sent ever larger and more complex spacecraft on ever longer journeys.*

On March 7, 1962, NASA launched Orbiting Solar Observatory 1, which was powered by solar photovoltaic cells and had a special midsection stabilized to point directly at the Sun. It was an unqualified success; but by early 1964 the space program was torn apart by arguments over funding, by the disastrous malfunction of a satellite at John F. Kennedy Space Center that killed three engineers, then by a third observatory's failure to make orbit when a rocket fired prematurely. At the end of 1965, the Advanced Orbiting Solar Satellite was canceled, and within four years the Nixon administration had terminated the program that had built the Saturn V Moon rocket—in one stroke depriving mankind of "a ubiquitous presence throughout the solar system."[26]

Fortunately, the case for using spacecraft for solar observation was compelling enough that NASA soon recovered, over the next decade launching Solar Observatory 3 as well as embarking on its most ambitious project, a crewed observatory in space, the Skylab Apollo Telescope Mount. An unmanned Skylab was launched in May 1973, with four main solar instruments—an ultraviolet spectrograph and spectroheliograph (cost, $40.9 million), an ultraviolet spectroheliometer ($34.6 million), a white-light coronagraph ($14.7 million), and an X-ray telescope (a comparatively cheap $8.3 million). Expenditures at this level mean that virtually all the material wealth that has gone into astronomy has been invested over the last forty years.

However, over that very decade of 1971–80, national priorities shifted away from science, for two prime reasons. First, economic: the cost of the Vietnam War, OPEC's vast increase in the price of oil, and the funds required for much-needed social and environmental programs raised doubts about the returns on

---

* *Explorer 1* was kept secret at the Jet Propulsion Lab, where it was nicknamed "Project Deal." As its manager, Jack Froehlich, a poker player, explained after *Sputnik*, "When a big pot is won, the winner sits around and cracks bad jokes, and the loser cries, 'Deal!'" Names are not always appropriate; in later years, when America was attempting to land men on the Moon, it called the program "Apollo."

the stream of federal money flowing into extraterrestrial technology. Second, political: scientists were often at the forefront of opposition to America's overseas and military policies, which raised questions about whether money spent on this community was fostering unpatriotic elements. Solar physics was particularly hard hit: an official report noted that by the end of the 1970s it appeared to be an isolated field of research, underfunded and understudied. Astronomers as a whole were "largely unaware of and maybe indifferent" to it.[27] Financial problems and cancellations continued, a situation exacerbated in January 1986 by the loss over Florida of the space shuttle *Challenger* with her entire crew. NASA postponed for four years the launch of *Ulysses*, a spacecraft for observing the solar output from above the Sun's poles, and abandoned altogether the development of the major solar telescope it had been planning. While America was foremost in reducing its programs, Europe was cutting back too: even as early as 1976, Greenwich had discontinued its sunspot photography after 102 years of continuous operation, and four years later the Swiss government closed its center for sunspot statistics. Researchers refused to be bowed, however, and from the 1970s onward came up with some remarkable discoveries. Most notable of all was what came to be known as "solar wind."

This requires going back a bit in time. In 1951, building on the hypothesis that the Sun discharged not just energy but matter, the German astronomer Ludwig Biermann (1907–1986) had suggested that the tails of comets were blown persistently away from the Sun, not, as had been assumed, by the known pressure of light but by an outward flux of particles, which deflected the ions in each comet's tail. Building on Biermann's insight, Eugene N. Parker (b. 1927), a distinguished theorist from Chicago, argued that such an efflux originated in a continuous million-mile-per-hour expansion of the corona, which he dubbed "solar wind" to emphasize its dynamic character. This wind travels inward from the Sun's north pole and outward from its south pole, argued Parker, the Sun's surface brightness being particularly sensitive to vertical currents. The wind's influence on the Earth might be feeble compared with that of solar radiation (it would not ruffle the hair on one's head), but, Parker went on, it affected the magnetic fields in interplanetary space, drawing lines of force out of the corona deep into the solar system. Among Parker's hypotheses was that the Earth's own magnetic field was "elongated" by the winds, which shaped it into something closer to a teardrop than a sphere.

Toward the end of the 1950s, the German geophysicist Julius Bartel ascribed the source of such wind to the Sun's "M regions"—*M* standing, appropriately, for "Mystery"—probably areas of unusually weak coronal emissions. Scientists at MIT decided to measure them—with a probe in *Explorer 1*, launched on February 25, 1961, which measured flux, speed, and direction. *Mariner 2*, launched a year later, established that the Sun discharged plasma continuously, at the speed of 248–434 miles per second, occasionally mounting to 775, temperatures increasing or decreasing along with speed. The corona indeed had sufficient energy to escape into space.

There followed a gap in research, possibly because of cutbacks, certainly because many scientists still did not accept that Parker was right. Then in August 1977 *Voyager 2* was launched, which discovered yet stronger winds as they crossed the orbits of planets ever farther from the Sun. It became clear that a constant surge of particles was indeed spilling relentlessly from the solar core. Parker was vindicated: matter was leaving the Sun "like streams of water whipping out from the whirling head of a lawn sprinkler."*

In 1978, in a report to the National Academy's Space Science Board, Parker declared, "Our daytime star is sufficiently near [to reveal] a variety of phenomena that at first sight defy rational explanation . . . but ultimately stimulate the theoretical understanding of new effects . . . elsewhere in physics and astrophysics." Solar physics, he concluded in a later essay, was the "mother of astrophysics."[28] Further,

> the temperature, luminosity, mass and radius of the Sun had to be known before there was a firm basis for arguing that the distant stars are suns. The gaseous nature of the Sun had to be established before the concept of the self-gravitating, self-supportive, luminous gaseous sphere could be developed quantitatively to give an idea on conditions in the deep interior of the Sun.[29]

In short, there would seem to be a natural progression in how we learn about our star. That progression has been hugely accelerated by half a century of discovery. By the time Parker penned these words, researchers had created three-dimensional models, mapped interplanetary magnetic realms, and begun to explore the dynamics of solar gales. *Voyager 1* (launched two weeks *after* its successor, on September 5, 1977) is still operating even as it draws away from the Sun, now 9.6 billion miles distant, having entered the "heliopause," where the solar wind is in balance with the surrounding interstellar medium.

---

* Karl Hufbauer, *Exploring the Sun* (Baltimore: Johns Hopkins University Press, 1993), p. 245. In the summer of 2005, I visited the main Beijing observatory and spoke to a lively young researcher, Chen Jie, who was completing a Ph.D. on coronal loops that cross the solar equator. She spoke reverently of Dr. Parker: "When they give a Nobel Prize to solar physics, it should go to him." As yet, it hasn't, possibly because he has not produced "one extraordinary piece of work," possibly because solar physics seems perpetually out of favor with the wise men of Stockholm.

Around 2015, it will have left the solar system altogether, to plunge on into interstellar space.*

Nearly every year now sees another craft sent aloft. Arguably the most important of all solar missions has been *SOHO* (December 1995), among its many services being early warning of mass ejections that can affect astronauts, and giving up to three days' notice of Earth-directed disturbances.[30] *TRACE* (Transition Region and Coronal Explorer, 1998) images the solar photosphere, transition region, and corona. *Stardust* (1999) returned to Earth in early 2006 with a million motes—weighing in total no more than a few grains of salt—picked up in three circuits of the Sun, while in 2000 a joint NASA/European project, Cluster 2, put into space four craft to analyze both how sunspots come into being and the interaction between solar winds and other magnetized particles.

In August 2003 the Spitzer Space Telescope blasted into orbit, following twenty-three years of preparation. The fourth and last of NASA's great observatories, it monitors the heat radiated by heavenly objects and records how primordial matter formed into galaxies: one galaxy, the Sombrero, is said to contain 800 billion suns.[31] Then, in the last months of 2006 alone, both *STEREO* (Solar Terrestrial Relations Observatory: two complementary NASA spacecraft that track massive solar electromagnetic storms) and Japan's now-named *Hinode* ("Sunrise": it is a Japanese custom not to name satellites until they are up and working) joined them in the heavens. *Hinode*'s telescope was soon sending back spectacular images of sunspots and solar storms, some of the most beautiful ever seen. More is yet to come: in 2010 NASA hoped to launch, on the back of its Atlas 5 rocket, the Solar Dynamic Observatory, designed to study the causes of the Sun's variability and its impact on Earth, and the first mission of the "Living with a Star" program.

We do not know—even approximately—just how many objects (say, of at least modest house size) circle the Sun, as we still lack the observational powers. Currently, some forty man-made research bodies are in orbit, fewer than six still functional, the rest useless hulks. TV and other satellites are legion, and there is already enough equipment out there to make controllers careful. For example, in September 2003 the fourteen-year-old *Galileo* probe, which

---

* As *Voyager* spacecraft leave the solar system (*Voyager 2* will follow *Voyager 1* in 2017), they carry a message for any intelligent alien life that may find them. At JPL, I was shown a copy of the golden record, no larger than a dinner plate, fixed to each craft's side. It contains eighty-seven images representing humanity: our bodies, our ways of life, and our interaction with our planet, as well as instructions for how to build a record player. Delegates from the United Nations have recorded greetings in almost every known national language, from "Please come to visit when you have time" to "May the honors of the morning be upon your heads." Music, from Beethoven to rock 'n' roll (Chuck Berry) and Australian Aboriginal chants to a chorus of crickets and frogs, finds its place—as well as the sound of a kiss.

brought off the exploration of Jupiter and its (then) sixty-three known moons as well as sending back crucial information about solar winds, was deliberately plunged into the giant planet's seething atmosphere so that it would not contaminate Europa, a Jovian moon believed capable of harboring life.*

Since 2004, the world has sent out at least eighty commercial satellites, and the next decade will see at least twenty more. A new generation of superrich entrepreneurs who grew up fascinated by space is pouring money into rockets of all types and sizes. Peter Diamandis, a founder of the Ansari X Prize, a $10 million incentive to put a pilot in space without government financing, says that given the energy and minerals to be found in the heavens, "the first trillionaires are going to be made in space." Economic considerations will play a bigger role than political ones in advancing the interplanetary frontier.[32] Paul G. Allen, a founder of Microsoft, paid for *SpaceShipOne,* the tiny craft that won the X Prize in 2004. Elon Musk, a founder of PayPal, is developing rockets through his company Space Exploration Technologies; Jeffrey P. Bezos, founder of Amazon.com, is developing rockets at a site in western Texas.

For the moment, however, NASA still holds the lead in research. When visiting the Jet Propulsion Lab in 2007, I was invited to sit in on a committee considering a solar monitoring device on the Moon. Plans are currently under consideration in Congress for a solar-powered, human-tended, continuously inhabited research outpost there, rising from either its north or south pole, where sunlight endures and water may exist.[33] But an even more dramatic venture, and one of the main reasons for my visit to Pasadena, is perhaps the most ambitious unmanned satellite project ever.

"JET PROPULSION LAB" is a proudly carried misnomer—the employees just like the title: "We're the only place like us that's not named after some person in the past," one staff member told me, counting off the Goddard, Johnson, Kennedy, George C. Marshall, and John C. Stennis space centers. "That's kinda cool." The whole place is kinda cool: perched on the edge of a national forest fourteen miles from downtown Los Angeles, it covers 177 acres and overlaps the town boundaries of Pasadena and La Cañada Flintridge, which vie with each other for primacy of address. Some five thousand full-time employees work in the facility, plus a few thousand contractors on any given day, so the

---

* London *Sunday Times*, September 7, 2003, News, p. 9. But it is not humanity that has cluttered the skies: billions of "stones in heaven" circle the Sun, from the mass of Jupiter to the uncountable populations of meteor swarms. Trillions of comets and about fourteen thousand known asteroids join them: thus the once tidy community of Sun, Moon, and five planets has fragmented into a cosmos formidably crowded.

place has the feel of a small town, with its own peculiar traditions (these once included the annual crowning of a "Miss Guided Missile," later "Queen of Outer Space," before the contest was discontinued without explanation in 1970) and an almost comically strict security code (although, to be fair, a drug-crazed young employee did sell an intelligence satellite program to the Soviets, an episode tracked in the book and film *The Falcon and the Snowman*).

I was to interview two of the scientists working on *Solar Probe Plus:* Dr. Neil Murphy, the group's supervisor, a Manchester (U.K.)-born astrophysicist in his mid-forties who, after working on the *Galileo* project, came out to Pasadena in 1999; and the man Murphy himself describes as "the smartest guy at JPL," Marco Velli, for the last four years its designated principal scientist, while remaining a tenured professor at the University of Florence. Murphy and I were set up in a small, windowless conference room deep within JPL's research area. "For hundreds of years we've been saying that history is much longer than we think," he told me.

> Then we said: How did the Sun get all its energy? There's a huge amount, and it's tied up in the magnetic field; it turns out that all the exciting stuff on the Sun is.
>
> So the next question is: We know the source of the Sun's energy: can we find out how it functions?
>
> What we're really looking for are trapped waves inside the Sun. We've been developing *Solar Probe* for about thirty years now, because the only way to understand the Sun's heating process is to get really close to it—there's no simple formula that describes how the dynamo inside the Sun works.[34]

Solar Probe has had its ups and downs, and has always been considered the thing that was just *beyond* what NASA could bring off; but Neil Murphy thinks the current blueprints will result in a funded mission to look at the plasma as it evolves. Astronomers use a "solar radius" to measure the size of planets and stars. Ten solar radii, he explained, are a little less than 4,360,000 miles—close enough that one would expect anything at that proximity to be fritzed in a millisecond. Somehow *Solar Probe Plus* has been designed to cope.

I was beginning to get out of my depth, and for a moment he relented. Plasma, he explained, is the "fourth state of matter," solids, liquids, and gases being the more familiar three. Formed from ionized atoms, it is the most common form of matter in the universe, making up the Sun, the other active stars, and interstellar gas. "The solar cycle is basically a process of stretching all this stuff around until it bursts through the surface. It's like twisting a rubber

*"We are capable of shutting off the sun and the stars," John Maynard Keynes warned in 1933,
"because they do not pay a dividend." In 2015, NASA's Solar Probe Plus should get within
4.3 million miles of the Sun—and bring us our biggest dividend yet.*

band: if you twist and twist, eventually this stuff will pop out, and that's what
provides the energy for a sunspot."

By this time there was no stopping Neil, his hands weaving about to illustrate what his satellite would be looking for. "One of the things that has evolved
in the last ten years," he continued,

> is our view of the Sun and the solar system as being all one large sys-
> tem. We've seen the effect of geomagnetic storms on sunlight and on
> the sunspot number. The main thrust of our group over that time is to
> have studied that solar wind, and how it interacts with the planets, and
> the effect of solar disturbances on geomagnetic environments—or
> what we call geospace.

He went on to describe the "tachocline" (from *tachos*, "fast," and *clene*,
"break"), the solid layer of the Sun that sits between the photosphere and that
region of the Sun's interior where rotation is uniform. Within the tachocline
everything also rotates at the same rate. This, Neil explained, was where the
plasma motions that generate the Sun's power come into being, via the process
of ionization. "If you take an ionized gas with a magnetic field in it, and push
the gas, the field moves with it. Ionized gases are almost perfect conductors,
and if the magnetic field is really strong, they'll move the plasma; but plasma is
highly energetic, and can push the magnetic field—which is particular to the
Sun, and, in physics, very unusual. What we are dealing with is how all these
particles interact once they've been magnetized like that."

So what was he hoping to discover? "Over the next fifty years we'll go through these stages of understanding what the processes are that drive the dynamo, and so on. Hopefully, we'll be able to gain some predictive power, to say, 'Given the situation, ten days from now we may see a sunspot appear.'" Nodding, I asked about solar winds. Neil leaned forward, grinning.

Outside the Sun's photosphere we've found some weird things happen. You have this transport of energy uphill, apparently because the temperature is going up, but we're not sure how the balances work yet. Two processes could take energy out of this magnetic field and accelerate the wind. One is when you take two magnetic fields pointing in the opposite direction to a plasma and you push them together and they break and re-form and liberate some energy doing that. And that's one of the processes responsible for explosive leaks of energy—putting energy into the solar wind.

The second one, to make a crude analogy, is like a microwave oven. The material flowing out from the Sun starts subsonically to go upward, but as it gets farther out it becomes supersonic—as well as being heated, it's being *accelerated*—and by the time it gets to Earth, it can be traveling from anything between three hundred and four hundred kilometers per second. So there's this process we don't understand; but we're now at the point of putting it all together.

Neil was about to leave for the Antarctic, to analyze the external structure of solar wind gusting there (later that morning I would watch footage with another JPL scientist, Paulett Liewer, of a comet being blown off course by the wind, both the comet's tail and the wind as clearly visible as the wisps of cloud in a medieval painting). But at this point came a knock on the door, and in walked a slight, well-groomed man a few years younger than his team leader: Marco Velli.

Marco told me that he had been involved with trying to reach the Sun for over thirty years, "and it finally looks as if we're getting there." The original plan, he explained, had been for *Solar Probe Plus* to get as near as three radii from the Sun's surface: approximately 1,297,356 miles. "Now we're talking about ten, but that still means the radiation will be unbearable." The satellite, he said, should be ready for launch in 2015. He handed over a hologram showing how *Solar Probe Plus* will appear in orbit: part of the great red furnace taking up two-thirds of the picture, with, bottom left, a cone of laser light, shining forth from a thick protective disk, behind it a yellow metal tower containing the recording devices. As one moved the card, the beam of light moved even deeper into the Sun's depths, while the star itself billowed out its winds and particles. Even as a mock-up, it made me catch my breath.*

What would photos from the satellite likely show? The Sun's mottled, grainy texture grows progressively denser as one moves toward its core, he said. The consistency around that core has been likened to thick yogurt; some of the magnetic fields found there are six thousand times more powerful than Earth's.

> *Solar Probe* is designed to go into the Sun as far as we can, to measure particles and also the structure of the solar atmosphere. We'll be asking: Where is all its energy stored? How can it be released so fast? Why is the wind the way it is? There is a seeming contradiction in the universe being so far away, requiring such complex techniques to explore it, while the Sun is right there, and we're looking at it all the time. In the next twenty years, we will come to know how the solar corona is powered, how solar wind is formed, and we will begin to learn the limits of the climate system, and of metereology, and how solar reactions are generated by the wind.[35]

That wind, he continues, does not blow uniformly, but in fast and slow streams: around the Sun's equator, it billows out at around 200–250 miles per second, decreasing from around the region of sunspots, while at the polar regions it travels much faster, about 500 mps. Marco, now as fired up as Neil had been, explained about plasma, how to define the photosphere, and what might cause the poles to switch magnetic forces, until it was time for me to put a final question: "Could you have chosen a more complicated subject to have immersed yourself in?" From out of Marco's small frame rolled a great basso profundo *Ho-ho-ho-ho,* like a department store Santa Claus. "I doubt it."

BY THEIR VERY nature, satellites orbiting high in the heavens are dramatic, but they pose major problems. For example, one carrying a large telescope needs special protection once it comes within ten radii of the Sun, and will be

---

* At the same time as the $750 million *Solar Probe Plus,* the European Space Agency is developing *Solar Orbiter,* to be launched in 2017. After traveling for three and a half years, covering some 75 million miles, it will enter into solar orbit at 20 million miles' distance, and is programmed to take the first-ever photos of the Sun from above each of its poles. The satellite's leading edge will rise to a temperature of 600°C (1,112°F), whereas the spacecraft's interior will remain at room temperature.

hugely expensive. In comparison, ground-based telescopes are achieving what satellites may only occasionally attain. Consequently, larger and larger telescopes are being developed for ground use: monsters such as the Quest camera and Samuel Oschin telescope on Mount Palomar, in Southern California, enabling astronomers, in the words of one writer, to "peel away cosmic history like the layers of an onion."*

These devices can pick up the most distant known quasar, 13.5 billion light-years away—and correspondingly many years ago. As of December 2009, some 347 planets had been detected outside our solar system, the smallest of which (unpoetically named Gliese 581) is five times as massive as Earth. Staring up through our atmosphere used to be like examining a passing plane from the bottom of a swimming pool, but a new technique called adaptive optics, developed by the U.S. military and now used extensively in astronomy, compensates for the fluctuations of air so that rather than having their view blurred by turbulence, stargazers get hard-edged pictures.

The latest examples of this breed are the Advanced Technology Solar Telescope, currently under construction at Sunspot, New Mexico, to be sited finally at Haleakala, in Hawaii, and Giant Magellan T, with a main lens 83.3 feet in diameter, being built in Chile for operation in 2013, which will have four times the light-gathering power of any previous telescope and should be able to view extrasolar planets directly. While such grand instruments will harvest information far beyond our solar system, they are also crucial to learning about the Sun. The ATST, for instance (brainchild of the Global Oscillation Network Group—GONG—a consortium of twenty-two institutions worldwide), is set to reveal the nature of the small fibrils of flux tubes within the chromosphere, believed to be the fundamental building blocks of its magnetic structure. The great fields tend to break up into separate compressed tubes of flux; the new telescope should explain why.

Such innovations are not the only ground-based technologies interrogating the Sun. Computers can model key parts of our system, from moons to solar winds, with increasing realism. In yet another new field, laboratory astrophysics, researchers create laboratory versions of plasmas whose behavior mimics phenomena such as the explosion of stars, the formation of galaxies, coronal ejections, and solar prominences—huge gaseous arches extending outward from the Sun's surface. Meanwhile, GONG has helped construct six

---

* Dennis Overbye, *The New York Times*, July 29, 2003, F4. While high science will be employing ever larger instruments, technological developments in the 1990s have made digital imaging of the sky possible for amateurs. Now, rather than using telescopes, many stargazers employ cameras that transfer the images to computer screens.

stations of velocity imagers around the world, so spaced that at least two of them can monitor the Sun at any given time.*

Though an accounting of all the current solar research ventures may sound a little like the chairman's annual report, beyond the dry facts is the reality that we are on the verge of huge breakthroughs. In 1952, the astrophysicist G. P. Kuiper could write confidently that "the golden age of solar physics has not yet been experienced."[36] Sixty years on, it is upon us. As early as 1979, Carl Sagan, too, was referring to a coming golden age:

> In all the history of mankind, there will be only one generation that will be first to explore the solar system, one generation for which, in childhood, the planets are distant and indistinct disks moving through the night sky, and for which, in old age, the planets are places, diverse new worlds in the course of exploration.[37]

"We are positioned for a new era of discovery about the Sun and the subtle ways it affects life on Earth," states GONG authoritatively. Carlos Frenk of Durham University, in northeastern England, told *National Geographic*, "It's no exaggeration to say that we're going through a period of change analogous to the Copernican revolution."[38] Then there were Neil Murphy's final words to me: "We're now in a position to pick up information beyond all previous imagining, data so complex that at last we should be able to deduce the nature of this intolerably inaccessible object! We're trying to find all the pieces of a puzzle that go together; and I think we will." "Indeed," concludes Eugene Parker, with a welcome acidity, "the activity of the Sun provides so many effects outside the realm of conventional laboratory physics that its contemplation is a humbling experience for the serious physicist, repeatedly demonstrating the incorrect nature of our best ideas and explanations."[39]

---

* Mauna Loa in Hawaii, Big Bear Observatory in Southern California, Idaipur on an island in Rajasthan in northwestern India, Learmouth on the North West Cape of Australia, Cerro Tololo in Chile (310 miles north of Santiago), and El Teide in the Canary Islands.

CHAPTER 30

UNDER THE WEATHER

"I've read somewhere that the sun's getting hotter every year," said Tom [Buchanan] genially. "It seems that pretty soon the earth's going to fall into the Sun—or, wait a minute—it's just the exact opposite. The sun's getting colder every year."

—F. Scott Fitzgerald,
The Great Gatsby[1]

Some say the world will end in fire
Some say in ice
From what I've tasted of desire
I hold with those who favor fire.

—Robert Frost, "Fire and Ice"[2]

OVER THE CENTURIES IT HAS BEEN ESTABLISHED THAT THE ODDS ARE heavily against our world's surviving forever. In *A Choice of Catastrophes* (1978), Isaac Asimov examined the range of disasters that threaten us,[3] defining five classes of peril: that the entire universe might so change that our world becomes uninhabitable; that the Earth might undergo a convulsion, making life impossible; that something (perhaps man-made) might destroy us; that civilization as we know it might be undermined, condemning what was left of humanity to a primitive existence; and that something might happen to the Sun. In all, he lists sixty-six different kinds of catastrophe, from expanding and contracting universes, black holes, and quasars through comets, meteorites, and asteroids and on to volcanoes, earthquakes, cosmic rays, pandemics, nuclear bombs, pollution, and depletion of resources. By the end of the catalogue one becomes almost light-headed, and it is a relief to turn to such age-old worries as wars and population explosions.

What is notable thirty-odd years on is that "hole in the ozone layer" and "global warming" are virtually absent from Asimov's list, the former appearing just once, the latter not at all. While both involve noxious gases, greatly increased heat, and catastrophic denouements, they are very different. The ozone layer has been, in the view of most scientists, to a large extent repaired; not so global warming.*

Ozone (from the Greek *ozein,* "to reek") is an unstable form of oxygen created when sunlight splits a standard oxygen molecule (O_2) into two single oxygen atoms that in turn combine with two more molecules of oxygen to form two molecules of ozone (O_3). It is found in the thin layer that encases the atmosphere some twelve to sixteen miles above the Earth—and on Earth itself, where it has a strong metallic odor that becomes evident whenever an electric discharge (such as lightning) passes through oxygen. Perhaps because of ozone's chlorinelike smell, it was for many years viewed as having cleansing properties; the Victorians spoke of "taking the ozone" at the seaside for their health (though in fact what they smelled was rotten seaweed, which is similarly pungent). Tubercular patients were often sent to mountain resorts, with their supposedly higher ozone levels; to attract customers such places would advertise themselves as "ozoniferous." In 1865 a letter in the London *Times* called the gas "Nature's Great Cleanser" and extolled its "bactericidal capabilities." Ozone is also found at ground level, but not in any bracing way. In city conditions, it causes smog, irritates the eyes, nose, and lungs, and has been linked to asthma. It also harms plants, insects, and fungi, and reduces by as much as half the capacity of trees to sop up greenhouse gases through photosynthesis. This is what one science writer calls its "Mr. Hyde side."[4]

On the other hand, the "ozone layer"—first detected in 1913 by the French physicists Charles Fabry and Henri Buisson—is the Dr. Jekyll side: it shields us from all kinds of ultraviolet light, and has done so for a billion years. Yet the ozone layer, which encircles the Earth far less densely than a film of mud on a tennis ball, is on average just three miles deep. It turns out to be fragile, too. In May 1985, scientists with the British Antarctic Survey detected a hole in the layer, and the following year American research confirmed the bad news: the polar ozonosphere thins seasonally, but the size of these gaps was something new—indeed, between 1978 and 1985, Earth's whole ozone shield eroded by 2.5 percent. It was soon predicted that before long these depletions would spread to over 30 percent of the globe's polar regions, allowing solar radiation to do incalculable damage.

---

* On March 3, 2006, the London *Times* published a letter in its "Q&A" column from a Mr. Thornton of the Scots border town of Galashiels, which asked: "When was the term 'global warming' first used?" Unlike nearly all other entries, this question was never answered. The first use I can find is in *Science* (August 8, 1975) by Wallace S. Broecker, a longtime researcher at Columbia University, who titled his article "Climate Change—Are We on the Brink of a Pronounced Global Warming?"—and the phrase caught on.

Lightning pierces the volcanic ash of the Icelandic volcano Eyjafjallajökull on
April 17, 2010. On average, over the entire Earth's surface, lightning strikes
approximately one hundred times per second—8,640,000 times a day.

It now appears that chlorofluorocarbons (CFCs), for decades hailed as one of the century's great inventions, may be the main cause of the hole. CFCs were the brainchild of an American mechanical engineer turned chemist and inventor, Thomas Midgley, Jr. (1889–1944), who ended up with over a hundred patents. (It has been remarked that Midgley "had more impact on the atmosphere than any other single organism in Earth history.")[5] In 1930, he developed a nontoxic refrigerant for household appliances, dichlorodifluoromethane, a chlorinated fluorocarbon that he dubbed Freon. CFCs replaced the various toxic or explosive substances previously used in refrigerator heat pumps. This wonder gas, odorless, colorless, and nonreactive, served not only as a coolant in refrigerators but also as a foam-blowing agent in fire extinguishers and a propellant in aerosols.

For more than four decades, CFCs seemed wholly beneficial. Their lack of reactivity confers upon them a life span of more than a hundred years, allowing them time to diffuse into the stratosphere. In the 1970s, three chemists (in 1995 to share the Nobel Prize), the Dutchman Paul Crutzen and the Americans Mario Molina and F. Sherwood Rowland, began a joint research project, simply as an academic exercise, to find out what happened to CFCs as they drifted upward; they discovered that, although they were stable near the Earth's surface, once in the stratosphere ultraviolet radiation broke off their chlorine atoms, which on their own are highly reactive and have the effect of decomposing the ozone layer. This occurs with particular severity around the south pole, where extreme weather patterns, supercold winters, and icy clouds concentrate the effect; but diffusion spreads the thinning as far equatorward as the middle latitudes.[6]

Although the first popular study of the damage was published in 1978, the same year as Asimov's book, governments long ignored the evidence.[7] At one point President Reagan's interior secretary suggested that if CFCs were indeed so destructive, anyone worried should just wear sunglasses and buy a hat: "People who don't stand out in the Sun—it doesn't affect them," he claimed.

Eventually, a United Nations protocol, signed in Montreal in 1987, severely restricted the production of CFCs. The agreement has been acclaimed as "far and away the most important piece of regulatory legislation on any environmental issue";[8] but given the chemical's long lifetime, the situation had to get worse before it got better. By January 1993, the Earth's average ozone layer concentration was the lowest on record (this remains the case); by March 1999, researchers were claiming that fish in Antarctica would soon be at risk of sunburn. Away from the poles, solar radiation was said to be causing other problems, visiting extreme deformities, such as extra hind legs, upon more than sixty species of frog and toad.[9]

Only in recent years, as the U.N. restrictions have begun to take effect, has the ozone layer shown signs of returning to its presumed natural state. Even so, many scientists maintain that it will take years before all CFCs are processed out of the atmosphere (several countries, indeed, still use them), and the layer is still about 10 percent thinner than it should be. Full recovery is not expected at the middle latitudes until about 2050, in the Antarctic until 2080. In 2002, satellite observations indicated that the Antarctic hole had shrunk from 9 million square miles to 6 million, the first dramatic reduction since the 1987 ban; but the hole evidently fluctuates in size, and as recently as September 2006, it matched its 2003 extent—exposing an area larger than the whole of North America. The patient is still convalescent.

Of course, this presumes that the diagnosis is correct in the first place. For the Sun seems to play a role here, too. In 2004 I went to talk to scientists at the Kitt Peak solar research center in Arizona, and spent a morning in the library of the local paper, the *Tucson Citizen*, which I discovered had, back in 1986, carried an AP report of satellite observations suggesting that it was not only CFCs but the Sun, too, that had been depleting the ozone. The article did not deny that CFCs carried particular responsibility—they plainly did—but it pointed out that an unusually intense climax of solar activity in late 1979 and early 1980 had set off a cascade of chemical changes that had reduced ozone worldwide, with a pronounced depletion over Antarctica, where, as solar activity subsided, ozone levels went up again. In other words, solar variations were affecting the ozone layer, and will continue to do so—a useful corrective to bear in mind as we turn to that other matter in which Isaac Asimov had displayed so little interest back in 1978: global warming.

PEOPLE HAVE LONG suspected that human activity could alter the climate (which word experts usefully define as "thirty years of weather"). But to what extent? And how does that relate to the various natural factors that have been identified—fluctuations in solar output, volcanoes belching smoke, dust, and sulfur, the effects of clouds ("still the climate expert's Achilles' Heel"),[10] and water vapor—and, over a longer term, the rise and erosion of mountain ranges (which divert wind patterns and ocean currents)—even changes in the composition of the air itself?

The first human-related warming effect may have been set in motion some ten thousand years ago, produced by the fires, clearing of vegetation, and soil disturbance of early farmers. That gases in the atmosphere might warm up the Earth was recognized in 1827 by Jean-Baptiste Fourier (1768–1830), a mathematician who had served under Napoleon in Egypt. Fourier pointed out the similarity between what happens in the atmosphere and in a greenhouse, as incoming sunshine warms up plants and soil faster than heat can escape either planet or glass-walled building. Were there no greenhouse gases, the Earth's temperature would average about −40°F (−40°C; the one temperature where the two scales coincide). However, up until the end of the nineteenth century, it remained unclear to what extent humans were responsible for these gases.[11] Then in 1896 the great Swedish chemist Svante Arrhenius (1859–1927) described how the burning of coal during and since the Industrial Revolution, which was releasing the trapped carbon from millions of years' worth of vegetation, might alter the carbon balance—and the temperature. The increase in CO_2 (which he called "carbonic acid gas," its name at the time) would trap solar infrared energy that would otherwise radiate back into space, thereby raising the planet's average temperature. This was what his much older contemporary John Tyndall (the man who explained why we see the sky as blue), picking up on Fourier, had a few years before dubbed the "greenhouse effect." "We are evaporating our coal mines," wrote Arrhenius—coal being about 70 percent carbon by weight. But such warnings found few supporters, most scientists arguing that whatever humans did was not significant enough to have any effect. Anyway, they added, most carbon dioxide came from volcanic emissions and other natural sources, not from human activity.

Over the next three decades, the argument did not greatly develop; but then, in the 1930s, scientists reported that the North Atlantic region and the United States had warmed significantly during the previous half-century, although they supposed this was just a phase in some natural cycle. The only voice suggesting otherwise was that of a professional coal engineer and amateur climatologist, Guy Stewart Callendar, who, standing before the Royal Meteorological Society in London in 1938, laid out the previous hundred years' worth of carbon dioxide measurements and warned that greenhouse heating should be a concern. Again, there was a period of silence. Then in the 1950s, researchers brought improved skills to bear on the question. In 1959, a California geochemist, Charles Keeling (1928–2005), showed that carbon dioxide levels rise and fall with the seasons, following the ebb and flow of land vegetation in the Northern Hemisphere (the Southern has far less land, hence less variability); but he also demonstrated that carbon levels were rising yearly, a trend that came to be known as the Keeling Curve.*

Over the next ten years, the study of ancient pollens and fossil shells confirmed that grave changes could take place within as little as a few centuries or even decades. In 1967, Roger Revelle (1909–1991), both figuratively and literally a giant of science—he was six foot four—the dynamic researcher who brought Keeling to join him at the Scripps Institution of Oceanography near San Diego, became the first person to make concerted measurements of carbon dioxide in the Earth's atmosphere, sending up weather balloons over the Pacific and plunging bottles down into it to measure radiocarbon in air and ocean respectively. Up until that time, scientists had believed the vast seas could absorb any increases in CO_2, but Keeling calculated that the amount being added by humans would take millennia to absorb. Within months, other scientists, assessing the gases released by burning tropical forests, by livestock and rice cultivation, and from town dumps and pipeline losses, were suggesting that average temperatures might rise significantly within the next century. However, the warming seemed a long way off, not a clear and present danger.

There was not even a consensus on what was actually happening over the relatively short term. Analysis of Northern Hemisphere weather statistics, for example, showed that a cooling trend had begun in the 1940s. The one point most scientists agreed on was that they scarcely understood this intricate system that appeared to respond to so many different inputs. It was apparently so

---

* Keeling employed a gas analyzer, which worked by passing an infrared beam through an air sample and measuring how much of the beam got through. The more carbon dioxide in the sample, the more it blocked the beam. Each analyzer cost around $20,000 then (about $145,000 now).

delicately balanced that almost any small perturbation might set off a great shift—improved computer models began to suggest how such jumps might come about, for example through a change in the circulation of ocean currents. However, the modelers who suggested this had to make many arbitrary assumptions, and some reputable scientists disputed the value of their results; others, raising questions about the effects of agriculture and deforestation on CO_2 levels, pointed out how little was known about the way ecosystems interact with climate.* One unexpected discovery was that the levels of certain other gases were rising, some of these (those ubiquitous CFCs) also degrading the ozone layer.

In the 1970s, as temperatures began to rise again, international scientific panels warned for the first time that humanity faced a serious threat: a molecule of CO_2 would in the course of its lifetime trap a hundred thousand times more heat than was released in producing it (indeed, the by-products of our industrial civilization trap almost a hundred times more energy than we actually use).[12] More than half the atmospheric CO_2 was coming from human activity, each of us on average shoveling an entire ton of carbon into the air each year. Al Gore summed up the coming crisis in his 2006 documentary *An Inconvenient Truth:* "The most vulnerable part of the world's ecological system is the atmosphere, because it's so thin. And we're in danger of changing its basic composition."†

That agenda-setting film was still a long way off at the time that scientists' warnings first caught the attention of a wider audience—during the stifling heat wave of 1988, then the hottest summer on record in the Northern Hemisphere, when the climatologist James E. Hansen (b. 1941) informed a sweating

---

* In *Emma,* Jane Austen describes an apple tree blooming at an odd time of year, which has always been viewed as a mistake on her part. Even Austen's brother Edward questioned her reputation for meticulous observation, writing: "I wish you would tell me where you get those apple-trees of yours that come into bloom in July." However, the weather for 1814–15, when Austen was writing *Emma,* was usually chilly, and H. H. Lamb in *Climate, History and the Modern World* cites the decade as the coldest since the 1690s. It was within the period known as the "Little Ice Age" (1350–1850), further chilled by volcanic dust.

† Although these figures have never been satisfactorily refuted by global warming deniers, a comment made by Oliver Morton remains telling: "It strikes me how unfortunate it is that English gives only a technical name to something as basic to life as carbon dioxide, something as fundamental as blood and breath. Because it is indiscernible without instruments and has thus never furnished our perceptual world, I'm forced to write about it under a name that, even if it doesn't alienate, certainly can't carry any freight of emotion. 'Water' is as rich in imagery as a word can be; 'oxygen,' coined just over two hundred years ago, has far fewer associations, but still has some general aura of necessity and energy and freshness. 'Carbon dioxide' is just a chemical. . . . It's salutary to remember that language itself can contrive to mask the richness and relevance of the world that science reveals."[13]

U.S. Senate committee that global warming threatened mankind. But the continuing uncertainties, and the sheer complexity of climate, made for vehement debate, with Hansen tagged alarmist in chief. Even so, governments were sufficiently worried to establish, that same year, a special Intergovernmental Panel on Climate Change (IPCC) under U.N. auspices, which issued reports in 1990, 1995, 2001, and 2007.

A weather balloon goes up during hail prevention operations in Georgia.

The first three of these surveys were all phrased with great caution, so that there might be some consensus among members: it was *much more likely than not*, they said, that our planet faced severe warming; but its cause could as easily be natural as man-made. The year following the 1990 appraisal, a U.S. National Academy of Sciences report separately declared that there was "no evidence yet" of dangerous climate change, a question-begging formulation: it was rumored that the Clinton administration had asked for much stronger wording. The 2001 IPCC report declared that 66 percent of warming was *probably* due to humans. The last time the world had been this warm was 50 million years ago, when, as Elizabeth Kolbert wrote in *The New Yorker* in 2005, "crocodiles roamed Colorado and sea levels were nearly three hundred feet higher."[14] By the time of the next survey, issued in 2007 and consisting of 1,572 pages that drew upon the work of more than two thousand scientists from 154 countries, both the tone and the figures had changed: the probability that human activities were responsible was assessed at 90 percent, and the panel was forthright in asserting that the late-twentieth-century phase had been the result of greenhouse gases, the effect of which was estimated to outweigh the Sun's influence by a factor of thirteen to one.* By the end of the year the panel had been awarded a Nobel Peace Prize, which it shared with Al Gore.

The tide of opinion had clearly turned. Speaking of the high levels of CO_2, as well as of other greenhouse gases such as methane and nitrous oxide (the Kyoto Protocol of 1997 covers no fewer than twenty-four such compounds), Mark Lynas, environmental correspondent of the British weekly *New Statesman*, declared: "That greenhouse gases have a warming effect, rather like an extra blanket around the globe, is indisputable, and has been established physics for over a hundred years."[†] Gernot Klepper of the Kiel Institute for World Economy, a member of the IPCC, has said bluntly, "The world is already at or above the worst case scenarios in terms of emissions."[15] Another expert, George Philander of Princeton University, recently told *National Geographic*, "We're now geological agents, capable of affecting the processes that determine climate."[16] Personal opinions all—but by far the majority of experts share them.

How can we quantify such gas increases? "In the 1780s," wrote Elizabeth Kolbert, "ice-core records show, carbon-dioxide levels stood at about 280 parts per million. Give or take ten parts per million, this was the same level that they had been at two thousand years earlier, in the era of Julius Caesar, and two thousand years before that, at the time of Stonehenge, and two thousand years before that."[17] When industrialization began to drive up these levels, they rose gradually at first, then much more steeply. When we started measuring, in the

---

* CO_2 doesn't have to be released into the atmosphere. Carbon capture and storage (CCS) procedures can absorb it, say from a power plant or similar large emitter, then inject it at high pressure into such handy spaces as depleted oil fields. The world's longest-existing CCS effort, maintained by the state-owned Norwegian energy company Statoil at a natural gas field in the North Sea, began in 1997. Columbia University is offering a $200,000 prize for ideas on how to keep CO_2 below an acceptable level, while in 2007 Virgin's Richard Branson offered $25 million for a workable plan to remove a billion tons of carbon dioxide per year. Meanwhile, a mystery exists: roughly that weight of carbon dioxide pours into the atmosphere every year, but nearly half, 43 percent, disappears—and nobody knows where it goes.

† Mark Lynas, *Six Degrees: Our Future on a Hotter Planet* (London: Fourth Estate, 2007), p. xix. In January 2008, Lynas became involved in an acrimonious debate over whether global warming had stopped. The *New Statesman* had run a website article by David Whitehouse (author of *The Sun: A Biography*) arguing that for the last ten years the Earth's average temperature had evened out and was no longer rising. Lynas contributed a fierce rebuttal castigating his fellow contributor. But the resultant debate on the Internet (more than twelve hundred replies on the *New Statesman*'s site, while twelve thousand continued on another website, making it the Internet's longest comment thread) excoriated Lynas for jumbled reasoning and misapplied statistics, and by some margin upheld Whitehouse's view. See the *New Statesman* website, www.newstatesman.com, for December 19, 2007 and January 14, 2008.

late 1950s, the ratio had reached 315 parts per million (ppm). In May 2005, the figure stood at 378; *Scientific American* of April 2007 gave it as 379. By February 2008, it was 383. It has been speculated that the maximum tolerable amount is 445. To keep concentrations even to that figure, the world must reduce emissions by 80 percent within forty years—an almost impossible aim, given the costs involved. In June 2009, James Hansen declared that CO_2 levels had already reached 385 ppm, and that CO_2 was being pumped into the air some ten thousand times faster than natural processes can remove it.[18] "So humans now are in charge of atmospheric composition." And that is just one greenhouse gas, albeit "the most general contaminant of all."[19] Other heat traps include water vapor (preeminently the single most common greenhouse gas on Earth), methane, chlorofluorocarbons, nitrous oxide, and ozone.

Even if these emissions stopped overnight, the world would warm a further 0.2°C by century's end, due to the release of energy already absorbed by the oceans: actual warming over the past 150 years has amounted to 0.6°C (just over 1°F), to around 59°. Removal of excess CO_2 will become less efficient as the planet heats up. The IPCC predicts that the Earth could be 3.5° to 8°C warmer by 2100, while in May 2009 researchers at MIT put the figure as high as 13.3°C.[20] One thinks of Kurt Vonnegut's despairing cry in his essay "Armageddon in Retrospect": "What can we do about global warming? We could turn out the lights, I guess, but please don't. I can't think of any way to repair the atmosphere. It's way too late"—an outburst that carries all the more weight when one learns his brother was one of the United States' leading meteorologists.[21] No wonder that the idea of constructing some huge planetary sunshade has recently migrated from the fringe to the mainstream.*

All of which brings us back to the Sun. How much is this "mildly variable star" responsible?[23] At a rough average, its rays strike the Earth's surface at about 288 watts per square meter—the power emitted by three good-sized incandescent lightbulbs. The Sun is about 15 percent broader than it was about 4.5 billion years ago, and thus is radiating about 25 percent more heat than at its birth. But in what is known as the "faint Sun paradox," the Earth had a warmer surface early in its life—when the atmosphere consisted mainly of

---

* According to the Soviet climate expert Mikhail Budyko, all we would have to do is dump gaseous sulfur dioxide into the stratosphere; it would form drops of sulfuric acid, which within months would be whirled around the globe by stratospheric winds, wrapping the planet in a white shroud. Since 2000, U.S. government sources have advocated such techniques as giant mirrors in space, reflective dust pumped into the atmosphere, enormous numbers of tiny shiny balloons, or tinkering with clouds to make them more reflective. The IPCC has labeled such ideas "speculative, uncosted and with potential unknown side effects." But in July 2009, the U.S. Department of Energy issued a report analyzing the possibility of using aerosol particles to reflect shortwave solar radiation back into space. Geoengineering seems to be the way of the future.[22]

carbon dioxide and water. The present-day "global dimming" is caused by various pollutants that prevent the energy from reaching us, either by reflecting radiation or by condensing more water droplets out of the air, to generate thicker, darker, more obscuring clouds.

From about 1960 to the early 1990s, the average amount of sunlight, in both duration and strength, reaching the Earth's surface has decreased by as much as 10 percent. In some regions such as Asia, the United States, and Europe, the drop has been even steeper: sunlight in Hong Kong has decreased by 37 percent.[24] The amount reaching us has diminished everywhere, not just over cities. Air travel may have exacerbated this effect: when commercial traffic was grounded for a couple of days after September 11, 2001, local temperature swings from day to night widened by several degrees, due to the absence of the jet contrails that normally absorbed sunlight and at night acted like a blanket (that metaphor again). So it might seem that the Sun was not responsible for global warming and—as Al Gore put it—"the science is all settled."

I WAS WILLING to accept that it was, but then I heard about Piers Corbyn, a climatologist who specializes in thirty-day, forty-five-day, and twelve-month forecasting. Corbyn is director of Weather Action, and the vigorous foe of the British Metereological Office, which he attacks without respite as being in the hands of those who mistakenly believe that the world is getting warmer solely because of greenhouse gases. The Met Office would probably like him to shut up or go away, but they cannot dismiss him as a freak, because too often, to their embarrassment, his forecasts have trumped their own.

Corbyn is in his mid-sixties, with a luxuriant beard and a mop of black hair, both speckled with gray, atop the kind of wiry frame one sees on a long-distance runner (which he used to be). One recent interviewer likened him to the BBC's Dr. Who. His office, located on Borough High Street, the long, slightly down-at-the-heel arterial road through South East London, is one of the smallest I have ever squeezed into, hardly affording room for two chairs amid all the papers, teacups, books, and printouts. On the wall hangs a framed *Guardian* profile headlined ALL THIS MAN'S CLOUDS HAVE SILVER LININGS. While still in high school, he published three papers on aspects of astronomy and meteorology as well as building his own weather station. After winning a scholarship to Imperial College, London, he received a first-class degree in physics and moved on to research work—successively on superconductivity, the mean matter density of the universe, galaxy formation, and solar activity.

Corbyn had for some time been interested in climate history, and by 1982 had begun to move into forecasting. Two years later came the miners' strike against the Thatcher government, and he was asked to predict whether it would be a cold winter. Yes, he reckoned, the turn of the year would be bitter—both his scientific view and his political hope, as he believed a cold winter would help the strike succeed (one of his brothers is Jeremy Corbyn, a fierce far-left member of Parliament). His political foresight fell short—the strike failed—but his weather predictions were spot-on: the end of 1984 saw a mild Christmas, but the new year began very cold. "By 12 January it was dire—my breath was freezing into icicles on my beard—a transcendental moment," Corbyn says, laughing.

By the summer of 1988, he felt confident enough to place bets on the weather with William Hill, one of the big three bookmakers in the United Kingdom and the only betting setup that would take them. It was offering odds of ten to one against that year's July being one of the ten wettest of the century. Corbyn put down his money, and won. Soon he was making more than £2,500 a year, but in 2000 William Hill stopped taking his wagers. Now, says Corbyn, it accepts bets on anything except the death of the monarch—which it is forbidden to do under law, since it gives the punter an interest in the sovereign's demise—and weather bets placed by Piers Corbyn, M.Sc., FRAS.

Plainly an evangelist for integrity in weather prediction, Corbyn is affronted by what he sees as the bad science behind most forecasts. "The global warmers' claim that CO_2 is or has been the main controller of climate fails when past data are examined," he explained, while energetically searching for a teabag. His research has suggested that water vapor, volcanoes, and upper-level clouds all play a greater role than CO_2.[25]*

Also of far more importance, he says, has been the motion of the magnetic toward the geographic North Pole, driving temperatures down sharply enough to cancel out much of their net rise in the last century. In fact, some scientists have been predicting that, far from being threatened by global warming, we may instead be facing an ice age, perhaps only a few hundred years from now. In 2005, the Russian astronomer Khabibullo Abdusamatov predicted that the Sun would reach a peak of sunspot activity in 2011, which would cause "dramatic changes" in temperatures—downward, not upward.[28]

---

* I am aware that I have given clouds—those ponds floating above our heads—short shrift; yet they play a vital role in modulating Earth's basic radiation balance. "All clouds have a split personality," scientist-reporter Richard Monastersky has observed. "They feature some characteristics that cool the Earth and others that warm it."[26] Or, as the dying monarch in Ionesco's *Exit the King* complains, "They don't listen—they do what they like." Relatively few data have been established about the ways clouds respond to other variables, but recently two Danish scientists revealed that the Earth's cloud cover varied in lockstep with a phenomenon known to be tied to the solar cycle. They called their discovery "a missing link in solar-climate relationships."[27]

During the last two million years, the planet has undergone more than twenty glacial advances and retreats, regions of the world having warmed and cooled differently under the influences of such factors as oceans, mountains, and winds. The last ice age began to break up about eleven thousand years ago, at which point the glaciers covering much of North America, Scandinavia, and northern Asia began to retreat to approximately their current positions.

One possible scenario for the onset of a new ice age would involve the shutting down of the Gulf Stream—which, unexpectedly, would be one consequence of global warming. This is because, as it flows, the North Atlantic Current, an extension of the Gulf Stream that cuts northeast across the Atlantic Ocean, surrounds the lands of the high latitudes of Europe with warm equatorial water, and releases heat equivalent to the annual output of a million medium-sized power stations into the overlying air. Together, warm water and warm winds are what keep Europe's climate so mild. One of the factors that drives this system involves vast tracts of seawater that freeze over each winter to form ice shelves, coincidentally releasing salt, which makes the surrounding water heavy enough to sink to the seabed, whence it flows southwest toward the Caribbean before warming again and rising. However, since 1977, the shelves have stopped forming as before (probably because of global warming), and the water has stopped sinking. By 2005, scientists were saying there was an almost one in two chance that the Gulf Stream would cease to flow. Europe's temperatures would plummet, with winter figures in Britain, for instance, dipping to −22°F (−30°C).[29] (Of course, as with almost every view on the subject of global warming, there is a counterview: in September 2009 the journal *Science* published fresh evidence that human activity was not only warming the globe, particularly the Arctic, but might even be delaying the inevitable descent into an ice age.)[30]

Another global-warming-related factor that could, paradoxically, help precipitate a new ice age involves the great volumes of frozen water in Antarctica, which measure several kilometers deep. As the ice pushes slowly oceanward, its progress is delayed by the sheets at the edges. A rise in sea level could disturb these edge sheets, however, and release vast glaciers into the sea, which could drive temperatures down dramatically. There are yet further scenarios for an ice age that do not involve global warming—by humans or otherwise. "Glacial cycles are initiated by slight, periodic variations in the Earth's orbit," writes Kolbert, which variations "alter the distribution of sunlight at different latitudes during different seasons and occur according to a complex cycle that takes a hundred thousand years to complete."[31]

Whether we should prepare ourselves for global warming or for a new ice age indirectly caused by a hotter Earth or by some other factor is almost impossible to forecast, given the current state of our technology. (Until the sixteenth century, "weather" and "whether" were interchangeable spellings.) The various computer models for global climates are—to date—insufficiently accurate. As one modeler complains, "We are starved for measurements. . . . When natural variability is so great, it is dangerous to attribute any particular change to any particular cause."[32] Nigel Calder, author of *The Weather Machine*, calls climatic change "one of the untidiest areas of modern science."[33] Or, as Joyce's Leopold Bloom so charmingly puts it, weather is "as uncertain as a child's bottom."[34]

THIS IS WHERE Piers Corbyn, for all his eccentricity (the British Met Office early on dubbed him "the mad scientist"), comes into his own. Unlike the Met Office, he believes that the Sun is central to climate change. While he is secretive about his methods, his forecasts are predicated on the assumption that solar particles and the magnetic links between Sun and Earth control our climate on all time scales, from days to hundreds of thousands of years. Solar particles are dramatically affected by alterations in the Earth's magnetic field, particularly shifts of the magnetic poles and the particles' excursions nearer the equator. Corbyn charts solar flares and other coronal mass ejections, eagerly ticking off each further influence. "There is the twenty-two-year cycle, the Moon, the Sun's magnetic field—these all affect the weather. One should also mention the shock waves in solar winds, known as 'red spikes,' and changes in the tilt of the Earth." The stronger the Earth's magnetic field, the more particles it collects from the solar wind; and the more particles it collects, the warmer the Earth becomes. The Sun's magnetic activity has more than doubled since 1901, Corbyn points out, reinforcing the field by a factor of 1.4 since 1964: and there is growing evidence that solar magnetic activity is reaching an eight-thousand-year high.

He certainly has made his fair share of incorrect forecasts, such as a white Easter in 1989 and "raging weather" in September 1997: both periods of serene warmth. But the record of the Met Office is considerably worse: since 1923 it has provided the BBC with forecasts, but the franchise has now been opened up to bidders after a particularly bad year in 2009: the Met predicted an "odds-on barbeque summer," which turned out so cold and wet the organization had to apologize, and then predicted a mild winter—cue the coldest January for twenty-three years. A recent poll claimed that 74 percent of people believe the Met Office forecasts are generally inaccurate.[35] Weather forecasters worldwide have hardly a better history.

In this increasingly contentious debate, one point in Corbyn's favor—to my mind—is that the IPCC's first report ignored solar variability, and its recent "summary for policymakers" still mentions sunlight only in passing, even then omitting the Sun's particle and magnetic effect.[36] As for the Copenhagen conference of December 2009, the emphasis was again on human-caused warming, and solar factors received hardly a mention—just that recent output has been at a low ebb.[37]

For Corbyn, however, everything the Sun does is vital in establishing what we can expect on Earth. The last time I saw him, in March 2009, he had embarked on two fresh areas of research about which he was very excited: first, the effects of the Moon on solar winds and on the Earth's magnetic field, which are sometimes powerful enough to disrupt the magnetic pathways from the Sun to the Earth; and second, what he dubs "virtual particle heating," in which charged solar particles change the poleward heat flow from the equator, and so affect the circulation of heat worldwide.

Corbyn's record in forecasting (for 2010 he predicts his success rate will be as much as 80 percent; as ABC News put it, "He's not perfect, but he's pretty good") suggests that he has it right, and that the IPCC has indeed underestimated the Sun's influence; just how seriously is hard to gauge. Sami Solanki, the director of the prestigious Max Planck Institute in Katlenburg-Lindau, Germany, sits on the fence: "The brighter Sun and higher levels of so-called 'greenhouse gases' have both contributed to the change in the Earth's atmosphere, but it is impossible to say which has had the greater impact." Timothy Patterson, director of the Ottawa-Carleton Geoscience Center of Canada's Carleton University, comes down more squarely on Corbyn's side: "Solar activity has overpowered any effect that CO_2 has had before, and it most likely will again. If we were to have even a medium-sized solar minimum," he goes on, using the term for a period of minimal sunspot activity, "we could be looking at a lot more bad effects than 'global warming' would have had."[38] To round off this list of heavy hitters, Willie Soon, a solar and climate scientist at the Harvard-Smithsonian Center for Astrophysics, wrote to me, "Bill Clinton used to sum up politics by saying, 'It's the economy, stupid!' Now we can fairly sum up climate change by saying, 'It's the Sun, stupid!' "[39]

Which side, then, to believe—those who sign on with the IPCC, for whom human-mediated global warming is "a fact like a rock you could hold in your hand,"[40] or those who talk of such alarms as scams? Are we due for another ice age or a planetary roasting? Or both? One recalls the great mathematical philosopher Alfred North Whitehead's reflection: "There are no whole truths. All truths are half-truths. It is trying to treat them as whole truths that plays the devil." Two extreme points of view are evident. One paints the bleak picture implicit in Roger Revelle's "We're at a hinge of history"; the other is close to total denial. One side rallies scientists who believe in the human-caused reasons for climatic disaster, bristling with charts, graphs, and research data, while the other side musters its own roster of experts, who deploy a different set of charts, graphs, and research to argue their way to a different conclusion— both sides growling that the other is direly wrong.*

What we can say with certainty is that the science is *not* all settled—how can it be, when there is still so much we do not know? Given all I have learned over the last eight years from the sun watchers in Tucson and on Mount Wilson, the Jesuit astronomer at Castel Gandolfo and the weather analysts in East Anglia, the physicists in Pasadena and scientists in China, Japan, India, Western Europe, Russia, and South Africa, it seems plain that nowhere amid this flux of information does there yet exist an adequate theory. We are still very much in the process of understanding climate. I do believe that we fail to control greenhouse gases at our peril; although, as Robert Kunzig points out in his study of the world's oceans, "Maybe a century from now, our fretting about carbon pollution will cause our great-grandchildren to chuckle, as we do now when we read the forecasts, made before the rise of the motor car, of city streets buried in horse manure."[41] Maybe it is even a form of self-importance to attribute such phenomena as climate shifts to human causes.

Yet one fact in all this debate rings clear and undeniable: the Sun so obviously embraces us, its atmosphere and wind surrounding us, its matter washing on our shores, that we must accept that our presiding star remains by far the major influence on our lives, and on our climate. Beyond that, we are all still searching.

---

* In November 2009 a hacker released thousands of emails and documents from the Climatic Research Unit at the University of East Anglia showing that researchers there had been covering up or tampering with evidence, but the outcry was overheated; there is plenty of other good evidence to consider.

CHAPTER 31

THE IMPOSSIBLE AND BEYOND

Law 1. When a distinguished but elderly scientist says that something is possible, he is almost certainly right. When he says that something is impossible, he is very probably wrong.

Law 2. The only way of discovering the limits of the possible is to venture a little way past them into the impossible.

Law 3. Any sufficiently advanced technology is indistinguishable from magic.

—Arthur C. Clarke[1]

One cannot choose but wonder.

—H. G. Wells, *The Time Machine*[2]

For nearly two thousand years, writers of what we now call science fiction have been conjuring up scenarios involving the Sun. In the earliest days of the genre their stories tended to reflect the values and concerns of their age, but had little in common with anything that we would today think of as science. As time passed and man's knowledge of his universe grew, writers of science fiction (never "sci-fi," they insist) began to anticipate—sometimes by hundreds of years—the discoveries of scientists, so that academic discipline and literary endeavor seemed to leapfrog each other in their speculations, on occasion building on each other in remarkable ways.

The first science fiction we know of was by Lucian of Samosata (A.D. 120–c. 180), a Syrian-Roman rhetorician, whose tale offers interstellar wars, kidnapping by extraterrestrials, and a trip to the Moon, in a pastiche of some of the more outlandish tales in the *Odyssey*. It is called "True History," because the whole story is nothing but lies. A character named Endymion describes how he had been carried away to the Moon while he slept, but has since escaped. He plans to make war on the People of the Sun, whose King Phaethon has refused to allow him to colonize Venus. In the ensuing titanic struggle, the

People triumph and Phaethon builds a high wall that prevents the light from his domain from ever reaching the Moon, thus casting it into permanent darkness.

Those other literary lions of ancient times, both of whom preceded Lucian—Cicero, in *The Dream of Scipio,* and Plutarch, in *Of the Face in the Moon Disk*—had also written works of speculation about celestial bodies, but these were nonfiction. To find anyone else among these early imaginers who wrote science fiction (at about the same time as Lucian, so far as we can judge), we have to travel around the globe to Zhang Heng (A.D. 78–139), the great Chinese court astronomer, who in his story "Meditation on the Mystery" described a journey beyond the Sun.[3] Many centuries later, Kepler would pick up again on the idea of space travel when he wrote the beginnings of a story called "The Dream" involving a boy's trip to the Moon.

The seventeenth century saw a proliferation of epic tales of interplanetary adventure,[4] a fad that, fueled by the explosion in scientific inquiry and the attention given it by writers of all kinds, continued full tilt into the eighteenth. In 1705, Daniel Defoe published *The Consolidator, or Memoirs of Sundry Transactions from the World in the Moon,* in which he anticipates a machine powerful enough to reach our then distant satellite.

Twenty years later, Jonathan Swift published *Gulliver's Travels,* in which, perhaps for the first time, even if only in jest, a writer contemplates the idea that both the Earth and the Sun are mortal. (One of the fears of the people of Lilliput, for instance, is that "the face of the Sun will by degrees be covered in its own effluvia, and give no more light to the world.") When Gulliver visits the great academy at Lagado, he finds that its members live in dread of a comet with a tail "ten hundred thousand and fourteen miles long" traveling too close to the Sun and acquiring "ten thousand times the heat of red-hot iron"— enough to burn up the Earth. "When they meet an acquaintance in the morning," Gulliver reports, "the first Question is about the Sun's Health: how he looked at his Setting and Rising, and what Hopes they have to avoid the Stroak of the approaching Comet."[5] But that is not the only troubling scenario; Swift goes on to satirize their fears about the possible demise of the Sun from sheer exhaustion of fuel, an extraordinarily prescient idea, albeit one Swift himself does not seem to take too seriously.

By the nineteenth century, speculative writers had started to consider the Sun a heavenly body like any other. William Herschel (1738–1822), although not meaning to write science fiction, left science well behind when he suggested that the Sun might be inhabited. E. Walter Maunder, of the Maunder Minimum, summarized what Herschel proposed:

He conceived that it was possible that its stores of light and heat might be confined to a relatively thin shell in its upper atmosphere, and that below this shell a screen of clouds might so check radiation downward that it would be possible for an inner nucleus to exist which should be cool and solid. This fancied inner globe would then necessarily enjoy perpetual daylight, and a climate which knew no variation from pole to pole. To its inhabitants the entire heavens would be generally luminous, the light not being concentrated into any one part of the vault; and it was supposed that, ignorant of time, a happy race might flourish, cultivating the far-spread solar fields, in perpetual daylight . . . but we now know that it corresponds in not a single detail to the actual facts.[6]

Accurate or not, this might seem a promising area for fiction, but writers left it alone. Jules Verne, whom one might have expected to have been curious about the Sun, actually had little to say about it, though he does mention it (if somewhat slightingly) in *Twenty Thousand Leagues Under the Sea* and its sequel, *The Mysterious Island* (1874), a romance rather than a work of science fiction; Verne writes about the effect of equinoxes, the contrasting natures of solar and terrestrial mean time, the rotation of the Earth, and using a gnomon to tell the time, but imaginative leaps about the star were outside his orbit.

A host of other nineteenth-century novelists tried their hands at science fiction at least once—Kipling ("With the Night Mail" and "As Easy as A.B.C."), Poe, James Fenimore Cooper, Twain, Melville, even Henry James—but for a major imaginative work that draws on the Sun one has to wait until 1895, when a London publisher paid £100 (no small sum then for a young man just out of college) for a story of some thirty-eight thousand words—one of the most memorable works of science fiction ever written: *The Time Machine*.

Herbert George Wells (1866–1946) had penned at least three previous versions of the story for various magazines, but in book form it mesmerized the reading public. No one had written a tale quite like it before. A product of its time, and of the distinct sensibility of its author, already a propagandist for his conviction that the future will ever be the direct outcome of existing social forces, it is as much dystopian socialist outpouring as science fiction. Wells imagines an inventor who discovers how to transport himself into the future and return, starting off in the seemingly stable days of the 1890s and ultimately traveling 30 million years forward in time. In this far-off world, the never named Time Traveler sees the last living things on a dying Earth, which "has come to rest with one face to the Sun," one hemisphere forever seared, the other lost to eternal cold and darkness. In the former, menacing crablike creatures wander bloodred beaches amid "intensely green vegetation." The Time Traveler jumps yet further forward, to see the Sun growing ever dimmer and more crimson:

> At last a steady twilight brooded over the earth, a twilight only broken now and then when a comet glared across the darkling sky. The band of light that had indicated the sun had long since disappeared; for the sun had ceased to set—it simply rose and fell in the west, and grew ever broader and more red. . . . At last, some time before I stopped, the sun, red and very large, halted motionless upon the horizon, a vast dome glowing with a dull heat, and now and then suffering a momentary extinction. At one time it had for a little while glowed more brilliantly again, but it speedily reverted to its sullen red heat. I perceived by this slowing down of its rising and setting that the work of the tidal drag was done. The earth had come to rest with one face to the Sun, even as in our own time the moon faces the earth. . . .
>
> So I travelled, stopping ever and again, in great strides of a thousand years or more, drawn on by the mystery of the earth's fate, watching with a strange fascination the sun grow larger and duller in the westward sky, and the life of the old earth ebb away.[7]

Wells was in fact expanding on a theme that has fascinated writers at least since Swift: the question of how long the Sun (and hence our planet) might last. Several works of Wells's time—such as the eminent astronomer Camille Flammarion's *Omega* (1893–94), George C. Wallis's "The Last Days of Earth" (1901), and William Hope Hodgson's *The House on the Borderland* (1908)—rest on the assumptions that the Sun is sustained by chemical combustion and that there will come a time, within the foreseeable future of civilization, when it will burn itself out.[8]

Winston Churchill, in his one novel, *Savrola* (1899), also envisioned a depleted Sun, but he posited this gradual process of extinction as an opportunity, a challenge mankind could meet, given certain moral and psychological qualities. In one episode, his protagonist (a highly idealized version of himself) gazes at Jupiter through a telescope and thinks of "the incomprehensible periods of time that would elapse before the cooling process would render life possible on its surface," which could conceivably lead to some kind of extraterrestrial Utopia. However, in the end he concludes that even if this came about, inexorably "the perfect development of life would end in death; the whole solar system, the whole universe itself, would one day be cold and lifeless as a burned-out firework."[9] This meditation might almost be a précis of the end of Wells's story—both melancholic conclusions written by ambitious depressives, two great men dedicated to improving the common lot.*

By the 1930s, it was known that the Sun produced heat by thermonuclear fusion and that it would one day die out. Clare Ashton, one of the few women then in the genre, produced *Phoenix*, a poignant story about a journey through

Danny Boyle's 2007 film Sunshine *imagined a dying Sun and a manned rocket sent to revivify it by blasting it with a thermonuclear bomb.*

space to rekindle the fading star—an idea used yet again in Gene Wolfe's series *The Book of the New Sun* (1980–83).[†] Although the solar surface temperature had been established in the 1890s, John Mastin also describes such a voyage in

---

* Churchill was a great fan of Wells's work, borrowing occasional phrases from it such as "the gathering storm" and in 1931 declaring that he could even "pass an exam" on his hero. In 1947 he paid an amusing tribute to Wells as a "seer. His *Time Machine* is a wonderful book, in the same class as *Gulliver's Travels*. It is one of the books I would like to take with me to Purgatory." Wells, for his part, created a caricature of Churchill in *Men Like Gods* (1923).

† Film and television have shared an interest in this idea. There has been little interest in the Sun per se (although in *Star Wars* Luke Skywalker sees across the vast desert landscape of his home planet *two* setting suns), films most understandably concentrating on space travel and transspacial rivalries. The premise of the 2007 movie *Sunshine*, directed by Danny Boyle of *Slumdog Millionaire* fame, is that the Sun is rapidly losing power, so a manned rocket is sent into space to set off a rejuvenating thermonuclear explosion: "Our Sun is dying; our purpose—to create a star within a star," as one scientist puts it. The scriptwriter Alex Garland kindly showed me around the set, a few acres of warehouses in southeast London: every detail, even the ship's oxygen gardens, had been checked with astronomers—a possible first.

Through the Sun in an Airship (1909), and H. Kaner set *The Sun Queen* (1946) on a sunspot.

From the 1930s on, science fiction practitioners were beginning to be recognized as something more than just enthusiastic amateurs: in the November 1945 editorial of the influential magazine *Astounding Stories of Super-Science,* the great science fiction editor John W. Campbell roundly declared of the previous decade: "The science-practitioners were suddenly recognized by their neighbors as not quite such wild-eyed dreamers as they had been thought, and in many soul-satisfying cases became the neighborhood experts."[10] Clifford D. Simak's "Sunspot Purge" (1940) and Philip Latham's "Disturbing Sun" (1959) correlated events on Earth with sunspot cycles. Robert Heinlein's six connected stories *The Man Who Sold the Moon* of 1950 ("daring adventures of the bold men of tomorrow . . . a world where energy from the Sun will be converted directly into power") ends with a chart of "Future History 1951–2600," forecasting weather control by 2075 and interstellar space travel by 2100, followed by consolidation of the solar system.[11]

The 1950s began a second golden age for science fiction, and both Cordwainer Smith (the pseudonym of the great counter-Chinese psychological warrior Paul Linebarger, 1913–1966) and Poul Anderson (1926–2001) belong to this period. In Smith's story "The Lady Who Sailed the *Soul*" (1960), he depicts voyagers crossing the heavens on the wings of the solar wind.* Anderson, equally happy writing within classic science fiction or in the new world of science fantasy, interested himself in plausible non-Earth-like planets, the benefits of space exploration, and faster-than-light travel beyond a solar system he regarded as "a hostile barren mess." In his tales "A Sun Invisible" (1966), "World Without Stars" (1966), and "Day of Burning" (1967), his space travelers visit other systems, while in *The Long Way Home* (1955) he envisions the possibility of a solar system that has become the seat of an interstellar empire, Earth turned into an Eden by global warming, with the Sun as an innocent bystander; but then the Sun is hardly ever a guilty participant.

A similar confidence in human ingenuity appears in Theodore L. Thomas's *The Weather Man* (1962), which features technicians who skim the solar surface

---

* As recently as November 2009, the science correspondent of *The New York Times* wrote: "About a year from now, if all goes well, a box about the size of a loaf of bread will pop out of a rocket some 500 miles above the Earth. There in the vacuum it will unfurl four triangular sails as shiny as moonlight and only barely more substantial. Then it will slowly rise on a sunbeam and move across the stars. . . . [It] will mark a milestone for a dream that is almost as old as the rocket age itself, and as romantic: to navigate the cosmos on winds of starlight." By the time you read this, science fiction should have become science fact.[12]

in "sessile boats" to modify its radiation—a characteristically optimistic vision by this chemical engineer turned patent lawyer. In Philip E. High's *Prodigal Sun* (1964), a sudden outpouring of intense solar radiation heads toward Earth, and scientists have to create a shield of gas in the upper atmosphere.

Other writers have imagined the Sun heating up rather than cooling down or disappearing. David Brin's *Sundiver* (1980), featuring a spectacularly close encounter between man and star, shows an exceptional grasp of modern science. In *One in 300* (1954), John D. MacDonald follows ten people who take off for Mars shortly after it has been discovered that the world is doomed: "Earth by this time

On May 19, 2005, NASA's Mars Exploration Rover Spirit *captured this extraordinary view as the Sun sank below the rim of the red planet's Gusev crater. As Mars is farther from the Sun than Earth is, the Sun appears only about two-thirds as large in the Martian sky.*

was dead, boiled sterile." (More fiction has been written about Mars than about any other heavenly body.) Robert Silverberg's "Thomas the Proclaimer," one of three separately authored novellas that make up *The Day the Sun Stood Still* (1972), has the Sun stopped in its path, to become a potential fly trap for spaceships.

None of these authors had much interest in exploring the Sun for its own sake, or making it the center of a story because of what was happening within it; this is true even of Hugh Kingsmill's "The End of the World" (1924) and Larry Niven's "Inconstant Moon" (1971), both based on the idea that the Sun might go nova. One exception is Ray Bradbury, who in his 1953 tale "The Golden Apples of the Sun" sends another *Icarus* thither "to touch it and steal part of it away forever."[13] Left with the job of bringing this venture to life on the page, Bradbury makes a game if not wholly convincing attempt, depicting a huge metallic cup, held by a strangely anthropomorphic hand, that reaches out and scoops the required amount of "precious gas." It proves equal to the task, and the story ends with the rocket heading home, to "take to Earth a gift of fire that might burn forever." Why, the captain asks, this whole exercise of journeying to the Sun, "playing tag, hitting and running"? This time Bradbury's prose gives us a vivid answer: "Because the atoms we work with our hands, on Earth, are pitiful; the atomic bomb is pitiful and small, and only the Sun knows what we want to know." This is not the only time that the Sun has been credited with consciousness, but it rates as one of the more compelling.

Somewhere between wild-eyed dreaming and scientific forecast lies the notion of igniting new suns to replace the current one. In 1957–58, Frederik Pohl and C. M. Kornbluth coauthored *Wolfbane*. The story is set in 2203, after a rogue planet, populated by machines known as Pyramids, has kidnapped the Earth and the Moon and taken them off into interstellar space. "The Sun grew too distant to be of use, and out of the old Moon the Pyramid-aliens built a new small sun in the sky—a five-year sun, that burned out and was replaced, again and again and endlessly again."[14] The novel begins with a character ruminating on whether the latest sun will be regenerated:

> In a week astronomers knew something was happening. In a month the Moon sprang into flame and became a new sun—beginning to be needed, for already the parent Sol was visibly more distant, and in a few years it was only one other star among many.
>
> When the inferior little sun was burned to a clinker *they* . . . would hang a new one in the sky; it happened every five clock-years, more or less. It was the same old moon-turned-sun; but it burned out, and the fires needed to be rekindled. The first of these suns had looked down on an Earthly population of 10 billion. As the sequence of suns waxed and waned there were changes; climactic fluctuation; all but immeasurable differences in the quantity and kind of radiation from the new source.

This is as far as Pohl and Kornbluth carry their notion, and the story soon takes more conventional turns; but it was an interesting variation on the theme of suns burning down—a subject that would be taken up and dealt with more ambitiously in works by Isaac Asimov and Arthur C. Clarke, the two "kings" of twentieth-century science fiction.

Isaac Asimov (1920–1992) was born in the Petrovichi shtetl of Smolensk Oblast, in the Soviet Union, but in 1923 his family moved to the United States, where his parents bought a succession of candy stores, most of which sold sci-fi in the heyday of the old-fashioned pulps. By the time he was eleven Asimov was composing his own stories (in all, he would write or edit more than five hundred books and an estimated nine thousand letters and postcards, publishing works in every major category of the Dewey Decimal System except philosophy). By the age of twenty-one he had written thirty-one stories. "My status . . . was as nothing more than a steady and (perhaps) hopeful third-rater."[15] Then, on March 17, 1941, he visited the offices of *Astounding*, whose editor, John Campbell, had an idea he wanted to put to his fledgling author:

> He had come across a quotation from an eight-chapter work by Ralph Waldo Emerson called *Nature*. In the first chapter, Emerson said: "If the stars should appear one night in a thousand years, how would men believe and adore; and preserve for many generations the remembrance of the city of God. . . ." Campbell asked me to read it and said, "What do you think would happen, Asimov, if men were to see the stars for the first time in a thousand years?"
>
> I thought, and drew a blank. I said, "I don't know."
>
> Campbell said, "I think they would go mad. I want you to write a story about that."[16]

The next day Asimov started "Nightfall," a twenty-eight-pager (13,300 words) that, twenty-three years later, the Science Fiction Writers of America would vote (by a considerable margin) the best of its kind ever written, and that has been so frequently reprinted that, as Asimov himself recognized, "It is now so well known that nothing like it can be published again" (though Asimov later wrote a novel-length treatment of the same story, fleshing out many details, including the final catastrophe). I confess to finding the characters tiresome and the plot lackluster, but along the way it does pose some interesting ideas, particularly about the possibility of the existence of planets that have multiple suns.

The action takes place in Saro City, on the planet Lagash, which over the ages has been bathed in perpetual light from six suns. These are in constant motion, so that at least one is always shining; but when the story opens, one by one, five of the six have passed out of sight. When Beta, the final sun, disappears, the whole planet will face mind-breaking night. The people of Lagash totally lack the mental equipment to deal with such a moment, and are poised to become the latest in a series of nine civilizations that have foundered at the height of their culture.[17]

Lagash's multiple suns have real-life counterparts, as we now know. In

UFO enthusiasts believe that flying saucers are particularly visible during solar eclipses.
The two circular blobs in this photograph are cited as evidence (as photographed by
Mildred Maier of Chicago on June 30, 1954).

2006 a professor of astrophysics reported that researchers continue to discover systems outside our own with at least one planet that has more than one star: one of the most recent, called HD 188753, lies 149 light-years away in the constellation Cygnus, and is heated by three suns.[18]

The theme of replenishing, or replacing, a dying sun—indeed, the idea of making a moon into a sun (as occurs in *Wolfbane*)—had earlier been taken up by Arthur C. Clarke (1917–2008) in his 1951 novel *The Sands of Mars.* In Clarke's story scientists are at work on "Project Dawn," which involves the ignition of the moon Phobos to burn for at least a thousand years and serve as a second sun for Mars, in the hope that the extra heat, together with the mass growth of oxygen-generating plants, will eventually make the Martian atmosphere breathable. The science of *The Sands of Mars* is in some ways ahead of *Wolfbane*'s, which is not surprising, given its author's stature.

Clarke was born in Somerset, in southwestern England, and like Asimov grew up on a diet of American sci-fi pulp. He served as a radar specialist involved in the early warning defense system during the Battle of Britain, and continued to be fascinated by spaceflight. His early stories appeared in fanzines between 1937 and 1945, his first professional sale coming in 1946, and from 1951 on he devoted himself full-time to writing. Many of his later books (he wrote or edited more than seventy) feature a technologically advanced but blinkered mankind confronted by a superior alien intelligence, although overall his writing is characterized by the optimistic view that science will enable humankind to explore the universe.*

Seven of his stories directly concern the Sun. One early tale, "The Star," was published in November 1955 after being entered (and altogether discounted) in a London *Observer* short story competition. The action begins on a spaceship ("No other survey ship has been so far from Earth"), whose mission is to visit the remnants of a supernova, and the crew finds "a single small world circling the star at an immense distance," which had once been host to intelligent life. The long-perished inhabitants have left behind a huge vault, evidence of "a civilization that in many ways must have been superior to our own." As he leaves the supernova, the chief astrophysicist, a Jesuit priest, reflects:

> I know the answers that my colleagues will give when they get back to Earth. They will say that the universe has no purpose and no plan, that since a hundred suns explode every year in our galaxy, at this very moment some race is dying in the depths of space. Whether that race has done good or evil during its lifetime will make no difference in the end: there is no divine justice, for there is no God. Yet, of course, what we have seen proves nothing of the sort.[19]

Armed with a shipful of computer-generated information, he calculates the exact date of the explosion that caused the supernova, which must have outshone "all the massed stars of the galaxy," and the story ends with the revelation that it must be the very same one that the shepherds and wise men saw above Bethlehem. That central conceit is clever enough, but what is most interesting is Clarke's early fascination with yet-to-be-mapped worlds and with the possibility of other suns.

In 1958 he published "A Slight Case of Sunstroke"—a minor offering, but one that uses the huge power of reflective sunlight as a plot device. In the capital of the Latin republic of Perivia, a crucial football match is in progress and passions are running high. Then the referee, obviously paid off by the visiting side, disallows a Perivian goal. A mistake: the "handsome souvenir program"

---

* Around 1972, after sharing a cab in Manhattan, Asimov and Clarke together published their "Treaty of Park Avenue," which proclaimed, tongue in cheek, that Asimov was required to insist that Clarke was the best *science fiction* writer in the world (reserving second place for himself), while Clarke had to insist that Asimov was the best *science* writer (reserving second place for himself). Thus the dedication in Clarke's book *Report on Planet Three* (1972): "In accordance with the terms of the Clarke-Asimov treaty, the second-best science writer dedicates this book to the second-best science fiction writer." But the implicit acknowledgment that Clarke was the better fiction writer is justified.

bought up by the home side's supporters is made of tinfoil, and in seconds fifty thousand makeshift mirrors are trained upon him:

> I never knew, until then, just how much energy there is in sunlight; it's well over a horsepower on every square yard facing the sun. Most of the heat falling on one side of that enormous stadium had been diverted into the single small area occupied by the late ref. . . . He must have intercepted at least a thousand horsepower of raw heat. . . . They play football for keeps in Perivia.[20]

Clarke's other solar-oriented stories are interspersed with his 2001 sagas: *2001: A Space Odyssey, 2010, 2060: Odyssey Three,* and *3001: The Final Odyssey,* all of them written between 1968 and 1997. The 1960 story "Summertime on Icarus" has a more serious tone, building on the original fable* to tell of a spaceship, the *Prometheus,* that tries to get close to the Sun by riding the near-orbiting asteroid Icarus, "the hottest piece of real estate in the solar system":

> Here was a unique opportunity for a research ship to get within a mere seventeen million miles of the sun, protected from its fury by a two-mile-thick shield of rock and iron. In the shadow of Icarus, the ship could ride safely round the central fire.[21]

Such overreaching seems destined to fail, but Clarke must have been in a forgiving mood, for even when Things Go Wrong he allows his main character, Astronaut Sherrard, to survive (with the help of a large metal-foil radiation screen)—though not before he comes terrifyingly close to being grilled alive, and we the readers have been given a lesson on the dangers of getting so close to ultimate power.

Clarke's next solar adventure, "The Wind from the Sun" (1964), described "a yacht race with a difference" in which half-square-mile sails run before the storm of radiation blowing from the Sun. Again, Clarke was ahead of most solar scientists:

> Deep beneath the surface of the Sun, enormous forces were gathering.
> At any moment the energies of a million hydrogen bombs might burst

---

* Daedalus, imprisoned with his son Icarus in a high tower by a vengeful King Minos, constructs two sets of wings, securing them to his own and his son's shoulders with wax. Daedalus warns Icarus that should he fly too high, the wax will melt. Father and son lift away. Exulting in his new-won freedom, the boy soars upward: sure enough, the blazing Sun softens the wax, and Icarus plunges headlong into the sea.

forth in the awesome explosion known as a solar flare. Climbing at millions of miles an hour, an invisible fireball many times the size of Earth would leap from the Sun and head out across space.[22]

Throughout his adult life, Clarke urged that humanity's destiny lay beyond the confines of Earth (hence the *2001* books). His work, such as his detailed forecast of telecommunication satellites, published in *Wireless World* magazine in 1945, two decades before one ever orbited the Earth, was often prophetic. Borrowing a phrase from the nineteenth-century psychologist-philosopher William James, he once suggested that exploring the solar system could serve as the "moral equivalent of war," giving an outlet to energies that might otherwise bring on nuclear holocaust.[23] If this seems far-fetched, consider that in 2004 George W. Bush proposed sending humans to Mars, and that Richard Branson is currently proposing space flights for paying passengers.*

For Clarke's most memorable solar story we have to go back to 1958 and "Out of the Sun." The narrator is aboard a spaceship within the (then believed) twilight zone of Mercury, part of a team exploring what goes on within the Sun at the height of a sunspot cycle: "From the far X-rays to the longest of radio waves, we had set our traps and snares; as soon as the sun thought of something new, we were ready for it. So we imagined." What follows is less a story than a meditation on man's attitude toward the Sun, and is still worth listening to:

> These great clouds of ionized gas moving far out from the sun are completely invisible to the eye and even to the most sensitive of photographic plates. They are ghosts that briefly haunt the solar system during the few hours of their existence; if they did not reflect our radar waves or disturb our magnetometers, we should never know that they were there.[24]

He picks up on his radar scanner a "tight little echo . . . outside all previous records of solar phenomena," which he reckons to be a gas cloud about five

---

* See Gerald Jonas's obituary of Arthur C. Clarke, *The New York Times*, March 19, 2008, C12. Virgin Galactic, an offshoot of Virgin Atlantic, is due to launch *SpaceShipTwo*, a six-passenger hybrid rocket, sometime in 2011. The Sun is some way off, however: the rocket will blast to just sixty-eight miles above the Earth, and passengers will enjoy four and a half minutes of weightlessness—and stupendous views—over a two-and-a-half-hour journey that will cost them $200,000 each. Branson promises that within ten years he will get the cost down to $40,000; already 340 seats have been booked.

hundred miles long and half that figure in width—and seeming to exhibit the characteristics of an intelligent being:

> Today . . . the idea no longer seems so strange to me. For what is life but organized energy? Does it matter *what* form that energy takes—whether it is chemical, as we know it on Earth, or purely electrical? . . . Only the pattern is important: the substance itself is of no significance. But at the time I did not think of this; I was conscious only of a vast and overwhelming wonder as I watched this creature of the sun live out the final moments of its existence.

As soon as the cloud hits Mercury in the center of its daylight side it will be dispersed—destroyed forever:

> Was it intelligent? Could it understand the strange doom that had befallen it? Perhaps, in those last few seconds, it knew that something strange was ahead of it . . . for it had begun to change. . . . It may be that I was looking into the brain of a mindless beast in its last convulsion of fear—or of a godlike being making its peace with the universe.

Whatever one makes of Clarke's flight of imagination, this is science fiction of the highest order—thrilling as a story (yet in total less than six pages), scientifically accurate within its limits, and addressing itself to questions that astronomers would find pertinent today. J. G. Ballard has asserted that sooner or later all the predictions of science fiction will come true,[25] a pardonable exaggeration; more accurate is Clarke's observation that "most technological achievements were preceded by people writing and imagining them." What we come to do to the Sun—or to escape from it—may well play out scenarios that science fiction has already described.

CHAPTER 32

THE DEATH OF THE SUN

If the sun should desert the day, / What would life be?
—COLE PORTER, "Do I Love You, Do I?"

Suns sink on suns, and systems systems crush
Headlong, extinct, to one dark centre fall.
And dark, and night, and chaos mingle all!
—ERASMUS DARWIN[1]

IN THE FINAL VOLUME OF C. S. LEWIS'S *THE CHRONICLES OF NARNIA*, that enchanted kingdom perishes when its sun explodes: "At last the sun came up. . . . They knew at once that this sun also was dying. It was three times—twenty times—as big as it ought to be, and very dark red . . . and in the reflection of that sun the whole waste of shoreless waters looked like blood."[2] So it must be for all worlds, and will be for our Sun, too. Before such a time, we may have to endure, if we can, several other horrors. Isaac Asimov, for instance, in his *A Choice of Catastrophes*, considers whether, with both Sun and Earth slightly on the move from their current orbit, they might collide. He dismisses that possibility; whether the Sun might crash into another star is another matter.

The Sun is thirty-two thousand light-years from the center of its galaxy of a hundred billion stars, which it orbits at about 155 miles per second, taking about 200 million years to complete a revolution, which means that it has completed twenty-four or twenty-five passages in its existence. In the 13.73 billion years of the universe's history, the stars of our galaxy seem to have arranged themselves into orbits that avoid any further collision; yet there remains the possibility of a maverick star or globular cluster of stars changing direction and, if not colliding with the Sun, passing by close enough to alter its orbit, which would surely affect life on Earth.[3]

The chance of any such calamity, Asimov admits, is "very small indeed," pointing out that we would anyway probably have a million years of warning. However, a drifting black hole (his next category of possible catastrophes) might give only a few years' notice: in 2005 scientists detected a hole the size of our solar system that has so far swallowed up the mass equivalent of 300 million stars the size of our Sun, although thankfully it is about 26,000 light-years (153,400 trillion miles) away, or about 63,000 times the distance between the Earth and the Sun. Should a mini–black hole strike the Sun, our star could probably ingest it without major repercussion; on the other hand, there is at least a chance that it could bring about either the Sun's terminal collapse or an explosion: not good news for Earthlings. Asimov goes on to consider clusters of antimatter and nonshining bodies (or "free planets"), both of which might wander into the Sun's path, to disastrous effect: but then, in theory, Britney Spears could win an Oscar. One feels he is ticking off anything at least physically possible, however unlikely.

In the decades since Asimov's book appeared, the possibility of Earth's being hit by asteroids has been rehearsed endlessly. There are enough in our system to worry about: between 1.1 and 1.9 million, with five thousand new ones being discovered every month.[4] Scientists have predicted that a thousand-foot asteroid will miss Earth in 2029 by between fifteen thousand and twenty-five thousand miles, then reorbit, just possibly to score a direct hit a few years later, probably around 2034.[5] The Earth's atmosphere is torn by various objects, ranging from those of basketball size (several a day) to car size (twice a year). Currently NASA's Large Synoptic Survey Telescope is monitoring the heavens for such missiles, and it is here that the Sun plays an unusual role: the telescope has a blind spot, since bodies that are either just ahead of or just behind Earth are easily lost in the solar glare. Sunlight also affects the *rotation* of asteroids less than about a mile in diameter at roughly average distance from the Sun: as they absorb its rays, their spin alters, and they accelerate on their axes "like pinballs in a breeze."[6] Many break up under the stress: but were an asteroid traveling toward Earth, to divert it we would need to change its velocity—according to one recent estimate—by a millimeter per second a decade in advance.*

ONE WAY OR ANOTHER, the question of the solar system's stability has fascinated and tormented astronomers for more than two hundred years. Some-

---

* In February 2005 *The Sun* magazine ran the front-cover headline: END TIMES SHOCKER—KILLER METEOR HEADS FOR U.S. Due to impact by 2012, the meteor "could easily fulfill the End Times prophecies of the Four Horsemen of Revelation." According to this powerful authority, "the five-thousand-foot-wide ball of rock and ice is not only massive enough to destroy the planet, it's wrapped in an unusual, four-layered cloud of dust and debris thousands of miles across." The magazine, however, was equal to the threat, advising its readers: "If you can afford it, *install an air purification system*."[7]

what to the embarrassment of contemporary experts, it remains one of the most unresolvably complicated calculations in celestial mechanics. But long before such calculations could be attempted, superstition and ignorance were advancing their own dire predictions. In Scandinavian myth, for instance, the Sun will eventually fail, to be followed by the three terrible years of Fimbulwinter, when the Earth is caught in the grip of a bone-chilling freeze, after which comes Ragnarok, when eternal darkness settles, and Sun and Moon are both devoured, heralding the end of all things. Such a belief touches a raw nerve: Charles Darwin, writing to a friend in 1865, voiced "his own pet horror . . . that of the Sun some day cooling and we all freezing."[8] The world's passing is often linked to the Sun's demise, only of course, in Darwin's day, the source of stellar energy was not known, since nuclear fusion had yet to be discovered. Now that we know how the Sun gets its power, and that its stores of hydrogen are finite, we can plot how the end may come.

The Sun's life is held in exquisite balance: while it is contracting under its gravitational field, the violence of thermonuclear reaction at its core exerts an expansive force, the two canceling each other out. It can retain its size only if these opposing forces are in sync, which requires an ever more rapid consumption of its hydrogen. At some point that fuel will run out. Our star's photosphere is made up of about 90 percent hydrogen, 9.9 percent helium, and a mix of sixty-seven elements, such as iron, calcium, and sodium, as well as eighteen chemical compounds, reflecting the makeup from which the Sun was originally formed (although nuclear reactions will steadily alter the mixture at the Sun's core, where plasma is compacted to such pressure and temperatures as to induce fusion). The larger any star is, the hotter it must be, and the more rapidly it must consume its hydrogen. More massive stars than the Sun run through their nuclear fuel far more rapidly than does our star, because they produce far higher temperatures.

This is all part of a process that has been going on for billions of years and will continue for many billions more. By at least 2.5 billion years ago, the Sun had contracted to a diameter of 186 million miles; only after that did it shrink to a size smaller than the Earth's orbit. At some point thereafter the Earth was formed (at just about the same time that the Sun began to radiate light). The Sun has continued its contraction, but not at a rate to signal its end anytime soon: it loses several million tons of matter every second, the mass equivalent of the energy it produces through thermonuclear reactions. Another few million tons are lost as solar wind and other particle emissions.

Although the Sun will eventually cool, it has first to go through a long period during which it actually grows hotter. As hydrogen converts into helium at its core, the resulting contraction will intensify the Sun's gravitational field and raise its inner temperature so that it will slightly expand. Eventually it will get so hot that it will initiate new thermonuclear reactions, the core's helium nuclei combining to form nuclei of the next higher elements, such as carbon, oxygen, magnesium, and silicon; but by now its essential balance will be lost to expansion. As the Sun swells, becoming at its maximum some 256 times as wide across as it is today, and 2,730 times as luminous, it will radiate more intensely from its extending surfaces: it will have become a red giant.

For the next few billion years, the Sun will fuse atoms quietly and steadily—only, rather than burning hydrogen, once its principal nuclear fuel starts to become depleted it will reorganize its structure so it can burn the next most plentiful fuel source: helium. To do this it needs an even higher core temperature, so its inner regions will shrink and get hotter still. However, at the same time its outer regions will expand greatly. During its last billion years, it will get 10 percent brighter still, searing the Earth's surface to 3,600°F (2,000°C), hot enough to start melting our planet. As Dennis Overbye paints the scene:

> Skimming over the flame tops of this giant, the bare, burned Earth would produce a bulge in the Sun. But friction would cause the bulge to lag as it tried to follow the Earth. The gravitational tug from the bulge would slow the Earth and would cause it to spiral inward, where friction from gases in the Sun's expanded atmosphere would slow it even more.
>
> Then it would go down [into the Sun].[9]

According to the scientific consensus, we can look forward to about 5.7 billion years of some kind of life on our planet—but not human life, which will have a far shorter span. Our long-range enemies will be heat and lack of carbon dioxide; half a billion years from now, the carbon dioxide concentration will be too low for most plants to continue photosynthesizing, and the Earth's average temperature will be about 120°F (49°C). Over the next billion years, after the extinction of plants, the atmosphere will fill with steam, as it did at Earth's birth, and be bombarded by intense sunlight. Two and a half billion years from now, the last water molecule will have disappeared, and the Earth will be a hot, dead world, covered by molten rock.

It could conceivably escape being engulfed—unlike Mercury and Venus, both of which will be swallowed up by the Sun once it reaches the greatest extension of its red gianthood. The most recent research suggests that although the Earth is at the borderline between being engulfed and surviving, the former fate is the more likely. But if it did remain outside the Sun's great swollen bulk, the enormous heat it would receive is likely to vaporize it.

Computer artwork depicting the Sun, around 5 billion years hence, heating a dying Earth. The oceans have evaporated, leaving salt-encrusted rocks. The Moon passes in front of the massively swollen Sun, already in its first red giant stage.

Should the Earth actually escape being both engulfed and vaporized, what would that mean for us? Would human life even be an option? One scenario is suggested by what happened to a planet in a neighboring galaxy. About twice a month we discover another extrasolar world; about 330 have been documented. In 2007, astronomers came across an Earthlike planet known as V 391 Pegasi, a gas giant at least three times as massive as Jupiter, orbiting about 150 million miles from a faint star in the constellation Pegasus. That star blew up as a red giant and lost half its mass, but did not destroy the planet; maybe, if V 391 Pegasi can survive, so might the Earth.*

As for other possibilities, as Isaac Asimov suggests, "There is, at least, ample warning. If humanity survives those billions of years, it will know for all those billions that it will have to plan an escape somehow. As its technological competence increases . . . an escape may become possible." So not only will we need to construct timetables for the various stages of the Sun's demise, but at the same time we must look for ways to migrate from our planet or to move it away from an expanding Sun. A particularly dangerous time for mankind would be at the end of the first red giant phase, when the Sun may become so luminous that its helium fuses in a gigantic explosive flash. Another period of jeopardy will occur as the Sun sheds its upper layers, thus diminishing its mass

---

* The British bookmakers William Hill, on learning that such a planet had been discovered, cut its odds on the existence of extraterrestrial life from a thousand to one to a hundred to one. For it to pay out, the British prime minister has to confirm the proof of such life officially, within a year of the bet's being placed.

and weakening its gravitational field, which will loosen its grip on the remaining planets (including ours). No longer so closely tethered to their orbital paths, they might intersect one another's orbits, or even collide with the Sun. At this point I am tempted to seek refuge in Woody Allen's take on forecasting the future: "I thought to myself that if our great golden star suddenly exploded, this planet would fly out of orbit and hurtle through infinity forever—another good reason to always carry a cell phone."[10]

But all is not necessarily lost. Human ingenuity is almost endless, and we are already researching the practicalities of moving to another planet in our solar system, or to one of its satellites—or moving outside the solar system altogether. One possibility, remote as it may be, is to use a series of small nuclear bombs to propel a starship: such charges would be exploded in sequence, so that the spacecraft would "ride" on the nuclear flash into either permanent orbit away from the Earth or somewhere more habitable. Another recent idea, outlined in a 2001 paper by Don Korycansky and Gregory Laughlin of the University of California, Santa Cruz, and Fred Adams of the University of Michigan, is that in the same way that space probes can boost their trajectory by playing gravitational pinball with Venus or Jupiter in order to be flung outward, so we, or some version of us, could bring about regular encounters between the Earth and a comet or asteroid, thus widening our orbit and propelling us farther from the Sun.[11] All that is required is for an asteroid about 62 miles (100 kilometers) in diameter to fly past us and transfer some of its orbital energy. The asteroid would then move out to encounter Jupiter, acquire more energy, and on a subsequent encounter pass it on to us. But this would be about as likely as moving an airplane by having a bumblebee go past it every ten years: even the scientists busy developing these scenarios say, "We are not advocating them as policy." One wonders what Arthur C. Clarke would have made of such ideas.

At least we can take solace from the fact that, skeptical as the scientists themselves are, we are talking of so many millions of years hence that the technologies that will be available then are impossible to predict. While the Earth may not survive indefinitely, mankind may outlive its destruction.

If we assume for the moment that we do manage to leave Earth on a nuclear-propelled starship, or some other vehicle of the future, where would we go? In March 2009 the Kepler space telescope was launched, to detect planets outside the solar system that have roughly the same size, conditions, and distance from their stars as Earth. By December of that year a planet—GJ 1214b–was found that was 2.7 times the size of Earth and was orbiting a star smaller and less luminous than our Sun; it was also comparatively near to our solar system—about forty light-years away—and was water-rich.[12]

Our means of making such a planet humanly habitable already has a name—"terraforming." And it may even be possible, since by the time the Sun has begun to expand sufficiently to start cooking the Earth, enough eons will have passed that we should have the technology to have established ourselves on hundreds of such "extrasolar" worlds. As H. G. Wells wrote, "There is no way back into the past. The choice is the Universe—or nothing." His fellow novelist Tom Wolfe, writing on the fortieth anniversary of man's landing on the Moon, asked: "When do we start building that bridge to the stars? We begin as soon as we are able, and this is that time. We must not fail in this obligation we have to keep alive the only meaningful life we know of."*

Not that we should underestimate the extraordinary challenges involved. Should we wish to settle on another planet in our solar system (highly unlikely, as none seems capable of supporting life), we know that Jupiter, Saturn, Uranus, Pluto, and Neptune will remain circling the Sun even after it has col-

* Tom Wolfe, "One Giant Leap to Nowhere," *The New York Times*, July 19, 2009, p. 11. Astronomers currently estimate that there are an astonishing 2×10^{11} planets within our galaxy that could support human life—more than 200 billion (although some experts more pessimistically believe that the Earth may be the only true example). But then the number of possible other universes is said to be 10^{100}.

An artist's impression of a manned base on Mars, modeled on similar bases in the Antarctic; only the artist has set the camp at the heart of a raging Martian dust storm, so casting an eerie and lurid light.

lapsed; and that although their orbits will expand, it will not be by much (less than double), so they will not be moving conveniently nearer to us. Even if such planet-hopping were to turn out to be viable, it is far more likely that by such a time we would have established structures in space capable of supporting large numbers of people, each community ecologically self-sustaining and independent. As the Sun grew hotter, they would adjust their orbits and slowly spiral away.

Such scenarios are not so much science fantasy as science forecast. The astronomer Carolyn Porco voiced the current orthodoxy when she wrote:

> Humanity's future need not be confined to mere survival on our home planet. Other worlds beckon, we know how to reach them and we will once more be outward bound. . . . This won't be a space race so much as a global exodus undertaken by an international community. . . . There could be no better way to say: the future is boundless, and it belongs to us.[13]

And if that doesn't seem visionary enough, Michio Kaku, in his book *Physics of the Impossible*, tells us that "physicists have now demonstrated that a law that prevents time travel is beyond our present-day mathematics. . . . Statements about what is possible and impossible have to take into account technologies that are millennia to millions of years ahead of ours."[14] In other words, we cannot rule out anything that we cannot prove impossible.

Though we may be able to dodge catastrophe—in what form, who can say, because the length of time involved is so great that the human race will likely have been succeeded (or will have replaced itself) by an entirely different species—the Sun will not. Eventually its surface will cool to a point where it glows red-hot rather than white-hot, as in its early days, and enters a second red gianthood. Once hydrogen fusion is no longer the main source of its energy, it can maintain itself for only a comparatively short additional period. It enters a brief unstable phase of diminishing energy returns as it depletes its helium and goes on to burn progressively heavier and scarcer elements—beryllium, boron, carbon, nitrogen, and oxygen—each one for a shorter period. The ability of this new giant to keep itself distended against the pull of gravity will falter, and it will begin to collapse, shrinking and expanding before shrinking for one last time. Over the span of a couple of hundred million years (a very brief period on the scale of most stellar lifetimes) the Sun will become lighter, its outer layers blowing into space in the form of a much denser solar wind. It will not be large enough to explode violently, so there will be no danger that in a few hours of fury the entire solar system will be scoured of life. It will simply contract into a white dwarf, a tiny shrunken cinder probably smaller than Mars, but with a density of two tonnes or more per cubic centimeter—a teaspoonful of its matter would weigh as much as a Rolls-Royce—leaving behind at most a thin film of its outer layer, making of itself a planetary nebula.

Once a white dwarf, it will be a dense ball of carbon and oxygen, no more than a dot of light even within its own system. Seen from the satellites of Jupiter (supposing mankind has pitched its tents there), it will be only 1/4000 as bright as the Sun appears to us now, and will deliver but a fraction of that energy as it fades away over the rest of time. Conceivably, long before the moment of white dwarfdom comes, future space settlers will have developed some form of hydrogen fusion power stations, and so would be independent of the Sun; more likely, they will have left the solar system altogether.

As for the Sun itself, as one writer has envisioned its last days, "At 100,000°C, this pinprick of white light . . . will continue to shine for billions of years, illuminating the giant gas clouds from the Sun's expelled outer layers to form a glorious, multi-colored nebula—our Sun's tombstone."[15]

Not quite: a final stage is yet to come in that long-drawn-out death. As the Sun cools, it will become a black dwarf, with the detritus of the planets circling it. It will not detonate into a supernova—again, it is too small; nor will it become a black hole through implosion or any other means. Our great, terrible, even lovable companion will simply dwindle away, a dark cinder of degenerate matter drifting in the vacuum, its life-giving journey done.

SUNSET: THE GANGES

When I, sitting, heard the astronomer, where he lectured with
  much applause in the lecture-room,
    How soon, unaccountable, I became tired and sick;
Till rising and gliding out, I wander'd off by myself,
  In the mystical moist night-air, and from time to time,
  Look'd up in perfect silence at the stars.

—WALT WHITMAN,
"When I heard the Learn'd Astronomer"[1]

Let us obtain the adorable splendor
of the Sun; may He arouse our minds.

—RIG VEDA[2]

*S*OLES OCCIDERE ET REDIRE POSSUNT," WROTE CATULLUS: SUNS CAN SET
and rise again. In English, the word "sunset" goes back at least to the 1440s.
Shakespeare, as one would expect, wrote of them; so too Milton, Byron,
Browning, Longfellow, Frost, Ginsburg, and Emerson, as well as many writers
of less fame. Different languages summon up different associations—the
French *soleil couchant*, for instance, or the German *Abendrot* (meaning "the red
glow of the evening sky"); but in nearly all cultures, the splendors that light up
the horizon at the Sun's going down are celebrated for their beauty and ex-
ploited for their symbolic potential.

Films, and sometimes books, end with hero and heroine departing into the
sunset. Deathbed sayings are often sunset references, from Tiberius' contemp-
tuous dismissal of his attendants ("You know when to abandon the setting and
hasten to the rising Sun!") to Jung's "Help me out of bed; I want to see the sunset."[3] Stephen Vincent Benét (1898–1943) won a Pulitzer Prize for his book-
length poem *John Brown's Body*, in which he describes Brown's majestically
plain address to the court upon being sentenced to death:

Here is the peace un-begged, here is the end,
Here is the insolence of the Sun cast off,
Here is the voice already fixed with night.

Whether seen over an ocean or across a vast prairie, from a great height or simply from inside a city apartment, sunsets are vivid experiences that move people ever and everywhere, sometimes to melancholy, sometimes to rapture. Winston Churchill, in his latter years, would travel to the Atlas Mountains to paint, because sunsets at the desert's edge are so exquisite. Thomas Hardy, when a boy, would sit on the stairs at his parents' home and watch as the evening Sun awoke a special intensity in the Venetian-red walls of the staircase.[4] Andy Warhol once filmed an entire sunset, with no action except the dying light and a jet's contrail drawing across the sky. During the imperial entertainments he put on in British East Africa between the wars, the prodigiously rich Sir Philip Sassoon, cousin of the poet, had the Union Jack hauled down from his mansion's roof because the colors clashed with the evening sky. Rangers in Yellowstone National Park are said to be "paid in sunsets," while every evening at the eighteenth hole of Pebble Beach Golf Links, in California, a bagpiper laments the close of day.[5] Perhaps no one on Earth, however, has seen a sunset as spectacular as those viewed from space. Astronauts have spoken of being overwhelmed by the prismatic radiances spreading across the face of the Earth.

Artists, of course, are particularly responsive. Van Gogh wrote of sunsets during the mistral:

Toward sunset it generally grows a little calmer, then there are superb sky effects of pale citron, and the mournful pines with their silhouettes standing out in relief against it with exquisite black lace effects. Sometimes the sky is red, sometimes of an extremely delicate neutral tone, and again pale citron, but neutralized by a delicate lilac.[6]

In 2007, a group of scientists produced a study of 554 paintings of sunsets, including many of Turner's 115 renderings, as well as works by Lorrain, Rubens, Rembrandt, Gainsborough, Hogarth, Copley, Degas, Caspar David Friedrich, Alexander Cozens, and Gustav Klimt, to help understand changes in climate. They found 181 artists who had painted sunsets between 1500 and 1900, then used a computer to work out the proportions of red and green along each picture's horizon.[7] As we now know, the magnificence of sunsets depends on cloud formation, or the level of particles in the air. The more airborne particles, the more sunlight scatters toward the red; so, the dustier the sky, the redder the sunset. Most of the paintings with the highest red/green ratios were painted in the three years following a documented eruption, there being fifty-four such "volcanic sunsets." Turner (all unknowing) recorded the effects of three: those of Tambora, in what is now Indonesia, in 1815; Babuyan, in the Philippines, in 1831; and Cosiguina, in Nicaragua, in 1835.[8] Though painters are not savants attempting precise depictions of nature, it is interesting to realize that such visions as *The Fighting Téméraire* (1838), once sharply criticized for its outrageous coloring, may be truer representations than once thought.

ONE OF SEVERAL sunset sites on the Internet publishes a list of the world's top ten spots for sunset viewing:

10. Key West, Florida
9. Ipanema Beach, Rio de Janeiro
8. The Maldives
7. Paradise Island, Bahamas
6. Natadola Beach, Fiji
5. Grand Canyon, Arizona
4. Great Pyramid, Giza, Egypt
3. Anchorage, Alaska
2. Kaunoao Beach, Hawaii
1. Oia, a village on Santorini, the shattered volcanic island off Turkey

I don't begrudge each its appeal, but I wanted to witness a sunset that *meant* something in addition to being visually spectacular. In October 2006 I set off, not for Hawaii or Los Angeles, but for the city of Varanasi, on the banks of the Ganges, taking a few Sun-related side trips along the way.

After landing in New Delhi, and visiting Jaipur and its famous ancient observatory, I flew on to Udaipur, about 250 miles farther south. October is the month of Diwali, the Hindu festival of light, and October 21, the day I arrived in Udaipur, the Hindu New Year. I had arranged that night to see the maharana, Arvind Singh Mewar, known as Sriji (noble leader), and received at my hotel a printed invitation for my audience, the envelope embossed with a Sun sporting a mighty, whirly mustache, accepted emblem of virility.

At 6:50 P.M. I presented myself at Shambhu Niwas Palace, the buildings ablaze with Diwali lights, and was led onto a large patio overlooking the famous lake of Udaipur, where Sriji was waiting. He was little more than five foot three, his outsized head adorned by a fittingly luxuriant mustache and full beard beneath lazily watchful eyes. Sixty-two years old (as he told me), he was wearing light-colored pants, an untucked striped shirt, and loafers, as if holidaying at some Caribbean resort. He asked a manservant for a gimlet (gin and lime), and looked disappointed when I opted to stay nonalcoholic.

"Now: what would you like to know?" Having learned that he had an interest in solar energy, I asked about his fourteen sun-powered vehicles for hire, from seven-geared pedal rickshaws with two seventy-five-watt panels to scooters with twelve-volt lead-acid batteries—and, on the river, a Surya-powered water taxi. Udaipur was already known as "India's solar state."

Sriji's response was brisk. "Oh, yes, all that. It's really not important." He turned instead to his family tree, whose roots go back to A.D. 569 and are ultimately "descended from the Sun," which is why his family crest shows the star so prominently. "For me the Sun is a human being," he said easily. Really? A mass of gases, human? Without missing a beat, he replied, "Well, maybe not a human *being*, but in terms of spirit, energy, yes, in what he gives to us, I *do* mean that the Sun connects in a special way. He is a god, a part of us. I'm not an intellectual, but it's what I feel, and I suppose what I'm saying is about how we *relate* to the Sun. People in the West seem to have lost that connection, to have lost their way.

"I think we need to redefine 'divinity,'" he went on. "Because the Sun *is* divine—only not in the way the word is used in your part of the world. It doesn't so much bring us light as take away the darkness. Each of us needs to find the Sun in ourselves."

The star had certainly endowed him with more than average energy. In his youth, he had been a fine cricketer, an able bat who represented Rajasthan. More recently, he had overseen the building of a dozen museums and libraries, put together a vintage car collection, formed a polo complex, managed a successful stud farm, pioneered local music initiatives, and, helped by a postgraduate course in hotel management, has doubled his chain of palaces into luxury hotels. He was chairman of three gem companies and several religious and educational trusts. In an era when most Indian princes are looking for a role and worrying about money (Indira Gandhi abolished the princedoms in 1971), he has both.

Three years later, in July 2009, the prime minister of India dedicated his country to the most ambitious solar energy development program in the world. All government buildings are to have solar panels by 2012, and 20 million households are to have installed solar lighting by 2020; to achieve these and similar goals, the government plans to invest 920 billion rupees (about $20 billion). But Sriji was the pioneer.

OVER THE NEXT FOUR DAYS, I visited Modhera, in Gujarat, to see one of the great sun temples, thence to Ahmedabad, Delhi, and finally, my ultimate destination, Varanasi, where I was met by my guide, Ravid ("It means 'Sun,'" he told me). Aged thirty-eight, he looked like a lead from a Bollywood movie, except he was too slim and unjowly, and his hair was too obviously treated with henna. One of the taller Indians I had met, he strode, gangly and loose-limbed, in sockless shoes, a silver bracelet on his right arm. Over the next three days he would oversee my education.

Varanasi is Hinduism's holiest city, and predates the modern metropolises of India—Chennai (Madras), Mumbai (Bombay), Kolkata (Calcutta), and New Delhi—by at least two thousand years, making it one of the most ancient in the world, as old as Jerusalem. It has a permanent population of 1.1 million. Here the faithful come to die, or their bodies are brought to be cremated (as was Mrs. Gandhi's), so absolving them of the burden of *samsara*, or reincarnation. No city is more symbolic of Indian culture. Called Banaras, or Benares (attempts of the ruling British at the Mughal version of the ancient Hindu name) until independence, it is also known as Kashi—from the Sanskrit root for "to shine," thus "the City of Light."[9] The wordplay underlines the relation of the name to enlightenment. It was here that India's most famous seeker, Siddhārtha Gautama, putting aside his life of luxury, preached his first sermon. Before long, he was acclaimed as the Buddha.

All the gods in the Hindu pantheon are manifest in Kashi—no need to go anywhere else, say the locals; everything the soul requires is present. But while

An 1831 photograph of Dasaswamedh Ghat (literally, "the riverfront of ten sacrificed horses"), one of the oldest and holiest Ganges landings in Varanasi

primarily a holy city—host to fifteen hundred temples—it is also renowned for its universities (foreign visitors have named it "the Athens of India") as well as for its courtesans, poets, thugs, brass shops, silk palaces, mangoes, *pan* leaves—even its sweets.

That first day, Ravid guided me to Lolarka Kund ("Trembling Sun"), one of the two shrines mentioned in that supreme Sanskrit epic of ancient India, the Mahabharata. It was set in a small square similar to those behind the main piazzas in Venice, but from this simple rectangle thirty-five startlingly steep steps descend to a bathing pool fifty feet down. The shrine was built up to its current size by a local ruler in the eighteenth century, after his bathing here had cured his leprosy. During the month of Bhadrapada, explained Ravid—sometime in August or September—the faithful flock in homage and in hope. Infertile women pray for their barrenness to be lifted, the pregnant for the good fortune of a son. To effect the latter, married couples must bathe together, the wife's sari tied to her husband's dhoti. As she bathes, the woman is expected to release some vegetables (a symbol of fertility), usually a variety of squash, into the water. Those suffering skin diseases leave their clothes behind. How many people come to so small a spot? "Twenty thousand, all in one day, at most in two."

Ravid was delighted to display his learning as he led me through the city's most cherished buildings. Hindu tradition is more than a lifetime's study—its complex myths, its manifold divine identities, its elaborate rituals, and its teachings about life and death. Those who lead a good life have brightness in their faces—*dhejas*. The Sun is compassionate, healing, friendly, Ravid continued, but Hindus believe that you must not follow a shadow, for it entails misfortune, nor tread on another's shadow, for it will drive him mad.

By now it was nearly six in the evening, and few people were about, just some children playing hopscotch, the odd mangy dog, and some monkeys. After sunset, the better class of person stays at home, while Untouchables scavenge for leftovers from the public wells or collect the warm droppings of cows to sell for fuel. It was time to return to my hotel.

The next morning I was up by 5:40, and was soon joined by Ravid, whose forehead was marked by a *roli* of red turmeric: different sects have different colors and signs, and red is for sun worshippers. We wound our way through the crumbling stone alleys, many no wider than footpaths, then walked along the row of ghats, or landings, whose staircases extend rootlike into the river. Some of the landings (eighty-four in all, stretching along just one riverbank) were almost deserted, others thronged with people, as we wended our way past waking fakirs and stallholders putting out their wares—bright orange garlands, ghee for lamps, costume jewelry, brass pots, trays and bowls, wooden toys, plastic water cans, and bottles of water. The evening before, I had read a late-nineteenth-century traveler's description:

> Up and down the ghâts, all day long, but especially in the early morn-
> ing, stream the endless course of pilgrims, ragged tramps, aged crones,
> horrible beggars, hawkers, Brahmin priests, sacred bulls and cows,
> Hindu preachers, wealthy *rajas* or bankers in gay palanquins, Fakirs,
> pariah dogs, and scoffing globetrotters. . . .[10]

Not much seemed to have changed in the hundred and ten years since.

At last we came in sight of the skiff Ravid had hired, and trod carefully over the hardened mud of the sandbank to board it. Our oarsman was about forty, small but sinewy and seemingly impervious to my presence: for him, just another day at the office. Although it was not yet dawn, there was light enough, and as we began to move slowly down the river I could see some people already at prayer, a few having their hair or beards cut by a street barber, others preparing themselves for immersion in the great Ganges. Before sunrise, the water is said to be not just warm, but *hot;* after sunrise, cooler. Generally the women come to bathe first, as early as 4 A.M., and will go to any of the ghats that line the water's edge.

Varanasi sits between two streams, the Varana ("Averter"), which flows into the Ganges on the north, and the lesser Asi ("Sword"), which joins the river on the south. Ravid was defensive about the Ganges's reputation for being heavily polluted. The great river had amoebalike bacteria in it, he said, and these eat harmful bacteria. *All* harmful bacteria? *Much* harmful bacteria.

We followed the river, its ghats alternating with the pavilions, temples, and terraces that crowd the banks, most of them hypnotically beautiful in the morning light, which at sunrise tinted them a uniform coral red. The local cactus, *Religios perseca,* seemed to sprout everywhere. As we passed Digpatiya Ghat, a chant rose from a gathering of priests. We slipped by an imposing twelfth-century ruin overrun by vagrants and goats. A wide expanse of riverbank had been covered with saris laid out to dry—a feast of color, the full length of each nearly twenty feet. Inevitably, Mark Twain had come by here, and for once wrote without aiming at comic effect:

> The Ganges front is the supreme show-place of Benares. Its tall bluffs
> are solidly caked from water to summit, along a stretch of three miles,
> with a splendid jumble of massive and picturesque masonry, a bewil-
> dering and beautiful confusion of stone platforms, temples, stair-
> flights, rich and stately palaces—nowhere a break, nowhere a glimpse

of the bluff itself; all the long face of it is compactly walled from sight by this crammed perspective of platforms, soaring stairways, sculptured temples, majestic palaces, softening away into the distances; and there is movement, motion, human life everywhere, and brilliantly costumed— streaming in rainbows up and down the lofty stairways, and massed in metaphorical flower-gardens on the miles of great platforms at the river's edge.[11]

As the Sun rose in the sky, I could see men and women of all ages practicing breath control, meditation disciplines, or the yogic sunrise ceremony, Surya Namaskar.*

Some men were already out hosing down the silt piled up by the seasonal flooding, while others raised hands cupped to hold the sacred water in supplication, or clutched the ceremonial *kusha* grass used in performing their rites. Most of the men were burly, certainly not slim: following my eyes, Ravid ex-

---

* Since my visit, the question of whether schoolchildren should be required by law to take up the Sun salutation has become a serious topic. In January 2007 the Hindu-Nationalist–led government of the state of Madhya Pradesh put forward a measure that required public school students to salute the Sun and recite certain chants in Sanskrit, but Muslims and Christians took issue, saying the chants were essentially a Hindu rite, violating the constitutional separation of faith and state. A court later ruled that neither the chants nor the salutation should be compulsory, but the issue remains contentious.[12]

A morning scene of the Ganges, showing youths worshipping the Sun with raised hands— unlike all other forms of prayer in Hindu India, which are offered with hands joined

plained that Varanasi is famous for its wrestlers. Everyone seemed caught up in his or her own world. An elder of seventy, completely bald, the three deep furrows on his forehead etched in yellow, gazed enrapt toward the horizon. Three women were busy washing one another using a plastic bag as a cloth, two girls splashing about joyfully. An older woman was doing her best to preserve her modesty while bathing in a minimal sari. A beautiful mother of about eighteen carried a baby, turning around seven times as the holy books prescribe, to cleanse herself and her child before the gods. Several bathers, up to their waists, were scouring their teeth with twigs from a holy tree. After their ablutions, most of those I watched crowded into temples along the river, their hands full of jasmine and sweets destined for the lap of a favored god, or else they carried the waters home in polished pots—made from the brass for which the city is celebrated.

One of the greatest Indian poets, Kabir (1440?–1518), although he lived in Varanasi, couldn't resist heaping disrespect on his hometown:

> *Going on endless pilgrimages, the world died,*
> *Exhausted by so much bathing.*[13]

Maybe so, but those who had braved the waters were not the only ones making use of the river. As we glided along, Ravid pointed out Manikarnika Ghat, the larger of the two main places of cremation (the only points along the river where photography is forbidden, although bodies are immolated almost everywhere). This burning ground has traditionally gone by yet another name: Jalasai Ghat, the "Sleeper on the Waters," and hosts a sacred fire said to have burned constantly for all history. Farther along, at the other main cremation center, Harishchandra Ghat, there now stands an electric crematorium where funeral pyres once blazed. But most prefer the old way.

All may be cremated, Ravid said, except snakebite victims, pregnant women, children, and the victims of smallpox. Cremation by flame costs 551 rupees, the mechanized version 151. The burnings go on without cease, each taking three to four hours: the great stacks of wood that line the shore, higher than houses, are forever being replenished and depleted. As we drifted slowly by, smoke wafted up from the low-lying pyres to curl about the temples that surround the sites. I could smell charring flesh.

When a body arrives—dead women draped in orange, men in white, explained Ravid—it is dipped into the Ganges, then taken to the funeral pyre, where it is sprinkled with sandalwood oil and garlanded. Tall bamboo poles dot the scene, each hung with a small basket filled with oil lamps to honor the ancestors of those passing through the fire. Wherever possible, the eldest or youngest son of the dead will take the central part in the ceremony. He will have had his head and eyebrows freshly shaved and will be clad in a seamless long white shirt, with the sacred thread, which usually hangs from the left shoulder, cast over the right. He will lead the mourners five times around the unlit pyre, counterclockwise (because, in death, everything is reversed), then he will light the pyre. Once the body is nearly consumed, the chief mourner performs the "rite of the skull," cracking the head with a bamboo stick to release the soul from entrapment. Then he turns to cast a pot of Ganges water over his left shoulder, dousing the embers, and strides away.

I wanted to venture beyond the formal line of ghats, so Ravid and I disembarked and walked on through the city. Before he left, I asked him to join me that evening so that during the rituals he could explain anything I didn't understand. He grinned. "Of course." Back at my hotel, I returned to Diana Eck's classic history *Banaras: City of Light*. A distinguished visitor to Varanasi, she writes, "described the dawning riverfront as 'one vast sun-temple,' and yet it is not the Sun alone which is honored. It is the Divine, of many names and forms, here physically manifest, present in warmth and light."[14] How much of this devotion centers on the Sun, and how much on other powers?

I had a chance to ask these questions a few hours later, when Ravid accompanied me to a meeting he had arranged with Dr. Sudharka Mishra, an eminent scholar of Sanskrit, at the professor's house. My host was a friendly if earnest figure, as intent as Ravid on enlightening my ignorance, and began by explaining something of the Indian attitude toward eclipses (as related in chapter 4), breaking off only to admonish one of his many children, who scampered about in delight at having such an unusual visitor.

Soon it was time for me to be instructed in the three levels of sun worship: the mythic, where the Sun is part of a ritualistic framework; the astronomical, where it is the king of stars (in southern India, for instance, the basis of the calendar); and the everyday, where it is an intrinsic part of normal life. "Without sun worship," Dr. Mishra concluded, "nothing is going to happen in Benares." The Sun has a soul, he said—*sutrama*. Inside each of us is a small map of the universe (somewhere near the bridge of the nose): proper meditation releases this knowledge, and the trillions of galaxies within us. Hinduism recognizes 330 million gods. . . .

I interjected that that was a crazy number, perhaps a metaphor for the many facets of God—"No!" he interrupted, suddenly leaning toward me. There really are that many, and Hinduism honors each one. In the vastness of the universe, is that so surprising? The West likes its cleanliness outside, not inside, he added, so Westerners do not understand. "We have no solution for you.

What is God? That is what you ask. Just as useful to demand, 'What is the color of pain?' There is no answer." Nor could I come up with one, although my mind went back to the white shrouds of the funeral pyres.

By the time Ravid and I took our leave and strolled together back into town, the Sun was about to set, and two of the main ghats were filling up, although the honoring service would not start until at least half an hour later, around 6:30. The largest landing, Dasaswamedh Ghat, already held about four hundred people, many of them having taken off their shoes, including the women, the majority dressed in their finest saris. Rows of pilgrim-priests huddled under bamboo umbrellas on low wooden *chaukis*.

Seven small altars, draped in fine orange silk (ruby silk being reserved for the central altar), were lined up at the river's edge, and acolytes passed from one to another, checking that all was in order: flowers, incense, joss sticks, the special fan each priest would use, conches, bells, combustibles, the *prasada* (consecrated food). I wanted to ask Ravid about all these, but he had wandered off into the melee.

I found a space directly behind the main altar, giving an unimpaired view. A large portrait had been placed on the altar, framed with white and red flowers: Arsalan Bapu, a local holy man. Beside me, a whippet-thin devotee of about fifty was already at prayer; I felt awkward stealing glances at him. Above the altars a large sign flapped in the wind: GAYA NIDHI SEVA, the major corporation that subsidizes the nightly ceremony. The stone steps afforded easy sitting, but many would still have to stand. Behind the altars, scores of tourist-laden boats had started to jockey for position, while minute leaf coracles bobbed their lamps of ghee and camphor, a long galaxy of lights under a perfect crescent moon. From somewhere along the river rose a rhythmic popping of firecrackers. The atmosphere was as charged as it is in a theater just before the curtain goes up. At last seven young priests, clad in orange shirts, light orange shoulder sashes, white pants, and strings of beads, took up their places at the altars.

Suddenly Ravid appeared by my side, evidently agitated. "You must come with me," he said sharply. I rose and followed. He took me to a small boat manned by the same oarsman we had had that morning. For a moment I thought he wanted to ensure a good vantage point, but after a bit he calmed down enough to explain: the year before, a bomb had been set off almost exactly where I had been sitting, and twenty worshippers had died. The government had whitewashed the atrocity as a simple matter of "some gas cylinders exploding," but locals were sure it was the work of provocateurs planted by Pakistani intelligence. No evidence was offered either way. The bomb was said to have been set for 6:45, when it would have killed several hundred people, but it exploded at 16:45.

Soon after I boarded our skiff, the seven priests, their black hair glistening under the lights, began to intone a great rumbling "Ohhhmmm,"* and I began to lose myself in the extraordinary tableau around me. Huge, vividly colored umbrellas, each a good eight feet across, lined the shores—maybe twenty on one side of the ghat, thirty on the other. More than a thousand spectators had gathered on the shore by now, on the river a further five hundred, the boats so densely packed that a young girl selling lamps skipped shoeless from one to another.

The priests eventually ceased their chanting and began to sound conch shells, turning east, north, south (for other gods), and finally west (to honor the Sun). The conches were meant to echo the "Om" resonance, Ravid explained. Next they raised seven-branched candelabra, symbolizing the world's hierarchy of species, and began to chant again, to a new rhythm. Once this part of the service was completed, they raised their outsized fans—like peacock feathers, only tasseled and made from cow hair, which Ravid explained repels small bees and flies as well as disease and bacteria. Then once more they made their four turns, as they invoked Vishnu, then Shiva.

Upon another surge of *Ohhhmmm*, less synchronized than before, there arose a clamor of bells, the priests chanted for peace in the world—*Om Shanti*—then the high priest cast flowers over the dark current as a final homage to his many gods, the Sun, the Moon, the planets, and the stars, and the ceremony was over. The great assembly began to break up, and we pushed off, first heading upriver where the cremation flames still smoldered, the rich reds of the pyres yielding new colors to the evening. Near the bank, the water was strewn with human ashes and bits of bone. By now it was nearly eight o'clock and the neighboring ghats were for the most part deserted, except for the occasional stray animal prowling for scraps.

In Geoff Dyer's recent novel set in Varanasi, his narrator offers a dyspeptic view of the rites I had just watched: "You didn't have to be a particularly discerning tourist to see that this was an exhausted pageant, drummed up for tourists, a *son et lumière* with a cast of hundreds. Any significance it was supposed to have had had been drained, possibly a long time ago or maybe just yesterday, or even now, right before our eyes."[15] While I do not deny that the tourists may have gone off to their beds happy but completely ignorant of the meaning of what they had just seen, I sensed a genuine reverence among the celebrants themselves. As with the Cuzco celebrations for the summer sol-

---

* "Om," sometimes written "Aum," is "the verbal symbol of the Supreme Reality," and is said to consist of five separate sounds, "A," "U," and "M," plus the nasalization (*bindu*) and the resonance (*nada*). One temple in Varanasi consists of five shrines, one for each of the notes.

stice, or the noisy crowds circling Stonehenge, or even that small group of travelers who watched the Moon pass across the Sun above the Antarctic ice, a feeling of awe, even something like fear, had filled the air as tribute was paid to forces that hold the power of life and death.

After a while our oarsman turned us about and pulled hard against the current toward the steps closest to my hotel. The air was chill. Above one of the main buildings, a banner read, in English, BANARAS IS BANARAS. BANARAS WAS BANARAS. BANARAS WILL BE BANARAS. We passed by Dasaswamedh Ghat, where the young priests and their helpers were clearing the fittings and instruments away, like orderlies in an operating theater. They would perform the same ceremony the following sunset, and the next. In their language, *kal* means both "yesterday" and "tomorrow."

ACKNOWLEDGMENTS

Chasing the Sun has taken almost eight years to research and write, during which time I have visited eighteen countries. That such trips were possible—vastly expanding what I first envisaged for the book—is due to the generous grant I received from the Alfred P. Sloan Foundation. To the foundation and particularly to its director, Doron Weber, I am extremely grateful.

Such an undertaking also required the help of an extraordinary range of people. To particularize: material on Japan was checked with Yoshiko Chikubu and by Junzo Sawa and Hamish Macaskill (of the English Agency, Japan); that on Peru by Marie Arana of *The Washington Post;* solstice customs by Professor Ronald Hutton of Bristol University; and North American dance rituals by Susan Gardner, my colleague at Kingston University, and by Wade Davis, explorer in residence at *National Geographic.* The chapters on eclipses were read by Thomas Crump, whose own book was so useful to me, and by Jay Pasachoff, Field Memorial Professor of Astronomy at Williams University.

For the chapters on ancient Babylonian, Egyptian, and Greek solar astronomy I received early advice and encouragement from Richard Parkinson, assistant curator of ancient Egyptian pharaonic culture at the British Museum; Christopher Walker and Michael Wright, two other BM alumni; Sir Geoffrey Lloyd, master of St. Catharine's College, Cambridge; and Dennis Rawlins, editor of *Dio.* I would single out for special thanks James Allen, ex-curator of Egyptian art at the Metropolitan Museum in New York: we came to an agreement that he would explain Egyptian culture to me if afterward I would give him a fencing lesson, so that well after the museum was closed to normal visitors, the mummies of ancient Egypt would look benignly down on us as our sabers clashed.

Chinese solar history was read early on by Nathan Sivin, professor of Chinese culture and the history of science at the University of Pennsylvania (though we did not always agree); Simon Winchester, recent biographer of Joseph Needham; and Christopher Cullen, head of the Needham Institute in Cambridge. The chapters from Copernicus to Newton were checked by Owen Gingerich, professor emeritus of astronomy and of the history of science at Harvard University. Brian Cathcart, professor of journalism at Kingston University and author of *The Fly in the Cathedral,* provided some invaluable advice on the physicists who helped build the atomic bomb.

On tree-ring details I was advised by Dr. Mike Baillie, of Queens University, Belfast. The chapters on how the Sun affects our bodies were read by Dr. Seth Orlow, chair of dermatology at NYU Langone, while Linda Prestgaard at the Norwegian embassy in New York checked details about Norway and other matters to do with SAD. David Davidar of Penguin Books told me about the Indian skin-bleaching techniques he employed for his novel *The House of Blue Mangoes.* The material on photosynthesis and other matters of natural history was read by Dr. Oliver Crimmen, head curator at London's Natural History Museum, and by Dr. Olivia Judson, the author and journalist. The details about animal migration were checked by Dr. Andre L. Martel and Pierre Poirier, of the Canadian Museum of Nature, and Henry Bouchard, of the U.S. Fish and Wildlife Service.

Details about solar navigation were corrected by Robert Ferguson, author of *The Vikings;* Professor Steven Walton; Dr. Richard Dunn, curator of the history of navigation at the National Maritime Museum, Greenwich; and Gloria Clifton, head of the Royal Observatory, Greenwich. The section on sundials benefited from the expertise of Dr. Frank King of Churchill College, Cambridge. My visit to Freiburg and the details on solar energy research generally were checked by Larisa Kazantseva of Nitol Solar and by Thomas Dresel of the Solar Energy Center in Freiburg, and my Spanish trip by José Solder of the Plataforma Solar de Almería.

The chapter on the Sun in art was read through by Turner's recent biographer James Hamilton (who "teased the text a bit"), and the passages on cinema by Kenneth Turan, film critic of the *Los Angeles Times.* For much of the material on classical music I am indebted to the late Gabriela Roepke, professor of opera at the New School. My musings on Nabokov were read by Azar Nafisi, author of *Reading Lolita in Tehran,* and on literature generally by two author friends, Clare Asquith and Betsy Carter. The chapter on politics and the Sun was read by Richard Bernstein, the author and sinologist, while Alex Cook of King's College, Cambridge, first alerted me to *The Ruins.* Tony Kasse corrected me on matters of Japanese culture, and Don Cohn of Columbia University was helpful on Mao and the Sun.

The labyrinths of quantum physics were explained to me by Andrew Blake, research associate at Cavendish Laboratory in Cambridge, and by Dr. Alan Walton, a senior Cambridge University physicist; Katie Yurkewicz, at CERN, was also an invaluable help. David Agle and Curtis D. Montano at the Jet Propulsion Lab gave me useful material on current projects and also on the early days of rocketry. Piers Corbyn, of Weather Action, gave me hours of his time, and the chapter on global warming was read by Dr. Willie Soon of Harvard University and by Dr. Robert Carter, professor at the Marine Geophysical Laboratory, James Cook University (both of whom, I fear, will not like all my conclusions). The history of science fiction and the Sun was checked by David Compton, editor and renowned science fiction author, and I was also helped by the screenwriter and novelist Alex Garland. The epilogue was read by Louise Nicholson, an authority on India, who in addition arranged my time in that country with such a sure hand.

Silke Ackermann of the Department of Prehistory and Europe was my "control" at the British Museum, leading me to expert after expert. During my researches I used almost a dozen libraries, from the one at Castel Gandolfo to that of the Munch Museum in Oslo. In particular, the Society Library in New York City and the London Library never let me down, while the New York Public Library was a home away from home, be it the Allen Room, the Wertheim Study, or the General Reading Rooms. David Smith at the NYPL has even won a profile in the pages of *The New York Times* for the quality of help he gives to struggling authors; for me, his business-card byline—"librarian to the stars"—seems especially appropriate.

Others who helped on one matter or another include: Professor Babette Babich of Fordham University's Department of Philosophy, Kai Cai, Dr. Nicholas Campion of Bath Spa University (on matters astrological), Fr. Juan Casanova, Lesley Chamberlain, Al Clement, Margaret Cook, Peter d'Epiro, Timothy Ferris, emeritus professor at the University of California, Berkeley, Dr. Valerie Fomin in St. Petersburg, Manuel Pérez García at the University of Almería, Carol Gaxiola, John Gerrard (who shared with me his *Watchful Portrait* [*Caroline*]), Professor Ai Guoxiang of the Chinese Academy of Science, Bob Hay, Jeff Hester, professor of physics and astronomy at Arizona State University, Miles Hilton-Barber (for telling me what it is like to fly a plane when you are blind), Chen Jie, who showed me around Beijing Observatory, Professor Phil Jones, longtime professor at the University of East Anglia, Chow Kii, Adam Kuper of Brunel University, Bill Livingston and John Leibacher of the National Solar Observatory in Tucson, Ved Mehta, Oliver Morton, Don Nicholson at the Mount Wilson Observatory in Pasadena (the one person in all my researches who called one of my questions "stupid"; I had asked how far his telescope could see), Colin Pearce, Alberto Righini, Mr. Tetsuya of the Watanabe National Astronomical Observatory in Tokyo, and Ingebjørg Ydstie of the Munch Museum in Oslo. I am also grateful to the astronomers at the observatory in Shanghai who spent time with me explaining their work, and the Mullard Space Science Laboratory in Dorking, Surrey.

The manuscript was read by Nick Webb, fresh from his first-class degree in astronomy; by Jonathan Weiner, author of *Beak of the Finch* and *Long for This*

World; and by Peter Petre, onetime science editor of *Fortune.* Two years after starting work on the book, I became aware that David Whitehouse, a professor of astronomy at Jodrell Bank and an award-winning science correspondent for the BBC, had published *The Sun: A Biography.* We went on to become friends, and one day he arrived at lunch with a large paper bag stuffed with books—the library he had used for his Sun book. 'They're for you," he said, then handed over a DVD—"It's got all my research," he added. It needs hardly adding how generous a gift this was, or how useful. He also kindly read an early complete draft of the book.

Friends and relations all did their bit—in many ways: John and Nina Darnton, Liza Darnton (for early research work on Carl Jung), Jamie Darnton (on some tricky Chinese etymology), Darrell McLeod, Sara Wheeler, Hilary Spurling, Elaine Shocas, Linden Stafford, Brian Brivati, my old university friend Colin Gleadell, Richard Oldcorn, Paul Pickering, Michael Johnson, Allen Kurzweil, David Bodanis (when not preoccupied reinterpreting the Ten Commandments), Bill and Linda Rich (for the insight about pirates), John de Stefano, my cello teacher, Laura Usiskin (for her research into matters ranging from Freemasonry to bird migration), Harry Hotz (for his knowledge of temperature ratios), Andrew Di Rienzo (for befriending me on a bus and teaching me some crucial physics thereafter), Ron Rosenbaum (for some Shakespeare tuition), Kevin Jackson, Jim Landis (not least for the foul-mouthed quotation by August Strindberg), Elizabeth Sifton (who told me of the Sun Museum in Riga), Ben Cheever (whom I forgot to mention last time), Peregrine Hodson (for making me climb Mount Fuji), Woodrow Campbell (my accomplice and photographer in Antarctica), Nancy Campbell (no relation), Mary Cunnane in Australia, my old tutor Dr. Michael Tanner of Corpus Christi College, Cambridge, Jackie Albers (for sharing her homework), and—not least—Toby, Mary, and Guy Cohen; it was Toby who one day put his hand on my shoulder and asked, "Daddy, why don't you make some money and write a *nice, short book?"* One day I will.

The head doorman of my apartment building, Bill Albers, made sure I was kept up to the mark with a constant flow of magazine items that he would slip under my door, along with pertinent questions about progress. Ann Godoff and Vanessa Mobley were both kind and helpful in the book's early days. At Random House, my editor, Beth Rashbaum, enormously improved the book with her powers of organization and her determination that nothing should appear that would not be clear to everyone; I am also extremely grateful for the skill and help of many others at Random, including Tim Bartlett, Tom Perry, Will Murphy, Barbara Bachman, London King, Angela Polidoro, Susan Kamil, and Gina Centrello, making my publication there feel like a family affair. The same is true of my British publishers, led by my longtime friend Ian Chapman, along with Mike Jones, Katherine Stanton, Rory Scarfe, Anna Robinson, Suzanne Babaneou, and Sally Partington. In Germany I am lucky to have as my friend and publisher Niko Hansen, whose patience and enthusiasm for the project never wavered (not in front of me, anyhow!). Catherine Talese has been the ideal illustration researcher, as well as a wonderful cheerleader and an indispensable guide to New York swimming pools.

At the Robbins office, Rachelle Bergstein, Karen Close, Mike Gillespie, Coralie Hunter, Katie Hut, and Ian King each read part or all of the manuscript, often more than once, and made valuable comments. Having them around and in support has been fun. Tim Dickinson, as he was on *By the Sword,* has been companion, sounding board, fount of knowledge both pertinent and not-too-pertinent (he would insist on the hyphens) but always highly interesting; I cannot thank him enough for all he did.

Last and supremely there is my literary agent, Kathy Robbins, aka Mrs. Cohen. Over the years I have become used to other authors thanking her in their books for her friendship, her counsel, her unrivaled humor, her ability to buoy up and push forward, her negotiating powers, her tenacity, her empathy, her peculiar ability to time both good and not-so-good news to maximum effect. I have experienced all these and more, but at least I can add one unique claim: it is she who, when woken in the small hours and told yet another extraordinary way in which chapter 4, or maybe 14, might begin, can say to me without a note of rancor, "That's *really* interesting, sweetheart," then turn over and at once return to a well-earned sleep.

Despite this roster of helpers, any mistakes in the book are of course my responsibility. One chases the Sun, but never catches up with it. I am comforted in this by a sentence I came across in Charles Darwin's autobiography (1882), where he writes: "Whenever I have found out that I have blundered, or that my work has been imperfect, and when I have been contemptuously criticized, and even when I have been over-praised, so that I have felt mortified, it has been my greatest comfort to say hundreds of times to myself that 'I have worked as hard and as well as I could, and no man could do more than this.'"

Richard Cohen, New York, July 2010

NOTES

PREFACE

1. *Mother Earth News*, no. 41, September–October 1976.
2. *Encyclopedia of Astronomy and Astrophysics* (New York: Nature Publishing, 2001), vol. 4, "Sun." Astronomers classify stars based on color, which relates to temperature, size, and longevity. Throughout the book, "Sun" appears thus, as astronomers and scientists prefer; when referred to in a nonspecific way, and often when mentioned in quoted matter, it is "sun." The word also appears lowercase in most compounds, such as "sun-glazed" or "sun worshipper."
3. Arthur Conan Doyle, *A Study in Scarlet*, 1897.
4. Ben Bova, *The Fourth State of Matter: Plasma Dynamics and Tomorrow's Technology* (New York: New American Library, 1974).
5. John Eddy, "Climate and the Role of the Sun," *Journal of Interdisciplinary History*, vol. 10, no. 4 (spring 1980), pp. 725–47.
6. See U.S. Scouting Service Project at www.usscouts.org, "Confidence and Team-Building Games." John Keats recalled a game at his school in which the boys whirled around the playground in a huge choreographed dance, trying to imitate the solar system: there is truly no new thing under the Sun.

SUNRISE: MOUNT FUJI

1. Wallace Stevens, "Notes Toward a Supreme Fiction," *The Collected Poems of Wallace Stevens* (New York: Random House, 1990).
2. Pink Floyd, "Time," from the album *Dark Side of the Moon*, Capitol Records, 1973.
3. Gateways are associated with the Sun in many cultures, Japanese gateways probably deriving from Indian *torana*, or "celestial gates"; the palace of the god-emperor at Kyoto is entered by "the Gate of the Sun." See William Lethaby, *Architecture, Mysticism and Myth* (New York: Braziller, 1975), pp. 186–89.

CHAPTER 1: TELLING STORIES

1. Max Müller, *Lectures on the Science of Language*, 1864 (reprinted New York: Kessinger, 2007).
2. John Donne, *Ignatius His Conclave*, 1611.
3. See the myth of Taios in Arthur Cook, *Zeus: A Study in Ancient Religion*, 3 vols. (Cambridge: Cambridge University Press, 1914–40), pp. 633–35 and 719ff.; see also W.K.C. Guthrie, *The Greeks and Their Gods* (Boston: Beacon Press, 1950), p. 211.
4. See Bruce Chatwin, "Art and the Image-Breaker," *The Morality of Things* (London: Cape, 1973), p. 179.
5. See Claude Lévi-Strauss, *Introduction to the Science of Mythology*, vol. 2, *The Origin of Table Manners* (New York: Harper and Row, 1978), p. 159.
6. Genesis XI:16–17. See also William Tyler Olcott, *Sun Lore of All Ages: A Collection of Myths and Legends Concerning the Sun and Its Worship* (New York and London: Putnam, 1914), pp. 38–39.
7. See John A. Crow, *The Epic of Latin America* (Berkeley: University of California Press, 1992), p. 33.
8. Jon R. Stone, ed., *The Essential Max Müller* (New York and Basingstoke: Palgrave Macmillan, 2002), pp. 155–56.
9. Max Müller, *India: What Can It Teach Us?* (New York: J. W. Lovell, 1883), p. 216.
10. W. J. Perry and G. E. Smith, *The Children of the Sun: A Study in the Early History of Civilization* (London: Metheun, 1930), p. 164.

"They're keeping me from practicing
my religion. I'm a sun-worshipper."

11. See Armin Kesser, "Solar Symbolism Among Ancient Peoples," *Graphics*, no. 100 (1962), p. 115.

12. Perry and Smith, *The Children of the Sun*, p. 141.

13. Ibid., p.160.

14. Carl Gustav Jung, *Memories, Dreams, Reflections* (New York: Pantheon, 1961), pp. 250–51. One is reminded of Rostand's Cyrano de Bergerac: "a lie is a sort of myth and a myth is a sort of truth."

15. Ibid., p. 252. Mountain Lake was the tribe's spokesman since he spoke the best English. He also told Jung that if the whites did not stop interfering with their religion, the Pueblos would make them regret their actions bitterly: they would stop helping Father Sun on his daily journey across the sky.

16. See Anthony Storr, *Jung* (New York: Routledge, 1973), p. 26.

17. Deirdre Bair, *Jung* (Boston: Little, Brown, 2004), p. 354.

18. Mircea Eliade, *Patterns in Comparative Religion* (New York: Meridian, 1963), pp. 124–51. A recent book, sadly available only in German, is particularly good on early solar myths: Dieter Hildebrandt, *Die Sonne* (Munich: Hanser Verlag, 2008).

CHAPTER 2: CELEBRATING THE SEASONS

1. John Donne, *A Nocturnall Upon St. Lucie's Day, Being the Shortest Day*, c. 1611. Donne believed December 13 to be the shortest day; in the Northern Hemisphere it is December 21 or 22.

2. Alan Furst, *Dark Star* (New York: Random House, 1991), p. 124.

3. Mark Twain, *The Adventures of Huckleberry Finn* (New York: Oxford University Press, 1999), p. 93.

4. Barry Lopez, *Arctic Dreams* (New York: Scribner, 1986), p. 20.

5. Macrobius, *The Saturnalia* (New York: Columbia University Press, 1969), book I, sec. 17, p. 114.

6. See *Cambridge Medieval History*, vol. IV, part 1 (Cambridge: Cambridge University Press, 1966), p. 43.

7. See J. M. Golby and A. W. Purdue, *The Making of Modern Christmas* (Athens: University of Georgia Press, 1986), pp. 123–24.

8. See Craig Harline, *Sunday: A History of the First Day from Babylonian Times to the Super Bowl* (New York: Doubleday, 2007), pp. 10, 17–24.

9. Ronald Hutton, *The Pagan Religions of the Ancient British Isles* (Oxford: Blackwell, 1991), p. 36.

10. John Brand, *Observations on Popular Antiquities*, ed. Henry Ellis (London, 1813), vol. I, pp. 238ff. The practice recalls the famous lines from Andrew Marvell's "To His Coy Mistress": "Thus, though we cannot make our Sun / Stand still, yet we will make him run"—though written with a very different end in view.

11. See *The Encyclopedia of Religion*, ed. Mircea Elinde et al., (New York: Free Press, 1987), vol. 14, pp. 139–40. See also John Matthews, *The Summer Solstice* (Wheaton, Ill.: Quest Books, 2002), pp. 20–21, 240–41, and 265–66.

12. Rune Hagan, private paper on midsummer witches, University of Tromsø.

13. Brand, *Observations on Popular Antiquities*, pp. 238–39.

14. Thomas Hardy, *The Return of the Native* (New York: Signet Classics, 1959), p. 23.

15. Emmanuel Le Roy Ladurie, *Carnival in Romans* (New York: Braziller, 1979).

16. See David Cressy, "The Fifth of November Remembered," *Myths of the English*, ed. Roy Porter (Cambridge: Polity Press, 1992), p. 78.

17. Mona Ozouf, *Festivals and the French Revolution* (Cambridge, Mass: Harvard University Press, 1988), pp. 158–67.

18. Jack M. Broughton and Floyd Buckskin, "Racing Simloki's Shadow: The Ajumawi Interconnection of Power, Shadow, Equinox and Solstice," in *Prehistoric Cosmology in Mesoamerica and South America* (Washington, D.C.: Smithsonian Occasional Publications in Mesoamerican Anthropology, 1988), pp. 184–89.

19. Clyde Holler, *Black Elk's Religion: The Sun Dance and Lakota Catholicism* (Syracuse, N.Y.: Syracuse University Press, 1995), p. 60.

20. See Robert Lowie, *The Crow Indians* (Lincoln and London: University of Nebraska Press, reprint 1983), p. 215: "Most characteristic was the intertwining of war and religion. The Sun Dance, being a prayer for revenge, was naturally saturated with military episodes."

21. Claude Lévi-Strauss, *The Origin of Table Manners* (New York: Harper and Row, 1978), pp. 222 and 212.

22. Joseph Epes Brown, *The Sacred Pipe: Black Elk's Account of the Seven Rites of the Oglala Sioux* (Norman: University of Oklahoma Press, 1995), p. 12; quoted in Ronald Goodman, *Lakota Star Knowledge: Studies in Lakota Stellar Theology* (Rosebud, S.D.: Sinte Gleska University Press, 1992). For the importance of scalping and its link with solar myths, see Lévi-Strauss, *Origin of Table Manners*, pp. 404–5.

23. Wade Davis, *One River* (New York: Simon and Schuster, 1996), pp. 79–82; see also Peter Matthiessen, *In the Spirit of Crazy Horse* (New York: Viking, 1983), p. 47.

24. Holler, *Black Elk's Religion*, p. 201.

25. See Friedrich Nietzsche, *The Birth of Tragedy* (New York: Anchor, 1956), p. 22.

CHAPTER 3: THE THREE THOUSAND WITNESSES

1. This poem was actually written in Italy in May 1948 and first published in *Horizon* that July; it then appeared in *Nones* (1951) and, revised, in the last chronological section of Auden's *Collected Shorter Poems, 1922–1957* (London: Faber, 1966).

2. See Caroline Alexander, "If the Stones Could Speak," *National Geographic*, June 2008, pp. 34–59.

3. Roger Deakin, *Wildwood: A Journey Through Trees* (London: Hamish Hamilton, 2007), p. 120.

4. C. P. Stacey, *American Historical Review*, vol. 56, no. 1 (October 1950), pp. 1–18.

5. Jacquetta Hawkes, *Man and the Sun* (New York: Random House, 1962).

6. Ronald Hutton, *The Pagan Religions of the Ancient British Isles* (Oxford: Oxford University Press, 1991), p. 36.

7. Ibid., p. 112.

8. See Thomas Barthel, *The Eighth Land* (Honolulu: University Press of Hawaii, 1978), pp. 248–49.

9. Bruce Chatwin, *The Songlines* (London: Penguin, 1988), p. 136.

10. Frank Delaney, *Ireland: A Novel* (New York: HarperCollins, 2005), pp. 51ff.

11. See Alexandre Dumas, *The Man in the Iron Mask* (Oxford: Oxford University Press, 1998), chapter 47, "The Grotto of Locmaria."

12. See John A. Eddy, *In Search of Ancient Astronomies*, "Medicine Wheels and Plains Indian Astronomy," pp. 147ff.

13. See Jay Atkinson, "America's Stonehenge: A Classic Whodunit and Whydunit," *The New York Times*, December 11, 2009, C38.

14. See Anthony Aveni, *Skywatchers* (Austin: University of Texas Press, 2001), p. 318; Gary Urton, *At the Crossroads of the Earth and the Sky* (Austin: University of Texas Press, 1981), p. 6; and R. T. Zuidema, "The Inca Calendar," in A. F. Aveni, ed., *Native American Astronomy* (Austin: University of Texas Press, 1988), p. 219. Archaeoastronomy has its critics, whose oldest quip is that the discipline's main achievement to date is to create a word of four consecutive vowels.

15. W. G. Auden, *Selected Poems* (New York: Viking, 1990).

16. Barry Cunliffe and Colin Renfrew, eds., *Science and Stonehenge* (Oxford: Oxford University Press, 1997), p. 572. The Great Wall had no major solar connections (although China is the other country to boast pyramids, such as those near Ch'ang, capital of the Western Han Dynasty of 206 B.C.–A.D. 220).

17. Martin Isler, *Sticks, Stones, and Shadows* (Norman: University of Oklahoma Press, 2001), p. 111.

Even so, as the orientalist Kate Spence writes, "The Egyptians were almost certainly unaware that their orientation method was precession-dependent and that it was therefore initially inexact, otherwise they would have used a more accurate bisection method" (*Nature*, vol. 412, August 16, 2001, p. 700). Isler's book has been particularly useful to the second half of this chapter.

18. See J. L. Heilbron, "Churches as Scientific Instruments," annual invitation lecture to the Scientific Instrument Society, Royal Institution, London, December 6, 1995, *SIS Bulletin* 48 (March 1996), 4–9.

19. See *The New York Review of Books*, December 15, 2005, p. 28.

20. See Leslie V. Grinsell, *Folklore of Prehistoric Sites* (London: David and Charles, 1976), p. 25.

21. See Joe Rao, "Sky Watch," *The New York Times*, July 9, 2006, p. 26.

CHAPTER 4: TERRORS OF THE SKY

1. Stuart Clark, *The Sun Kings: The Unexpected Tragedy of Richard Carrington and the Tale of How Modern Astronomy Began* (Princeton, N.J.: Princeton University Press, 2007), p. 15.

2. Dorothy Jean Ray, "Legends of the Northern Lights," *Alaska Sportsman*, April 1958.

3. Robert A. Henning, quoted in S.-I. Akasofu, *Aurora Borealis: The Amazing Northern Lights* (Anchorage, Alaska: Alaska Geographic Society, 1979), p. 5.

4. Leonard Huxley, *Scott's Last Expedition: The Personal Journals of Captain R. F. Scott, R.N., C.V.O., on His Journey to the South Pole* (London: Murray, 1941), p. 257.

5. Ernest W. Hawkes, *The Labrador Eskimo* (Ottawa: Government Printing Bureau, 1916), p. 153.

6. The explorer Samuel Hearne, quoted in Barry Lopez, *Arctic Dreams* (New York: Scribner, 1986), p. 235.

7. 2 Maccabees 5:1–4, written about 176 B.C.

8. Sir Walter Scott, *The Lay of the Last Minstrel*, ii, 8.5 (General Books, 2010).

9. R. M. Devens, *Our First Century, Being a Popular Descriptive Portraiture of the One Hundred Great and Memorable Events of Perpetual Interest in the History of Our Country, Political, Military, Mechanical, Social, Scientific, and Commercial: Embracing Also Delineations of All the Great Historical Characters Celebrated in the Annals of the Republic; Men of Heroism, Statesmanship, Genius, Oratory, Adventure and Philanthropy* (Chicago: C. A. Nichols, 1878), pp. 379ff.

10. See Warren E. Leary, "Five New Satellites with a Mission of Finding a Source of Color in Space," *The New York Times*, January 23, 2007.

11. Salvo De Meis, *Eclipses: An Astronomical Introduction for Humanists* (Rome: Istituto Italiano per l'Africa e l'Oriente, 2002).

"Listen, pal, it's a free country. You
worship the sun god your way, and I'll
worship the sun god my way."

12. See *The Eclipse Chasers*, three-part series originally broadcast on BBC 4 on March 31, April 9, and April 16, 1996.
13. The NASA claim may seem far-fetched, but see Cally Stockdale, "Agnihotra Ancient Healing Fire Practice," www.rainbownews.co.nz, August–September 2006; Dr. Shantala Priyadarshini, "Homa Fire Ceremony," AHC magazine, February 3, 2005; and "NASA and Cow Dung," available on the Internet. See also Hitesh Devnani, "Longest Solar Eclipse of the Century Shrouds Asia," *The Epoch Times* (July 23, 2009), A1, A3.
14. See Lévi-Strauss, *The Origin of Table Manners*, p. 42.
15. See J. P. McEvoy, *Eclipse: The Science and History of Nature's Most Spectacular Phenomenon* (London: Fourth Estate, 1999).
16. See Rev. A. H. Sayce, *Astronomy and Astrology of the Babylonians* (San Diego: Wizards Bookshelf, 1981), p. 29.
17. F. R. Stephenson, *Historical Eclipses and Earth's Rotation* (Cambridge: Cambridge University Press, 1997), chapter 11, "Eclipse Records from Medieval Europe," p. 226.
18. Joseph Needham, *Science and Civilisation in China* (Cambridge: Cambridge University Press, 1959), vol. 3, p. 188.
19. Niccolò Machiavelli, *Discorsi*, 1:56.
20. Plutarch, *Life of Pericles* (New York: Bolchazy-Carducci, 1984), 35.1–35.2.
21. Dio Cassius, *History of Rome*, book LX, chapter 26. He does not say how Claudius acquired his knowledge.
22. See BBC Radio 4 broadcast on Lawrence, narrated by John Simpson, October 16, 2005.
23. Shigeru Nakayama, *A History of Japanese Astronomy* (Cambridge: Harvard University Press, 1969), p. 51.
24. *Iliad*, book XVII, lines 266–68: "Nor would you say that the Sun was safe, or the Moon, for they were wrapt in dark haze in the course of the combat"; and *Odyssey*, book XX, lines 356–57: "The Sun has utterly perished from heaven and an evil gloom is overspread." Salvo de Meis makes the point in his *Eclipses: An Astronomical Introduction for Humanists* (p. 21) that such lines may reflect poetic license: "They contain no faithful descriptions of an eclipse, neither information which could lead to an acceptable dating."
25. William Smith Urmy, *The King of Day* (New York: Nelson and Phillips, 1874).
26. Luke 23:44.
27. See Stephenson, *Historical Eclipses*, p. 344.
28. J. A. Cabaniss, *Agobard of Lyons, Churchman and Critic* (Syracuse: Syracuse University Press, 1953), p. 18.
29. Barbara K. Lewalski, *The Life of John Milton* (Oxford: Blackwell, 2000), pp. 210, 259, and 278.
30. Quoted in Derek Appleby and Maurice McCann, *Eclipses: The Power Points of Astrology* (Northampton, Mass.: Aquarian Press, 1989), p. 11.
31. *Diary of Samuel Pepys*, ed. M. Bright, 1879, vol. VI, p. 208.
32. Thomas Crump, *Solar Eclipse* (London: Constable, 1995), pp. 115–17. Halley invited observations from those members of the public "with a pendulum clock of which many people are furnished" (De Meis, *Eclipses*, p. 167).

CHAPTER 5: THE FIRST ASTRONOMERS

1. Ralph Waldo Emerson, *Journals and Miscellaneous Notebooks* (Cambridge, Mass.: Belknap Press, 1982), November 14, 1865.
2. Gustave Flaubert, *Dictionary of Accepted Ideas* (New York: New Directions, 1954).
3. J. Norman Lockyer, *The Dawn of Astronomy: A Study of Temple Worship and Mythology of the Ancient Egyptians* (New York: Dover, 2006; first published 1894).
4. See Rev. A. H. Sayce, *Astronomy and Astrology of the Babylonians* (San Diego: Wizards Bookshelf, 1981), p. 145; reprinted from vol. 3, part I of *Transactions of the Society of Biblical Archaeology*, 1874.
5. See John Britton and Christopher Walker, "Astronomy and Astrology in Mesopotamia," in Christopher Walker, ed., *Astronomy Before the Telescope* (London: British Museum Press, 1996), p. 42.
6. See J. G. Macqueen, *Babylon* (New York: Praeger, 1964), p. 212.
7. See *Cambridge Ancient History*, vol. 3, part 2 (Cambridge: Cambridge University Press, 1969), p. 285.

8. Noel M. Swerdlow, *The Babylonian Theory of the Planets* (Princeton: Princeton University Press, 1998). I found this book very useful, but it received a critical mauling on publication. The world of Babylonian studies is a fierce one. It is worth recalling what was told me by Christopher Walker, the great expert on Babylonian astronomy: "We have just the first few filaments of the spider's web. To my mind, the debate as to what the Babylonians knew is still wide open; we are all operating in a vacuum."

9. J. P. McEvoy, *Eclipse: The Science and History of Nature's Most Spectacular Phenomenon* (London: Fourth Estate, 1999), pp. 63–64.

10. See Ronald A. Wells, "Astronomy in Egypt," in Walker, ed., *Astronomy Before the Telescope*, p. 29.

11. See Jacquetta Hopkins Hawkes, *The First Great Civilizations: Life in Mesopotamia, the Indus Valley, and Egypt* (London: Hutchinson, 1973), pp. 230–31; see also Otto E. Neugebauer and Richard A. Parker, *Egyptian Astronomical Texts, Vol. III: Decans, Planets, Constellations and Zodiacs* (Providence, R.I.: Brown Egyptological Studies 3, 1969).

12. See Clemens Alexandrinus, *Stomata*, Lib. VI. See also *Encyclopedia of Religion*, p. 145.

13. See James P. Allen, *Genesis in Egypt: The Philosophy of Ancient Egyptian Creation Accounts* (New Haven, Conn.: Yale University Press, 1988), pp. 8 and 12.

14. Ibid., pp. 13–14. Atum's seed becomes all life: "Atum is the one who developed growing ithyphallic, in Heliopolis. / He put his penis in his grasp / That he might make orgasm with it." Among the Greeks, the Stoic Zeno, too, chose a biological model. At first there is nothing except God, then God creates a difference within himself, such that he is "contained" in moisture. This is the living "sperm" that produces the cosmos.

CHAPTER 6: ENTER THE GREEKS

1. See Jacques Brunschwig and Geoffrey E.R. Lloyd, eds., *Greek Thought: A Guide to Classical Knowledge* (Cambridge: Harvard University Press, 2000), p. 269.

2. See Christopher M. Linton, *From Eudoxus to Einstein: A History of Mathematical Astronomy* (Cambridge: Cambridge University Press, 2004), p. 15.

3. D. R. Dicks, *Early Greek Astronomy to Aristotle* (New York: Norton, 1970), p. 29.

4. Plato, *The Apology, Crito and Phaedo of Plato* (New York: Merchant, 2009), 26D.

5. See Will Durant, *The Life of Greece* (New York: Simon and Schuster, 1939), pp. 337–38, 627, and 161.

6. See Thomas Heath, *A History of Greek Mathematics*, vol. 1 (New York: Dover, 1981), pp. 176 and 178. Sir Thomas Heath was permanent secretary for administration of the British Treasury in the 1920s and in his spare time a formidable historian of mathematics.

7. See William Tarn, *Hellenistic Civilization* (London: Arnold, 1966), p. 299, and Michael Fowler, "How the Greeks Used Geometry to Understand the Stars," at http://galileoandeinstein.physics.virginia.edu/lectures/greek_astro.htm.

8. Aristotle, c. 340 B.C./1960, 15, 23. This dualism lasted until the time of Galileo.

9. Durant, *Life of Greece*, p. 337.

10. See Otto E. Neugebauer, *Studies in Civilization*, p. 25.

11. Nathan Sivin and Geoffrey Lloyd, *The Way and the Word: Science and Medicine in Early China and Greece* (New Haven and London: Yale University Press, 2002), p. 101.

12. Timothy Ferris, *Coming of Age in the Milky Way* (New York: Anchor, 1989), p. 41. In a later footnote (p. 65) Ferris writes: "One could write a plausible intellectual history in which the decline of sun worship, the religion abandoned by the Roman emperor Constantine when he converted to

Christianity, was said to have produced the Dark Ages, while its subsequent resurrection gave rise to the Renaissance."

13. Dennis Rawlins, "Astronomy and Astrology: The Ancient Conflict," *Queen's Quarterly* 91/4 (Winter 1984), pp. 969–89.

14. See Robert R. Newton, *The Crime of Claudius Ptolemy* (Baltimore: Johns Hopkins University Press, 1997), pp. 76ff.

15. Colin A. Ronan, *The Astronomers* (London: Evans, 1964), p. 91.

16. Ibid., p. 95.

17. See "The Universe of Aristotle and Ptolemy," at http://csep10.phys.utk.edu/astr161/lect/retrograde/aristotle.html.

18. Frederick Nietzsche, *The Will to Power*, in *Collected Works*, new edition, ed. O. Levy (London: T. N. Foulis, 1964).

19. Franz Cumont, *The Oriental Religions* (New York: Dover, 1956; first published 1911), p. 134.

CHAPTER 7: GIFTS OF THE YELLOW EMPEROR

1. Quoted by Joseph Needham in *Science and Civilization in China* (Cambridge: Cambridge University Press, 1959), vol. 3, p. 171.

2. Zou Yuanbiao, "Da Xiguo Li Madou," in *Yuan xue ji* 3/39, quoted in Jonathan D. Spence, *The Memory Palace of Matteo Ricci* (London: Faber, 1985), p. 151.

3. See Simon Winchester, *Joseph Needham: The Man Who Loved China* (New York: HarperCollins, 2008), pp. 17–21. Winchester adds, "Also a chain-smoker and devastating womanizer."

4. See *The New York Times*, March 27, 1995.

5. See Horace Freeland Judson, "China's Drive for World-Class Science," MIT's *Technology Review* January 2006. When Needham's first volume appeared, eminent American academics published indignant reviews, calling rubbish the notion that Chinese science and technology played any consequential part in history and warning that the book was part of the Red Menace.

6. See C. Ronan, "Astronomy in China, Korea and Japan," in Christopher Walker, ed., *Astronomy Before the Telescope* (London: British Museum Press, 1996), pp. 245–68.

7. Quoted in Needham, *Science and Civilization in China*, p. 193.

8. See, for example, Article 110 in *The T'ang Code: Vol. II, Specific Articles*, trans. Wallace Johnson (Princeton, N.J.: Princeton University Press, 1997).

9. Needham, *Science and Civilization in China*, p. 406.

10. Ibid., p. 227.

11. See "Ancient Chinese Cosmology," at www.astronomy.pomona.edu/archeo/china/china3.html.

12. Vincent Cronin, *The Wise Man from the West* (New York: Dutton, 1955), p. 232.

13. Needham, *Science and Civilization in China*, p. xlv.

14. Christopher Cullen, "Joseph Needham on Chinese Astronomy," *Past and Present* no. 87 (May 1980), p. 42. Cullen ends his essay: "Compilers of works of popularization and 'cross-disciplinary syntheses' who try to treat Needham's work as a bran-tub will find a few mousetraps awaiting their fingers as they rummage." Point taken.

15. Cronin, *The Wise Man from the West*, p. 11. Needham lists forty outstanding inventions over thirty thousand years up to 1500 B.C., and 287 inventions from then until A.D. 1700.

16. Xue Shenwei, *Journal of Ancient Civilizations*, vol. 2 (Changchun, Jilin Province: IHAC Northeast Normal University, 1987), "Brief Note on the Bone Cuneiform Inscriptions," pp. 131–34.

17. Author's conversation with Christopher Walker, Department of Ancient Near East, British Museum, January 12, 2006.

18. See Ivan Hadingham, "The Mummies of Xinjiang," *Discover*, vol. 15, no. 68 (April 1994); see also Drauenhelm, "Ulghur-I Ulgur Mummies Say 'No' to China," *East Bay Monthly*, vol. 29, no. 3 (December 1998); Edward Wong, "The Dead Tell a Tale China Doesn't Care to Listen To," *The New York Times*, November 19, 2008, A6; Elizabeth Wayland Barber, *The Mummies of Ürümchi* (New York: Norton, 1999) and Nicholas Wade, "A Host of Mummies, A Forest of Secrts." *The New York Times*, March 16, 2010, D1.

19. Cronin, *Wise Man from the West*, p. 100.

20. Ibid., pp. 192–93.

21. Needham, *Science and Civilization in China*, pp. 239, 220.

22. See Edward Rothstein, "A Big Map That Shrank the World," *The New York Times*, January 20, 2010, C1 and C7.

23. See "Eunuchs, Sleuths and Otherwise," http://cruelmusic.blogspot.com/2008/04/eunuchs -sleuths-and-otherwise.html, April 21, 2008.

24. Cronin, *Wise Man from the West*, pp. 140–42. This is largely Cronin's account, with just small changes.

25. Fan Tsen-Chung, *Dr. Johnson and Chinese Culture* (London: The China Society, 1945), p. 9.

CHAPTER 8: THE SULTAN'S TURRET

1. Qur'an, Surah 6, 76–78.

2. Arthur Koestler, *The Sleepwalkers* (London: Hutchinson, 1959), p. 105.

3. See article on Arab astronomy by J. J. O'Connor and E. F. Robertson, www-history.mcs.st -andrews.ac.uk/Biographies/Sinan.html.

4. See Mohammad Ilyas, *Islamic Astronomy and Science Development: Glorious Past, Challenging Future* (Petaling Jaya, Selangor Darul Ehsan, Malaysia: Pelanduk Publications, 1996), p. 22.

5. M.A.R. Khan, *A Survey of Muslim Contributions to Science and Culture* (Ashraf, Lahore: Internet edition), p. 14. See also David A. King, *Astronomy in the Service of Islam* (Brookfield, Vt.: Variorum, 1993), p. 246.

6. Ilyas, *Islamic Astronomy*, p. 253. See also "Aspects of Applied Science in Mosques and Monasteries" in Science and Theology in Medieval Islam, Judaism, and Christendom international symposium, Madison, Wisconsin, April 15–17, 1993.

7. David A. King, "Islamic Astronomy," in Christopher Walker, ed., *Astronomy Before the Telescope* (London: British Museum Press, 1996), p. 160.

8. Ilyas, *Islamic Astronomy*, pp. 1–2. The office of the *muwaqqit*, or mosque astronomer, was developed during the thirteenth century in Egypt. Most Egyptian and Syrian astronomers of consequence for the next two centuries were *muwaqqit*.

9. See also Otto Neugebauer, "The Early History of the Astrolabe," in *Astronomy and History: Selected Essays* (Berlin: Springer, 1983), p. 279. A nice description of the astrolabe also appears in Edward Rutherfurd's novel *London* (London: Century, 1997), p. 340.

10. Abdul-Qasim Said ibn Ahmad, *The Categories of Nations*, quoted in Neugebauer, "The Transmission of Planetary Theories in Ancient and Medieval Astronomy," in Christopher Walker, ed., *Astronomy and History*, pp. 3 and 129.

11. See Christopher M. Linton, *From Eudoxus to Einstein: A History of Mathematical Astronomy* (Cambridge: Cambridge University Press, 2004), p. 85.

12. Otto Neugebauer, *A History of Ancient Mathematical Astronomy* (Berlin: Springer, 1975), p. 127.

13. See David Pingree, "Astronomy in India," in Christopher Walker, ed., *Astronomy Before the Telescope* (London: British Museum Press, 1996), pp. 123–42.

14. See John Henry, *Moving Heaven and Earth* (Cambridge: Icon Books, 2001), p. 41.

15. Saint Augustine, *Works*, quoted in Linton, *From Eudoxus to Einstein*, p. 113.

16. John David North, *Stars, Minds, and Fate: Essays in Ancient and Medieval Cosmology* (London: Hambledon Press, 1989), p. 403.

17. Neugebauer, *A History of Ancient Mathematical Astronomy*, p. 943. He was known to friends and students alike as "the Elephant."

18. See *Encyclopaedia Britannica*, 11th edition (1910), vol 3, p. 153.

19. See Koestler, *The Sleepwalkers*, p. 110. Much of this paragraph paraphrases his analysis.

CHAPTER 9: THE EARTH MOVES

1. Nicolaus Copernicus, *On the Revolutions of the Heavenly Spheres*, 1543; reprinted in two vols, ed. Christina J. Moose (Pasadena, Calif.: Salem Press, 2005).

2. Martin Luther, *Works*, vol. 22, c. 1543.

3. Groucho Marx, *The Groucho Letters* (London: Sphere, 1967), p. 10.

4. Thomas A. Bailey, *The American Pageant: A History of the Republic*, 13th ed. (Boston: Houghton Mifflin, 2006), chapter 1.

5. William Manchester, *A World Lit Only by Fire* (Boston: Little, Brown, 1992), p. 89.

6. See Christopher M. Linton, *From Eudoxus to Einstein: A History of Mathematical Astronomy* (Cambridge: Cambridge University Press, 2004), p. 116.

7. Dava Sobel, *Galileo's Daughter* (New York: Penguin, 2000), p. 59.

8. See Thomas Heath, *A History of Greek Mathematics*, p. 301. The omitted passage runs: "Philolaus believed in the mobility of the Earth, and some even say that Aristarchos of Samos was of that opinion." One also wonders what Copernicus would have made of the assertion by Leonardo da Vinci that "the Sun does not move. . . . The Earth is not in the center of the circle of the Sun, nor in the center of the universe" (*Notebooks* I, pp. 310 and 298).

9. See Joshua 10:12–13 and Ecclesiastes 1:4–5.

10. Patrick Moore, in Christopher Walker, ed., *Astronomy Before the Telescope* (London: British Museum Press, 1996), p. 10.

11. See Garrett Hardin, "Rewards of Pejoristic Thinking" (1977). See www.garrethardinsociety.org/art_rewards_pejoristic_thinking.html.

12. Otto Neugebauer, "On the Planetary Theory of Copernicus," in Otto Neugebauer, ed., *Astronomy and History*, p. 103. This essay sets forth in detail the many ways in which Copernicus adhered to Ptolemaic calculations. There is also an interesting discussion of the extent to which Copernicus may have plagiarized Arab theorems in Dick Teresi, *Lost Discoveries* (New York: Simon and Schuster, 2002), pp. 3–5.

13. Arthur Koestler, *The Sleepwalkers* (London: Hutchison, 1959), p. 202.

14. Tycho Brahe, "An Oration on Mathematical Principles," quoted in Linton, *From Eudoxus to Einstein*, p. 155.

15. John Robert Christianson, *On Tycho's Island: Tycho Brahe and His Assistants, 1570–1601* (New York: Cambridge University Press, 2000), p. 82.

16. Koestler, *The Sleepwalkers*, p. 302.

17. Johannes Kepler, *New Astronomy* (Cambridge: Cambridge University Press, 1992), introduction. The book's subtitle, "Celestial Physics," implies a program of investigation that struck Kepler's contemporaries as paradoxical. When the book was first published it was almost universally ignored. For more on this, and a contrary view, see Owen Gingerich, *The Book Nobody Read* (New York: Walker, 2004).

18. Gingerich, *The Book Nobody Read*, pp. 168, 47. In his entertaining account, Gingerich not only writes off what he calls "the legend of epicycles upon epicycles," but is dismissive of Koestler's theory of scientific advance: "Some critics said his book should have been entitled *The Sleepwalker*, because he had only one good example of a scientist groping in the dark, Kepler himself. A few even more perceptive critics said he had zero examples, an opinion borne out by recent scholarship" (p. 49). Maybe some scientists bridle at being likened to sleepwalkers, but the fact remains that many important discoveries come out of left field—and that was Koestler's point.

19. Bertholt Brecht, *Galileo* (New York: Grove, 1966), p. 51.

20. Brian Clegg, *Light Years: An Exploration of Mankind's Enduring Fascination with Light* (London: Piatkus, 2001), p. 74.

21. See Dennis Overbye, "A Telescope to the Past as Galileo Visits U.S.," *The New York Times*, March 28, 2009, A8. But Galileo's colleague Cesare Cremonini, a professor of natural philosophy, was paid two thousand florins a year: mathematicians were still regarded as inferior to philosophers.

22. Timothy Ferris, *Coming of Age in the Milky Way* (New York: Anchor, 1989), p. 84.

23. Clegg, *Light Years*, pp. 54–55.

24. Noel M. Swerdlow, "Galileo's Discoveries with the Telescope and Their Evidence for the Copernican Theory," *The Cambridge Companion to Galileo*, ed. Peter Machamer (Cambridge: Cambridge University Press, 1998), pp. 244–45.

25. See *Discoveries and Opinions of Galileo*, trans. and ed. Stillman Drake (New York: Doubleday, 1957), pp. 87–144.

26. H. L. Mencken, *A Treatise on Right and Wrong* (New York: Knopf, 1934), p. 279.

27. Koestler, *The Sleepwalkers*, p. 483.

28. Ludovico Geymonat, *Galileo Galilei: A Biography and Enquiry into His Philosophy of Science*, trans. Stillman Drake (New York: McGraw-Hill, 1965), p. 73.

29. John Milton, *Paradise Lost*, book VIII, ll. 167ff.

30. See George Sim Johnston, "The Galileo Affair," www.catholiceducation.org/articles/history/

world/wh0005.html. Roger Bacon, Giordano Bruno, and other scholars either imprisoned or put to death for their views would hardly agree with Whitehead; still, his main point holds: Galileo got off lightly.

31. See Richard S. Westfall, *Essays on the Trial of Galileo* (Vatican Observatory Foundation, 1989), p. v; and Richard Owen and Sarah Delaney, "Vatican Recants with a Statue of Galileo," London *Times*, March 4, 2008, p. 11, which quotes the current pope, while still a cardinal, making a speech in 1990 in which he described Galileo's trial as fair. The Vatican later said he had been misquoted.

CHAPTER 10: STRANGE SEAS OF THOUGHT

1. David Brewster, *Life of Sir Isaac Newton* (London: John Murray, 1831), vol. 2, chapter 27.
2. Leo Tolstoy, *War and Peace*, trans. Richard Pevear and Larissa Volokhonsky (New York: Vintage, 2008), p. 606.
3. Richard S. Westfall, *Never at Rest: A Biography of Isaac Newton* (Cambridge: Cambridge University Press, 1983), p. 10.
4. Ibid., p. 581.
5. Albert Einstein, "What Is the Theory of Relativity?" London *Times*, November 28, 1919.
6. John Herivel, *The Background to Newton's* Principia (Oxford: Clarendon Press, 1965), p. 67.
7. François-Marie Arouet de Voltaire, *An essay upon the civil wars of France [. . .] and also upon the epick poetry of the European nations*, reprinted in *Letters on England* (New York: Prentice Hall, 2000), p. 75.
8. William Stukeley, *Memoirs of Sir Isaac Newton's Life* (London: Taylor & Francis, reprint 1936), p. 20.
9. Quoted in David Whitehouse, *The Sun: A Biography* (West Sussex: Wiley, 2005), pp. 71–72.
10. See R. L. Gregory, *Eye and Brain: The Psychology of Seeing* (London: Weidenfeld, 1966), p. 16.
11. Mircea Eliade, *Patterns in Comparative Religion* (New York: Meridian, 1963), p.128. Of all the memory aids for the colors of the rainbow, my favorite remains "Richard Of York Gained Battles In Vain."
12. See Gerald Weissmann, "Wordsworth at the Barbican," *Hospital Practice* (1987) 22:84, quoted in Raymond Tallis, *Newton's Sleep*, p. 21. The original source is the *Diary* of Benjamin Haydon.
13. See Joseph Needham, *Science and Civilization in China*, vol. 3 (Cambridge: Cambridge University Press, 1959), p. 219.
14. Peter Dizikes, "Word for Word on the Web, Isaac Newton's Secret Musings," *The New York Times* June 12, 2003.
15. John Maynard Keynes, "Newton, the Man," *The Royal Society Newton Tercentenary Celebrations* (Cambridge: Cambridge University Press, 1947), p. 27.
16. John Milton, *Paradise Lost*, book 3, ll. 606–9.
17. James Gleick, *Isaac Newton* (New York: Pantheon, 2003). Gleik's concise study is the source for many of the biographical details in this chapter.
18. Donald Fernie, *The Whisper and the Vision: The Voyages of the Astronomers* (Toronto: Clarke, Irwin, 1976), p. 10.

NARCISSISM: THINKS WORLD REVOLVES AROUND IT

19. See Richard Holmes, *The Age of Wonder* (London: Harper, 2008), p. 17.
20. Ted Kooser, *Delights and Shadows* (Port Townsend, Wash: Copper Canyon Press, 2004), p. 62.
21. See Karl Hufbauer, *Exploring the Sun: Solar Science Since Galileo* (Baltimore: Johns Hopkins University Press, 1993), p. 33.
22. Fanny Burney, *Diary*, September 1786, pp. 169–70.
23. See Stuart Clark, *The Sun Kings: The Unexpected Tragedy of Richard Carrington and the Tale of How Modern Astronomy Began* (Princeton, N.J.: Princeton University Press, 2007), p. 35.
24. Christopher M. Linton, *From Eudoxus to Einstein: A History of Mathematical Astronomy* (Cambridge: Cambridge University Press, 2004), p. 360.
25. Pierre-Simon Laplace, *Oeuvres Complètes* (Paris: 1884), vol. VI, p. 478, and vol. VII, p. 121.

CHAPTER 11: ECLIPSES AND ENLIGHTENMENT

1. James Joyce, *Ulysses: The Corrected Text* (New York: Random House, 1986), p. 136, ll. 364–70.
2. J. C. Beaglehole, *The Life of Captain James Cook* (Stanford: Stanford University Press, 1974), p. 87.
3. *Sky and Telescope*, April 2000, pp. 63–65.
4. See Agnes M. Clerke, *A Popular History of Astronomy During the Nineteenth Century* (London: A. & C. Black, 1893), p. 74.
5. Mabel Loomis Todd, *Total Eclipses of the Sun* (Boston, 1894), p. 110.
6. Mark Twain, *A Connecticut Yankee in King Arthur's Court* (New York: Barnes and Noble, 1995), pp. 86, 34–35.
7. See John B. Duff and Peter M. Mitchell, eds., *The Nat Turner Rebellion* (New York: Harper and Row, 1971), p. 19.
8. William Styron, *The Confessions of Nat Turner* (New York: Random House, 1967), p. 348.
9. Paul D. Bailey, *Wovoka, the Indian Messiah* (Los Angeles: Westerlore Press, 1957), pp. 93–94.
10. Virginia Woolf, *The Diary of Virginia Woolf*, Vol. III, *1925–30*, ed. Anne Olivier Bell (New York: Harcourt, 1980), pp. 142–43.
11. Isaac Asimov, "New Stars," in *The Relativity of Wrong* (New York: Doubleday, 1988), p. 138.
12. Stephen Jay Gould, "Happy Thoughts on a Sunny Day in New York City," in *Dinosaur in a Haystack* (New York: Harmony Books, 1985), p. 3.
13. John Updike, *The Early Stories, 1953–75* (New York: Knopf, 2003), pp. 645–46.
14. See Jay M. Pasachoff, "Solar-Eclipse Science: Still Going Strong," *Sky and Telescope*, February 2001, pp. 40–45.
15. Albert Einstein, "Grundlage der allgemeinen Relativitätstheorie" ("The Foundations of the Theory of General Relativity"), *Annalen der Physik* vol. 49, (1916), pp. 769–822.
16. See Christopher M. Linton, *From Eudoxus to Einstein: A History of Mathematical Astronomy* (Cambridge: Cambridge University Press, 2004), p. 450.
17. A. S. Eddington, *Obituary Notices, Fellows of the Royal Society* no. 8, vol. 3, January 1940, p. 167.
18. Paul Johnson, *Modern Times* (London: Weidenfeld, 1999), pp. 2–4. See also the opening chapter of Nicholas Mosley, *Hopeful Monsters* (New York: Vintage, 2000), p. 26, where the episode is re-created.
19. See Ronald W. Clark, *Einstein: The Life and Times* (London: Hodder, 1973), p. 225; see also Arthur I. Miller, *Empire of the Stars: Obsession, Friendship and Betrayal in the Quest for Black Holes* (Boston: Houghton Mifflin, 2005), pp. 51–52.
20. Thomas Crump, *Solar Eclipse* (London: Constable, 1995), p. 134; I have drawn in some detail from his description of the expedition.

CHAPTER 12: THE SUN DETHRONED

1. Tommaso Campanella, *A Dialogue Between a Grand Master of the Knights Hospitallers and a Genoese Sea Captain: The City of the Sun* (Berkeley: University of California Press, 1981).
2. Woody Allen, "Strung Out," *The New Yorker*, July 28, 2003, p. 96.
3. See *Catholic Encyclopedia* on CD-ROM, entry for Secchi.
4. See *The New York Times*, Science section, July 11, 2006.
5. Mark Haddon, *The Curious Incident of the Dog in the Night-time*, pp. 12–13.
6. Ron Cowen, "The Galaxy Hunters," *National Geographic*, February 2003, p. 16.

7. See A. E. Housman, *Selected Prose*, ed. J. Carter and J. Sparrow (Cambridge: Cambridge University Press), 1961.

8. Martin Gorst, *Measuring Eternity: The Search for the Beginning of Time* (New York: Broadway, 2001), p. 11. See also Marcia Bartusiak, *The Day We Found the Universe* (New York: Pantheon, 2009).

9. Stephen W. Hawking, *A Brief History of Time* (New York: Bantam, 1988), p. 7.

10. Gorst, *Measuring Eternity*, p. 95. I have made significant use of Gorst's excellent account in this section.

11. Timothy Ferris, *Coming of Age in the Milky Way* (New York: Anchor, 1989), p. 246.

12. Gorst, *Measuring Eternity*, p. 187.

13. Ibid., p. 188.

14. Sir J.F.W. Herschel, *Treatise on Astronomy* (London: Longman, 1833), p. 212.

15. George Gamow, *The Birth and Death of the Sun: Stellar Evolution and Subatomic Energy* (New York: Viking, 1949), p. 12.

16. For a discussion of how "outsiders" contributed to solar physics from 1910 on, see Karl Hufbauer, *Exploring the Sun*, pp. 81ff.

17. Cecilia H. Payne, *Stellar Atmospheres* (Harvard Observatory Monograph No. 1, Cambridge, Mass., 1925), p. 185. Eddington's own monograph, *The Internal Constitution of the Stars*, published a year later, became an instant classic.

18. See Owen Gingerich, "The Most Brilliant PhD Thesis Ever Written in Astronomy," Harvard Smithsonian Center for Astrophysics, p. 16, available at www.harvardsquarelibrary.org/unitarians/payne2.html.

19. See Gamow, *Birth and Death of the Sun*, p. v.

20. See Finn Aaserud, *Redirecting Science: Niels Bohr, Philanthropy, and the Rise Of Nuclear Physics* (Cambridge: Cambridge University Press, 1990), p. 2.

21. David Kaiser, "A × B ≠ B × A: Paul Dirac," *London Review of Books*, February 26, 2009, p. 21.

22. Letter to Bohr, April 21, 1932, quoted in Aaserud, *Redirecting Science*, p. 55.

23. Aaserud, *Redirecting Science*, p. 7.

24. See "The Elements" (2007); see seedmagazine.com/content/article/cribsheet_8_the_elements/.

25. See Michio Kaku, *Physics of the Impossible: A Scientific Exploration into the World of Phasers, Force Fields, Teleportation, and Time Travel* (New York: Doubleday, 2008), p. xv.

26. See Richard Rhodes, *Dark Sun* (New York; Simon and Schuster, 1995), p. 222.

27. Reginald Victor Jones, *Most Secret War: British Scientific Intelligence, 1939–1945* (London: Hamish Hamilton, 1978); published in the United States as *The Wizard War: British Scientific Intelligence, 1939–1945*.

28. A full account of how Einstein's letter was drafted and how it reached the White House is given in Walter Isaacson, *American Sketches* (New York: Simon and Schuster, 2009), pp. 149–55.

29. See also Jeremy Bernstein, "The Secrets of the Bomb," *New York Review of Books*, May 25, 2006, p. 41.

30. See John Hersey, *Hiroshima* (Harmondsworth: Penguin, 1946), publisher's note, p. viii.

31. See the obituary of Charles Donald Albury (1920–2009) published in *The Miami Herald* from June 4 to June 29, 2009.

32. See Bethe's obituary in the London *Times*, March 8, 2005.

CHAPTER 13: SUNSPOTS

1. William Smith Urmy, *The King of Day* (New York: Nelson and Phillips, 1874), pp. 82–83.

2. See Joseph Needham, *Science and Civilization in China* (Cambridge: Cambridge University Press, 1959), p. 435. Needham mentions several Arab observations, including those in 840 (mistaken for a transit of Venus), 1196, and 1457.

3. Fr. Juan Casanovas, "Early Observations of Sunspots," San Francisco: Astronomical Society of the Pacific Conference Series, Proceedings of a Meeting Held in Puerto de la Cruz, Tenerife, Spain, October 2–6, 1996, vol. 118, pp. 3 and 19.

4. Ellen M. McClure, *Sunspots and the Sun King* (Chicago: University of Illinois Press, 2006), p. 1.

5. Heinrich Schwabe, *Astronomische Nachrichten*, vol. 20, no. 495 (1843).

6. Simon Mitton, *Daytime Star: The Story of Our Sun* (New York: Scribner, 1981), p. 122.

7. Sten F. Odenwald, *The 23rd Cycle: Learning to Live with a Stormy Star* (New York: Columbia University Press, 2001), p. 54.

8. See essay on Schwabe and Wolf by Dr. David P. Stern; contact education@phy6.org.
9. Stuart Clark, *The Sun Kings: The Unexpected Tragedy of Richard Carrington and the Tale of How Modern Astronomy Began* (Princeton, N.J.: Princeton University Press, 2007), p. 23.
14. See ibid., p. 142.
10. See *Encyclopedia of Astronomy and Astrophysics*, p. 3203.
11. See Mitton, *Daytime Star*, p. 130. See also William Livingston and Arvind Bhatnagar, *Fundamentals of Solar Astronomy* (New Jersey: World Scientific Publishing, 2005).
12. W. Livingston, J. W. Harvey, O. V. Malenchenko, and L. Webster, "Sunspots with the Strongest Magnetic Fields," *Solar Physics*, vol. 239, nos. 1–2, December 2006, pp. 41–68.
13. See F. E. Zeuner, *Dating the Past* (London: Sutton, 1952), p. 19.
14. Harlan True Stetson, *Sunspots and Their Effects* (New York: Whittlesey House [McGraw-Hill], 1937), p. 43.
15. Samuel Johnson, *Rasselas, Poems, and Selected Prose*, ed. Bernard H. Bronson (New York: Henry Holt, 1965), pp. 592–93.
16. Stuart Clark, "Quiet Sun Puts Europe on Ice," *New Scientist*, April 14, 2010.
17. See J. A. Eddy, "Climate and the Role of the Sun," in J. A. Eddy, ed., *The New Solar Physics* (Boulder, Col.: Westview Press, 1978), pp. 11–34; Andrew E. Douglass, *Dictionary of American Biography*; and *Nature*, vol. 431, p. 1047.
18. See Spencer R. Weart, *The Discovery of Global Warming* (Cambridge, Mass., and London: Harvard University Press, 2003), p. 131.
19. Nigel Calder, *The Weather Machine* (London: BBC, 1974), p. 131.
20. See William F. Ruddiman, "How Did Humans First Alter Global Climate?" *Scientific American*, March 2005, pp. 46–53; also Donald Goldsmith, "Ice Cycles," *Natural History*, March 2007, pp. 14–18.
21. See William James Sidis, "A Remark on the Occurrence of Revolutions," *Journal of Abnormal Psychology* 13 (1918), pp. 213–28. See also Martin Gardner, *Mathematical Carnival* (New York: Knopf, 1977).
22. See *Citizen Cohn*, scripted by Nicholas von Hoffman, HBO film for television, Breakheart Films/Spring Creek Productions, 1992; von Hoffman's book of the same title (New York: Doubleday, 1988), p. 180; and Wayne Phillips, "Harassing Feared by 'Voice' Suicide," *The New York Times* March 7, 1953, p. 10.
23. Urmy, *King of Day*, p. 79.
24. See Odenwald, *The 23rd Cycle*, pp. 8ff. and 70.
25. See Roger Guesnerie, *Assessing Rational Expectations: Sunspot Multiplicity and Economic Fluctuations* (Cambridge, Mass.: MIT Press, 2001), p. 1.
26. Valmore C. La Marche, Jr., and Harold C. Fritts, "Tree-Rings and Sunspot Numbers," *Tree-Ring Bulletin*, vol. 32 (1972), pp. 19–33.
27. See Mike Baillie, "Tree-Rings Focus Attention on Global Environmental Events That Could Be Caused by Extraterrestrial Forcing," School of Archaeology and Palaeoecology, Queen's University, Belfast. See also P. D. Jones and E. Mann, "Climate over Past Millennia," *Review of Geophysics*, May 2004, p. 21.
28. See Paul Simons, "Summer 1816," London *Times*, October 1, 2008, p. 9.
29. See *Science News*, April 10, 2010, p. 8.
30. See Paul Simons, "Weather Eye," London *Times*, April 10, 2010, p. 85.

CHAPTER 14: THE QUALITIES OF LIGHT

1. Quoted in Michael Sims, *Apollo's Fire: A Day on Earth in Nature and Imagination* (New York: Viking, 2007), p. 29.
2. Graham Greene, *The Power and the Glory* (London: Penguin, 1940), p. 70.
3. Galileo Galilei, *Dialogue Concerning the Two Chief World Systems* (Philadelphia: Running Press, 2005).
4. See Katie McCullough, "Speed of Light Formal Report (The Foucault Method)," http://njas.org/projects/speed_of_light/cache/2/lightspeedformal.htm.
5. Michio Kaku, *Physics of the Impossible* (New York: Doubleday, 2008), p. 19.
6. See Dorothy Michelson Livingston, *The Master of Light* (Chicago, 1973), and Joel Achenbach, "The Power of Light," *National Geographic*, October 2001, p. 15.
7. John Lloyd and John Mitchinson, *The Book of General Ignorance* (London: Harmony, 2007), p. 57.

8. See David Attenborough, *The Private Life of Plants* (Princeton, N.J.: Princeton University Press, 1995), p. 102.

9. Stuart Clark, *The Sun Kings* (Princeton, N.J.: Princeton University Press, 2007), p. 94.

10. See Duncan Steel, *Marking Time*, pp. 145–50. See also Mick O'Hare, ed., *Why Don't Penguins' Feet Freeze? And 114 Other Questions* (London: Profile, 2006), pp. 154–55.

11. Clark, *Sun Kings*, p. 96.

12. See Richard M. Bucke, *Cosmic Consciousness* (Secaucus, N.J.: University Books, 1961), quoted in Peter Matthiessen, *The Snow Leopard* (New York: Viking, 1978), p. 98.

13. Elizabeth Wood, *Science for the Airplane Passenger* (Boston: Houghton Mifflin, 1968), p. 60.

14. *Encyclopaedia Britannica*, 10th edition, "Meteorology," pp. 278–79.

15. *Natural History*, October 2005, p. 11.

16. Anthony Holden, *Big Deal* (London: Abacus, 2002), pp. 49 and 120. In Charles Bock's breakout first novel, *Beautiful Children* (New York: Random House, 2008), Las Vegas is "The neon. The halogen. The viscous liquid light. Thousands and millions of watts, flowing through letters of looping cursive and semi-cursive, filling then emptying, then starting over again. Waves of electricity, emanating from pop-art facades, actually transforming the nature of the atmosphere, creating a mutation of night, a night that is not night—daytime at night."

17. See Christopher Hibbert, *The Great Mutiny: India, 1857* (London: Penguin, 1980), p. 147, and Bernard Cornwell, *Sharpe's Rifles: Richard Sharpe and the French Invasion of Galicia, January 1809* (London: Penguin, 1989), p. 86.

18. Edith Nesbit, *The Wouldbegoods* (London: Puffin, 1996), chapter 3, "Bill's Tombstone."

19. See Air Vice Marshal J. E. Johnson, *The Story of Air Fighting* (London: Chatto, 1964), pp. 24–25.

20. See Geoffrey Bennett, *Coronel and the Falklands* (Edinburgh: Birlinn, 2000), pp. 13 and 28.

21. Sam Shepard, "A Short Life of Trouble," originally published in *Esquire*, 1987, reprinted in Jonathan Cott, ed., *Bob Dylan: The Essential Interviews* (New York: Wenner Books, 2006), p. 365; John Dos Passos, *The Fourteenth Chronicle: Letters and Diaries of John Dos Passos*, ed. Townsend Ludington (Boston: Gambit, 1973), p. 567.

22. For a discussion of how shadow and glare have affected baseball in the United States, see John Branch, "Postseason's Afternoon Start Times Put Shadows in Play," *The New York Times*, October 7, 2009, B17.

23. C. P. Snow, *Variety of Men* (New York: Scribner's, 1967), p. 22.

24. See Edward Marjoribanks, *For the Defence: The Life of Sir Edward Marshall Hall* (New York: Macmillan, 1929), pp. 147 and 408–12.

25. A. Roger Ekrich, *At Day's Close: Night in Times Past* (New York: Norton, 2005); see also *The New Yorker*, May 30, 2005.

26. Anisha Vranckx, in an interview with the author, October 2006. Ms. Vranckx, of Greek extraction but now based in Udaipur, is a Romany, and has twice represented Gypsy interests at the United Nations.

27. W. Sangster, *Umbrellas and Their History* (London: Cassell, 1871), p.15.

28. Edmund C. P. Hull, *The European in India, or The Anglo-Indian's Vade Mecum* (London: 1871), p. 61.

29. Robert Louis Stevenson, *The Strange Case of Dr. Jekyll and Mr. Hyde and Other Tales of Terror*, ed. Robert Mighall (London: Penguin, 2002), p. 23.

30. See J. K. St. Joseph, "Air Photography and Archaeology," *Geographical Journal*, vol. 105, no. 1–2 (January–February 1945), pp. 47–59.

31. G.W.F. Hegel, *Philosophy of Right*, trans. T. M. Knox (New York: Oxford University Press, 1967; first published 1820).

32. Jill Nelson, *Sexual Healing* (Chicago: Agate Publishing, 2003); see review of this first novel by Beverly Lowry, *The New York Times Book Review*, June 22, 2003, p. 20.

33. Vladimir Nabokov, *Speak, Memory* (New York: Putnam, 1960), p. 81; see also Margaret Wertheim, "The Shadow Goes," *The New York Times*, June 20, 2007, and Charles Champlin and Derrick Tseng, *Woody Allen at Work: The Photographs of Brian Hamill* (New York: Abrams, 1995), p. 29, photo caption showing Allen on a beach in Southampton, Long Island, during the shooting of *Interiors* (1978).

34. A belief promulgated by Rabbi Jacob Emden (1697–1776), who cited the biblical verse (Joshua 1:8) "You will study [the Torah] day and night," leaving room for secular studies during hours that are neither truly day nor truly night.

"Call it 'Astronomy Beyond the Visible-Light Spectrum.'"

35. See John Coldstream, *Dirk Bogarde* (London: Weidenfeld, 2005), p. 517.
36. Carl Gustav Jung, *Memories, Dreams, Reflections* (New York: Pantheon, 1961), p. 269.

CHAPTER 15: BENEATH THE BEATING SUN

1. Rudyard Kipling, "Dear Auntie, Your Parboiled Nephew," holograph letter to Edith Macdonald, June 12, 1883, in *Early Verse by Rudyard Kipling, 1879–89* (New York: Oxford University Press, 1986), p. 19.
2. Thurston Clarke, *Equator* (New York: Avon, 1988), pp. 11, 137, and 143.
3. Jonathan Spence, *The Memory Palace of Matteo Ricci* (London: Faber, 1985), pp. 37 and 84.
4. Leo Tolstoy, *War and Peace*, trans. Richard Pevear and Larissa Volokhonsky (New York: Vintage, 2008), p. 1013.
5. Ruth Prawer Jhabvala, *Heat and Dust* (London: John Murray, 1975), p. 71.
6. Noël Coward, *The Complete Illustrated Lyrics* (New York: Overlook, 1998).
7. Rudyard Kipling, *Kim* (London: Penguin, 1987), pp. 110 and 330.
8. See David Grann, *The Lost City of Z* (New York: Doubleday, 2009), p. 56.
9. See Sara Wheeler, *Too Close to the Sun: The Audacious Life and Times of Denys Finch Hatton* (New York: Random House, 2007), pp. 55, 69, and 95.
10. Richard Collier, *The Sound of Fury: An Account of the Indian Mutiny* (London: Collins, 1963), pp. 120 and 14–15.
11. Ibid.
12. Robert Burton, *The Anatomy of Melancholy* (New York: New York Review of Books, 2001), p. 378.
13. Jonathan Weiner, *The Next One Hundred Years* (New York: Bantam, 1990), p. 154.
14. Barack Obama, *Dreams from My Father* (New York: Three Rivers Press), 2004.
15. Babette Babich, "Reflections on Greek Bronze and 'The Statue of Humanity,'" *Existentia*, vol. 17 (2007), p. 436.
16. William Shakespeare, sonnet 62, line 10. There are 253 references to the Sun in Shakespeare.
17. Miranda Seymour, "White Shoulders," review of Deborah Davis's biography *Strapless: John Singer Sargent and the Fall of Madame X* in *The New York Times Book Review*, September 28, 2003, p. 19.
18. Paul Fussell, *Abroad: British Literary Travelling Between the Wars* (Cambridge: Cambridge University Press, 1980), p. 137. Several of the details on these pages come from this study, notably the chapter "The New Heliophily," pp. 137–41.
19. F. Scott Fitzgerald, *Tender Is the Night* (London: Grey Walls Press, 1953), p. 83. Sun worship also appears in *The Beautiful and Damned*. He called the beach at Antibes "a bright tan prayer mat."
20. See the review of Emily W. Leider's *Dark Lover: The Life and Death of Rudolph Valentino*, "We Lost It at the Movies," by Barry Gewen, *The New York Times Book Review*, May 11, 2003, p. 15.

21. See Bennetta Jules-Rosette, *Josephine Baker in Art and Life* (Chicago: University of Illinois, 2007).

22. Mary Blume, "Vive the Vacation: The French Look Back," *International Herald Tribune*, July 20, 2006, p. 20.

23. Stephen Potter, *Some Notes on Lifemanship* (New York: Henry Holt, 1950), pp. 18–20.

24. Thurston Clarke, *Ask Not: The Inauguration of John F. Kennedy and the Speech That Changed America* (New York, Henry Holt, 2004), pp. 41–42. Clarke may have chosen not to stress the effects of the Sun in his travel book, *Equator*, but he makes his point here.

25. Gore Vidal, *Palimpsest: A Memoir* (New York: Random House, 1995), p. 355.

26. John Fowles, *Wormholes: Essays and Occasional Writings* (New York: Henry Holt, 1998), pp. 284–87.

27. Jane Austen, *Persuasion* (Cambridge: Cambridge University Press, 2006), p. 110.

28. Thomas More Madden, *On Change of Climate* (London: 1874), p. 32.

29. See R. A. Hobday, "Sunlight Therapy and Solar Architecture," *Medical History* (1997) 42:455–72. See also Richard Hobday, *The Light Revolution: Health, Architecture, and the Sun* (Forres, U.K.: Findhorn Press, 2006), p. 12.

30. A. Ransome, *The Principles of "Open-Air" Treatment of Phthisis and of Sanatorium Construction* (London: Smith, Elder, 1903), pp. 72–73.

31. See Thomas Stuttaford, London *Times*, February 2, 2006, section 2, p. 9.

32. See Martin Burgess Green, *Children of the Sun* (New York: Stein and Day, 1976), p. 6. Green openly acknowledges taking his title from W. J. Perry and Grafton Elliot Smith's 1930 study.

33. Fiona MacCarthy, London *Times*, December 19, 1991.

34. "The Solar Revolution" is a phrase of the British critic John Weightman's (1915–2004).

35. See Robert Mighall, *Sunshine: One Man's Search for Happiness* (London: Murray, 2008), p. 46.

36. Stephen Spender, *World Within World* (London: Hamish Hamilton, 1951; reissued by Faber, 1997).

37. Fowles, *Wormholes*, p. 285.

38. Quoted in Fussell, *Abroad*, p. 136.

39. See Mighall, *Sunshine*, p. 39.

40. Albert Camus, *Lyrical and Critical Essays* (New York: Knopf, 1968), pp. 352, 80–83. After his death, his longtime publisher, Blanche Knopf, wrote an appreciation in the *Atlantic* (February 1961, pp. 77ff.) entitled "Albert Camus in the Sun": "Camus was born of the Sun and always had a yearning to be in it."

41. See *Tucson Star*, August 19, 1986.

CHAPTER 16: SKIN DEEP

1. See Joel L. Swerdlow, "Unmasking Skin," *National Geographic*, November 13, 2002, pp. 36–63.

2. Alice Hart-Davis, "Take More Cover," London *Evening Standard*, May 20, 2009, p. 29.

3. See Steve Jones, *In the Blood* (London: HarperCollins, 1996), p. 194.

4. Jeffrey Gettleman, "Albinos, Long-Shunned, Face Deadly Threat in Tanzania," *The New York Times*, June 6, 2008, pp. 8 and 20; see also Donald McNeil, "Bid to Stop the Killing of Albinos," *The New York Times*, February 17, 2009, D5, and "The Enduring Curse of African Witchcraft," *The Week*, April 11, 2009, p. 16.

5. See G. P. Studzinski and D. C. Moore, "Sunlight: Can It Prevent as Well as Cause Cancer?" *Cancer Research* (1995), 55:4014–22, and Paul Vitello, "Skin Cancer Up Among Young; Tanning Salons Become Target," *The New York Times*, August 14, 2006, B1. See also Kaur Mandeep, M.D., and others, *Journal of the American Academy of Dermatology*, part 1, July 2004.

6. See *The New Yorker*, December 6, 2004, p. 52.

7. G. K. Chesterton, *The Annotated Innocence of Father Brown*, ed. Martin Gardner (New York: Dover, 1997), p. 215.

8. Mick O'Hare, ed., *Why Don't Penguins' Feet Freeze? And 114 Other Questions* (London: Profile, 2006), pp. 4–5.

9. Jane E. Brody, "A Second Opinion on Sunshine: It Can Be Good Medicine After All," *The New York Times*, June 17, 2003, F7. See also *Science News*, April 24, 2010, p. 9.

10. See Michael Holick and Mark Jenkins, *The UV Advantage: New Medical Breakthroughs Reveal Powerful Health Benefits from Sun Exposure and Tanning* (New York: iBooks, 2003). In 2007 the London *Independent* gave up its whole front page to tell its readers, "A new study reveals that vitamin D, unlocked by sunlight on our skin, can ward off colds and flu. It also halves the risk of cancer, combats heart disease and fights diabetes. Now scientists are calling it the 'wonder vitamin' " (April 14).

11. Jones, *In the Blood*, pp. 177–97.

12. See J. L. Cloudsley-Thompson, "Time Sense of Animals," in J. T. Fraser, ed., *The Voices of Time: A Cooperative Survey of Man's Views of Time as Expressed by the Sciences and by the Humanities* (New York: George Braziller, 1966), p. 299.

13. Morag Preston, "Children of the Moon," London *Times, Saturday Magazine*, date unknown.

14. Lisa Sanders, "Perplexing Pain," *The New York Times Magazine*, November 1, 2009, pp. 24–25.

15. See Elizabeth Kostova, *The Historian* (New York: Little, Brown, 2005), p. 55; also p. 148, "He is not going to jump on you in broad daylight, Paul"—discussing a modern-day vampire.

16. See Jones Denver, *An Encyclopedia of Obscure Medicine* (New York: University Books, 1959).

17. See also Nick Lane, "New Light on Medicine," *Scientific American*, January 2003, vol. 288, no. 1, p. 38, and "The Straight Dope," *Washington City Paper*, "Did Vampires Suffer from the Disease Porphyria—or Not?"

18. See Mel Sinclair, http://www.eczemasite.com.

19. John Updike, *Self-consciousness: Memoirs* (New York: Knopf, 1989), pp. 42–78.

20. See "Winter Depression," *Harvard Mental Health Letter*, November 2004, p. 4.

21. See Abigail Zuger, "Nighttime, and Fevers Are Rising," *The New York Times*, September 28, 2004, F5.

22. Nigel Hawkes, "Why We Get So Gloomy in Winter," London *Times*, February 17, 2004, p. 5. See also Jane Brody, "Getting a Grip on the Winter Blues," *The New York Times*, December 5, 2006, F7. Ms. Brody had articles appear on Sun-related matters for four consecutive months that year, none of which questioned the SAD lobby's findings.

23. Rick Atkinson, *The Day of Battle: The War in Sicily and Italy, 1943–1944* (New York: Henry Holt, 2007), p. 55.

24. Ken Chowder, "Copenhagen," *The New York Times Magazine*, November 21, 2004, p. 92.

25. O. Lingjaerde and others, *Journal of Affective Disorders* 33 (1995), pp. 39–45.

26. Ted Reichborn-Kjennerud, *Patients with Seasonal Affective Disorder: A Study of the Clinical Picture, Personality Disorders, and Biological Aspects* (Oslo: University of Oslo Press, 1997), p. 119.

27. Judith A. Perry, David H. Silvera, Jan H. Rosenvinge, Tor Neilands, and Arne Holte, "Seasonal Eating Patterns in Norway: A Non-clinical Population Study," *Scandinavian Journal of Psychology* (2001), 42:307–12.

28. Andrés Magnússon and Jóhann Axelsson, "The Prevalence of Seasonal Affective Disorder Is Low Among Descendants of Icelandic Emigrants in Canada," *Archives of General Psychiatry*, December 1993, vol. 50, pp. 947ff.

29. Tim Brennen, Monica Martinussen, Bernt Ole Hansen, and Odin Hjemdal, "Arctic Cognition: A Study of Cognitive Performance in Summer and Winter at 69°N," *Applied Cognitive Psychology* vol. 13 (1999), pp. 561–80.

CHAPTER 17: THE BREATH OF LIFE

1. Quoted in *The New York Times*, November 16, 2002.

2. Galileo Galilei, *Dialogue Concerning the Two Chief World Systems*, (Philadelphia: Running Press, 2005).

3. Bill Bryson, *A Short History of Nearly Everything* (New York: Broadway, 2003), p. 298.

4. Oliver Morton, *Eating the Sun: How Plants Power the Planet* (London: Fourth Estate, 2007), pp. xvii and 56. See also Peter H. Raven, *Biology*, 7th edition (New York: McGraw-Hill, 2007); David Williams, *Lessons from Joseph Priestley: The 2004 Essex Hall Lecture* (London: Lindsay Press, 2004).

5. See Georg Hartwig, *The Subterranean World* (New York: Scribner, 1871), p. 88, and Mark Twain, *The Innocents Abroad, or The New Pilgrim's Progress* (London: Hotten, 1870): "Our author makes a long, fatiguing journey to the Grotto del Cane on purpose to test its poisoning powers on a dog—got elaborately ready for the experiment, and then discovered that he had no dog."

6. Stephen Hales, *Vegetable Staticks* (1727).

7. Thomas S. Kuhn, *The Structure of Scientific Revolutions* (Chicago: University of Chicago Press, 1962), pp. 53–56.

8. Jan Ingenhousz, *Experiments upon Vegetables, Discovering Their Great Power of Purifying the Common Air in the Sun-shine, and of Injuring It in the Shade and at Night* (London, 1779).

9. Nancy Y. Kiang, "The Color of Plants on Other Planets," *Scientific American*, April 2008, pp. 48–55.

10. See John Maddox, *What Remains to Be Discovered* (New York: Free Press, 1998), p. 147.

11. See Morton, *Eating the Sun*, pp. 168–75.

12. See Mikolaj Sawicki, "Myths About Gravity and Tides," *Physics Teacher* 37, October 1999, pp. 438–41.

13. *Excursions: The Writings of Henry David Thoreau*, vol. 9 (Boston: Houghton Mifflin, 1893), p. 292.

14. David Attenborough, *The Private Life of Plants* (Princeton: Princeton University Press, 1995), pp. 45ff.

15. Morton, *Eating the Sun*, p. 222.

16. Attenborough, *Private Life of Plants*, p. 1.

17. Vernon Quinn, *Stories and Legends of Garden Flowers* (New York: Frederick Stokes, 1939), pp. 116–17. In alchemy, both the lily and the holm oak are associated with the Sun, the lily embodying purity, the holm oak strength and glory.

18. See Peter Tompkins and Christopher Bird, *The Secret Life of Plants* (New York: Harper, 1973), pp. 58 and 170–71.

19. Dan Morgan, *Merchants of Grain* (New York: Viking, 1979), p. 16.

20. See William J. Broad, "CIA Revives Data Sharing on Environment," *The New York Times*, January 5, 2010, A1.

21. See "Night Schools for Fish," *The Washington Post*, March 30, 2009, p. A5.

22. See C. Claiborne Ray, "Birds of a Feather," *The New York Times*, October 3, 2006, F2, and Natalie Angier, "Some Blend In, Others Dazzle," ibid., F1.

23. See D. V. Alford, *Bumblebees* (London: Davis-Poynter, 1975), p. 75.

24. Frisch's account is authoritative but a little leaden; here I have paraphrased the excellent description in David Attenborough, *Discovering Life on Earth* (Boston: Little, Brown, 1981), p. 103.

25. See Edward O. Wilson, *The Insect Societies* (Boston: Harvard University Press, 1971).

26. See M. Dacke and others, "Built-in Polarizers Form Part of a Compass Organ in Spiders," *Nature*, September 30, 1999, pp. 470ff.

27. Wilson, *Insect Societies*, p. 216.

28. See Norman R. F. Maier and T. C. Schneirla, *Principles of Animal Psychology* (New York: McGraw-Hill, 1935), p. 188.

29. See Leland Crafts, Theodore C. Schneirla, Elsa E. Robinson, and Ralph W. Gilbert, "Migration and the 'Instinct' Problem," in ibid., pp. 25–39.

30. See also Kenneth P. Able, ed., *Gatherings of Angels: Migrating Birds and Their Ecology* (Ithaca, N.Y.: Cornell University Press, 1999), pp. 14–15, and David Attenborough, *The Life of Birds* (Princeton N.J.: Princeton University Press, 1998), p. 62.

31. See *Nature Neuroscience Reviews*, February 2005, and *The New York Times*, February 1, 2005, F1.

32. See Frank P. Gill, *Ornithology* (New York: W. H. Freeman, 1995), chapter 13, and P. Berthold, *Bird Migration: A General Survey* (New York: Oxford University Press, 1993), pp. 156–57.

33. See David Attenborough, *Life on Earth*, DVD, vol. 7, "Victors of the Dry Land," and *Life on Earth* (Boston: Little, Brown, 1979), p. 152.

34. See David Attenborough, *The Trials of Life: A Natural History of Animal Behavior*, DVD, vol. 6, "Homemaking."

35. See James Lovelock, *Gaia: A New Look at Life on Earth* (Oxford: Oxford University Press, 1979), and *The Ages of Gaia* (Oxford: Oxford University Press, 1988).

36. Morton, *Eating the Sun*, p. 256.

CHAPTER 18: THE DARK BIOSPHERE

1. Robert Kunzig, *Mapping the Deep: The Extraordinary Story of Ocean Science* (New York: Norton, 2000), p. 207.
2. James Hamilton-Paterson, *The Great Deep: The Sea and Its Thresholds* (New York: Henry Holt, 1992), p. 165.
3. E. A. Wallis Budge, tr. of Pseudo-Callisthenes, 1933; see Hamilton-Paterson, *Great Deep*, p. 167.
4. See David Grann, "The Squid Hunter," *The New Yorker*, May 24, 2004, pp. 56–71.
5. W. J. Broad, "Deep Under the Sea, Boiling Founts of Life Itself," *The New York Times*, September 9, 2003, F1–4.
6. See Julie A. Huber, David A. Butterfield, and John A. Baross, *FEMS Microbiology Ecology*, vol. 43 (2003), pp. 393–409.
7. See Kunzig, *Mapping the Deep*, p. 53. In *The Ecological Theater and the Evolutionary Play* (New Haven: Yale University Press, 1965), the famous Yale ecologist G. Evelyn Hutchinson suggested that it might be possible in theory for organisms to make a living off the internal heat of the Earth rather than off sunlight, but he had no idea how they might do it.
8. David Attenborough, *Discovering Life on Earth* (Boston: Little, Brown, 1981), p. 22.
9. See Peter Whitehead, *How Fishes Live* (London: Phaidon, 1975), pp. 111–12.
10. Rachel Carson, "The Sunless Sea," in *Great Essays in Science*, ed. Martin Gardner (New York: Prometheus, 1994), pp. 287–304.
11. See Hamilton-Paterson, *The Great Deep*, pp. 112–13.
12. See Peter Herring, *The Biology of the Deep Ocean* (Oxford: Oxford University Press, 2002), chapter 8, "Seeing in the Dark."
13. Hamilton-Paterson, *The Great Deep*, p. 203.
14. Charles Q. Choi, "Thousands of Strange Sea Creatures Discovered," LiveScience.com, November 22, 2009, www.livescience.com/animals/091122-deep-sea-creatures.html.
15. See M. F. Moody and J. R. Parris, "Discrimination of Polarized Light by Octopus," *Nature* 186, 839–40 (1960), and Nadav Shashar, Phillip S. Rutledge, and Thomas W. Cronin, "Polarization Vision in Cuttlefish—A Concealed Communication Channel?" *Journal of Experimental Biology* 199 (1996), 2077–84.
16. See Bruce Robison, "Life in the Ocean's Midwaters," *Scientific American*, July 1995, p. 60.
17. "Deep-Sea Fish Sees Red," *BBC Wildlife*, August 1998, p. 59.
18. See J. T. Fraser, ed., *The Voices of Time: A Cooperative Survey of Man's Views of Time as Expressed by the Sciences and by the Humanities* (New York: George Braziller, 1966), p. 309.
19. See N. Angier, "Out of Sight, and Out of a Predator's Stomach," *The New York Times*, July 20, 2004, F4. See also Sonke Johnsen, "Transparent Animals," *Scientific American*, February 2000, pp. 80–90, and K. Madin and D. Kovacs, *Beneath Blue Waters: Meetings with Remarkable Deep-Sea Creatures* (New York: Viking, 1996).
20. Kunzig, *Mapping the Deep*, p. 8.
21. See Piers Chapman, "Ocean Currents," www.waterencyclopedia.com/Mi-Oc/Ocean-Currents.html.
22. See William J. Broad, "Rogue Giants at Sea," *The New York Times*, July 11, 2006, F1 and F4.
23. See Stephen H. Schneider, ed., *The Encyclopedia of Climate and Weather* (Oxford: Oxford University Press, 1996), vol. 2, p. 734.
24. William Whewell, *The Philosophy of the Inductive Sciences: Works in the Philosophy of Science, 1830–1914* (New York: Continuum, 1999).
25. Broad, "Rogue Giants at Sea," F4.
26. See Kenneth Chang, "Findings," *The New York Times*, August 8, 2006.
27. See *Dictionary of National Biography 1910–11*, Darwin, Sir George Howard, supplement to main dictionary.
28. *Encyclopaedia Britannica*, 11th ed., p. 944, "Tides." It is a very Darwin family deprecation, though: Charles Darwin is constantly saying that subjects are too difficult to tackle.

CHAPTER 19: THE HEAVENLY GUIDE

1. Herman Melville, *Moby-Dick* (New York: Random House, 1930), p. 716.
2. Letter to Thomas Wentworth Higginson, 1862, in *Selected Letters*, ed. T. Johnson (Cambridge, Mass.: Harvard University Press, 1971).

SUNRISE ←
SUNSET →
NOON ↑

ZIEGLER

3. Daniel Boorstin, *The Discoverers* (New York: Random House, 1983), p. 46.

4. Ibid., p. 219.

5. *Konungs Skuggsja, or The King's Mirror,* trans. Laurence Marcellus Larsen (New York: Twayne Pil.,1917). In another place and time, Samuel Pepys would set up an examination for naval officers that included navigation, and naval schoolmasters were put on board ships to instruct the crew in mathematics: see also J. T. Fraser, ed., *The Voices of Time: A Cooperative Survey of Man's Views of Time as Expressed by the Sciences and by the Humanities* (New York: George Braziller, 1966), pp. 216–17.

6. Robert Ferguson, *The Hammer and the Cross: A New History of the Vikings* (London: Allen Lane, 2009), p. 62.

7. See "The Viking Sunstone: Is the Legend of the Sun-Stone True?" at www.polarization.com/viking/viking.html, and Bradley E. Schaefer, "Vikings and Polarization Sundials," *Sky & Telescope*, May 1997, p. 91.

8. See Kenneth Chang, "Etched in Lava, Diary of the Earth's Magnetic Field Shows a Temporary Calm," *The New York Times*, May 16, 2006, F3.

9. Timothy Ferris, *Coming of Age in the Milky Way* (New York: Anchor, 1989), p. 128.

10. Miguel de Cervantes, *Don Quixote*, trans. J. M. Cohen (Harmondsworth: Penguin, 1950), part II, chapter 29, p. 58. In the original Catalan, "Quixote" translates as "horse's ass."

11. See Jonathan Spence, *The Memory Palace of Matteo Ricci* (London: Faber, 1985), pp. 66–69.

12. Fernand Braudel, *The Mediterranean and the Mediterranean World in the Age of Philip II*, vol. 1, trans. Sian Reynolds (London: Collins, 1972), p. 104.

13. See Michael Grant, *The Ancient Mediterranean* (New York: Penguin, 1988), p. 146.

14. Braudel, *The Mediterranean and the Mediterranean World*, p. 138.

15. *Encyclopaedia Britannica*, 11th ed., p. 284, "Navigation."

16. See V. Gordon Childe, *What Happened in History* (Harmondsworth: Penguin, 1946), pp. 246–47.

17. See Richard Hakluyt, Epistle Dedicatorie to Charles Howard, second edition of *Principal Navigations* (1598); see also Derek Howse, *Greenwich Time and the Discovery of Longitude* (London: Oxford University Press, 1980), p. 12, and Lisa Jardine, *Ingenious Pursuits: Building the Scientific Revolution* (New York: Random House, 1999), pp. 137–38 and 159.

18. See David Grann, "The Map Thief," *The New Yorker*, October 17, 2005, p. 68.

19. See John Burt, *History of the Solar Compass* (Detroit: O. S. Gulley's Presses, 1878).

20. See Harry M. Geduld, ed., *The Definitive Time Machine* (Bloomington: Indiana University Press, 1987).

21. Melville, *Moby-Dick*, pp. 740–43.

22. From time to time in the writing of this book, I have consulted Wikipedia, but with considerable trepidation. Its account of the longitude problem, however, is hard to improve on, and I am duly grateful.

23. William Watson, *Ode on the Coronation of King Edward VII, 1902* (Whitefish, Mont.: Kessinger, 2009).

CHAPTER 20: OF CALENDARS AND DIALS

1. Sir Hermann Bondi, quoted in Paul Davies, *About Time: Einstein's Unfinished Revolution* (New York: Simon and Schuster, 1996), p. 1.

2. E. J. Bickerman, *The Ancient History of Western Civilization* (New York: Harper & Row, 1976).

3. Umberto Eco, "Times," in *The Story of Time*, ed. Kristen Lippincott (London: Merrell, 1999), p. 11.

4. See *Encyclopaedia Britannica*, 11th edition.

5. Allan R. Holmberg, quoted in Charles C. Mann, *1491: New Revelations of the Americas Before Columbus* (New York: Knopf, 2005), and in David Grann, *The Lost City of Z* (New York: Doubleday, 2009), p. 30.

6. See *The Cambridge History of Islam*, p. 98, and Mohammad Ilyas, *Islamic Astronomy and Science Development: Glorious Past, Challenging Future* (Petaling Jaya, Selangor Darul Ehsan, Malaysia: Pelanduk Publications, 1996), pp. 97–107.

7. See Tim Weiner, "Hailing the Solstice and Telling Time, Mayan Style," *The New York Times*, December 23, 2002.

8. See R. T. Zuidema, "The Inca Calendar," in Anthony F. Aveni, ed., *Native American Astronomy* (Austin: University of Texas Press, 1977).

9. See Duncan Steel, *Marking Time: The Epic Quest to Invent the Perfect Calendar* (New York: Wiley, 1999), p. 10.

10. See Christopher Hirst, "A Thousand Years of Tinkering with Time," *The Week*, March 8, 2008, pp. 44–45.

11. J.R.R. Tolkien, *The Lord of the Rings*, pp. 1140ff.

12. William Hazlitt, "On a Sun-Dial," first published in *New Monthly Magazine*, October 1827, reprinted in G. Keynes, ed., *Selected Essays of William Hazlitt* (New York: Random House, Nonesuch Press, 1930), p. 345.

13. See Geoffrey Chaucer, *The Parson's Prologue*; also the Introduction to *The Man of Law's Tale*, when he calculates the hour as ten o'clock by the same method. The original runs:

> Foure of the clokke it was tho, as Igesse,
> For ellevene foot, or litel moore or lesse,
> My shawe was at thilké tyme, as there,
> Of swiche feet as my lengthé parted were
> In sixe feet equal of proporcioun.

14. Joseph Needham, "Time and Knowledge in China and the West," in J. T. Fraser, ed., *The Voices of Time* (New York: George Braziller, 1966), p. 106.

15. See Albert E. Waugh, *Sundials: Their Theory and Construction* (New York: Dover, 1973), pp. 4–5.

16. J. V. Field, "European Astronomy in the First Millennium: The Archaeological Record," in Christopher Walker, ed., *Astronomy Before the Telescope* (London: British Museum, 1996), p. 121.

17. Dava Sobel, "The Shadow Knows," *Smithsonian*, January 2007, p. 91. There are two distinctive terms for sundials: the "style," a straightedge whose shadow is a straight line on a plane dial, which shadow normally identifies a *direction* on the dial; and the "nodus," a small ball or disk whose shadow identifies a *point* on the dial. See Gerhard Dohrn–van Rossum, *History of the Hour: Clocks and Modern Temporal Orders* (Chicago: University of Chicago Press, 1996). The gnomon as a whole serves as a style and its tip as a nodus.

18. *Horam non possum certam tibi dicere: facilis inter philosophos quam inter horologia convenient.* Seneca the Younger (attrib.), *Apocolocyntosis*, stanza 2, line 2.

19. *Julius Caesar*, Act II, scene i, ll. 4ff.; *Cymbeline*, Act II, scene ii; *Richard II*, Act V, scene v. See also *As You Like It*, Act II, scene vii, ll. 20–22.

20. Frank W. Cousins, *Sundials: A Simplified Approach by Means of the Equatorial Dial* (London: John Baker, 1972), p. 9.

21. Dava Sobel, "The Shadow Knows," p. 91.

22. *Henry VI, Part III*, Act II, scene v, ll. 21–30. As so often with Shakespeare, these lines bear a second reading; "quaint" also signifies female genitalia.
23. Hazlitt, "On a Sun-Dial," p. 336.
24. See Tad Friend, "The Sun on Mars," *The New Yorker*, January 5, 2004, pp. 27–28.

CHAPTER 21: HOW TIME GOES BY

1. James Joyce, *Ulysses: The Corrected Text* (New York: Random House, 1986), p, 47.
2. *The Taming of the Shrew*, Act IV, scene ii, l. 197.
3. Anthony Burgess, *One Man's Chorus* (New York: Carroll and Graf, 1998), pp. 120–23.
4. Jackson, *Book of Hours*, p. 15.
5. Daniel Boorstin, *The Discoverers* (New York: Random House, 1983), p. 40.
6. *History Magazine*, February–March 2008, p. 12.
7. Lisa Jardine, *Ingenious Pursuits: Building the Scientific Revolution* (New York: Random House, 1999), p. 133.
8. Virginia Woolf, *Mrs. Dalloway* (London: Hogarth Press, 1925, reprinted Granada, 1979), p. 91. The novel was originally entitled "The Hours"; in a book where so much is given a precise time and place, critics disagree about the hour when it opens—9 A.M. or 10 A.M.—while the hour of 11 A.M. to noon fills nearly a quarter of the novel.
9. Valentine Low, "The King of Clocks," *The Week*, January 10, 2009, p. 37.
10. See Peter James, *Dead Simple* (London: Macmillan, 2005), p. 318.
11. Clark Blaise, *Time Lord: Sir Sandford Fleming and the Creation of Standard Time* (New York: Pantheon, 2000), p. 170.
12. T. E. Lawrence, *Seven Pillars of Wisdom: A Triumph* (London: Cape, 1973), p. 573.
13. See Blaise, *Time Lord*, pp. 69, 129, and 35.
14. Burgess, *One Man's Chorus*, p. 123.
15. Blaise, *Time Lord*, pp. 142ff.
16. Mark M. Smith, *Mastered by the Clock: Time, Slavery, and Freedom in the American South* (Chapel Hill: University of North Carolina Press, 1997).
17. See "Q+A," *The New York Times*, May 8, 2007, F2.
18. See Jackson, *Book of Hours*, pp. 188–89.
19. See David Prerau, *Seize the Daylight: The Curious and Contentious Story of Daylight Saving Time* (New York: Thunder's Mouth Press, 2005), and Michael Downing, *Spring Forward: The Annual Madness of Daylight Saving Time* (Emeryville, Calif.: Shoemaker & Hoard, 2005).
20. Anahad O'Connor, "Really?" *The New York Times*, March 10, 2009, D5.
21. Gail Collins, "The Great Clock Plot," *The New York Times*, August 23, 2007, A21.

"Now that we can tell time, I'd like to suggest that we begin imposing deadlines."

22. See Cecil Adams, "The Straight Dope: Why Is India 30 Minutes Out of Step with Everybody Else?" *Washington City Paper,* June 5, 1981, p. 18.

23. Ibid.

24. See Marlise Simons, "Synchronizing the Present and Past in a Timeless Place," *The New York Times* (international edition), September 12, 2005.

25. See www.yeswatch.com.

26. See *The Week,* February 21, 2009, p. 17.

27. Christopher Hirst, "A Thousand Years of Tinkering with Time," *The Week,* March 8, 2008, p. 44.

CHAPTER 22: THE SUN IN OUR POCKET

1. Jonathan Swift, *Gulliver's Travels,* Part III, "A Voyage to Laputa."

2. More recently, researchers at MIT and the University of Arizona have reached similar conclusions: see Ian Sample, "Doubts Cast on Archimedes' Killer Mirrors," *The Guardian,* October 24, 2005.

3. Frank T. Kryza, *The Power of Light: The Epic Story of Man's Quest to Harness the Sun* (New York: McGraw-Hill, 2003), p. 53.

4. Ibid., p. 30.

5. Stuart Clark, *The Sun Kings: The Unexpected Tragedy of Richard Carrington and the Tale of How Modern Astronomy Began* (Princeton, N.J.: Princeton University Press, 2007), p. 60.

6. See www.californiasolarcenter.org/history_solarthermal.html.

7. See Charles Smith, "Revisiting Solar Power's Past," *Technology Review,* July 1995.

8. George Gamow, *The Birth and Death of the Sun: Stellar Evolution and Subatomic Energy* (New York: Viking, 1949), p. 1.

9. See Mary Archer, "Hello Sunshine," Royal Institute Proceedings no. 66, p. 10.

10. See Eisuke Ishikawa, *O-edo ecology jijo* (*The Edo Period Had a Recycling Society*) (Tokyo: Kodansha, 1994): see also "Japan's Sustainable Society in the Edo Period (1603–1867)", Japan for Sustainability newsletter, April 6, 2005, www.energybulletin.net/node/5140.

11. Mary Jo Murphy, "Becoming the Big New Idea: First, Look the Part," *The New York Times,* August 24, 2008, p. 4. As recently as 1976, *Nature* thought it necessary to inform its readers, "The Sun can be used for a variety of energy-related purposes," in an editorial titled "Is the Sun Being Oversold?" (May 20, 1976).

12. Arthur C. Clarke, *Astounding Days* (New York: Wiley, 1984), p. 203.

13. *The New York Times,* August 3, 2006, C1. Tidal and wave projects are currently under way in Rhode Island, Cantabria (Spain), Daishan (China), northern Portugal, and at least three sites in the United Kingdom.

14. See Elizabeth Kolbert, "The Car of Tomorrow," *The New Yorker,* August 11, 2003, p. 40.

15. See Henry Alford, "Solar Chic," *The New Yorker,* September 24, 2007, p. 128.

16. *Gourmet,* January 2005, p. 8; *The New York Times Magazine,* August 21, 2005, p. 51.

17. Paul D. Spudis, "Why We're Going Back to the Moon," *The Washington Post,* December 27, 2005.

18. See Elizabeth Kolbert, "The Climate of Man—Part III," *The New Yorker,* May 9, 2005, p. 57, and W. Wayt Gibbs, "Plan B for Energy," *Scientific American,* September 2006, p. 84—an issue devoted to alternative sources of power; but then the magazine devoted an issue to energy problems as far back as 1913. *Plus ça change.* See also Chris Smyth, "'Sailing' on Sunlight May Help Polar Observations," London *Times,* September 11, 2009, p. 26.

19. See Craig Whitlock, "Cloudy Germany a Powerhouse in Solar Energy," *The Washington Post,* May 5, 2007, A1 and A14.

20. Keith Bradsher, "China Tries a New Tack to Go Solar," *The New York Times,* January 9, 2010, B1 and B4; and Keith Bradsher, "China Leading Race to Make Clean Energy," *The New York Times,* January 31, 2010, p. 1.

21. James T. Areddy, "Heat for the Tubes of China," Marketplace, *The Wall Street Journal,* March 31, 2006, p. 1. The article's title is a play on a famous 1930s book (and later film), *Oil for the Lamps of China,* by Alice Tisdale Hobart.

22. *Mother Earth News* 41 (September–October 1976).

23. See *The New York Times,* September 1, 2003.

24. "(Solar) Power to the People Is Not So Easily Achieved," *The New York Times,* January 23, 2008, B2, and K. Zweibel, J. Mason, and V. Fthenakis, "Solar Grand Plan," *Scientific American,* January 2008, p. 66.

25. See "As Earth Warms, the Hottest Issue Is Rethinking Energy," *The New York Times*, November 4, 2003.

26. See Neil deGrasse Tyson, "Energy to Burn," *Natural History*, October 2005, pp. 17–20.

27. Oliver Morton, *Eating the Sun: How Plants Power the Planet* (London: Fourth Estate, 2007), p. 395.

28. Since I wrote this chapter, these words were quoted back at me—by Al Gore, addressing the Democratic Convention in Denver in August 2008.

CHAPTER 23: THE VITAL SYMBOL

1. Fyodor Dostoyevsky, *The Brothers Karamazov* (London: Penguin, 1967), p. 628.

2. William Blake, *A Vision of the Last Judgment*, c. 1810.

3. See "The Netherlands: Anne Frank's Chestnut Tree to Be Cut Down," Associated Press, November 15, 2006.

4. Hilary Mantel, *Wolf Hall* (New York: Holt, 2009), p. 78. (Edward's words come from a different, contemporary source.)

5. See Ian Thompson, *The Sun King's Garden: Louis XIV, André Le Nôtre and the Creation of the Gardens at Versailles* (New York: Bloomsbury, 2006), p. 48.

6. From the records of André Félibien, quoted in Orest Ranum, "Islands of the Self," in *Sun King: The Ascendency of French Culture During the Reign of Louis XIV*, ed. David Lee Rubin (London: Folger, 1992), pp. 30–31.

7. Thompson, *The Sun King's Garden*, p. 751.

8. See John Haffert, *The Peacemaker Who Went to War* (New York: Scapular Press, 1945), p. 196.

9. Craig Smith, "Nearly 100, LSD's Father Ponders His 'Problem Child,' " *The New York Times*, January 7, 2006.

10. See Mircea Eliade, "A South American," in *Occultism, Witchcraft, and Cultural Fashions: Essays in Comparative Religions* (Chicago: University of Chicago Press, 1978).

11. See Olivia A. Isil, *When a Loose Cannon Flogs a Dead Horse There's the Devil to Pay* (New York: Ragged Mountain Press, 1996).

12. Joanna Pitman, *On Blondes* (London: Bloomsbury, 2003), p. 12.

13. Raymond Chandler, *Farewell, My Lovely* (London: Penguin, 1940), pp. 76–77.

14. Pitman, *On Blondes*, p. 13.

15. James Joyce, *Ulysses*, p. 50, ll. 240–42. Joyce famously makes reference to the Sun in Episode 14, "The Oxen of the Sun" (ll. 383–428), when the introductory Latin verses, the maternity hospital, the nurses, and Dr. Horn are all symbols of the fertility that the oxen of the Sun embody.

16. See R. R. Brooks, *Noble Metals and Biological Systems* (New York: CRC Press, 1992), pp. 13, 99, 111, 116, 121, 199, and 297.

17. Lyndy Abraham, *A Dictionary of Alchemical Imagery* (New York: Cambridge University Press, 1998), p. 130.

18. Victor Hugo, *The Hunchback of Notre Dame* (New York: Oxford University Press, 1999), p. 235.

19. See Sarah Arnott, "What Is So Special About Gold, and Should We All Be Investing in It?" *The Independent*, September 10, 2009, p. 34.

20. See David Grann, *The Lost City of Z* (New York: Doubleday, 2009), pp. 148–49.

21. See Ben Goldberry, *The Mirror and Man* (Charlottesville: University Press of Virginia, 1985), p. 37.

22. See Sabine Melchior-Bonnet, *The Mirror: A History* (New York: Routledge, 1994), p. 46.

23. Ibid., p. 147.

CHAPTER 24: DRAWING ON THE SUN

1. See Paul Simons, "Windmills Foretell a Thaw in the Air," London *Times*, March 14, 2006, p. 64.

2. See *Encyclopaedia Britannica*, 15th ed., 1974, p. 695, column 1.

3. Graham Reynolds, *Turner* (London: Thames and Hudson, 1969), p. 12.

4. See Anthony Bailey, *Standing in the Sun* (London: Sinclair-Stevenson, 1997), p. 102.

5. Ibid., p. 248.

6. James Hamilton, *Turner: A Life* (London: Hodder, 1997), p. 216. All unsourced quotations on Turner come from Hamilton's book.

7. See Bailey, *Standing in the Sun*, p. 385.

8. Ibid., pp. 343–44.

"Can Hiram call you back? He's adjusting our solar panels."

9. See the video recording *American Light: The American Luminist Movement 1850–75* (Washington, D.C.: Camera Three Productions and National Gallery of Art, c. 1980), and Katherine Mansthorne and Mark Mitchell, *Luminist Horizons: The Art and Collection of James A. Suydam* (New York: National Academy Museum, 2006).
10. London *Times*, December 5, 1926. See also Valerie J. Fletcher, "The Light of Art," in *Fire of Life: The Smithsonian Book of the Sun* (New York: Norton, 1981), p. 192.
11. See Reinhold D. Hohl, "The Sun in Contemporary Painting and Sculpture," *Graphis*, vol. 18, no. 100, p. 228.
12. *The Complete Letters of Vincent van Gogh to His Brother* (New York: New York Graphic Society, 1958), vol. 3, pp. 374, 3, 25–27, 202, 205.
13. Robert Mighall, *Sunshine: One Man's Search for Happiness* (London: Murray, 2008), p. 235; see also John Updike, "The Purest of Styles," *The New York Review of Books*, November 22, 2007, p. 16.
14. Van Gogh, *Complete Letters*, pp. 204, 298.
15. For this recollection of a Paul Nash painting I am indebted to my friend Nicola Bennett. She also reminds me that in *Guernica*, Picasso paints a lightbulb shedding jagged light onto horses and bulls, a scene that one might expect to be depicted outdoors—only it is too awful to take place under the healing Sun.
16. Hilary Spurling, *Matisse the Master: A Life of Henri Matisse—The Conquest of Colour, 1909–1954* (London: Hamish Hamilton, 2005). The quotations here are drawn from the second volume of the biography, and my discussion of Matisse draws heavily on her observations.
17. See Spurling, *Matisse the Master*, pp. 216–17. Spurling dates their first meeting as January 16, 1918.
18. See Michael Kimmelmann, "The Sun Sets at the Tate Modern," *The New York Times*, March 21, 2004.
19. Trip Gabriel, "David Hockney: Acquainted with the Light," *The New York Times*, January 21, 1993.
20. Tim Adams, "David Hockney: Portrait of the Old Master," London *Observer*, November 1, 2009.
21. Michael Sims, *Apollo's Fire: A Day on Earth in Nature and Imagination* (New York: Viking, 2007), p. 79.

CHAPTER 25: NEGATIVE CAPABILITIES

1. Le Corbusier, *Towards a New Architecture* (New York: Praeger, 1960), p. 16.
2. Robert Leggat, *A History of Photography*, Internet document, http://www.rleggat.com/photohistory, 15–32.
3. See Peter Galassi, *Before Photography: Painting and the Invention of Photography* (New York: Museum of Modern Art, 1981), p. 11.
4. Roland Barthes, *Camera Lucida: Reflections on Photography* (New York: Hill and Wang, 1982), p. 15.

5. This photograph, and the insight about how it captures light, are taken from Geoff Dyer, *The Ongoing Moment* (New York: Pantheon, 2005), pp. 131–32.

6. Robert Frank, *The Americans* (University of Michegan: Scalo, 1959). A French edition appeared the previous year.

7. See Mark Olshaker, *The Instant Image: Edwin Land and the Polaroid Experience* (New York: Stein and Day, 1978), p. 11.

8. Quoted in Paul Delany, *Bill Brandt: A Life* (Stanford: Stanford University Press, 2004), p. 281.

9. Patrizia Di Bello, "From the Album to the Computer Screen: Collecting Photographs at Home," in James Lyons and John Plunkett, eds., *Multimedia Histories: From the Magic Lantern to the Internet* (Exeter: University of Exeter Press, 2007), p. 58.

10. See Ronald R. Thomas, "Making Darkness Visible," in Carol T. Christ and John O. Jordan, eds., *Victorian Literature and the Victorian Visual Imagination* (Berkeley: University of California Press, 1995), pp. 134–56.

11. See David Bowen Thomas, *The Origins of the Motion Picture* (London: His Majesty's Stationery Office, 1964), pp. 7–21. See also Vito Russo, "Adventures in CyberSound": "A History of Motion Pictures," www.acmi.net.au/AIC/ENC-CINEMA.html, and *Edison: The Invention of the Movies*, 4-disc boxed set, Museum of Modern Art/Kino International, 2005 (see www.kino.com/edison). The first movie kiss occurred in 1896, taking up fifty feet of film and nineteen seconds.

12. David Thomson, *A Biographical Dictionary of Film* (London: Deutsch, 1975), p. 602, entry on Vincent Price.

13. Michael Sims, *Apollo's Fire: A Day on Earth in Nature and Imagination* (New York: Viking, 2007), p. 27.

14. Ibid., p. 140.

15. Holland Cotter, "Full Constant Light," *The New York Times*, December 26, 2008, p. C39.

16. John Tauranac, *The Empire State Building* (New York: St. Martin's Press, 1995), p. 51.

17. Quoted in Jacques Guitton, ed., *The Ideas of Le Corbusier on Architecture and Urban Planning* (New York: George Braziller, 1981), p. 104.

CHAPTER 26: TALK OF THE DAY

1. Nikolay Rimsky-Korsakov, *My Musical Life* (New York: Knopf, 1923), pp. 141–42.

2. See Richard Donington, *Opera and Its Symbols* (New Haven: Yale University Press, 1990), pp. 125ff.

3. See Bryan Magee, *The Tristan Chord: Wagner and Philosophy* (New York: Henry Holt, 2000), pp. 216–21.

4. See Michael Tanner, *Wagner* (London: HarperCollins, 1996), p. 101.

5. William Mann, *The Operas of Mozart* (New York: Oxford University Press, 1977), p. 597.

6. A. G. Mackey, *An Encyclopedia of Freemasonry and Its Kindred Sciences: Comprising the Whole Range of Arts, Sciences, and Literature as Connected with the Institution* (New York: Masonic History Company, 1912), p. 765.

7. Alex Ross, "Apparition in the Woods," *The New Yorker*, July 9 and 16, 2007, p. 55.

8. *The Beatles Anthology* (New York: Chronicle Books, 2000).

9. Interview with Tony Burrell, "Best of Times, Worst of Times," London *Sunday Times Magazine*, June 11, 2006. See also Ted Anthony, *Chasing the Rising Sun* (New York: Simon and Schuster, 2007), p. 230.

10. Recording of the Sun "singing" can be heard on the Internet at a number of sites including http://solar-center.stanford.edu/singing/singing.html.

11. *The Diaries of Kenneth Tynan*, ed. John Lahr (New York: Bloomsbury, 2001), p. 76.

CHAPTER 27: BUSIE OLD FOOLE

1. Henry Wadsworth Longfellow, *Evangeline*, Part I, section iv.

2. Dick Francis, *Smokescreen* (New York: Harper, 1972), p. 6.

3. Vladimir Nabokov, *Ada: A Family Chronicle* (New York: Vintage, 1990), p. 45.

4. See, for example, Vladimir Nabokov, *Pale Fire*, in *Novels 1955–62* (New York: Library of America, 1996), p. 460: "The pen stops in midair, then swoops to bar / A cancelled sunset or restore a star, / And thus it physically guides the phrase / Toward faint daylight through the inky haze."

5. Vladimir Nabokov, *Speak, Memory* (New York: Putnam, 1966). The many solar references in

Speak, Memory are often delightful, without either the perverse sensual element or the doom-laden significance they possess in *Lolita*. For instance, as a boy Nabokov plays with some wax, "using a candle flame (diluted to a deceptive pallor by the sunshine that invaded the stone slabs on which I was kneeling)" (p. 58), and on another occasion he notes the lurid gleam cast by the Sun at the end of a rainy day on the fresh-picked mushrooms lying on a garden table (p. 44).

6. Brian Boyd, *Vladimir Nabokov: The Russian Years* (Princeton, N.J.: Princeton University Press, 1990), pp. 434–37 and 44.

7. Brian Boyd, *Vladimir Nabokov: The American Years* (Princeton, N.J.: Princeton University Press, 1990), pp. 159, 297, and 306.

8. See Vladimir Nabokov, *Lolita* (London: Penguin, 2006), pp. 7, 8, 182, 42, 68, 44, 184, 48, 60, 63, 65, 66, 66, 79, 86, 107, 109, 176, 125, 101, 215, 141, 142, 241, 244, 250, 269, 270, 300, 292 (a "green Sun," being over in an exquisite instant, may be a Nabokov reference to orgasm), 324, 320, 326, 334 (the phrase "burning like a man" is terrifying as an image, but not easy to explain), and 348.

9. *Iliad*, 1. 472. See also D. R. Dicks, *Early Greek Astronomy to Aristotle* (Ithaca: Cornell University Press, 1985), pp. 29–32.

10. Dolores L. Cullen, *Chaucer's Pilgrims: The Allegory* (Santa Barbara, Calif.: Fithian Press, 2000), p. 15; see also Ameerah B. P. Mattar et al., "Astronomy and Astrology in the Works of Chaucer," www.math.nus.edu.sg/alasken/gem-projects/hm/astronomy_and_astrology_in_the_works_of_Chaucer.pdf, and Owen Gingerich, "Transdisciplinary Intersections: Astronomy and Three Early English Poets," *New Directions for Teaching and Learning*, vol. 1981, issue 8, pp. 67–75. Chaucer's name, incidentally, means "shoemaker."

11. John Milton, "Ode on the Morning of Christ's Nativity," ll. 229–33.

12. Andrew Marvell, "Upon Appleton House," ll. 661–64.

13. Robert Herrick, "To the Virgins, to Make Much of Time," ll. 5–8.

14. See John Milton, *Paradise Lost*, Book 3, ll. 571–87. Other notable mentions in *Paradise Lost* are found in Book 1, ll. 594–99, when the Sun in eclipse is used as a simile for Satan's blasted appearance; Book 3, ll. 608–12, a reference to the Sun generating precious stones in the ground; Book 4, ll. 32–39, when Satan addresses the Sun; and Book 9, l. 739, when Satan tempts Eve at high noon (in *Paradise Regained* [2:292] he tempts Christ at the same ominous time of day). There are also references in *Paradise Lost*, Book 6, ll. 479–81, and *Comus*, ll. 732–36, to gems that are said to shine in the dark because they store solar fire within them.

15. Frances Amelia Yates, *The Art of Memory* (London: Routledge, 1966), p. 153.

16. See Alberto Manguel and Gianni Guadalupi, *Dictionary of Imaginary Places* (New York: Harcourt, 2000), p. 632.

17. *Romeo and Juliet*, Act III, scene v, ll. 12–24. Another favorite poem about lovers and the Sun is from Catullus, *Ode to Lesbia*, V, the lines beginning "Though the Sun each night expires, Morning renovates its fires. . . ."

18. Sir George More, *A Demonstration of God in His Workes: Agaynst all such as eyther in word or life deny there is a God*, 1597, quoted in John Stubbs, *Donne: The Reformed Soul* (London: Viking, 2006), p. 178.

19. John Donne, "An Anatomy of the World, the First Anniversary." For a longer discussion of Donne's anti-Copernican views, see Arthur Koestler, *The Sleepwalkers* (London: Hutchinson, 1959), pp. 214ff.

20. Samuel Taylor Coleridge, letter to Humphry Davy, July 15, 1800, in *Collected Letters*, vol. 1 (London: Oxford University Press, 2002), p. 339.

21. Alfred Edward Housman, letter to *The Times Literary Supplement*, December 12, 1928. For his own poetry mentioning the Sun, see *Collected Poems*, "The West," and *Last Poems*, "The Sun Is Down"; "The Welsh Marches," no. 28 of *A Shropshire Lad* (published in 1896), "The vanquished eve, as night prevails," and no. 10, "The . . . golden wool of the ram." Best of all is "Revolution," "the golden deluge of the morn."

22. David Herbert Lawrence, *Sons and Lovers* (1913), ed. Helen and Carl Baron (Cambridge: Cambridge University Press, 1992), p. 464.

23. D. H. Lawrence, *Collected Stories* (New York: Knopf, 1994), pp. 981, 983, 990, 996, and 998. See also *The Letters of D. H. Lawrence*, ed. George J. Zytaruk and James T. Boulton (Cambridge: Cambridge University Press, 1982), p. 481, and N. H. Reeve, "Liberty in a Tantrum: D. H. Lawrence's *Sun*," *Cambridge Quarterly*, vol. 24, no. 3, pp. 209–20.

24. D. H. Lawrence, *Apocalypse* (London: Penguin, 1996), pp. 27ff.

25. Edward Brunner, "Harry Crosby: A Biographical Essay," *Modern American Poetry*, www.english .illinois.edu/MAPS/poet/a_f/crosby/crosby.htm 2001.

26. Paul Fussell, *Abroad: British Literary Travelling Between the Wars* (Cambridge: Cambridge University Press, 1980), p. 139.

27. Brunner, "Harry Crosby."

28. The details about the Black Sun in Mandelstam are taken from an Internet essay of April 30, 2003, by "language hat," www.languagehat.com/archives/003376.php and www.languagehat .com/archives/2009_01.php. Later contributors to the site quoted Nerval's writing about a Black Sun in "El Desdichado," "the Black Sun of melancholy," and Blake in "Marriage of Heaven and Hell": "By degrees we beheld the infinite Abyss, fiery as the smoke of a burning city; beneath us at an immense distance was the Sun, black but shining."

29. See Simon Jenkins, "Betjeman's Discreet, Dignified Muse Makes Today's Look Like Mere Groupies," *The Guardian*, April 18, 2008, p. 34.

30. Among novels offering solar passages of note are *Moby-Dick* (dying sperm whales turn their heads toward the Sun in chapter 116) and *The Wind in the Willows*, in which Rat and Mole are granted an epiphany at sunrise. Playwrights, too, have recorded the Sun as an overwhelming presence, from Tennessee Williams to Peter Shaffer, whose *Royal Hunt of the Sun* (1964) attempted to recapture the world of the Inca and the Spanish conquest—and whose *Equus* would have delighted Lawrence. *Hyperion* (1797–99), the first novel by the German lyric poet Friedrich Hölderlin, is in large part an analysis of the Sun's role in Western philosophy. Colette (1873–1954) reveals in *Sido* that during her childhood her mother would reward good behavior by letting her see the dawn.

    The division into poets and novelists is an artificial one, but the list of fiction writers would continue: John O'Hara ("Against the Game"), William Faulkner (*Light in August* and "That Evening Sun"), Patrick White and Doris Lessing (both particularly good on tropical relentlessness), Elizabeth Bowen (*A World of Love*), William Golding (*The Scorpion God*), Frank O'Hara, Garrison Keillor (both *Wobegon Boy* and his parody "Casey at the Bat"), Updike again (stories such as "Deaths of Distant Friends" and "Leaves"), Norman Rush (*Mating* is about a Utopian community in the Kalahari that survives on solar power), and J. G. Ballard (especially *Empire of the Sun*). Among poets one adds Dylan Thomas, in *Under Milk Wood*, Matthew Arnold ("Is it so small a thing / To have enjoy'd the Sun"), Walt Whitman ("Give Me the Splendid Silent Sun"), Ezra Pound, Louis MacNeice ("The Sunlight on the Garden"), Wilfred Owen ("Exposure," "Futility"), Wallace Stevens ("Sunday Morning"), Philip Larkin ("Solar" and "High Windows"), Thom Gunn ("Sunlight"), Richard Eberhart ("This Fevers Me"), Simon Armitage (whose collection *Tyrannosaurus Rex Versus the Corduroy Kid* offers light as a metaphor for poetry), John Updike again ("Seven Stanzas at Easter," "Cosmic Gall"), Lisa Jarnot (*Ring of Fire*), and Amy Clampitt (particularly "The Sun Underfoot Among the Sundews," "A Baroque Sunburst," "What the Light Was Like," and "Winchester: The Autumn Equinox"). Deliberately omitted here, mainly for reasons of space, is the tradition of the "pastoral," in which the writer looks back to an idealized past when either it is forever summer or the characters are bathed in sympathetic light. Novelists such as Evelyn Waugh (in *Brideshead Revisited*), L. P. Hartley (*The Go-Between*), George Orwell (*Coming Up for Air*), and, supremely, Marcel Proust have all contributed to this form of solar deference, but it is an indirect form of homage. There also exists a galaxy of bad writing about the Sun; one has some sympathy with the bartender in the TV series *Cheers* who picks up his girlfriend's book and remarks, "*The Sun Also Rises*—well, that's real profound."

31. See footnote 2 to the Introduction to *Paul Verlaine: Selected Poems*, trans. Martin Sorrell (Oxford World's Classics, 1999).

32. See Charles Nicholl, *Somebody Else: Arthur Rimbaud in Africa, 1880–91* (London: Cape, 1999), pp. 356 and 439.

33. See Graham Robb, *Rimbaud* (New York: Norton, 2000), p. 93. Robb makes the point (p. 386) that Rimbaud exaggerated the heat of the Sun.

34. André Gide, *The Immoralist*, trans. Dorothy Bussy (Harmondsworth: Penguin, 1960), pp. 25, 51, and 26. Gide's sun-worship is discussed at length in Robert Mighall, *Sunshine: One Man's Search for Happiness* (London: Murray, 2008), pp. 86–88. For the remarks on Gide and Camus I have drawn on an unpublished essay by the novelist Lesley Chamberlain.

35. Elizabeth Strout, *Olive Kitteridge* (New York: Random House, 2008), pp. 176, 3, 80, 8, 103, 202–3, and 270.

CHAPTER 28: THE RISING STAR OF POLITICS

1. Peter Morgan, *Frost/Nixon* (London: Faber, 2006), p. 66.

2. "Mr. Bush received one D in four years [at Yale], a 69 in astronomy [during his freshman year] " (*The New York Times,* June 8, 2005, A10). By comparison, John Kerry had four Ds in his freshman year, but in this area comedians prefer to target Republicans. Thus, during Ronald Reagan's presidency, the David Letterman show of August 12, 1987, featured "Top Ten Surprises in the President's Speech"; at No. 1: "Hysterical shouts of 'We're hurtling towards the sun!' made poor closing statement."

3. See Nicholas Campion, "Prophecy, Cosmology, and the New Age Movement," unpublished manuscript, chapter 3, p. 52.

4. Friedrich Nietzsche, *Ecce Homo: How One Becomes What One Is* (London: Penguin, 1992), p. 3.

5. Friedrich Nietzsche, *The Gay Science* (1882, reprinted Cambridge: Cambridge University Press, 2001). Its first translated title was *The Joyous Wisdom,* but "the gay science" was a well-known phrase at the time, derived from a Provençal expression for the technical skill required for poetry-writing. In Book 3, paragraph 108, Nietzsche first proclaims that "God is dead."

6. From an unpublished article, "Philosophy Under the Sun," by Lesley Chamberlain.

7. Translated from the draft of a letter in the Munch Museum quoted in Patricia Gray Berman, "Monumentality and Historicism in Edvard Munch's University of Oslo Festival Hall Paintings," a dissertation submitted to New York University, 1989, p. 62.

8. Quoted to me in a letter from James Landis, author of a novel about Strindberg.

9. Edvard Munch, Notebook OKK reg. no. N55, Munch Museum.

10. Sue Prideaux, *Edvard Munch: Behind the Scream* (London: Yale University Press, 2005), p. 276.

11. This is one of at least eight alternative texts Munch wrote in Norwegian and in German, as well as a version in French. The texts, written from 1895 to 1930, are reproduced in Reinhold Heller, *Edvard Munch: The Scream* (London: Allen Lane, Penguin, 1973), pp. 105–6.

12. See Marion Meade, quoted in David Grann, *The Lost City of Z* (New York: Doubleday, 2009), p. 41.

13. H. P. Blavatsky, *Isis Unveiled: A Master Key to the Mysteries of Ancient and Modern Science and Theology* (Pasadena, Calif.: Theosophical University Press, 1877, reissued 1931); pp. 302, 502, 271, and 258.

14. Cornelius Tacitus, *Germania,* ch. 4, quoted in Simon Schama, *Landscape and Memory* (New York: Knopf, 1995), p. 82.

15. Ibid., p. 117.

16. Richard Noll, *The Jung Cult: Origins of a Charismatic Movement* (Princeton, N.J.: Princeton University Press, 1994), p. 104. Noll attacks Jung again in *The Aryan Christ: The Secret Life of Carl Jung* (New York: Random House, 1997), branding him an ambitious charlatan. The case is well argued, to the point that Noll has become a demon figure among Jungians.

17. See Nicholas Goodrick-Clarke, *The Occult Roots of Nazism* (New York: New York University Press, 1985), p. 5.

18. George L. Mosse, *The Crisis of German Ideology* (New York: Howard Fertig, 1964), p. 59.

19. Noll, *Jung Cult,* pp. 80–81.

20. See Servando González, *The Riddle of the Swastika: A Study in Symbolism,* privately published by the author, Box 9555, Oakland, Calif., 94613. See also the section on "Psyche and Swastika" in Geoffrey Cocks, *Psychotherapy in the Third Reich* (New York: Oxford University Press, 1985), pp. 50–86.

21. See Richard J. Evans, *The Third Reich in Power, 1933–1939* (New York: Penguin, 2005).

22. See Carl Gustav Jung, *Symbols of Transformation* (New York: Bolingen Foundation, 1976, first published 1912), pp. 121–31 and 171–206.

23. Noll, *Jung Cult,* p. 133.

24. Max Müller, *Lectures on the Science of Language, Delivered at the Royal Institution of Great Britain, February–May 1863,* second series (New York: Scribner, 1869), p. 520.

25. Nicholas Goodrick-Clarke, *Black Sun: Aryan Cults, Esoteric Nazism and the Politics of Identity* (New York: New York University Press, 2002), p. 4. See also Wilhelm Landig's lurid and racially repellent trilogy of novels beginning with *Götzen gegen Thule* (1971) and Goodrick-Clarke's *The Occult Roots of Nazism.*

26. David H. James, *The Rise and Fall of the Japanese Empire* (London: Allen and Unwin, 1952), pp. 50 and 181.

27. John Gunther, *The Riddle of MacArthur: Japan, Korea and the Far East* (New York: Harper, 1950), pp. 107–8.
28. See John W. Dower, *Embracing Defeat: Japan in the Wake of World War II* (New York: Norton, 1999), p. 310. For a fuller discussion, see pp. 308–18. See also Herbert P. Bix, *Hirohito and the Making of Modern Japan* (New York: HarperCollins, 2000), pp. 560–62. The official translation appears in U.S. Department of State, *Foreign Relations of the United States, 1946*, vol. 8, pp. 134–35.
29. See Dower, *Embracing Defeat*, p. 307.
30. See Norimitsu Onishi, "Wanted: Little Emperors," *The New York Times*, March 12, 2006, p. 4.
31. Vincent Cronin, *The Wise Man from the West* (New York: Dutton, 1955), p. 97.
32. *Basic Theories of Traditional Chinese Medicine* (Beijing: Academy Press, 1998), p. 23.
33. Joseph Ratzinger, *The Spirit of the Liturgy* (San Francisco: Ignatius Press, 2000), pp. 24, 54, 42, 68, 69, 82, 96, 101, 103, 107, 107, 109, and 128–29.

CHAPTER 29: OVER THE HORIZON

1. George Ellery Hale, letter to H. M. Goodman, March 5, 1893, quoted in H. Wright, *Explorer of the Universe: A Biography of George Ellery Hale* (New York: Dutton, 1966), p. 102.
2. Quoted in Helen Dukas and Banesh Hoffmann, eds., *Albert Einstein, The Human Side: New Glimpses from His Archives* (Princeton: Princeton University Press, 1979), p. 104. A whole industry exists, busily making up fake Einstein quotations, but this one appears genuine: on his fiftieth birthday, March 14, 1929, Einstein was showered with gifts, and as a way of thanking everybody, composed a humorous doggerel. These are its (translated) final two lines.
3. See Thomas Levenson, "Einstein's Gift for Simplicity," *Discover* magazine, September 30, 2004, http://discovermagazine.com/2004/sep/einsteins-gift-for-simplicity.
4. Dennis Overbye, "The Next Einstein? Applicants Would Be Welcome," *The New York Times*, March 1, 2005, F4.
5. Ruth Moore, *Niels Bohr* (Cambridge, Mass: MIT Press, 1985), p. 127.
6. Brian Cathcart, *The Fly in the Cathedral: How a Group of Cambridge Scientists Won the International Race to Split the Atom* (New York: Farrar, Straus, and Giroux, 2005), p. 70. (Boris Kachka, Roger Straus's current biographer, incidentally told me, "I've lost count of the number of times people have referred to him as 'the Sun King.'")
7. Timothy Ferris, *Coming of Age in the Milky Way* (New York: Anchor, 1989), p. 262.
8. See Elizabeth Kolbert, "Crash Course," *The New Yorker*, May 14, 2007.
9. John N. Bahcall, "How the Sun Shines," www.nobelprize.org/noble_prizes/physics/articles/fusion/index.html, June 29, 2000.
10. Ibid. Ironically, of course, the accepted solution to the neutrino problem was announced within the year.
11. Quoted in Kolbert, "Crash Course."
12. Brian Clegg, *Light Years* (London: Piatkus, 2001), p. 3.
13. See John N. Bahcall, "Neutrinos from the Sun," *Scientific American*, vol. 221, no. 1, July 1969.
14. Ibid., pp. 28–37.
15. John Maddox, *What Remains to be Discovered* (New York: Free Press, 1998), p. 86.
16. Professor Marco Velli, author interview at the Jet Propulsion Lab, December 6, 2007. Velli added, with some feeling, "Solar physics has always been 'in between,' a subgroup of a subgroup. Some of the famous physicists of history haven't always given the right consideration to solar research."
17. Tom Stoppard, "Playing with Science," *Engineering and Science*, Fall 1994, p. 10. He calls *Hapgood* "a play which derived from my belated recognition of the dual nature of light—particle and wave."
18. Karl Hufbauer, *Exploring the Sun: Solar Science Since Galileo* (Baltimore: Johns Hopkins University Press, 1993), p. 123.
19. Quoted in ibid., pp. 125–60.
20. See Michael J. Neufeld, *Von Braun: Dreamer of Space, Engineer of War* (New York: Knopf, 2007).
21. J. N. Wilford, "Remembering When the U.S. Finally (and Really) Joined the Space Race," *The New York Times*, January 29, 2008, F3.
22. "The Space Age," *Science Times*, September 25, 2007, F1.

23. Quoted in Michael Sims, *Apollo's Fire: A Day on Earth in Nature and Imagination* (New York: Viking, 2007), p. 14.
24. Wilford, "Remembering When the U.S. Finally (and Really) Joined the Space Race." The first American attempt, *Vanguard TV3*, launched on December 6, 1957, was an embarrassing failure, shutting down a few feet off the launching pad and at once nicknamed "Flopnik."
25. See the online history of the Jet Propulsion Lab, editor and lead author Franklin O'Donnell, 2002, www.jpl.nasa.gov/jplhistory.
26. Carolyn Porco, "NASA Goes Deep," *The New York Times*, February 20, 2007, A19.
27. A. K. Dupree, J. M. Beckers, and others, "Report of the Ad Hoc Committee on the Interaction Between Solar Physics and Astrophysics, [June 18,] 1976." See Hufbauer, *Exploring the Sun*, pp. 191–92.
28. See E. N. Parker, C. G. Kennel, and L. J. Lanzerotti, eds., *Solar System Plasma Physics* (Amsterdam: North-Holland, 1979), 1:3–49.
29. E. N. Parker, "Solar Physics in Broad Perspective" in *The New Solar Physics: Proceedings of an AAAS Selected Symposium 17*, ed. John Eddy (Boulder, Colo.: Westview Press, 1978), p. 3.
30. See Alex Wilkinson, "The Tenth Planet," *The New Yorker*, July 24, 2006, pp. 50–59.
31. See Adam Frank, "How Nature Builds a Planet," *Rochester Review*, Summer 2006, pp. 15–21, and Dennis Overbye, "Dusty Planet-Forming Process May Be Playing Out in Miniature," *The New York Times*, February 8, 2005, F3. As recently as February 2008, the discovery was reported of a smaller version of our system five thousand light-years across the galaxy, with outer giant planets and room for smaller inner ones: two planets, one about two-thirds the mass of Jupiter, the other about 90 percent the mass of Saturn, orbit a reddish star of about half the Sun's mass (see *The New York Times*, February 15, 2008, p. A20). It did not even make the front page.
32. "The Space Age," *The New York Times*, February 11, 2010, F8.
33. Porco, "NASA Goes Deep," A19.
34. Neil Murphy, interview with the author, December 6, 2007.
35. Marco Velli, interview with the author, December 6, 2007.
36. Gerard Peter Kuiper, *The Solar System*, vol. 1, *The Sun* (Chicago: University of Chicago Press, 1953).
37. Prospectus of the Advanced Technology Solar Telescope, published by GONG, 2006.
38. *National Geographic*, p. 10.
39. Parker, "Solar Physics in Broad Perspective," p. 1.

CHAPTER 30: UNDER THE WEATHER

1. F. Scott Fitzgerald, *The Great Gatsby* (New York: Scribner, 1925, p. 124).
2. Robert Frost, "New Hampshire" (first published in *Harper's Magazine*, 1920), *The Poetry of Robert Frost* (New York: Henry Holt, 2002).
3. Isaac Asimov, *A Choice of Catastrophes* (New York: Simon and Schuster, 1978).
4. Michael Sims, *Apollo's Fire: A Day on Earth in Nature and Imagination* (New York: Viking, 2007), p. 125. Further, "it has been known for a decade or more that at levels routinely encountered in most American cities ozone burns through cell walls in lungs and airways. Tissues redden and swell." David V. Bates, "Smog: Nature's Most Powerful Purifying Agent," *Health & Clean Air*, fall 2002 newsletter, http://healthandcleanair.org/newsletters/issue.html.
5. J. R. McNeill, *Something New Under the Sun: An Environmental History of the Twentieth-Century World* (New York: Norton, 2001), p. xxvi.
6. See Elizabeth Kolbert, *Field Notes from a Catastrophe: Man, Nature, and Climate Change* (New York: Bloomsbury, 2006), p. 183. See also note 14 below.
7. Harold Schiff and Lydia Dotto, *The Ozone War* (Garden City, N.Y.: Doubleday, 1978).
8. Aaron Wildavsky, *But Is It True? A Citizen's Guide to Environmental Health and Safety Issues* (Cambridge, Mass.: Harvard University Press, 1995), p. 334.
9. See Andrew R. Blaustein, *Scientific American*, February 2003, p. 60.
10. Jonathan Weiner, *The Next One Hundred Years* (New York: Bantam, 1990), p. 102.
11. See Sir John Houghton, *Global Warming* (Cambridge: Cambridge University Press, 1997), p. 12.
12. See *The New Yorker*, November 20, 2006, p. 69.
13. Oliver Morton, *Eating the Sun: How Plants Power the Planet* (London: Fourth Estate, 2007), p. 371.
14. Elizabeth Kolbert, "The Climate of Man—II," *The New Yorker*, May 2, 2005, p. 70. Kolbert's *Field*

Notes from a Catastrophe (New York: Bloomsbury, 2006) is an eloquent extension of her *New Yorker* articles.

15. See Andrew C. Revkin, "U.N. Report on Climate Details Risks of Inaction," *The New York Times*, November 17, 2007, A1. For several of the details of this history, see also Spencer R. Weart, "The Discovery of Global Warming," at http://www.aip.org/history/climate.

16. *National Geographic*, September 2004, p. 10.

17. Elizabeth Kolbert, "The Climate of Man—III," *The New Yorker*, May 9, 2005, p. 54.

18. See Elizabeth Kolbert, "The Catastrophist," *The New Yorker*, June 29, 2009, p. 42.

19. Weiner, *Next One Hundred Years*, p. 70.

20. Gary Rosen, "More Heat than Light," *New York Times Magazine* July 8, 2007, p. 20; see also Bill McKibben, "Carbon's New Math," *National Geographic* October 2007, pp. 33–37.

21. Kurt Vonnegut, *Armageddon in Retrospect* (New York: Putnam, 2008), p. 26.

22. See John Tierney, "The Earth Is Warming? Adjust the Thermostat," *The New York Times*, August 11, 2009, D1.

23. See Weiner, *Next One Hundred Years*, p. 73.

24. See K. Chang, "Globe Grows Darker as Sunshine Diminishes 10% to 37%," *The New York Times*, May 13, 2004.

25. See Cecil Adams, "The Straight Dope," *Washington City Paper*, April 20, 2007, p. 18.

26. Mark Lynas, *Six Degrees: Our Future on a Hotter Planet* (London: Fourth Estate, 2007), p. 6.

27. See Stuart Clark, *The Sun Kings: The Unexpected Tragedy of Richard Carrington and the Tale of How Modern Astronomy Began* (Princeton: Princeton University Press, 2007), p. 180.

28. "The Sun Also Sets," *Investor's Business Daily*, February 7, 2008.

29. See Jonathan Leake, "So, Are We Going to Freeze or Fry?" London *Sunday Times*, December 2005, p. 16; see also Richard N. Cooper, *International Approaches to Global Climate Change*, World Bank Research Observer 15:145–72 (http://wbro.oxfordjournals.org/cgi/content/abstract/15/2/145.

30. See A. C. Revkin, "Global Warming is Delaying Ice Age, Study Finds," *The New York Times*, September 4, 2009.

31. Kolbert, *Field Notes from a Catastrophe*, p. 32.

32. Wildavsky, *But Is It True?* pp. 352–53.

33. Nigel Calder, *The Weather Machine* (London: BBC, 1974), p. 76.

34. James Joyce, *Ulysses*, p. 75.

35. See *The Week*, February 20, 2010, p. 13.

36. See IPCC, *Summary for Policymakers, Climate Change 2007: The Physical Science Basis. Contribution of Working Group 1 to the Fourth Assessment Report of the Intergovernmental Panel on Climate Change* (Cambridge and New York: Cambridge University Press, 2007).

37. See Elizabeth Kolbert, "The Copenhagen Diagnosis: Sobering Update on the Science," *Yale Environment 360*, November 24, 2009, available at http://e360.yale/edu/content/feature/msp?id=2214.

38. "The Sun Also Sets," *Investor's Business Daily*, February 7, 2008.

39. Willie Soon, letter to the author, May 7, 2009.

40. Weart, *The Discovery of Global Warming* (Boston: Harvard University Press, 2008), p. 196.

41. Robert Kunzig, *Mapping the Deep*, pp. 318–19.

CHAPTER 31: THE IMPOSSIBLE AND BEYOND

1. Arthur C. Clarke, *Astounding Days: A Science Fictional Autobiography* (London: Gollancz, 1989), p. 207.

2. H. G. Wells, *The Time Machine* (New York: Airmont, 1964), p. 125.

3. See Joseph Needham, *Science and Civilization in China* (Cambridge: Cambridge University Press, 1959), p. 440.

4. See Marjorie H. Nicolson, *Science and Imagination: Collected Essays on the Telescope and Imagination* (Ithaca: Cornell University Press, 1965), p. 71.

5. Jonathan Swift, *Gulliver's Travels*, part 3, chapter 2.

6. E. Walter Maunder, *Are the Planets Inhabited?* (London: Harper & Bros., 1913), chapter 3.

7. Wells, *The Time Machine*, p. 111.

8. For a full list of such stories, see John Clute and Peter Nicholls, eds., *The Science Fiction Encyclopedia*, 2nd ed. (New York: St. Martin's Press, 1993), pp. 1177–78.

9. See P. K. Alkon, *Winston Churchill's Imagination* (Lewisburg, Pa.: Bucknell University Press, 2006), p. 150.
10. Quoted in Clarke, *Astounding Days*, p. 224.
11. Robert Heinlein, *The Man Who Sold the Moon* (New York: Signet, 1950).
12. Dennis Overbye, "Setting Sail into Space, Propelled by Sunshine," *The New York Times*, November 10, 2010, D1.
13. Ray Bradbury, *The Golden Apples of the Sun: Classic Stories 1* (London: Bantam, 1990), pp. 154–60.
14. Frederik Pohl and C. M. Kornbluth, *Wolfbane* (New York: Ballantine, 1959), pp. 28 and 11.
15. Isaac Asimov, *In Memory Yet Green* (New York: Doubleday, 1979), pp. 295–96.
16. Isaac Asimov, "Nightfall," in *The Best of Isaac Asimov* (London: Sphere, 1973), pp. 37, 40, and 60.
17. Isaac Asimov and Robert Silverberg, *Nightfall* (London: Pan, 1990), p. 352.
18. See Charles Liu, "My Three Suns," *Natural History*, October 2006, pp. 70–71. On July 19, 2005, *The New York Times* reported that astronomers had discovered a planet with three suns.
19. Arthur C. Clarke, *The Collected Stories of Arthur C. Clarke* (New York: Tor, 2000), pp. 517–21.
20. Clarke, "A Slight Case of Sunstroke," in *Collected Stories*, pp. 687–92.
21. Clarke, "Summertime on Icarus," in *Collected Stories*, pp. 724–32.
22. Clarke, "The Wind From the Sun," in *Collected Stories*, pp. 828–42.
23. See *Saga* magazine interview with Nina Myskow, quoted in *The Week* (U.K. edition), July 7, 2007, p. 10.
24. Clarke, "Out of the Sun," in *Collected Stories*, pp. 652–57.
25. J. G. Ballard, 1987, quoted by John Strausbaugh, "Aiming for Life's Jugular in Deadly Verbal Darts," *The New York Times*, December 1, 2004, E9.

CHAPTER 32: THE DEATH OF THE SUN

1. Erasmus Darwin, *The Botanic Garden*, part 1, canto IV (Air), ll. 380–83.
2. C. S. Lewis, *The Last Battle* (New York: Macmillan, 1956), p. 148.
3. Isaac Asimov, *A Choice of Catastrophes* (New York: Simon and Schuster, 1978), pp. 92–130.
4. See Dennis Overbye, "Sun Might Have Exchanged Hangers-On with Rival Star," *The New York Times*, February 12, 2004, A25.
5. London *Times*, April 18, 2005, p. 17.
6. Kenneth Chang, "Prediction Proved: Light Speeds Up an Asteroid as It Spins," *The New York Times*, March 13, 2007.
7. *The Sun*, February 28, 2005, pp. 8–10.
8. Quoted in Michael Sims, *Apollo's Fire: A Day on Earth in Nature and Imagination* (New York: Viking, 2007), p. 157.
9. Dennis Overbye, "Kissing the Earth Goodbye in About 7.59 Billion Years," *The New York Times*, March 11, 2008.
10. Woody Allen, "Strung Out," *The New Yorker*, July 28, 2003, p. 96.
11. Dennis Overbye, "Scientists' Good News: Earth May Survive Sun's Demise in 5 Billion Years," *The New York Times*, September 13, 2007.
12. See "Scientists Spot Nearby 'Super-Earth' Planet," Sphere News, December 17, 2009.
13. Carolyn Porco, "NASA Goes Deep," *The New York Times*, February 20, 2007, A19.
14. Michio Kaku, *Physics of the Impossible: A Scientific Exploration into the World of Phasers, Force Fields, Teleportation, and Time Travel* (New York: Doubleday, 2008), pp. xvi–xvii.
15. Stuart Clark, "The Death of the Sun," *Focus*, January 1, 2010, p. 34.

SUNSET: THE GANGES

1. Walt Whitman, *Leaves of Grass*.
2. Gayatri mantra, from Rig Veda (London: Penguin, 1982), book III, 62, 10.
3. C. Bernard Ruffin, *Last Words: A Dictionary of Deathbed Quotations* (London and North Carolina: McFarland, 1995).
4. Florence Emily Hardy, *The Early Life of Thomas Hardy* (New York: Macmillan, 1928), p. 19.
5. See Eric Blehm, *The Last Season* (New York: HarperCollins, 2006).

6. *The Complete Letters of Vincent van Gogh to His Brother* (New York: New York Graphic Society, 1958), p. 238. Others who have written supremely well about sunsets are Jules Verne, in *20,000 Leagues Under the Sea* (London: Penguin, 2004), p. 239; Mark Twain, in *Life on the Mississippi* (London: Penguin), p. 396; Vladimir Nabokov, in *Speak, Memory* (New York: Putnam, 1906), p. 213; and T. E. Lawrence, in *Seven Pillars of Wisdom: A Triumph* (London: Cape, 1973), p. 575.

7. Michael Connolly, *The Closers* (New York: Little, Brown, 2005), p. 270.

8. See David Adam, *The Guardian*, October 1, 2007, and the AP report by Kate Schuman, November 28, 2007.

9. See Diana I. Eck, *Banaras: City of Light* (New York: Columbia University Press, 1962), pp. 14–15.

10. W. S. Caine, *Picturesque India: A Handbook for European Travelers* (London: Routledge, 1890), p. 302.

11. Mark Twain, *Following the Equator: A Journey Around the World* (New York: Dover, 1989), p. 496.

12. See Somini Sengupta, "Debate in India: Is Rule on Yoga Constitutional?" *The New York Times*, January 26, 2007.

13. See Charlotte Vaudeville, *Kabir* (Oxford: Clarendon Press, 1974), p. 267.

14. Diana C. Eck, *Banaras: City of Light*, p. 182.

15. Geoff Dyer, *Jeff in Venice, Death in Varanasi* (New York: Pantheon, 2009), p. 173.

"That can't be good."

INDEX

Page numbers in italics indicate illustrations.

Abbasid Dynasty, 103
Abdusamatov, Khabibullo, 471
Abul-Wafa, 101
abyssal floor, 262
Academos, 76
Achaemenian Later Avestan calendar, 290
Ada (Nabokov), 399
Adams, Fred, 495
Adams, William Grylls, 327
Aereopagitica (Milton), 133
aerial perspective, 359n
Aeschylus, 267
Aesop, 4
Africa: albinism in, 232–33; Dancing
    Stones solar monument, Kenya, 28;
    early spread of Islam in, 102; heat in,
    215–16, 325, 416; in *King Solomon's
    Mines*, 158–59; native timekeeping
    systems in, 288, *289*, 300; Sun myths,
    3, 10, 208, 213. *See also* Egypt; Sahara
    Desert
Agulhas Current, 267
Airy, George, 316
Ajumawi Indians, 209
Akhenaten (Egyptian ruler), 38, *67*
Alaska, 316n
Albers, Bill, 244, 248
albinism, *232*, 232–33
alchemy, 84, *145*, 145–46
Alexander the Great, 51, 65, 70, 77, 79, 259
Alexandria (Egypt), *72, 80*, 82, 102, 108,
    110, 294, 300
Alfonso X (king of Castile), 112
algebra, 65n
Alighieri, Dante, 405
Allen, Paul G., 453
Allen, Woody, 168, 383
Alloway, Lawrence, 374
Almagest (Ptolemy), 83, 104, 111, 113, 117,
    118, 120
Almería, Spain, 334–35

Also Sprach Zarathustra, 395, 420, 421, 422,
    425
Alter, David, 205
alternative energy. *See* solar energy
altitude sickness, xxv–xxvi, xxvii, xxviii,
    xxx
Amaterasu (Japanese sun goddess), xxvi,
    xxix, 356, 428
The American Claimant (Twain), 196–97
American Revolution, 199n
American Stonehenges, 34
Anaxagoras, 52, 70, 73, 74
Anaximander, 73, 303
Anaximenes, 73–74
"ancient lights" doctrine, 339n
Anderson, Poul, 481
André Chenier (opera), 394
Andromeda, 170
Angkor Wat, Cambodia, 33
animal kingdom: role of sunlight in
    migration, 256; role of sunlight on
    activity, 254–57; sun-avoiding species,
    257
annular eclipses, 46, 49, 50, 51n, 156
Ansari X Prize, 453
Anson, George, 285
Antarctica: and global warming, 472; and
    ozone layer, 463, 464; solar eclipse,
    45, 153–55, *154*, 157
ante meridiem, 300, 315
Antiochus I (Babylonian ruler), 65
Antonioni, Michelangelo, 384
Antony and Cleopatra (Shakespeare), 249
Apollinaire, Guillaume, 411
Apollo, Sun as, xxix, 234, 347, 348, 387,
    399, 410, 420n
Apollo program (NASA), 449n
Arabs. *See* Islamic civilization
Arapaho Indians, 24
Arcadia (Stoppard), 69, 136
Archer, Mary, 330

Archilochus of Paros, 42
Archimedes, 70, 111, 323, 323–24
architecture, 384–87
Aristarchus of Samos, 70, 79–80, 117, 119
Aristocles, 75
Aristophanes, 74–75
Aristotle: Arab translations of work, 111;
    background, 76–79, 369n;
    commentaries by Averroës, 112;
    counted among ancient astronomers,
    70; medieval interest in, 114–15;
    Prime Mover concept, 78, 84;
    understanding of light, 139; views on
    comets, 124; views on heavens as
    incorruptible, 122, 131, 189
Arnold, Matthew, 410
Arrhenius, Svante, 464
arushá, 3
Aryabhatta, 103
Asam, Aegid Quirin, 359
Ashton, Clare, 480
Ashurbanipal, 63n
Asimov, Isaac, 146, 163, 460, 483–85,
    486n, 490–91, 494
Assassins (Islamic sect), 106–7
Assyrian Empire, 63n
asteroids, 491, 495
astroarchaeology, defined, 35
astrolabes, 105–6, 113, 279, 280
astrology vs. astronomy, 61–62, 82, 84, 99,
    109, 112–13
astronauts, 270n, 452, 487, 500
astronomical twilight, 213
astronomy: in ancient Greece, 69–85; in
    Babylon, 59–65, 62; beginnings,
    59–68; in China, 86–99; compared
    with astrology, 61–62, 82, 84, 99, 109,
    112–13; early instruments for, 105–6;
    in India, 109; Islamic contributions,
    100–108; Khayyám's contributions,
    101; naked-eye, 62, 122, 280–82; as
    navigation aid, 275–76, 283;
    significance in calendar-making,
    289–95; solar physics, xxiii, 177, 181,
    197, 444, 445, 447–51, 459; Vatican's
    heavenly interest, 112–13, 169,
    187–89, 405, 406
astrophysics, 203, 451, 458–59
Aswan Dam, 38n
Atget, Eugène, 379–80, 381
atomic bomb, 179–82, 183
atomic clock, 317, 317, 320, 321
atomic theory, 440–41; Einstein's role, 439
Attenborough, David, 251, 261
Atum (Egyptian Sun god), 67, 68
aubades, 407–8
Auden, W. H., 28, 35, 415–16
Augustine, Saint, 18, 112, 277, 308

Augustus (Roman emperor), 84, 295
aurora australis (southern lights), 42
aurora borealis (northern lights), 42, 43, 44
auroras, 42–43, 43, 44, 45
Austen, Jane, 225, 229, 386n, 466n
Austerlitz, Battle of, 206
Autolycus, 113
autumnal equinox, 15–16, 23, 35
Averroës, 112
Avicenna, 101
Aztecs: calendars, 289–90, 293; as
    Children of the Sun, 9, 10; solar
    monuments, 34–35; Sun myths, 4; use
    of mirror imagery, 356

Babylon: as birthplace of astronomy,
    59–65, 62; calendars in, 289–90, 291;
    mathematics in, 63–65; relocation of,
    65; study of eclipses, 48; Sun myths,
    60; Table of Portents, 42
Bacon, Francis, 235
Bacon, Roger, 114, 115, 128, 296, 324
Baghdad, as capital of Abbasid Dynasty,
    103, 106
Bahcall, John, 442, 444
Bailey, Paul, 160
Bailey, Thomas, 117n
Baillie, Mike, 200–201
Baily, Francis, 157
Baily's Beads, 157
Baker, Josephine, 222
Bapu, Arsalan, 509
Barbauld, Anna, 246n
Barberini, Maffeo, 132
Baring-Gould, S., 211n
Barnard, Émile, 368
Barnes, William, 378n
Barrie, James M., 208–9
Barrymore, Lionel, 382–83
Bartel, Julius, 450–51
basal cell carcinoma, 231, 233
Bastiat, Frédéric, 408n
Beardsley, Aubrey, 222
Beaumont and Fletcher, 211n
Becquerel, Alexandre-Edmond, 330
Becquerel, Henri, 174
Bede, Venerable, 296
bees, role of sunlight on activity, 255
Beethoven, Ludwig van, 394
Beg, Ulugh, 107–8
Belknap, Jeremy, 44
Bell Labs, 330
Bellarmine, Robert, 130n
Bend Sinister (Nabokov), 400
Benedict, Saint, 308, 359, 360
Benedict XIV, Pope, 134
Benedict XVI, Pope, 134, 433–35, 434
Benét, Stephen Vincent, 499–500

Bengalis, Sun myths, 6
Bergman, Ingrid, 222
Bermuda Triangle, 267
Bernhard, Oskar, 226–27
Bethe, Hans, 179, 182, 184
Betjeman, John, 416
Bezos, Jeffrey P., 453
Bible, 51–52, 62, 63n, 112, 119, 123, 130, 169, 171–73, 205n, 239, 358. *See also* Genesis, Book of
Bickerman, E. J., 288
Bien, Reinhold, 299–300
Biermann, Ludwig, 450
Big Ben, 312
Bigelow, Mrs. Albert Smith, 415
biological clock, 236n. *See also* circadian rhythm
birds: role of sunlight on activity, 254, 256; in Sun myths, 3
al-Biruni, 101, 104, 117
black dwarf, Sun as, 498
Black Hole of Calcutta, 246n
black holes, 169, 444, 490–91
black smokers, 260
black sun, 343, 414–15, 427
Blaise, Clark, 313, 314
Blake, William, 345, 365, 395, 409
Blavatsky, Madame, 423, 425n, 427
Bliss, Arthur, 374
blondness, 353
Bohr, Niels, 184, 440
Bondi, Hermann, 288
bonfires, in solstitial celebrations, 19–22
The Book of Hours, 309–10
Boorstin, Daniel, 273, 277, 310
Bose-Einstein condensate, 204
bosons, 441, 442
Bova, Ben, xxi
boy scouts, xxiii
Boyd, Brian, 400–401
Boyle, Danny, 480
Bradbury, Ray, 482–83
Bradley, James, 203
Brahe, Tycho, 120–25, 127, 129, 188, 310
Bram, France, 385
Branson, Richard, 468n, 488
Brassaï, 379
Braudel, Fernand, 281
Brecht, Bertholt, 127, 133
Breyer, Franziska, 332, 333
Brighter Than a Thousand Suns (Jungk), 176
Brin, David, 482
British Empire, colonial weather extremes, 216–17
Brodsky, Josef, 416
Broecker, Wallace S., 461
Brooke, Rupert, 228
The Brothers Karamazov (Dostoyevsky), 345

Brougham, Henry Peter, 225–26
Brown, Dan, 277n
Brown, Joseph Epes, 26
Browning, Robert, 410, 499
Bruno, Giordano, xix, 132
Bryson, Bill, 169, 245, 445
Buddha, 503
buffalo, 26n
Buffon, Georges-Louis de, 172–73, 324
Buisson, Henri, 461
Bunsen, Robert, 205
Burgess, Anthony, 308, 313
Bürgi, Jost, 310
"burn daylight," 351
Burnet, Thomas, 172
Burney, Fanny, 150
burning mirrors, 323, 323–24
Bush, George W., 418
butterflies, role of sunlight on activity, 254
Byron, George Gordon (Lord Byron), 410, 411, 499

Calder, Nigel, 473
calendars: astronomical calculations, 289–90; early makers, 290–95; lunar, 289–90; systems for creating, 288
California Current, 266
caliphs, defined, 102n
Callendar, Guy Stewart, 465
Calvert, Edward, 365
Calvin, Melvin, 248
camera lucida, 149
camera obscura, 149, 376
Cameron, Julia Margaret, 378
Camillo, Giulio, 407
Campanella, Tommaso, 168, 407
Campbell, John, 283, 484
Camus, Albert, 229–30, 417
Canaletto, 362
cancer, 231–33, 237, 253n
Candlemas bear, 22
The Canterbury Tales, 404. *See also* Chaucer, Geoffrey
carbon, in photosynthesis, 245
carbon capture and storage (CCS), 468n
carbon dioxide, 245, 246, 247, 464, 465, 466, 468–69, 470, 471, 493
carbonic acid gas, 246n, 464
carcinomas, 231, 233
Carnac Stones solar monument, Brittany, France, 33
Carrington, Richard, 192–93, 200
Carson, Rachel, 262, 263
Carter, Jimmy, 338
Casanovas, Juan, 187–89
Castel Gandolfo, Italy, 187–89
Cathcart, Brian, 440
cathedrals, European, 39–40

Cat's Cradle (Vonnegut), 399
Catullus, 499
Caux, Salomon de, 325
Cervantes, Miguel de, 65n, 280, 298n
cesium clocks, 317n
Cézanne, Paul, 367
Chadwick, James, 179
Chaldeans, 60–63
Chaliapin, Feodor, 396
Challenger space shuttle, 450
Chamberlin, Thomas, 174
Chandigarh, India, 385–86
Chandler, Raymond, 351–52, 353
Chanel, Coco, 221–22
Chaplin, Charlie, xx, 427, 428
Charlemagne, 52
Charles I (king of England), 304–5
Charles II (king of England), 136, 220, 283
Chatwin, Bruce, 31
Chaucer, Geoffrey, 269, 302, 404
Chávez, Hugo, 319
chemosynthesis, 261
Chen Jie, 451n
Chesterton, G. K., 234
Cheyenne Indians, 24
chi (Chinese term), 91–92
Chichén Itzá, Mexico, 35, 35
Children of the Sun, 8–10, 228
Chile, Sun myths, 3
China: arrival of foreigners, 94–99;
    astronomical history, 86–99;
    calendars in, 289–90, 291; cosmology
    in, 91–92; cultural longevity, 10;
    Cultural Revolution, 431–33;
    expansion and foreign influence, 93;
    Imperial Observatory in Beijing, 98;
    Indian astronomy in, 109; key
    inventions for exploration, 96; Li Ki
    (Book of Rites), 89–90; method of
    tracking planets, 88–89; national flag,
    433n; sea travel by Chinese, 275, 277;
    as solar-power leader, 339; study of
    eclipses, 48, 49; study of sunspots,
    189; Sun as medical symbol, 433; Sun
    myths, 1, 4, 87; symbolic authority of
    Sun, 431–33; timekeeping in, 302; use
    of mirror imagery, 356; water-driven
    astronomical clock tower, 94
chlorofluorocarbons (CFCs), 462–63
chlorophyll, 245, 248, 261
Christian IV (king of Denmark), 124
Christianity: church response to
    heliocentric system, 118–20, 132, 133,
    134; early rituals, 16–17; origin of
    halo, 17; relationship of Christ's
    birthday to winter solstice, 17–18; Sun
    as God theory, 419–20. *See also* Jesus

Christ; Prime Mover (Aristotle);
    Roman Catholic Church
Christie, Agatha, 229
Christmas, relationship to winter solstice,
    17–18, 296, 434
The Chronicles of Narnia (Lewis), 117n,
    490
chronometers, 286, 287
Church of England, 144n, 171
Churchill, Winston, 182, 479, 480n, 500
Cicero, 71, 119, 477
cinematographers, 383
circadian rhythm, 240–41, 319n
civil twilight, 213
Clark, Stuart, 193, 205
Clarke, Arthur C., 336, 476, 485–89, 495
Clarke, Thurston, 214–15
classical music, Sun in, 388–96
Claude. *See* Lorrain, Claude
Claudius (Roman emperor), 50, 303
Claudius Ptolemaeus. *See* Ptolemy
Clavius, Christopher, 157, 296, 300
Clegg, Brian, 128, 442n
Clement IV, Pope, 114
Clement VIII, Pope, 187
clepsydras, 302
climate change, 464–75
clocks: chiming, 304; Chinese water-driven
    astronomical clock tower, 94;
    mechanical, 310–14; pendulum, 304;
    ship's, 285, 287; weight-driven, 304
clouds, 471n
Cluster 2 space mission, 452
coal, 338n, 464
Cockcroft, John, 179
Codrington, Admiral, 311
Cohn, Roy, 199
Coleridge, Samuel, 409–10
Collier, Richard, 217
Collins, Gail, 319
color, 204–5
color perspective, 359n
Columbus, Christopher, 117, 158, 160, 277,
    283n, 354
comets, 124
compasses, 91, 96, 277–78, 279, 284, 286
Comte, Auguste, 169, 205
Conan Doyle, Arthur, xxi
concave reflectors, 323, 324
The Confessions of Nat Turner (Styron), 160
A Connecticut Yankee in King Arthur's Court
    (Twain), 159–60
Constable, John, 361, 368
Constantine (Roman emperor), 16, 17n
continental shelf, 261–62
Cook, James, 148, 155–56, 287
Cool Hand Luke (movie), 384

Cooper, James Fenimore, 162, 478
Copenhagen (play), 184
Copernicus, Nicolaus: background, 118; belief in heliocentric universe, 116, 118–20, 127, 130, 132–33; church response to heliocentric system, 118–20, 132, 133, 134; comparison of his system with Ptolemaic system, 118–20, 121, 122, 132–33, 280, 407, 439; first takes up Ptolemy's tables, 83; Galileo's endorsement, 130, 132–33; in library at Castel Gondolfo, 188; models Earth's passage around Sun, 119; portrait, 129; significance among scientists, 127; Tycho Brahe's alternative system, 122–23
Copley, John, 500
Corbyn, Piers, 470–71, 473–74
coronagraphs, 448, 449
coronas, 155, 157–58, 169, 201, 450–51
Corot, Jean-Baptiste-Camille, 364
Cortez, Hernándo, 160, 354
Cotter, Holland, 383–84
Cottingham, E. T., 166
Council of Tours, 17
cow dung, 47n
Coward, Noël, 216, 222
Cowley, Malcolm, 414
Cozens, Alexander, 500
Crane, Hart, 415
Creation: dating, 171–73; Milton's account, 406; relationship of Sun and Moon in accounts, 6
cremation ceremony, at Varanasi, India, 503, 507–8
crops, Soviet, 253
Crosby, Harry, 412, 414–15
Crow Indians, 24
Crowley, Aleister, 423
Crump, Thomas, 55, 167
Crusaders, 106, 324
Crutzen, Paul, 462
Cullen, Christopher, 290n
Cullen, Dolores, 404
Cumont, Franz, 85
The Cunning Little Vixen (opera), 394
Curie, Irène, 179
Curie, Marie and Pierre, 174–75
currents, ocean, 266–68
curved reflectors, 323, 324
Cymbeline (Shakespeare), 304
Cyrano de Bergerac, 190, 411

The Da Vinci Code, 187, 277n
Daedalus, 487n
Daguerre, Louis, 377
Daisyworld, 258

Dancing Stones solar monument, Namoratunga, Kenya, 28
Darius (king of Persia), 51
dark biosphere, defined, 260–61
dark matter, 169
Darwin, Charles, 119n, 120, 171, 173–74, 492
Darwin, Erasmus, 490
Darwin, George, 269, 270–71
Das Rheingold (opera), 390
Davis, Raymond, 444
Davis, Wade, 26
Davy, Humphry, 249
daylight saving time, 317–19
de' Dondi, Giovanni, 310
De thermis, 437
"dead of night," 301
Death Valley, California, 214
Debussy, Claude, 395
Defoe, Daniel, 477
Degas, Edgar, 500
Delacroix, Eugène, 370
Delaney, Frank, 32–33
Delisle, Joseph-Nicolas, 147
Delius, Frederick, 394
Democritus, 150
dendrochronology: current thinking, 200–201; Eddy's study of tree rings, 197; origin of science, 195, 195–96
Deng Xiaoping, 432
depression, 240, 241, 243
Descartes, René, 137, 141
d'Holbach, Paul Henri Dietrich, 172n
Diamandis, Peter, 453
Dickens, Charles, 410
Dickinson, Emily, 275, 410
Dicks, D. R., 70
digital photography, 380–81
Diocletian, 288
Diogenes Laertius, 71
Dionysius Exiguus, 288
The Divine Comedy (Dante), 405–6
Dixon, Jeremiah, 148
Doldrums, 266
Dolphin, David, 238–39
Don Quixote (Cervantes), 65n, 280, 336
Donne, John, 3, 14, 300, 407, 408–9
Dos Passos, John, 207n
Douglass, Andrew, 195–96
Dracula, 237–38
Dresel, Tom, 332, 333
Dryden, John, 283
du Maurier, Daphne, 44n
Dubček, Alexander, 418
duels, 301n
Duncan, Isadora, 392
Durant, Will, 72, 80–81

Durrell, Lawrence, 229–30
dusk, 212–13
Dvořák, Antonin, 395
Dyer, Geoff, 510
Dylan, Bob, 207n
Dyson, Frank, 165, 166

eagles, in Sun myths, 4
Earth: auroras, 43; characteristics of core, 175n–176n; circumnavigation, 285; comparison of Copernican and Ptolemaic systems, 118–20, 121, 122, 132–33, 280, 406–7, 439; Copernicus models passage around Sun, 119; distance from sun, xix; effect of sunspots, 196–201; environmental threats, 460–75; estimating age, 110; future, 493–95; Gagarin's description, 448n; hottest temperatures, 214–18; incidence of sunshine, 218; Kepler's views on planetary movement, 125, 126; migrating from, 494–95; orbital variations, 15–16, 36n, 197–98; in relation to rest of universe, 115; shape of, 15, 64, 94, 96, 116–17; solar system stability, 490–98; solar-influenced atmosphere, 198–99; source of internal heat, 175n
Easter, 290, 295, 296, 299, 435
Easter Island moai, 31
Easy Rider (movie), 384
eclipses: Antarctica (2003), 153–55; associated phenomena, 156–58; characteristics, 45–46; Chinese records, 90; Cook's pursuit, 155–56; in creation stories, 51; defined, 45, 46; descriptions of experience, 162, 162–64, 165; following, 155–56; in history, 50–52; and portrait of Saint Benedict, 359, 360; predicting, 48–49, 50; relationship of light to gravity, 164–67; Ricci's Chinese experience, 97; scientific study, 164–67; statistics, 156; superstitions and fears, 46–48, 49–50, 52, 55, 158–62
Eco, Umberto, 288
Ecumenical Maps, 282
Eddington, Arthur, 165–67, 176–77
Eddy, John, xxii, 197–98, 200
Edison, Thomas, 342, 382, 383n
Edo, Japan, 331n
Edward IV (king of England), 346–47
Edward VIII (king of England), 228, 355n–356n
Egypt: advances in science and math, 65, 66–68; calendars in, 65, 289–90, 292–93; cultural longevity, 10; Great Sphinx, 28; pyramids, 36–39, 37; site

of first solar power station, 328; Sun myths, 4, 6, 9, 15; timekeeping in, 300, 302, 303; use of mirror imagery, 356
Einhard, 189
Einstein, Albert: and absolute time, 313; and atomic bomb, 179, 180, 180–81, 182n; and equivalence principle, 320; general theory of relativity, 164–65, 166, 439–40; and hydrogen bomb, 184; and light, 202, 203, 204, 244; special theory of relativity, 439; view of Newton, 137
Eisenhower, Dwight, 448
El Niño, 267–68
electricity. See solar energy
electromagnetic radiation, 169. See also Maxwell, James Clerk
Elements (Euclid), 83
Elgar, Edward, 395
Eliade, Mircea, 10–11, 349–50
Eliasson, Olafur, 373
Eliot, George, 7
Eliot, T. S., 415
Elizabeth I (queen of England), 219–20, 273, 297
Emerson, Ralph Waldo, 59, 484, 499
Empedocles, 202
Empire State Building, New York City, 384
Eneas, Aubrey, 327–28
Epicurus, 70
Epiphany (January 6), 17–18
Equator (Clarke), 214–15
Equatorial Counter Current, 266
equinoxes: autumnal, 15–16, 23, 35; cultural role, 16–22; defined, 15; and Earth's orbit, 15–16; and Egyptian pyramids, 39; and Gregorian calendar, 296, 299; vernal, 15–16, 28, 35, 40, 289, 296, 299, 435
Ernst, Max, 370
Essen, Louis, 317
Ethiopia, calendar system in, 288, 289
Euclid, 70, 83, 111, 113
Eudoxus of Cnidus, 70, 77, 78–79
Explorer spacecraft, 449, 450–51
eye, human, 204n, 232, 234–35, 241

Fabricius, 131, 188
Fabry, Charles, 461
Fantascope, 382
Fatima, 349
Faulkner, William, 313n
al-Fazari, Muhammad, 105
Fechner, Gustav, 234
fencing, 208
Fenton, Roger, 378, 381
Ferguson, Robert, 278
Fermi, Enrico, 179, 441, 444n

fermions, 441, 442
Fernández de Oviedo, Gonzalo, 354
Ferris, Timothy, 82, 169, 174, 279–80
Fidelio (opera), 394
The Fighting Téméraire (Turner painting), 363, 363–64
filmmaking, role of Sun in, 382–84
Finnegans Wake (Joyce), 441n
Finsen, Niels, 227–28
fire clocks, 302
fires, in solstitial celebrations, 19–22
fish, role of sunlight on activity, 254
Fisher, Kitty, 220
fission bomb, 181, 182, 184
Fitzgerald, Edward, 101
Fitzgerald, Ella, 391n
Fitzgerald, F. Scott, 222, 460
Fitzroy, Augustus Henry, 220
Fizeau, Hippolyte, 203
flags, national, Sun on, 345–46, 428, 433n
Flammarion, Camille, 479
Flandrin, Jules, 358
Flaubert, Gustave, 59
Fleming, Ian, 322, 323
Fleming, Sandford, 314–15, 316
Fletcher, Valerie, 367
flower clocks, 252–53
flying saucers, 485
Follini, Stefania, 236n
Ford, Eileen, 234
Ford, Henry, 342
fossil fuels, 336
Foucault, Léon, 193, 203
Foulques, Guy de, 114
Fourier, Jean-Baptiste, 464
Fowles, John, 225, 229
foxhunting, xxiii
France: Carnac Stones solar monument, Brittany, 33; French Revolution of 1789, 23, 64n, 199n; Unité d'Habitation, Marseille, 385. *See also* Louis XIV (king of France)
Francis, Dick, 399
Frank, Anne, 345
Frank, Robert, 380
Franklin, Benjamin, 317
Fraunhofer, Josef, 204–5
Fraunhofer Institute of Solar Energy Systems, Freiburg, Germany, 332
freckles, 221
Frederick II (king of Denmark), 123, 124
Frederick the Great (king of Prussia), 311
Freemasons, 393, 424
Freiburg, Germany, 331–34
French Revolution of 1789, 23, 64n, 199n
Frenk, Carlos, 459
Freon, 462
Freud, Sigmund, 134

Friedrich, Caspar David, 500
Friedrich, David, 365
Frisch, Karl von, 255
Frisch, Otto Robert, 179, 181
Froehlich, Jack, 449n
From Here to Eternity (Jones), 225
Frost, Robert, 460, 499
Frost, Terry, 353
Furst, Alan, 14
fusion energy, xxi, 178, 184, 341, 441, 480, 492, 497
Fussell, Paul, 414

Gagarin, Yuri Alekseyevitch, 448n
Gaia Theory, 258
Gainsborough, Thomas, 365, 500
galaxies, 170–71
Galen, 111
Galileo: death, 134; endorses Copernican system over Ptolemaic, 130, 132–33; observation about Sun, 244; portrait, *129*; recants heliocentrism, 133; register in library at Castel Gondolfo, 188; role in development of telescope, 127–30; and Roman Catholic Church, 130, *131*, 132–33, 134; study of sunspots, xx, 130, 131, *188*, 189, 190, 409; study of tides, 269; views on absolute time, 286; views on light, 202; visited by John Milton, 54
Galileo space probe, 452–53
Gamow, George, 177, 178–79, 329, 440
Gandhi, Indira, 502, 503
Gandhi, Mohandas, 423
Ganges River, 501, 503, 505–6, *506*, 507–8
Garbo, Greta, 222
Garfunkel, Art, 239
Garland, Alex, 480n
Gautreau, Amélie, 221
Gauvain, Henry, 231
Gehry, Frank, 387
Gell-Mann, Murray, 441n
general relativity, 439–40
Genesis, Book of, 6, 171, 172, 405
Genghis Khan, 106
Geographia (Ptolemy), 83, 117, 282
geometry, 65n
George III (king of England), 150
George V (king of England), 355n
geothermal energy, 323
Gerardus Mercator Rupelmundanus. *See* Mercator
Germany, as solar-power leader, 331–34, 339
Germinal (French Republican calendar month), 23
ghee (clarified butter), 47n
Gide, André, 416–17

Gilbert, William, 137
Gingerich, Owen, 127
Giordano, Umberto, 394
Gish, Lillian, 382–83
Gleick, James, 146
global dimming, 470
Global Oscillation Network Group (GONG), 458–59
global warming, xxii–xxiii, 461n, 464–75
gnomons, 40, 64, 301–2, 302, 303
Goddard, Robert, 445–46
Godfrey, Thomas, 283
Goethe, Johann Wolfgang von, 142n–143n, 361, 394, 425–26
Gogarty, Oliver St. John, 307
gold, 353–55, 356. See also yellow color
Goldberg, Leo, 447
The Golden Cockerel (opera), 389
Goldsmid, Johann, 131
Goldwyn, Samuel, 306
Goodell, Jeff, 338n
Gorbachev, Mikhail, 418
Gore, Al, 253, 466, 468, 470
Gorst, Martin, 171–72
Gould, Stephen Jay, 163
Gounod, Charles, 391n
Grann, David, 263, 283n
Grant, Cary, 222
gravity, 139, 144, 164–67
The Great Dictator (movie), xx, 427
Great Pyramid, Egypt, 36–37, 37
Great Sphinx, Egypt, 28
The Great Wave, Sète (photograph), 381
Greece: ancient statuary, 219n; calendars in, 289–90, 293–94; sea travel by ancient Greeks, 275, 276–77; Sun myths, 3, 7; timekeeping in, 303
Greed (movie), 384
Green, Charles, 148
Green, Henry, 229–30
Green, Martin, 228
Greene, Graham, 202
greenhouse effect, 464, 465, 468, 469
Greenwich Mean Time (GMT), 316
Greenwich Observatory. See Royal Observatory, Greenwich
Gregorian calendar, 290, 296–300
Gregory Nazianzen, Saint, 18
Gregory of Tours, 189
Gregory XIII, Pope, 187, 273, 296. See also Gregorian calendar
Griffith, D. W., 382–83
Groundhog Day, 23
Groves, Leslie R., Jr., 181
Gu Yanwu, 90
Gulf Stream, 266, 267, 472
Gulliver's Travels (Swift), 286, 311, 477
Gunther, John, 428–29

Gurdjieff, George Ivanovich, 422
Gutenberg, Johannes, 117
Guy, Thomas, 171
Guys and Dolls (musical), 391n
Guzmán, Abimael, 239
Gyldenstierne, Knud, 125n

Haddon, Mark, 170n
Hadley, James, 149
Hadley, John, 283
Haggard, H. Rider, 158
Hahn, Helena Petrovna, 423n
Hahn, Otto, 179
Hakluyt, Richard, 282
Hale, George Ellery, 193, 439
Hales, Stephen, 247
Hall, Conrad, 384
Halley, Edmond, 55, 138, 143–44, 147, 172
Halley's comet, 90, 144n
halo, Christian, 17
Hamilton, James, 363–64
Hamilton-Paterson, James, 259
Hamlet (Shakespeare), 119n, 125n, 249
Hammurabi, 60
Han Dynasty, China, 93, 290
Hansen, James E., 466–67, 469
Hansen, Truls Lynne, 242
Hanukkah lights, 20
Hanway, Jonas, 211n
Hapgood (Stoppard), 445
Harakhty (Egyptian Sun god), 4
Haraldson, Olav, 278–79
Harbor Scene (Claude painting), 359
Harding, Warren G., 319
Hardy, G. H., 208
Hardy, Thomas, 21–22, 410, 500
Harkness, William, 148
Harmonics (Ptolemy), 82–83
Harriot, Thomas, 131, 188, 234
Harris, Richard, 25
Harrison, John, 287
Hawking, Stephen, 169, 172n, 444
Hay, David and Joan, 281n
Hay, Harold, xix, 340
Haydn, Joseph, 395
Hazlitt, William, 300, 305–6
Healey, Matthew, 355n
Heaney, Seamus, 352
heat. See temperatures
Heezen, Bruce, 259
Hegel, G.W.F., 410
Heinlein, Robert, 481
Heisenberg, Werner, 184
heliocentric universe, 83, 116, 117, 118–20, 127, 130, 132–33, 134
Heliodorus, 63
heliography, 376–77, 378, 380
heliopause, 451

Helios (in Greek mythology), 7, 410
Helios (prototype flying wing), 336
"Helios" Overture, 395, 397
helioseismology, 398
heliospectroscope, 169
heliostats, 334, 339n
helium, 59, 164, 178, 341, 441, 492–93, 494, 497
Henlein, Peter, 310
Henri IV (king of France), 190n
Henry I (king of England), 51
Henry VI (king of England), 305
Hepburn, Audrey, 11
Heracleides of Pontus, 79
Heraclitus, 70, 71
Herbert, Alan, 229–30
Herod, 52
Herodotus, 15, 37, 51, 61, 63, 65, 70
Herrick, Robert, 406
Herschel, John, 151, 170, 171, 177, 325, 378
Herschel, William, xx, 149–51, 171, 196, 477
Hesiod, 15, 411
Hesse, Hermann, 228
Higgins, Godfrey, 420
High, Philip E., 482
Hill, William, 471
Hilton-Barber, Miles, 300n
Himmler, Heinrich, 427
Hindus: calendars, 289–90; contributions to solar astronomy, 108–10; Sun worship, 33–34, 506, 506, 508–9; Varanasi as holy city, 47, 501, 503, 503–11; views on eclipses, 47. See also India
Hinode (Japanese spacecraft), 452
Hippalus, 276–77
Hipparchus, 80, 81, 83, 85
Hippocrates, 239, 242
Hirohito (Japanese emperor), 429–31
Hitler, Adolf, xx, 353, 424–25, 428
Hobbes, Thomas, 133
The Hobbit (Tolkien), 237n
Hobday, Richard, 385n
Hockney, David, 373–75
Hodgson, William Hope, 479
Hofmann, Albert, 349
Hogarth, William, 297, 500
"hole in the sea," 267
Holick, Michael, 235–36
Holler, Clyde, 26–27
Hollywood, California, 222, 383
Holmes, Richard, xxi
Holmes, Sherlock, xix, xxi, 312
Holst, Gustav, 395–96
Holy Trinity, 144n
Homer, 7, 15, 51n, 275, 403
Hooke, Robert, 138, 143–44

Hopi Indians, 4–5
Hopper, Dennis, 384
horary plants, 252
horizon, defined, 276
horologue, defined, 310
Horrocks, Jeremiah, 147
horsepower, defined, 327n
horses, in Sun myths, 3
hour, as unit of time, 300–303
Housman, A. E., 171, 410
Houyi (mythic Chinese ruler), 1, 4
Hoving, Lucas, 392
Huayna Capac (Inca conqueror), 6
Hubble, Edwin, 170
Hufbauer, Karl, 451n
Huitzilopochtli (Aztec Sun god), 4
Hülagu Khan, 106–7
Humboldt, Alexander von, 191–92
Hunt, Leigh, xx–xxi
Hutton, Ronald, 18, 31
Huxley, Aldous, 411
Huygens, Christopher, 143, 203
Hven (island), Denmark, 123–24
hydrogen, xx, 178, 193, 245, 247n, 248, 261, 266, 341, 492, 493, 497, 498
hydrogen bomb, xx, 184, 487
hydrothermal vents, 260
Hypatia, Lady, 110
Hyperion (Keats), 410
Hyperion, Sun as, 410

Ibn al-Haytham, 324, 376
Ibn Rushd, 112
Ibsen, Henrik, 420–21
Icarus (ballet), 392
Icarus (in Greek mythology), 487n
ice ages, 172, 197, 198n, 471–72
Icelandic volcano, 462
Iliad (Homer), 308, 403
Impression: Sunrise (Monet painting), 366
Impressionists, 365–66
Incas: calendars, 289–90, 293; Inti Raymi festival, 11–13; solar monuments, 34–35; Sun myths, 6, 12, 13; views on gold, 354
incense clocks, 302
India: astronomy in, 107, 108–10; colonial English in, 216, 217; and darkness, 210–11; development of mathematical tools, 110; heat in, 216; Jantar Mantar observatory, 36; reckons Earth's age, 110; Rig Veda, xix, 7, 109, 205n, 499; solar monuments, 33–34; Sun myths, 3; Udaipur as solar state, 502; untouchables, 210–11, 504; views on eclipses, 46–47. See also Hindus
Indians. See North American Indians
infrared radiation, 151

Ingenhousz, Jan, 247
Innocent IV, Pope, 94
Inquisition, 133
insects, role of sunlight on activity, 255–56
Intergovernmental Panel on Climate Change (IPCC), 467–68, 469, 474
International Earth Rotation Service, 320
International Energy Agency, 334
Internet Time, 321
Inti (ancient Peruvian Sun god), 3
Inti Raymi (Peruvian festival), 11–13
Inuit people: Sun myths, 5; views on auroras, 44
Inuk calendar, 312
IPCC. See Intergovernmental Panel on Climate Change (IPCC)
Ireland, Liamh Greine solar monument, 31–33, 32
Iris (opera), 394
iron: in deep sea, 260–61; Sahara Desert ore deposits, 279n
Irving, Henry, 237
Islamic civilization: and alchemy, 145; calendars in, 291–92, 299; contributions to astronomy, 100–108; embrace of learning, 102–3; gauging latitude, 277; invention of astronomical instruments, 105–6; need for knowledge of Sun, 102, 104–5; in Spain, 111–12
Isler, Martin, 39

Jabir, 101
Jackson, Kevin, 309
jaguars, in Sun myths, 4
Jai Singh II, Maharajah, 305
Jains, 109
Jaipur, India, 501
James, Henry, 83n, 212–13, 478
James I (king of England), 53
Jamison, Judith, 392
Janáček, Leoš, 394
Janssen, Pierre Jules-César, 164, 316
Jantar Mantar observatory, India, 36
Japan: adoption of Western calendar, 298; during Edo Period, 331n; Hinode space satellite, 452; Minamoto-Taira civil war, 50–51; national flag, 345, 428; as solar-power leader, 331n, 339; Sun ideographs, 428n; Sun myths, 428; surrender at end of Second World War, 429–31; use of mirror imagery, 356
Japanese Mountain Association, xxv
Jardine, Lisa, 310
Jean Grey (comic-book superheroine), 411
Jefferson, Thomas, 305, 419

Jesuits, arrival in China, 95–99
Jesus Christ: birth as basis for counting time, 288; eclipse alleged on day of crucifixion, 52; relationship of birthday to winter solstice, 17–18; Sun's brilliance as image for Transfiguration, 348. See also Christianity
Jet Propulsion Laboratory (JPL), 446, 447–48, 452n, 453–54. See also Murphy, Neil; Velli, Marco
Jews: calendars, 289–90, 292; symbolism of menorah, 348
Jhabvala, Ruth Prawer, 216
Jimmu (first emperor of Japan), xxix
John, Elton, 239
John of Worcester, 189
John Paul II, Pope, 134, 433
Johnson, Samuel, xxx, 36, 196n
Johnston, Edna, 403
Joliot, Frédéric, 179
Jones, James, 225
Joyce, James, 153, 353, 379n, 415, 423
Julian calendar, 290, 294–95, 296
Julius Caesar, 17, 288, 294–95
Julius Caesar (Shakespeare), 304
Jung, Carl, 10–11, 213, 426, 428, 499
Jungk, Robert, 176
Jupiter, 43, 119, 121n, 130, 453, 496
Justinian's law, 339n
Jyotisa Vedanga, 109

Kaaba, Mecca, 104
Kabir (Indian poet), 507
Kaku, Michio, 203, 497
Kan Te, 90
Kandinsky, Wassily, 370
Kaner, H., 481
Kant, Immanuel, 151–52
Kaplan, Raymond, 199
Kardashev, Nikolai, 341
Keats, John, 157, 410
Keeling, Charles, 465
Keenan, Larry, 350
Kemp, Clarence, 326–27
Kennedy, Bobby, 224n
Kennedy, Jackie, 224n
Kennedy, John F., 224, 224
Kentridge, William, 393
Kenya, Dancing Stones solar monument, Namoratunga, 28
Kepler, Johannes: background, 125; calculates world's creation date, 51; and camera obscura, 376; and Galileo, 130n, 133; portrait, 129; represented in library at Castel Gondolfo, 188; third law, 147; view about Sun's

corona, 157; views on planetary movement, 120, 125–27, *126*; views on tides, 269; writes about space travel, 477
Kepler space telescope, 495
Kerouac, Jack, 380
Keynes, John Maynard, 144n, 354–55, 454
Khan, M.A.R., 103
Khayyám, Omar, 101
al-Khayyámi, Ghiyath al-Din Abu'l-Fath Umar ibn Ibrahim Al-Nisaburi. *See* Khayyám, Omar
Khepri (Egyptian Sun god), 4
Khrushchev, Nikita, 418
al-Khwarizmi, 101, 104
al-Kindi, 104
King, David, 104–5
King Lear (Shakespeare), 53, 209
King Solomon's Mines (Haggard), 158–59
Kingsmill, Hugh, 482
Kinsey, Alfred, 423
Kiowa Indians, 26
Kipling, Rudyard, 214, 216, 478
Kircher, Athanasius, 282–83, 376
Kirchhoff, Gustav, 203, 205
Kitt Peak Observatory, Arizona, 463
Klee, Paul, 370
Klepper, Gernot, 468
Klimt, Gustav, 500
Ko Hung, 142n
Koch, Robert, 226
Koestler, Arthur, 100–101, 115, 120
Kolbert, Elizabeth, 467, 468, 472
Konso people, Africa, 300
Koontz, Dean R., 237n
Kooser, Ted, 149
Korcycansky, Don, 495
Kornbluth, C. M., 483
Kremer, Gerard de, 282
Kryza, Frank, 324–25
Kublai Khan, 95
Kuhn, Thomas, 247n
Kuiper, G. P., 459
Kunzig, Robert, 259, 265
Kurosawa, Akira, 384
Kuroshio Current, 266, 267
Kyoto Protocol, 468

La Boheme (opera), 389–90
La Niña, 268
Labrador Current, 266
LACIE (Large Area Crop Inventory Experiment), 253
Ladurie, Emmanuel Le Roy, 22
Lakota Sioux Indians, 24, 26–27
Lalande, J. J., 148, 170
Lamb, H. H., 466n

Land, Edwin, 380
Laplace, Pierre-Simon, 152
Large Synoptic Survey Telescope, 491
Latham, Philip, 481
latitude: defined, 276; gauging, 277; role of Sun in determining, 276; in Viking navigation, 278
Lauds (time for Christian devotion), 308, 309
Laughlin, Gregory, 495
Laughter in the Dark (Nabokov), 400
Lavoisier, Antoine, 247n
Lawrence, D. H., xxii, *411*, 411–14
Lawrence, T. E., 50, 313n
Le Corbusier, 376, 384–87
Le Gray, Gustave, *381*
Le Morte d'Arthur (Malory), 406
Le Moulin de la Galette, Montmartre, 366
Leadbetter, Charles, 155
leap seconds, 320
Lee, Christopher, 322
Lenin, Vladimir, 298
Leo, Alan, 423
Leo the Great, Pope, 18, 39
Leo X, Pope, 118
Leonardo da Vinci, 128, 324, 359n
Lévi-Strauss, Claude, 24
Lewis, C. S., 117n, 490
Lhuyd, Edward, 172
Li Ki (Book of Rites), 89–90
Liamh Greine, Irish prehistoric monument, 31–33, *32*
Liewer, Paulett, 456
Lifemanship (Potter), *185*, 223–24
light: early beliefs about, 202–3; particle vs. wave debate, 143, 203; speed of, 203–4, 442n. *See also* photosynthesis; sunlight
lighthouses, 281
lightning, 461, *462*
light-year, defined, 151n
Lilius, Aloysius, 296
Limbourg brothers, 309
linear perspective, 359n
Linebarger, Paul, 481
Linnaeus, Carolus, 251, 252–53
lion, in Sun myths, 3
Lipperhey, Hans, 128
literature, Sun in, 399–417
Little Ice Age, 191, 197n, 198, 466n
Little Red Book (Mao's), *431*, 432
Liu Shaoqi, 432
"Living with a Star" program, 452
Livy, 83n
Lockyer, Norman, 59, 164
lodestone, 277
Lolita (Nabokov), 401–3, 417

Longfellow, Henry Wadsworth, 208, 399, 499
longitude, 284–85
Loos, Anita, 222
Lopez, Barry, 14
The Lord of the Rings (Tolkien), 117n, 210, 298n–299n
Lorenz, Konrad, 255
Lorrain, Claude, 359, 360, 361, 500
Louis IX (king of France), 282
Louis XIV (king of France), 311, 324, *346*, 347–48, 357, 418
Lovelock, James, 258
Lu Gwet-djen, 86
Lu Hao-dong, 433n
Lucian of Samosaia, 476–77
lumière, defined, 378
Lumière, Louis and Auguste, 382
lunar calendars, 289–90
lunar eclipses, defined, 45
Luther, Martin, 116, 119–20
Luxor Resort and Casino, Las Vegas, 206
Lycidas (Milton), 54
Lyell, Charles, 173
Lynas, Mark, 468

Maalouf, Amin, 106n
MacArthur, Douglas, 215, 429
Macaulay, Thomas, 410
MacDonald, John D., 482
Machiavelli, Niccolò, 49
Machu Picchu, xx, 12n, 35
Macrobius, 16
Madama Butterfly (opera), 390
Maddox, John, 444
Magellan, Ferdinand, 279
The Magic Flute (opera), xx, 393, 395
"magic hour," 383
magic lantern, 376
magnetic compass, 91, 96, 277–78, 279, 284, 286
magnetic field: Earth's, 137, 176n, 192, 197, 279n, 473, 474; role in auroras, 43, 45; Sun's, 157, 194, 195, 454, 455–56, 457, 473
Maier, Mildred, 485
Mailer, Barbara (Norris Church), *413*
Mailer, Norman, *413*
Mairan, Jean-Jacques Dortous de, 173
Mallarmé, Stéphane, 411
Malory, Thomas, 406
Mamluk sultans, 107–8
al-Ma'mun, 103–4
A Man Called Horse (movie), 25
The Man with the Golden Gun (Fleming), 322
Manchester, William, 117
Mandelstam, Osip, 415n

Manhattan Project, 181–82
Mann, William, 393
Manon Lescaut (opera), 390
al-Mansur, 103
Mantel, Hilary, 346–47
Mao Zedong, 431–33
maps, as navigation aid, 282–84
Marat, Jean-Paul, 239
Marcus Aurelius, 17
Marey, Étienne-Jules, 382
Maria, countess of Coventry, 220
Marian, J. J. de, 251n
marigolds, 252
Mariner 2 spacecraft, 451
Mars, 119, 198, 248n, *482*, 496
Marseille, France, 385
Martin, Bert, 388
Martyr, Justin, 17
Marvell, Andrew, 406
Marx Brothers, 116
Mascagni, Pietro, 394
Mason, Charles, 147
Masonic system, 393, 424
Massenet, Jules, 394
Massey, Gerald, 423
Mastin, John, 480–81
al-Masudi, 101
mathematics: in ancient China, 93; in Babylon, 63–65; in India, 110; Islamic contributions, 102–8; Khayyám's contributions, 101; place/value numerical system, 64; at time of French Revolution, 64n; vocabulary of, 65n
Matisse, Henri, 358, 370–73
Maunder, Edward Walter, 194, 195–96, 197, 198, 477–78
Maurice of Nassau (German prince), 128
Maxwell, James Clerk, 203
Maya civilization, 9, 10, 34–35, 35, 293
Mayakovsky, Vladimir, 396
Mayer, Julius Robert, 247–48
McClure, Ellen, 190n
McEvoy, James, 65
"mean" sun concept, 291, 310
measurement, ancient, 66n
Mecca, 102, 104, 105, 292
MEDEA (Measurements of Earth's Data for Environmental Analysis), 253
Mede-Lydian battle, *50*, 51
Medicine Mountain, Wyoming, 34
Mehta, Ved, 300n
Meitner, Lise, 179
melanin, 232–33
melanoma, 231, 233
Melchior-Bonnet, Sabine, 357
Melville, Herman, 275, 478
Mencken, H. L., 133

Menelaus (king of Sparta), 275–76
Ménestrier, Claude-François, 348
menorah, 348
Mercator, Gerardus, 282
Mercury, 43, 119
meridians, 91, 105, 106, 121, 276, 287, 296, 316. See also prime meridian
The Merry Wives of Windsor (Shakespeare), 351
Mesopotamia: climate, 63; Sun myths, 7
Messier, Charles, 170n
Meton of Athens, 74–75, 294, 403
Mewar, Arvind Singh (Sriji), 501–2
Mexico: burning sun-ball contest, 21; solar monuments, 34–35
Michelangelo, 358
Michelson, Albert, 203–4
Midgley, Thomas, Jr., 462
Mid-Oceanic Ridge, 262
midsummer, 19–22, 29, 30, 31
A Midsummer Night's Dream (Shakespeare), 20
Mighall, Robert, 369
Milankovic, Milutin, 197–98
miliaria, 218
Milky Way, 130, 150, 170, 171, 206
millennium, 288, 289
Milton, John, 54–55, 133–34, 146, 406–7, 499
Mindanao Trench, 262
Ming Dynasty, China, 95–96, 290
minute, defined, 311
Miró, Joán, 370
mirrors: relationship to Sun, 356–57; as source of heat, 324; as weapons, 323, 323–24
Mishra, Sudharka, 47, 508–9
Mithras, 387, 420, 426, 434
Miyagawa, Kazuo, 384
Moby-Dick (Melville), 275, 284
Molina, Mario, 462
Monastersky, Richard, 471n
Monet, Claude, 366, 366–67
money, association with Sun, 355n
Montalbini, Maurizio, 236n
Montezuma (Aztec emperor), 34, 354
months, naming on Julian calendar, 295
Monty Python, 117n
Moon: as basis for calendars, 289–94; building installations there to harvest solar energy, 338; question of reliability, 290; relationship to Sun among ancients, 6
Moore, Jonas, 286
Moore, Patrick, 120
More, George, 408–9
Morgan, Dann, 253
Morgan, J. P., 414

Morley, Edward W., 204
Morton, Oliver, 250, 258, 342, 466n
Moses, 348
Mouchot, Augustin, 325, 326
Mount Fuji, xxv–xxxi
"Mouth of Truth" (Bocca delia Verita), 11
movies, role of Sun in, 382–84
moving images, 381–84
Moyle, Dorothy Mary, 86
Mozart, Wolfgang Amadeus, xx, 393–94
Mrs. Dalloway (Woolf), 311
Mrs. Sheridan (Gainsborough painting), 365
Much Ado About Nothing (Shakespeare), 219
Muhammad, 102, 103, 104, 291–92
Müller, Friedrich Max, 3, 7–8, 8
Müller, Johann, 118
Müller, Max, 403, 426–27
Munch, Edvard, 375, 420–22
muons, 445
Murnau, F. W., 238, 383
Murphy, Neil, 454–56, 459
music, Sun in, 388–98
Musk, Elon, 453
Muybridge, Eadweard, 382
Myers, Beverly Graves, 97n

Nabokov, Vladimir, 399–400, 400, 417
naked-eye astronomy, 62, 122, 280–82
Namoratunga, Kenya, Dancing Stones, 28
Napoleon (Bonaparte), 37, 152, 190n, 206, 215, 464
Napoleon III, 325
NASA (National Aeronautics and Space Administration): creation by Eisenhower, 448–49; exploratory spacecraft, 45, 451–53. See also Jet Propulsion Laboratory (JPL)
NASCAR, 207–8
Nash, Paul, 370
Native Americans. See North American Indians
nautical twilight, 213
navigation: astronomy viewed as aid, 275–76, 283; determination of longitude, 284–87; "hugging the shore," 281; instruments for, 276, 277–82; maps for, 282–83; in Moby-Dick, 275, 284; in sixteenth-century, 281
Nazianzen, Gregory. See Gregory Nazianzen, Saint
Nazis, 411, 422, 425–28. See also swastikas
Needham, Joseph, 86–87, 88, 89, 91, 92, 93, 97, 290, 302
Nefertiti (Egyptian ruler), 38

Neptune, 496
Nergal (Mesopotamian Sun god), 7
Nero (Roman emperor), 84
Nesbit, E., 206, 411
Neugebauer, Otto, 69–70, 109, 120
neutrino detectors, 443, 444
neutrinos, 442, 443–45
New Year, 8, 18, 23, 89n, 160, 290, 293, 294, 295n, 298, 299, 434
New York City solar monuments, 40–41
Newgrange, Ireland. See Liamh Greine, Irish prehistoric monument
Newton, Isaac: background, 136–38, 439; debate with Huygens over whether light is particles or waves, 143, 203; feud with Hooke, 143, 144; friendship with Halley, 144; and gravity, 139, 144, 409; interest in light, 139–43, 140, 409; interest in Sun, 138; portrait, 138; religious views, 144n; represented in library at Castel Gondolfo, 188; response to criticism, 143; significance of apple, 139; views on alchemy, 145–46; views on tides, 269
Nicholas IV, Pope, 114
Nicholas of Cusa, 117
Nielsen, Carl, 395, 397
Niépce, Joseph Nicéphore, 249, 376–77
Nietzsche, Friedrich, 84, 411, 420, 421, 422, 424, 426
night, 210–11, 213
Nightingale, Bernard, 69
1984 (Orwell), 315n
Nineveh, 63n
Niven, Larry, 482
Nixon, Richard, 418
No. 10 Downing Street, London, 419
noctilucence, 213
Noll, Richard, 424, 426
noon, 301, 309
Norse wheel, 18
North, John, 112
North American Indians: and eclipses, 47, 160–62; medicine wheels, 34; Pueblo views on Sun, 10; sun dances, 23–27
North Celestial Pole, 279n
North Equatorial Current, 266
North Geographical Pole, 279n
North Geomagnetic Pole, 279n
North Magnetic Pole, 279n
north poles, 279n
northern lights (aurora borealis), 42, 43
Northern Pole of Inaccessibility, 279n
Norway, views on auroras, 44
Nosferatu: A Symphony of Horror (movie), 238
Notorious (movie), 231

nuclear fission, 179, 180–81, 182, 184
nuclear fusion, xx, 178, 184, 341, 441, 480, 492, 497

Obama, Barack, 215, 218–19, 341
occultism, 423, 424
oceans: Clarke's views on energy from, 336; continental shelf, 261–62; currents, 266–68; deep waters, 262–65; factors in current formation, 267; layers, 267; monster waves, 267–68; navigating, 275–87; role in global warming, 469; statistics, 26; tides, 268–71, 323, 479
Octavian (Roman emperor), 84
Odyssey (Homer), 403–4
Oenopides of Chios, 74
Olbers' Paradox, 170n
Olympia (movie), 384
Omega (watch company), 337
On Colour (Newton), 142, 143
On the Heavens (Aristotle), 114–15
On the Philosophy of Mythology (Müller), 7–8
opera, Sun in, 388–96
opera glasses, 130n
Oppenheimer, J. Robert, 180, 181, 182, 184
Orbiting Solar Observatory program, 449
Oresme, Nicole, 115
The Origin of Species (Darwin), 119n, 173
Orloff, Valerii, 199n
orreries, 280
Orwell, George, 315
Ott, John Nash, 253n
Overbye, Dennis, 493
oxygen, 245, 246–47, 260, 261, 461
Oyashio Current, 266
ozone layer, 460, 461–64

painters, and Sun, 358–75
Paiute Indians, 160–62
Pale Fire (Nabokov), 400
Palmer, Samuel, 365
Palomar Observatory, 458
Papago Indians, 24
Paradise Lost (Milton), 54, 133–34, 406–7
Paris, as City of Light, 378–79
Paris Commune, 199n
Parker, Dorothy, 222
Parker, Eugene N., 450, 451, 459
Parmenides, 70, 73
Pascal, Blaise, 126
Passement (French astronomer), 320
The Passenger (movie), 384
Passover, 52, 295
Patmore, Coventry, 410
Patrick, Saint, 18
Patterson, Timothy, 474

Paul, Saint, shipwreck on Malta, 276
Paul III, Pope, 119
Payne, Cecilia, 177–78
Peck, Gregory, 11
Péguy, Charles, 207
Peierls, Rudolf, 181
Pelamis machine, 336
Pepys, Samuel, 29, 55
Periander (tyrant of Corinth), 354
Pericles, 49
Perry, Jack, 317
Perry, Judith, 242
Perry, William James, 8–10
Persian calendar, 292
perspective, in art, 359n
Peru: Inti Raymi festival, 11–13; solar
    monuments, 35; Sun myths, 3
Peter of Maricourt, 279
Peter Pan, 208–9
peyote, 26n
Pherekydes of Syros, 70
Philander, George, 468
philosopher's stone, 146
Phoebus, Sun as, 410
phoenix legend, 410–11
photography: oblique, 212; and Sun,
    376–84; time-lapse, 253n. See also
    moving images
photons, xx, 42, 203, 335, 440, 442n, 443
photoperiodism, 252
photosynthesis, 244–45, 247, 248–54, 261
phototropism, 249–50
photovoltaic cells, 330–31
photovoltaic panels, 335
Piaf, Édith, 391n
Piazzi, Giuseppe, 170
Picasso, Pablo, 374, 375
Pickford, Mary, 382–83
Pied Piper, 288
pilot book, 278
Pink Floyd, xxv, xxx
Pissarro, Camille, 367
pitchblende, 174–75
Pitman, Joanna, 353
Pizeau, Armand-Hippolyte-Louis, 193
"plain sailing," 282
Planck, Max, 440
plane astrolabes, 105–6
plane charts, 282
planets: auroras, 43; Copernicus models
    passage around Sun, 119; discovery of
    Uranus, 150; leaving Earth to go
    somewhere else, 495–97; outside solar
    system, 458, 495; solar-influenced
    atmosphere, 198–99. See also Earth
The Planets (orchestral suite), 395
plants, nonphotosynthetic, 251n. See also
    photosynthesis

plasma: creating in laboratory, 458;
    defined, 455
Plataforma Solar de Almería (PSA), 335
Plateau, Joseph, 382
Plato: anecdote about Thales, 71n; Arab
    translations of work, 111; as
    astronomer, 63, 70, 75–76, 78, 84;
    views on vision, 139, 202
Plautus, 304
Pliny, 17, 48–49, 51, 71, 116, 219n, 324
Plutarch, 49, 71, 119, 477
Pluto, 496
Poe, Edgar Allan, 410, 478
Poetics (Aristotle), 369n
Pohl, Frederik, 483
polarizing light, 380
pole star, role in ancient astronomy, 91
Polo, Marco, 94–95
Polynesians, as astronomers, 275
Pompey (Roman leader), 294
Pope, Alexander, 143
popular music, Sun in, 396–97
Porco, Carolyn, 497
porphyria, 237–39
Porter, Cole, 222, 490
Portillo, Alfonso Sevilla, 335
post meridiem, 300, 315
Potter, Harry, 411
Potter, Stephen, 185, 223, 239
Poussin, Nicolas, 359
Priestley, Joseph, 245–47
prime meridian, 287, 316
Prime Mover (Aristotle), 78, 84
printing press, 117, 118, 134, 282, 325–26,
    326
prisms, 140–41
Prokofiev, Sergey, 396
Protestants, 119–20
protons, 177, 441–42, 445
psoriasis, 239–40
psychedelic drugs, 349–50
Ptolemy: Arab translations of his work,
    104, 111–12; background, 82–84, 85;
    charts rediscovered, 282; comparison
    of his system with Copernican system,
    118, 120–21, 132–33, 280, 406–7,
    439; counted among ancient
    astronomers, 70; doubts about his
    theories of planetary motion, 111–12;
    experimentation with light, 139;
    system for judging eclipses, 49; works,
    82–83, 103, 111, 113, 117, 118, 120,
    282
Puccini, Giacomo, 389–90
Pudor, Heinrich, 424
Pueblo Indians, 10
Pushkin, Alexander, 415n
pyramid, derivation of term, 38n

Pyramid of the Sun, Mexico, 34
pyramids, Egyptian, 36–39, 37
Pythagoras, 70, 71–73, 72, 84

quadrants, 98, 106, 121, 280, 283, 284
quantum physics, 203, 440–41; Einstein's
    role, 439
quantum tunneling, 441
quarks, 441, 442
quasars, 458
Quest camera, 457
Qur'an, 100, 102–3, 105

Ra (Egyptian Sun god), 4, 6, 9, 39
radio astronomy, 448
radio signals, 199n
radioactivity, 174–76, 179
radioisotopes, 175n
radium, 174–75
Raeburn, Henry, 365
railroads, and timekeeping, 313, 314–15
rainbows, 141–42
Raleigh, Walter, 131
Ramadan, 299
Ramses II (Egyptian ruler), 38
Rashômon (movie), 384
Rasselas (Samuel Johnson), 36, 196n
Ratzinger, Joseph Cardinal. See Benedict XVI,
    Pope
Ravel, Maurice, 395
Rawlins, Dennis, 82
Ray, John, 172
Ray, Man, 415
al-Razi, 101
Reagan, Ronald, 338
Rebecca (du Maurier), 44n
red dwarfs, 206
red giant, Sun as, xx, 493–94, 494, 497
red spikes, 473
reflected light, 206–7. See also mirrors
reflecting telescopes, 142, 149–50
Regulus (Turner painting), 362
Rembrandt van Rijn, 359, 500
renewable energy, solar radiation as
    source, 322–42
Renoir, Pierre-Auguste, 366, 372
retina. See eye, human
Revelle, Roger, 465, 474
Revkin, Andrew C., 338n
Reynolds, Freewheelin' Frank, 350
Reynolds, Joshua, 220
Rhodes (Greek island), 81
Ricci, Matteo, 97–99
Richard II (Shakespeare), 304
Richelieu, Cardinal, 310
Riefenstahl, Leni, 384
Rig Veda, xix, 7, 109, 205n, 499

Rimbaud, Arthur, 416
Rimsky-Korsakov, Nikolai, 389
The Ring of the Nibelungs (opera), 390–91
Robinson Crusoe, 211
rockets: after WWII, 447, 449, 452–53;
    future, 480n, 481n, 483, 488n; pre-
    WWII, 445–46, 446; WWII research,
    447
Roeg, Nicolas, 384
Roepke, Gabriela, 388–89, 390, 391, 393,
    394, 395, 396
Roget, Peter Mark, 381–82
Roman, Colin, 83
Roman Catholic Church: arrival of Jesuits
    in China, 95–99; and Copernicus,
    118–19, 120, 132, 133, 134; and
    Galileo, 132–33, 134; heavenly interest
    in astronomy, 112–13, 169, 187–89,
    405, 406; Ratzinger's treatise on solar
    imagery in Christianity, 433–35;
    Vatican Observatory, 187–89. See also
    Christianity
Roman Empire: astronomical knowledge
    in, 82; calendars in, 294–95;
    harnessing of sunlight, 324; Sun
    myths, 16; timekeeping in, 300–301;
    use of mirror imagery, 356
Roman Holiday (movie), 11
Romans, France, 22, 23
Romeo and Juliet (Shakespeare), 351, 408
Roméo et Juliette (opera), 391n
Rømer, Ole, 203
Rooney, David, 320, 321
Roosevelt, Franklin D., 181
Rosenkrantz, Frederik, 125n
Rosenthal, Norman, 241
Rosicrucianism, 424
Ross, Alex, 394
Ross, John, 263
Rossetti, Dante Gabriel, 101
Rotch, Josephine, 415
Rothschild, Lord, 439
Rouault, Georges, 370
Rowland, F. Sherwood, 462
Royal Observatory, Greenwich, 283, 287,
    312, 316
Rubáiyát of Omar Khayyám, 100, 101, 166–67
Rubens, Peter Paul, 500
Rubinstein, Arthur, 396
Ruders, Poul, 397–98
Rule a Wife and Have a Wife (Beaumont and
    Fletcher), 211n
Ruskin, John, 360, 362–63, 364
Russell, Bertrand, 411
Russell, Henry Norris, 178
Rutherford, Ernest, 175–76, 178, 441
Ruysdael, Jacob van, 359

Sabine, Edward, 191
Sabinianus, Pope, 309
SAD (seasonal affective disorder),
    240–43
Sagan, Carl, 169, 459
Sahara Desert: conditions, 214, 215–16,
    218; foraging ants in, 254; iron ore
    deposits, 279n
Saint Bridget's Day, 22
Saint Vitus' dance, 19n
Samrat Yantra, 305
Samson Agonistes (Milton), 54–55
Samuel Oschin telescope, 458
sandglasses, 302
Sandia Laboratories, 335
Sargent, John Singer, 221
Sarton, George, 101
Sartre, Jean Paul, 417
Sassoon, Philip, 500
satellites, planetary: defined, 130n; of
    Jupiter, 130
satellites, unmanned, 448, 449–53, 454,
    455–59
Saturn (planet), 119, 121n, 496
Saturn (Roman god), 16
Saussure, André de, 86
Sawai Jai Singh II, Maharaja, 36
Sayers, Dorothy L., 162
Scandinavia, Sun myths, 492
Schama, Simon, 424
Scheele, Carl Wilhelm, 247n
Scheiner, Christoph, 131, 188, 192
Schikaneder, Johann Josef, 393
Schopenhauer, Arthur, 422, 428
Schräder, Otto F., 426n
Schreck, Max, 238
Schrödinger, Erwin, 440
Schwabe, Samuel Heinrich, 190–91, 192
Schwarz, T. H., 356n
science fiction, 476–89
Scientific Revolution, 146, 247n, 409
scientist, origin of term, 269
Scott, Robert F., 43–44
Scott, Walter, 44
Scriabin, Alexander, 374, 422
The Sea Around Us (Carson), 262, 263
sea clocks, 287
sea cucumbers, 260
sea travel: astronomy as navigational aid,
    275–76, 283; determination of
    longitude, 284–87; earliest, 275;
    instruments for navigation, 276,
    277–82; maps for navigation, 282–83.
    See also oceans
seafloor, 262–65
seasonal affective disorder (SAD),
    240–43

seasons: and Gregorian calendar, 299;
    question of when they begin, 289. See
    also equinoxes; solstices
seaweed, 265n
Sebastian I (king of Portugal), 215
second, defined, 320
Seleucus, 65
Sendero Luminoso, 418
Seneca, 44n, 303
Sennacherib (Assyrian ruler), 61
Seurat, Georges, 367–68
Sevilla Portillo, Alfonso, 335
sextants, 98, 108, 121, 280, 283
shade, 210, 211
shadows, 208–11, 212, 213
shah of Iran, 292
Shakespeare, William, 20, 53, 119n, 125n,
    219, 298n, 304, 305, 307, 310, 351,
    407, 408, 411, 499
Shattering Suns (Sun sounds), 397
Shaw, George Bernard, 439
Shelburne, Lord, 246
Shelley, Percy Bysshe, 410
Shen Kua, 277
Shepard, Sam, 207n
shepherd's sundial, 252
Sherriff, R. C., 211
Shirley, Rodney W., 283n
Shih Shen, 90
Shoshoni Indians, 24
Shovell, Clowdesley, 285–86
Shugar, Daniel, 340
Shuman, Frank, 328
Sibelius, Jean, 394
Sidey, Hugh, 224n
Sidis, William James, 199n
Sidney, Philip, 407
silicon photovoltaic cells, 330–31
Silk Road, 95n
Silverberg, Robert, 482
Simak, Clifford D., 481
Simloki (Soldier Mountain), 209
Sims, Michael, 383
Sister Angelica (opera), 390
Sistine Chapel, 358
Skene, James, 361–62
skin cancers, 231–33, 237
skin color: in black and white movies, 222;
    prejudice against darker skins,
    218–19; as signifier of standing, 219;
    suntans, 221–25, 224; whiteners,
    219–21, 225
sky, viewed as blue, 205
Skylab, 449
skyscrapers, 384, 385
slow glass, 442n
Smith, Cordwainer, 481

Smith, Grafton Elliot, 8–10
Smith, Mark, 315
Smith, Willoughby, 330
Snow, C. P., 180
The Snow Maiden (opera), 389
Sobel, Dava, 119, 169, 285, 303, 304
Socrates, 75
Soddy, Frederick, 175–76
SOHO space mission, 452
Sokurov, Aleksandr, 430n
Solanki, Sami, 474
solar cells, 330–31
solar constant, 169
solar disk, 57. *See also* wheel, relationship
    to solstice
Solar Dynamic Observatory mission, 452
solar eclipses: ancient study, 48–49, 50;
    Antarctica, 2003, 153–55; associated
    phenomena, 156–58; Baily's Beads,
    157; characteristics, 45–46; Chinese
    records, 90; Cook's pursuit, 155–56;
    coronas, 155, 157–58, 169, 201;
    danger to human eye, 235; defined, 45;
    descriptions of experience, 162–64;
    following, 155–56; Hindu view, 47;
    illustration, 96; photos, 162, 165; and
    portrait of Saint Benedict, 359, 360;
    predicting, 48–49; relationship of light
    to gravity, 164–67; Ricci's Chinese
    experience, 97; scientific study,
    164–67; statistics, 156; superstitions
    and fears, 49–50, 52, 55, 158–62
solar energy: cost-effectiveness, 340–41;
    early inventions, 327–29; extent of
    development, 331–42; first devices
    employing, 325–27, 326; government
    spending on, 338–40; in Japan, 331n;
    R&D centers, 335, 338
Solar Energy Research Institute, 338
Solar Orbiter (European Space Agency),
    457n
solar particles, 473, 474
solar physics, xxiii, 177, 181, 197, 444, 445,
    447–51, 459
solar plexus, 352
Solar Probe Plus, 454, 454–57
solar prominences, 157, 164, 169, 458
solar retinopathy, 234
solar roof panels, 327, 338, 340
solar spectrum, 151, 169, 204–5, 245
solar water heaters, 326, 329, 330, 339
solar wind, 450–51, 452, 453, 455, 456–57,
    473, 474, 481n
Solar-H, 248
solar-powered aircraft, 336, 337
SolarRegionFreiburg, Germany, 332
Soldier Mountain, California, 209
Soler, José Martínez, 335

SOLO (Solar and Heliospheric
    Observatory), 189
solstices: celebrations and rituals, 18–22;
    cultural role, 16–22; defined, 14, 15;
    and Earth's orbit, 15–16; and Egyptian
    pyramids, 39; and Gregorian calendar,
    296, 299; relationship of Christmas to,
    17–18, 296, 434; relationship of wheel
    to, 18–19
songs, Sun in, 396–97
Sons and Lovers (Lawrence), 412
Soon, Willie, 474
Sophocles, 69
Sorrell, Martin, 416
The Sorrows of Young Werther (opera), 394
Sosigenes of Alexandria, 294
sounds, Sun representations, 397–98
South America: heat in, 214–15; Inti Raymi
    festival, 11–13; solar monuments,
    34–35; Sun myths, 3, 6, 12, 13. *See also*
    Mexico; Peru
South Equatorial Current, 266
South Pacific (play), 352
southern lights (aurora australis), 42
Soviet crops, 253
SpaceShipOne spacecraft, 453
Spain, as intellectual center at end of Dark
    Ages, 111
Specola Vaticana. *See* Castel Gandolfo,
    Italy
spectroheliographs, 193, 449
spectroheliometers, 449
spectroscopy, 169, 205
Spence, Jonathan, 280
Spender, Stephen, 228–29
Spenser, Edmund, 406
Speusippus, 76
spinning wheels, 18, 19, 203n
Spinoza, Baruch, 142
spiritualism, 423
Spitzer Space Telescope, 452
spring equinox. *See* vernal equinox
Spurling, Hilary, 371, 372
Sputnik, 448
squamous cell carcinoma, 231, 233
squid, deep-sea, 263, 264
Sri Lanka, 407, 418
Sriji (Arvind Singh Mewar), 501–2
SSP (space-based solar power), 338
St. Mark's Church-in-the-Bowery, New
    York, 40
stade, defined, 66n
Stalin, Joseph, 239
standard time, 315–16
star catalogues, 90, 170
star spectroscope, 169
Star Wars (movie), 480n
Stardust spacecraft, 452

Statue of Liberty, *387*
steam engine, 325, 326, 327n, 328
Steel, Duncan, 296
Steiner, Rudolf, 423, 426n
STEREO (Solar Terrestrial Relations Observatory), 452
Stetson, H. True, 196n
Stevens, Wallace, xxv
Stevenson, Robert Lewis, 211, 222
Stoke, Roger, 310
Stoker, Bram, 237–38
Stonehenge, *29*, 29–31; meaning of name, 28n
Stoppard, Tom, 69, 125n, 130n, 136, 445
Strabo of Amasia, 61
Strassmann, Fritz, 179
Strauss, Richard, 395, 422
Stravinsky, Igor, 395, 396
Strindberg, August, 421, 422
Strout, Elizabeth, 417
Stukeley, William, 30
Sturt, Charles, 216
Styron, William, 160
Su Song, 94
sub-untouchables, 210–11
Suckling, John, 407
Sumerians, 59–60
summer solstice: climbing Mount Fuji, xxv–xxxi; defined, 14; midsummer celebrations, 18–22; role of Meton of Athens, 74–75
Sun: as alternative energy source, 322–42; in ancient cultures, 3–7; as aphrodisiac, 229–30; in art, 358–75; auroras, 42–45; characteristics, xix–xx, 168, 169, 205, 206, 442, 492; corona, 155, 157–58, 169, 201, 450–51; defined, xix; earliest ideas about, xx, xxi–xxii; early scholars, 7–11; Earth's distance, xix, 146–47; eclipses, 45–55, 153–67; estimating age, xix, 169, 177; first astronomers, 59–68; first photograph, 193n; future, xxii, xxiii, 459, 490–98; Inti Raymi festival in Peru, 11–13; in language, 351–52; in literature, 399–417; methods of determining direction by, 279n, 303n; in music, 388–98; on national flags, 345–46, 428, 433n; and photography, 376–84; role in climate change, 469–70, 471, 473–75; role in tides, 268, 269–70, 323, 479; in science fiction, 476–89; seasonal aspects, 14–27; as self-renewing resource, 322–42; sound representations, 397–98; structural monuments, 28–41; sunspots, 187–201; as symbol, 345–57; variety of

myths about, 3–7, 350–51. *See also* astronomy; sunlight
The Sun (movie), 430n
sun arcs, 383
sun dance, 23–27
"The Sun Gets Married" (Aesop's fable), 4
Sun King, 190n, 195. *See also* Louis XIV (king of France)
Sun Quartets (Haydn), 395
Sun Rising Through Vapor (Turner painting), 361
sunbathing, *185*, 223, *224*, 228–30, 400, 401
Sunday, as holy day, 17
Sunday Afternoon on the Island of La Grande Jatte (Seurat painting), 367–68
sundials, *41*, 302–6, *305*
sundowners, 352
sunfish, 254
sunflowers, 250–51, 368
sunlight: "ancient lights" doctrine, 339n; and cancer, 231–33, 237; as curative, 226–28; and deep ocean, 262–65; effect of too little, 235–36; global dimming, 470; health benefits, 228–29; as human right, 339–40; impact on military operations, 206–7; passing sunbeam through prism, 204–5; and photovoltaic cells, 330–31; role in animal activity, 254–57; in sports, 207–8. *See also* photosynthesis; ultraviolet light
sunrise, 383, 395
sunset: cultural significance, 499–501; danger to eyes, 235; in moviemaking, 383; in Varanasi, India, 504, 509–11; world's ten top spots for viewing, 501
Sunshine (movie), *480*, 480n
sunshine, incidence on Earth, 218
sunshine laws, 352
sunspots, 98, 187–201; activity cycles, 194; activity peak predicted for 2011, 200, 471; characteristics, 193–94; Chinese observation, 90; and climate variations, 197n, 198, 201, 471, 474; early observations, 130–32; effect on Earth, 196–201; Galileo's studies, 130, 131–32, 134; Greenwich discontinues photography, 450
sunstone, 278–79
suntans, 221–25, *224*
supernovas, 122, 123
superstition, and eclipses, 49–50, 52, 55, 158–62. *See also* Sun, variety of myths about
Sutherland, Graham, 370
swastikas, *343*, 425
Swatch, 321

Swerdlow, Noel, 65, 128–29
Swift, Jonathan, 286, 322, 477
synesthesia, 374–75
Syntaxis mathematica (Ptolemy), 82
Szilard, Leo, 179–80, 181

tachocline, 455
Tacitus, 423–24
Taj Mahal, 385n
Talbot, William Fox, 377, 378
The Taming of the Shrew (Shakespeare),
        297–98, 307
Tamur the Lame (Arabic king), 107
Tang Dynasty, 109
tanned skin, 221–25, 224, 233
tanning salons, 233
Taoism, 92
Tauranac, John, 384
Tavener, John, 395
Taylor, E.G.B., 281n
Taylor, Stephen, 397
Teardrop Park, New York City, 339n
telescopes: adaptive optics for, 458;
        Bacon's role in development, 128;
        Galileo's role in development, 127–30;
        impact on study of sunspots, 190–95;
        modern, for learning more about
        Sun, 458; Newton's, 142; origin of
        term, 130n; popularity, 149–50;
        satellite compared with ground-based,
        457–58
telespectroscope, 169
Teller, Edward, 184
Tellier, Charles, 327, 328
temperatures: estimate of Sun's surface,
        169; hottest on Earth, 214–18; impact
        on military, 215–16
Tennyson, Alfred Lord, 410
terraforming, 496
Tertullian the Convert, 84
Tesibius, 302
Teton Sioux Indians, 24
Tetrabiblos (Ptolemy), 82
Tezcatlipoca (Aztec Sun god), 4
Thacker, Jeremy, 286
Thales (astronomer), 51, 52, 70–71
Theodosius, 113
Theophrastus of Eressos, 79, 188
Theosophical Society, 423, 425n
Thermidor (French Republican calendar
        month), 23
thermonuclear reactions, 184, 441,
        443–45, 480, 492, 493
Third Reich. *See* Nazis
Thomas, Theodore L., 481–82
Thomson, David, 383
Thomson, William, 269–70
thorium, 175, 175n, 176

Thus Spake Zarathustra, 395, 420, 421, 422,
        425
Tiberius (Roman emperor), 44n, 84, 499
tides, ocean, 268–71, 323, 479
tiger, in Sun myths, 3
time zones, 315
timekeeping, 307–21; among ancients,
        300–303; individual timepieces, 310;
        need for measuring time with
        precision, 308–9; sundials, 302–6.
        See also calendars
time-lapse photography, 253n
Tintin, 159
Titicaca, Lake, solar markers, 35
Todd, Mrs. D. P., 157
Toffana, Giulia, 219
Tolkien, J.R.R., 117n, 210, 298n–299n.
        See also The Hobbit (Tolkien)
Tolstoy, Leo, 136
Tosca (opera), 390
tower of Babel, 62
TRACE (Transition Region and Coronal
        Explorer), 452
transparent animals, 265
Treaty of Verdun, 53
tree rings, 195, 195–96, 197, 200–201
Tristan und Isolde (opera), 391–93
Tromso, Norway, 241–43
Truman, Harry, 182
tsunamis, 201
tuberculosis, 226, 227
Turandot (opera), 388, 390
Turner, Georgia, 388
Turner, Joseph Mallord William, 358,
        360–64, 363, 371, 500, 501
Turner, Nat, 160
al-Tusi, Nasir al-Din, 106–7
Twain, Mark (Samuel Langhorne
        Clemens), 159–60, 173, 196–97, 362,
        478, 505–6
Twelfth Night (Shakespeare), 219
twenty-four-hour clock, 315
twilight, 212–13
Two Gentlemen of Verona (Shakespeare), 219
Tycho. *See* Brahe, Tycho
Tynan, Kenneth, 398
Tyndall, John, 142n, 205, 464
Tyson, Neil deGrasse, 41, 341

Udaipur, India, 501–2
UFOs, 485
ultraviolet light, 227, 233, 234, 235, 237,
        249n, 461
Ulysses (Joyce), 153, 307, 353, 473
Ulysses spacecraft, 450
umbrellas, 211
undead, 238n
Unité d'Habitation, Marseille, France, 385

United Nations, 316, 463
United States: aurora borealis of Nov. 14, 1837, 44; hottest temperatures, 214; national intelligence, 253–54; solar monuments, 34; space program, 448–53
untouchables, 210–11, 504
Updike, John, 164, 239–40, 442
uranium, 174
Uranus, 150, 496
Urban VIII, Pope, 132, 133
urbanism, 385–86
Urmy, William, 187
U.S. Naval Observatory, 316, 319, 320
U.S. Scouting Service Project, xxiii
Ussher, James, 171, 172, 173
UTC (Coordinated Universal Time), 316–17

Valentino, Rudolph, 222
vampire myth, 237–39
Van Gogh, Vincent, 368–70, 374, 500
Varanasi, India, 47, 501, 503, 503–11
Varro Reatinus, 83n
Vatican Observatory. See Castel Gandolfo, Italy
Vauban neighborhood, Freiburg, Germany, 333
Vedic religion, 47, 108–9. See also Hindus; Rig Veda
Vega, Garcilaso de la, 6
Velli, Marco, 454, 456–57
Vendémiaire (French Republican calendar month), 23
Venice, 127–28, 354, 356, 362
Venus: Copernicus models passage around Sun, 119; solar-influenced atmosphere, 198; transit between Earth and Sun, 146–49
Verlaine, Paul, 416
vernal equinox, 15–16, 28, 35, 40, 289, 296, 299, 435
Verne, Jules, 316, 478
Vespasian (Roman emperor), 9
Vespers (time for Christian devotion), 309
Vespucci, Amerigo, 283n
Vidal, Gore, 224
Vigilius, Pope, 39
Vikings, 278
Virgil, 189
vitamin D, 235–36
Vitruvius, 305
Voice of America, 199
volcanoes: Eyjafjallajökull, Iceland, 462; harnessing, 341; possible Sun connection, 201; and sunsets, 501; undersea, 259–60
Volk movement, 423–25, 426, 427n

Volney, Constantin-Francois, 419, 435
Voltaire, 297
von Braun, Werner, 447
von Liebig, Justus, 357
von Moltke, Helmuth, 313
von Sachs, Julius, 249
von Stoheim, Erich, 384
Vonnegut, Kurt, 399, 469
Voyager spacecraft, 451–52

Wagner, Richard, 390–92, 422
Waiting for Godot (play), 321
Wald, George, 248n
Walkabout (movie), 384
Wallach, Eli, 376
Wallis, George C., 479
Walt Disney Concert Hall, Los Angeles, 386, 387
Walton, Ernest, 179
Wang Mang, 93
Wanika (East African people), 208
Warhol, Andy, 500
Washington, George, 215, 305
watches (portable clocks), 287, 311, 313; as reckoners, 303n
watches (time periods), 287, 292, 301
water clocks, 66, 302
water heaters, solar, 326, 329, 330, 339
Watson, William, 287
Watt, James, 327n
waves, monster, 267–68. See also tides, ocean
The Waves (Woolf), 163n
Weather Action, 470. See also Corbyn, Piers
The Weather Project (art installation), 372n, 373
Wedgwood, Josiah, 219
weekends, origin of, 295n
Weiner, Jonathan, 218
Weizmann Institute, 335
Wells, Herbert George, 478–79, 480, 496
Wells, H. G., 180, 248n, 476
wheel, relationship to solstice, 18–19
Whewell, William, 269n
white dwarf, Sun as, xx, 498
"white glare," 386n
Whitehead, Alfred North, 134, 474
Whitehouse, David, 468n
Whitman, Walt, 228, 499
Wilde, Oscar, 308, 312–13, 314, 417
Willett, William, 318
Willsie, Henry E., 328
Wilson, E. O., 255–56
Wilson, Jack, 160
Wilson, Tom, 235
winds, in navigation, 276–77. See also solar wind

Winter Landscape with Two Windmills (Ruysdael painting), 359
winter solstice: defined, 14; relationship of Christmas to, 17–18, 296, 434; rituals in history, 20–22; when to celebrate, 17–18
The Winter's Tale (Shakespeare), 252
Wöhrbach, Otto, 332
Wolf, Rudolf, 191
Wolfe, Gene, 480
Wolfe, Nero, 352
Wolfe, Tom, 496
Wolff, Geoffrey, 414
Wollaston, William, 204
Wood, Elizabeth, 205n
Wood, Florence and Kenneth, 403n
Wood, John, Jr., 386n
Wood, Pat, 28
Woodward, John, 172, 173
Woody, Todd, 338n
Woolf, Virginia, 162–63, 311
Wooster, Bertie, xix
Wovoka, 160–62
Wren, Christopher, 139, 143–44
Wright, Thomas, 150
wristwatches. *See* watches (portable clocks)
Wu Hsien, 90

X Prize, 453
Xavier, Francis, 96–97
xeroderma pigmentation (XP), *236,* 236–37
Xerxes of Persia, 51
Xi Chu, *89*
Xihi (Chinese goddess), 4
Xue Shenwei, 95n

yang (Chinese term), 91, 92
Yates, Frances, 407
years, systems for measuring, 288
Yeats, William Butler, 411, 416, 423
yellow color, 352–53. *See also* gold
yellow dwarf, Sun as, xix
yin (Chinese term), 91, 92
Yucatán, 35
Yule (Danish tradition), 18
Yule log, 20

zenith, defined, 276
Zeno of Elea, 70
Zhang Heng, 93–94, 477
ziggurats, 62
Zoopraxiscope, 382
Zoroaster, 393
Zou Yuanbiao, 86

ABOUT THE AUTHOR

RICHARD COHEN is the author of *By the Sword,* a three-thousand-year history of swordplay. The former publishing director of Hutchinson and of Hodder & Stoughton and the founder of Richard Cohen Books, he was for two years program director of the Cheltenham Festival of Literature. He is a visiting professor of creative writing at the University of Kingston-upon-Thames, has written for *The New York Times* and most leading London newspapers, and has appeared on BBC radio and television. He lives in New York City.